Plant Taxonomy and Biosystematics

Plant Taxonomy and Biosystematics
Classical and Modern Methods

Edited by

T.S. Rana
K.N. Nair
D.K. Upreti

CSIR – National Botanical Research Institute
Lucknow – 226 001, Uttar Pradesh

and

NEW INDIA PUBLISHING AGENCY

New Delhi – 110 034

NEW INDIA PUBLISHING AGENCY
101, Vikas Surya Plaza, CU Block, LSC Market
Pitam Pura, New Delhi 110 034, India
Phone: + 91 (11)27 34 17 17 Fax: + 91(11) 27 34 16 16
Email: info@nipabooks.com
Web: www.nipabooks.com

Feedback at feedbacks@nipabooks.com

ISBN: 978-93-83305-41-4

The contributors/editors/institute and the publisher thankfully acknowledge Dr. Stephen Sharnoff the original copyright holder for figure 18A&B in Chapter 5.

Composed, Designed and Printed in India

सीएसआईआर-राष्ट्रीय वनस्पति अनुसंधान संस्थान
CSIR-National Botanical Research Institute
(वैज्ञानिक तथा औद्योगिक अनुसंधान परिषद, नई दिल्ली)
राणा प्रताप मार्ग, लखनऊ – 226 001, उ.प्र., भारत
(Council of Scientific & Industrial Research, New Delhi)
Rana Pratap Marg, Lucknow - 226 001, U.P., India

डॉ. चन्द्र शेखर नौटियाल
निदेशक
Dr. C.S. Nautiyal
Director

Foreword

Taxonomy is the science of describing, naming and classifying organisms, and therefore, it is an important tool in understanding the basic tenets of biological diversity. Unfortunately, the ongoing efforts to survey, inventory and documenting the biological resources have been limited by the scarcity of taxonomic expertise in many groups of organism. Without taxonomists it is impossible to describe and name new species, survey biological diversity and identify biodiversity hotspots or other areas of special conservation concern.

There are only few active centers in India, where researchers are being trained in plant taxonomy. It is observed that the taxonomic researches are not attracting reasonable funding to carry forward the Linnaean legacy; therefore there is a constant declining trend in taxonomic research as well as taxonomists the world over. This concern has been expressed not only in India but the world over by eminent biologists. Considering the increasing demand for more trained taxonomists to resolve the issues of biodiversity inventorying, conservation, monitoring and management, it is important that taxonomy-based institutes in India take stock of the situation, and find out appropriate solutions by organizing training courses on various aspects of traditional and contemporary plant taxonomy.

CSIR-NBRI is among the few research centers in India, where plant taxonomy and systematic research has been actively pursued within a multidisciplinary framework. Plant systematic research at NBRI covers a broad spectrum of areas including classical taxonomic revisions, monographs, floristic inventories of all groups of plants from algae to angiosperms, and plant molecular systematics. The Institute is known for its outstanding contributions to enriching the knowledge base on plant diversity and is a National Repository of Indian flora designated by the National Biodiversity Authority of India. The Institute is also a recognized Lead Botanic Garden under the Ministry of Environment and Forests, and a DUS Test Centre under the Protection of Plant Varieties and Farmers Rights Authority (PPVFRA), Ministry of Agriculture, Government of

Phones : (Off.) (0522) 2205848, EPABX Phone : (0522) 2297800 to 2297999 Fax : (0522) 2205839, Post Box No. 436, Lucknow
e-mail : director@nbri.res.in, csnnbri@yahoo.com, Website: http://www.nbri.res.in

India. It is, therefore, befitting that NBRI has taken the initiative to organize training programs in Plant Taxonomy and Biosystematics for the benefits of student, researchers and young faculties in taxonomy and allied subjects. I am happy that the scientists at Plant Diversity and Systematics Group of NBRI has brought out a manual on "Plant Taxonomy and Biosystematics (Classical and Modern Methods)".

The manual includes chapters written by experts on various groups of plants, microbes, fungi, and other aspects on contemporary tools and techniques being used in plant systematics. I am confident that this manual will be a valuable primer for those who intend to take up plant taxonomy and systematics as an active area of their scientific research. I congratulate the authors and editors of the manual for bringing out such a publication, which I am sure, will serve as an excellent reference and study material for students, researchers and teachers in Plant taxonomy, systematics and related discipline in plant sciences.

Dated 25-01-2014

C.S. Nautiyal

Phones : (Off.) (0522) 2205848, EPABX Phone : (0522) 2297800 to 2297999 Fax : (0522) 2205839, Post Box No. 436, Lucknow
e-mail : director@nbri.res.in, csnnbri@yahoo.com, Website: http://www.nbri.res.in

Preface

It is very often observed that taxonomic researches are not attracting reasonable funding to carry forward the Linnaean legacy; therefore there is a constant declining trend in taxonomic research as well as taxonomists the world over. Considering the increasing demand for more trained taxonomists to resolve the issues of biodiversity inventorying, conservation, monitoring and sustainable management of bio-resources, CSIR-National Botanical Research Institute, Lucknow has taken the initiative to publish a manual on " **Plant Taxonomy and Biosystematics (Classical and Modern Methods)"** for young students and faculties, who have active interest in pursuing plant systematics as their academic as well as research pursuits.

The basic aim of this manual is to provide useful resource materials for training young students and faculties working in the area of plant systematics. The manual provides updated information on basic as well as applied aspects of plant systematics on various groups of plants like Algae, Lichens, Bryophytes, Pteridophytes, Gymnosperms and Angiosperms.

Chapter 1 to 3 describe the various approaches and methods to study microbial and fungal diversity, which is basically a very useful precursor to the students and young researchers. Chapter 4 and 5 deal with the multi-dimensional approaches in Lichen systematics. The book progresses upwards through the plethora of information on the diversity and systematics of Algae, Bryophytes, Pteridophytes and Gymnosperms (Chapter 6-10).

Chapter 11 to 15 cover methodological details on plant identification, approaches and methods of writing Flora, revision, monograph and development of herbarium. This information is very important for the students and young faculties who intend to pursue their researches in plant taxonomy. Chapter 14 and 15 particularly provide all the relevant information on the International Code of Plant nomenclature including cultivated plants. These chapters *per se* are very significant for the amateur as well as serious readers of plant taxonomy.

Plant taxonomy and biosystematics is a dynamic subject, as it derives information from various other disciplines like palynology, seed morphology, pharmacognosy, molecular biology, etc. We have, therefore, broaden the scope of this book by including the chapters on palynology, seed morphology, molecular systematics, biostatistics, ecological and remote sensing methods for diversity analyses, and pharmacognostical tools for identification of herbal drugs (Chapter 16-22). The knowledge and information on these applied aspects of biology in relation to taxonomy will certainly infuse the interest in readers.

Chapter 23 and 24 describe the various methods of characterization and evaluation of ornamental and medicinal plants. The last chapter (25) of the book provides the information about CSIR-NBRI Botanic Garden and its various repositories, which could be of great interest to the readers from the perspectives of ornamental plants.

We trust that this manual will be immensely useful to Taxonomists, Foresters, Conservationists, Herbalists, Agriculturists, Floriculturists, Naturalists, Environmental Biologists and policy makers.

We express our gratitude to Dr. C.S. Nautiyal, Director, CSIR-National Botanical Research Institute, Lucknow for encouragement and unrelenting support. We are also thankful to all the authors of this manual for contributing the articles and responding timely to our request. We sincerely, acknowledge the friendly co-operation and assistance of our colleagues especially Mr. Baleshwar, Dr. Bhaskar Datt and Dr. L.B. Chaudhary, at various stages of preparation of the manuscript.

In-spite of our genuine efforts, there might still be some inadvertent errors or omissions in this Manual. We shall greatly appreciate if the users of this Manual very kindly bring such instances to our notice.

T.S. Rana
K.N. Nair
D.K. Upreti

Contents

Contributors

Aarti Kumari
Plant Molecular Virology Laboratory
CSIR-National Botanical Research Institute
Lucknow – 226 001, Uttar Pradesh

A. Mishra
Plant Microbe Interaction
CSIR-National Botanical Research Institute
Lucknow – 226 001, Uttar Pradesh

A.K. Asthana
Bryology Laboratory
Plant Diversity, Systematics and Herbarium Division
CSIR-National Botanical Research Institute
Lucknow – 226 001, Uttar Pradesh

A.P. Singh
Pteridology Laboratory
Plant Diversity, Systematics and Herbarium Division
CSIR-National Botanical Research Institute
Lucknow – 226 001, Uttar Pradesh

Arti Garg
Botanical Survey of India
Central Regional Centre
Allahabad – 211 002, Uttar Pradesh

A.K.S. Rawat
Pharmacognosy and Ethnopharmacology Division
CSIR-National Botanical Research Institute
Lucknow – 226 001, Uttar Pradesh

A.K. Goel
Botanic Garden
CSIR-National Botanical Research Institute
Lucknow – 226 001, Uttar Pradesh

Baleshwar
Plant Diversity, Systematics and Herbarium Division
CSIR-National Botanical Research Institute
Lucknow – 226 001, Uttar Pradesh

Bhaskar Datt
Plant Diversity, Systematics and Herbarium Division
CSIR-National Botanical Research Institute
Lucknow – 226 001, Uttar Pradesh

C.S. Nautiyal
Division of Plant-Microbe Interactions
CSIR-National Botanical Research Institute
Lucknow – 226 001, Uttar Pradesh

D.K. Upreti
Lichenology Laboratory
Plant Diversity, Systematics and Herbarium Division
CSIR - National Botanical Research Institute
Lucknow – 226 001, Uttar Pradesh

D. Narzary
Molecular Systematics Laboratory
Plant Diversity, Systematics and Herbarium Division
CSIR-National Botanical Research Institute
Lucknow – 226 001, Uttar Pradesh

Kiran Toppo
Algology Laboratory
Plant Diversity, Systematics and Herbarium Division
CSIR-National Botanical Research Institute
Lucknow – 226 001, Uttar Pradesh

K.N. Nair
Molecular Systematics Laboratory
Plant Diversity, Systematics and Herbarium Division
CSIR- National Botanical Research Institute
Lucknow – 226 001, Uttar Pradesh

K.S. Mahar
Molecular Systematics Laboratory
Plant Diversity, Systematics and Herbarium Division
CSIR-National Botanical Research Institute
Lucknow – 226 001, Uttar Pradesh

Kanak Sahai
Seed Biology Laboratory
Plant Diversity, Systematics and Herbarium Division
CSIR-National Botanical Research Institute
Lucknow – 226 001, Uttar Pradesh

L.B. Chaudhary
Plant Diversity, Systematics and Herbarium Division
CSIR-National Botanical Research Institute
Lucknow – 226 001, Uttar Pradesh

M.R. Suseela
Algology Laboratory
Plant Diversity, Systematics and Herbarium Division
CSIR-National Botanical Research Institute
Lucknow – 226 001, Uttar Pradesh

M. A. Usmani
Algology Laboratory
Plant Diversity, Systematics and Herbarium Division
CSIR-National Botanical Research Institute
Lucknow – 226 001, Uttar Pradesh

Manoj Semwal
CSIR–Central Institute of Medicinal and Aromatic Plants
Lucknow – 226 015, Uttar Pradesh

P.B. Khare
Pteridology Laboratory
Plant Diversity, Systematics and Herbarium Division
CSIR-National Botanical Research Institute
Lucknow – 226 001, Uttar Pradesh

P.S. Chauhan
Division of Plant Microbe Interactions
CSIR-National Botanical Research Institute
Lucknow – 226 001, Uttar Pradesh

Poonam C. Singh
Division of Plant Microbe Interactions
CSIR-National Botanical Research Institute
Lucknow – 226 001, Uttar Pradesh

Rajani Verma
Division of Plant Microbe Interactions
CSIR-National Botanical Research Institute
Lucknow – 226 001, Uttar Pradesh

S.K. Raj
Plant Molecular Virology Laboratory
CSIR-National Botanical Research Institute
Lucknow – 226 001, Uttar Pradesh

Susheel Kumar
Plant Molecular Virology Laboratory
CSIR-National Botanical Research Institute
Lucknow – 226 001, Uttar Pradesh

S.P. Singh
Division of Plant Microbe Interactions
CSIR-National Botanical Research Institute
Lucknow – 226 001, Uttar Pradesh

Shipra Pandey
Division of Plant Microbe Interactions
CSIR-National Botanical Research Institute
Lucknow – 226 001, Uttar Pradesh

Sanjeeva Nayaka
Lichenology Laboratory
Plant Diversity, Systematics and Herbarium Division
CSIR - National Botanical Research Institute
Lucknow – 226 001, Uttar Pradesh

M.M. Pandey
Pharmacognosy and Ethnopharmacology Division
CSIR-National Botanical Research Institute
Lucknow – 226 001, Uttar Pradesh

R.K. Roy
Botanic Garden
CSIR-National Botanical Research Institute
Lucknow – 226 001, Uttar Pradesh

Rameshwar Prasad
Botanic Garden
CSIR-National Botanical Research Institute
Lucknow – 226 001, Uttar Pradesh

R.C. Nainwal
Distant Research Centers
CSIR-National Botanical Research Institute
Lucknow – 226 001, Uttar Pradesh

Shilpi Singh
Botanic Garden
CSIR-National Botanical Research Institute
Lucknow – 226 001
Uttar Pradesh

Saurabh Sachan
Central National Herbarium
Howrah – 711 103, West Bengal

S. Verma
Plant Diversity, Systematics and Herbarium Division
CSIR-National Botanical Research Institute
Lucknow – 226 001, Uttar Pradesh

S.A. Ranade
PMB (Genomics) Laboratory
CSIR-National Botanical Research Institute
Lucknow – 226 001, Uttar Pradesh

Soumit K. Behera
Plant Ecology and Environmental Sciences Division
CSIR-National Botanical Research Institute
Lucknow – 226 001, Uttar Pradesh

S.K. Tewari
Distant Research Centers
CSIR-National Botanical Research Institute
Lucknow – 226 001, Uttar Pradesh

S.K. Sharma
Distant Research Centers
CSIR-National Botanical Research Institute
Lucknow – 226 001, Uttar Pradesh

Shweta Singh
Distant Research Centers
CSIR-National Botanical Research Institute
Lucknow – 226 001, Uttar Pradesh

T.S. Rana
Molecular Systematics Laboratory
Plant Diversity, Systematics and Herbarium Division
CSIR-National Botanical Research Institute
Lucknow – 226 001, Uttar Pradesh

Chapter – 1

Methods and Tools to Assess the Soil Microbial Community

P.S. Chauhan, A. Mishra and C.S. Nautiyal

1. Introduction

Microbial diversity is becoming increasingly essential with the requirement to preserve the integrity, function and long-term sustainability of natural and managed ecosystems. However, our knowledge of the diversity of microbes remains obscure due to lack of appropriate approaches to evaluate the contribution of different components of the soil microbes to ecosystem. This lack of approaches together with the complex and heterogeneous matrix that soil microbes are embedded in, poses great challenges to understand the dynamics of microbial communities in soils (Nautiyal *et al.* 2010; Chauhan *et al.* 2013).

In classical terms, diversity is described as the total number of species present, i.e. species richness or species abundance, and the distribution of individuals among those species i.e. species evenness or species equitability. However, the incorporation of these two components into diversity indices has led to much controversy among ecologists (Atlas 1984). The Shannon-Weaver index (Shannon and Weaver 1949) is probably the most widely used index for measuring diversity of a microbial community. But this index, as many others that have been proposed, has its own limitations. A major difficulty is that both richness and evenness play a role in determining the value of the index. Therefore, any diversity index is a single value and thus cannot indicate the total make-up of a community. It is recommended that evenness, richness and diversity values should be calculated in order to obtain an objective assessment of community structure (Atlas 1984). From molecular point of view, diversity often refers to the number of different sequence type present in a habitat e.g. a soil (Ogram 2000).

Measuring this type of diversity does not give information about the frequency of individual species, but it can give valuable information about changes in the community structure and species richness in response to changes in the physico-chemical properties of the soil and soil management practices (Nautiyal *et al.* 2010).

A number of modern techniques have been performed to assess the microbial diversity in the soil. Despite the enormous genetic diversity within each of the bacterial divisions, the high frequency with which these groups are recovered from soils of diverse origin argues that many bacteria within each of these groups have special characteristics that make them specially suited to life in the soil (Sait *et al.* 2002). However, high diversity is beneficial to ecosystem performance since it guarantees a certain functional redundancy and enhances productivity, stability and resilience (Chauhan *et al.* 2011).

Soil as a microhabitat

It has been acknowledged that the wealth of genetic complexity within soil by far exceeds that of any other habitat on earth (Torsvik *et al.* 2002). Progress in testing contemporary ecological hypotheses in soil has been limited by the difficulty of accurately measuring species richness and evenness and by the lack of information about the species really involved in the measured microbial activities. Soil is a complex microhabitat (Fig.1) for the following distinctive properties (Nannipieri and Badalucco 2003).

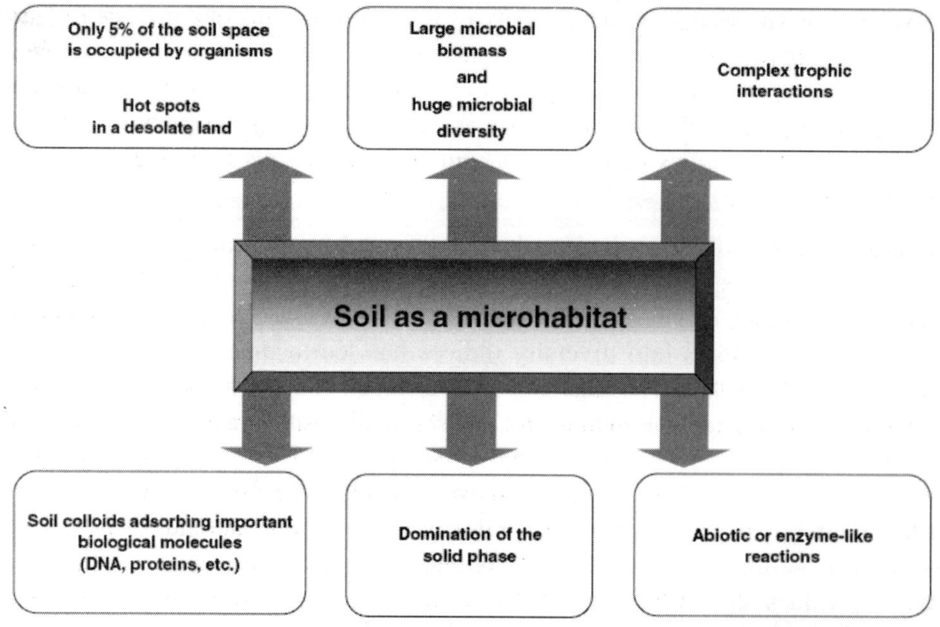

Fig. 1. Schematic representation of the main characteristics of soil as a microhabitat (Nannipieri *et al.* 2003).

The microbial population in soil is very diverse. Soils are reported to contain 10^5–10^8 bacteria, 10^6–10^7 actinomycetes and 10^5–10^6 fungal colony forming unit g^{-1} (Cresswell and Wellington, 1992). Likewise, according to a recent estimate, 1 g of soil may harbor up to 10 billion bacteria of possibly 4000–7000 different species and a biomass density of 300–30,000 kg ha^{-1} (Rosello-Mora and Amann 2001). Bacteria that are able to grow on agar plates typically account for only around 1% of the bacteria in soil (Hobbie and Fletcher 1988).

The chemical, physical and biological characteristics of these microhabitats differ in both time and space. Even if the available space is extensive in soil, the biological space, that is, the space occupied by living microorganisms, represents a small proportion, generally less than 5% of the overall available space (Ingham *et al.* 1985). Another peculiarity is the presence of 'hot spots', zones of increased biological activity, zones with accumulated particulate organic matter and the rhizosphere (Nautiyal *et al.* 2008).

According to Hattori (1973), almost 80–90% of the microorganisms inhabiting soil are on solid surfaces. Mechanisms by which these microorganisms interact with soil surfaces have been much studied. Another distinctive characteristic of soil as a microhabitat is the property of the solid phase to adsorb important biological molecules such as proteins and nucleic acids. The adsorption of organic compounds by soil colloids retards their microbial degradation; the location of potential substrates inside pores or microaggregates reduces their accessibility to soil microorganisms.

The surfaces of soil mineral components can themselves catalyze many reactions. Clay minerals and Mn (III and IV) and Fe (III) oxides catalyze electron transfer reactions, such as the oxidation of phenols and polyphenols with formation of humic substances (Ruggiero *et al.* 1996). Other abiotic reactions catalyzed by soil minerals include deamination, polymerization, polycondensation and ring cleavage.

2. Microbial Ecology and Microbial Habitats

Even after three centuries of research, soils are still considered the last great frontiers for biodiversity research and are home to an extraordinary range of microbial groups. The period between 1890 and 1910 is considered as the first 'Golden Age' of soil microbiology, however, the recent explosion in microbiology research should finally break open the complexities of the soil biota 'black box'. The current excitement among soil biologists can largely be attributed to developments in modern molecular techniques which are not only helping to unravel the exceptionally high microbial diversity but also developing new concepts in microbial ecology or challenging some old ones. The term 'functional microbial ecology' is gaining momentum as it is now possible to determine, in detail, population dynamics through functional diversity to the molecular regulation of key processes, including the intractable micro-scale level microbial habitat structure (Fig. 2).

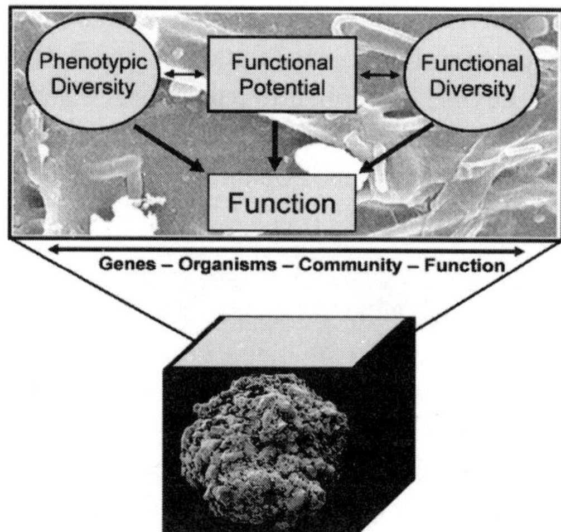

Fig. 2. Functional microbial ecology: a new approach to soil microbial community analysis and the opening of the soil biota 'black box' (Gupta *et al.* 2008).

A range of molecular, biochemical, physiological, microscopic and isotopic techniques are now available to detect, identify and study *in situ* various functional groups of microbiota and their links to functions. The progress achieved in the development of various methods has now reached to a level that they could be and are being used for a meaningful (reproducible/reliable) assessment of the diversity (taxonomic and functional) and functioning of soil biota. Soil structure and microbial habitat are intimately related in terms of genesis and functioning of soils. Because of the high microbial diversity, complex spatial arrangement of biota and their resources, and physical protection of microorganisms, it has been difficult to understand the nature and dynamics of microbial habitat. Methodological restrictions in soil microbiology, have prevented the development and testing of unifying concepts for microbial ecology that integrate the diversity, physiology and ecology of organisms with habitat properties. Modern microbial techniques show a great promise in linking the ecological concepts and theoretical framework into ecosystem level functions, which in turn would allow us to apply soil microbiology principles to practical applications.

3. Approaches to Measure Microbial Diversity in Soil

Microbial diversity in soil has been analyzed by various methods including culturable and unculturable part. There is strong evidence that most of the soil bacteria observed under the microscope are viable and active, but are unable to form visible colonies on agar plates (Amann *et al.* 1995). A combination of both culturable and unculturable approaches is required to get more accurate measure of the extent of microbial diversity in soil. It is still difficult with current techniques to study true diversity since we do not know what is present and we have no way of determining the accuracy of our extraction or detection methods. However, the methods usually available for studying microbial diversity can be discussed in the following three categories (Fig. 3).

- Genetic diversity
- Structural diversity
- Functional diversity

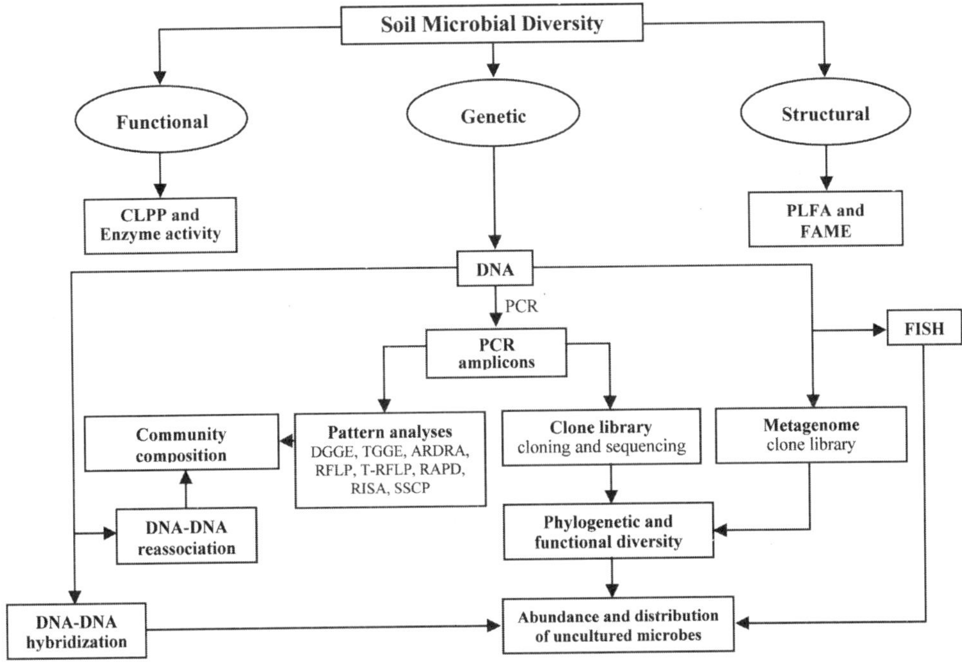

Fig. 3. Methods of accessing microbial diversity. *CLPP* community level physiological profiling, *PLFA* phospholipid fatty acids, *FAME* fatty acid methyl ester, *PCR* polymerase chain reaction, *DGGE* denaturing gradient gel electrophoresis, *TGGE* temperature gradient gel electrophoresis, *ARDRA* amplified rDNA restriction analysis, *RFLP* restriction fragment length polymorphism, *T-RFLP* terminal restriction fragment length polymorphism, *RAPD* random amplification of polymorphic DNA, *RISA* ribosomal intergenic spacer analysis, *SSCP* single-stranded conformation polymorphism, *FISH* fluorescence *in situ* hybridization.

3.1. Genetic diversity

Significant advances in equipment design, automation, and decreases in prices of equipment have arisen as a result of the Human Genome project and the widespread commercial interest in genetic engineering. Consequently these technologies are now within the reach of most research laboratories and they are being quickly taken up. The genetic diversity of soil microorganisms is an indicator of the genetic resource. Methods for determination of the genetic microbial diversity are as mentioned below.

3.1.1. 16S rDNA

Genetic diversity of bacteria is most commonly studied by diversity of the 16S rDNA genes, which occur in all bacteria and which show variation in base composition among species. 16S rDNA genes are thus used for phylogenetic affiliation of Eubacteria and Archaea and large databases exist on sequences of 16S rDNA (e.g. (www.ncbi.nlm.nih.gov/BLAST/) and http://rdp.cme.msu.edu/html/). It consists of variable and conserved regions, and this has facilitated the design of primers in the conserved regions for targeting the majority of members of defined groups of bacteria. Several comparable methods have been developed to examine the diversity of 16S rDNA sequences in total DNA extracted from soil microbial communities.

3.1.2. Denaturing gradient gel electrophoresis (DGGE)/temperature gradient gel electrophoresis (TGGE)

Denaturing gradient gel electrophoresis (PCR-DGGE) (Muyzer *et al.* 1993) and Temperature gradient gel electrophoresis (PCR-TGGE) (Heuer and Smalla 1997) are widely used methods for estimation of microbial community diversity. They are based on variation in base composition and secondary structure of fragments of the 16S rDNA molecule. DNA is extracted from soil samples and amplified using PCR with universal primers targeting part of the 16S or 18S rRNA sequences. The 5′-end of the forward primer contains a 35–40 base pair GC clamp to ensure that at least part of the DNA remains double stranded. This is necessary so that separation on a polyacrylamide gel with a gradient of increasing concentration of denaturants (formamide and urea) will occur based on melting behavior of the double-stranded DNA. If the GC-clamp is absent, the DNA would denature into single strands. On denaturation, DNA melts in domains, which are sequence specific and will migrate differentially through the polyacrylamide gel (Muyzer 1999). Theoretically, DGGE can separate DNA with one base-pair difference (Miller *et al.* 1999). TGGE uses the same principle as DGGE except the gradient is temperature rather than chemical denaturants. DGGE/TGGE has been used to assess the diversity of bacteria and fungi in the rhizosphere (Smalla *et al.* 2001), caused by changes of nutrient addition (Iwamoto *et al.* 2000) and addition of anthropogenic chemicals (Whiteley and Bailey 2000). The partial community level fingerprints derived from DGGE/TGGE banding patterns have been analyzed for diversity studies based on the number and intensity of the DNA bands as well as similarity between treatments.

3.1.3. Terminal restriction fragment length polymorphism (T-RFLP)

Terminal restriction fragment length polymorphism (T-RFLP) (Liu *et al.* 1997) is an alternative method for examining diversity of 16S rDNA sequences of microbial communities. It is also based on PCR amplification of 16S rDNA with specific primers. The primers are labeled with a fluorescent tag at the terminus resulting in labeled

PCR-products. The products are cut with several restriction enzymes, one at a time, which result in labeled fragments that can be separated according to their size on agarose gels. Recently, the potential of the T-RLFP method to discriminate soil bacterial communities in cultivated and non-cultivated soils has been demonstrated (Buckley and Schmidt 2001).

3.1.4. Single strand conformation polymorphism (SSCP)

Another technique that relies on electrophoretic separation based on differences in DNA sequences is single strand conformation polymorphism (SSCP). Like DGGE/TGGE, this technique was originally developed to detect known or novel polymorphisms or point mutations in DNA (Orita *et al.* 1989). When DNA is denatured into single strands each strand folds up into a configuration based not only on size but also on sequence. This feature can separate single stranded DNA on an agarose gel according to folding and secondary structure (Lee *et al.* 1996). SSCP has been used to study bacterial or fungal community diversity (Stach *et al.* 2001).

3.1.5. Restriction fragment length polymorphism (RFLP)/amplified ribosomal DNA restriction analysis (ARDRA)

Restriction fragment length polymorphism (RFLP), also known as amplified ribosomal DNA restriction analysis (ARDRA) is another tool used to study microbial diversity that relies on DNA polymorphisms. In the study by Liu *et al.* (1997), PCR amplified rDNA is digested with a 4-base pair cutting restriction enzyme. Different fragment lengths are detected using agarose or non-denaturing polyacrylamide gel electrophoresis in the case of community analysis (Tiedje *et al.* 1999). This method is useful for detecting structural changes in microbial communities but not as a measure of diversity or detection of specific phylogenetic groups (Liu *et al.* 1997).

3.1.6. Ribosomal intergenic spacer analysis (RISA)

A final example of a method based on PCR amplification of rDNA fragments is ribosomal intergenic spacer analysis (RISA) in which the region between 16S and 23S rDNA is amplified by primers targeting conserved regions of 16S and 23S rDNA. The amplified fragments vary considerably in size between species but also between the ribosomal operons within a single cell. The fragments are separated according to size by electrophoresis. The RISA method has been used in soil and rhizosphere to monitor effects of deforestation (Borneman and Triplett 1997).

3.2. Structural diversity

The structural or phenotypic analyses concept can be applied at a community levels, hence, they offer an integrated picture as to the physical, chemical and biotic state of the soil microbial community.

Phospholipid fatty acids (PLFA) and Fatty acid methyl ester (FAME)

A method that does not rely on culturing of microorganisms is fatty acid methyl ester (FAME) analysis. Fatty acids are ubiquitous in their distribution and occupy a wide range of structural functions in both prokaryotes and eukaryotes. They are the basic components of many lipids, including phospholipids and the total amount of phospholipids in cells is correlated to biomass. Phospholipid fatty acids (PLFAs) are stable components of the cell wall of most microorganisms. They are polar lipids specific for subgroups of microorganisms, e.g. gram-negative or gram-positive bacteria, methanotrophic bacteria, fungi, mycorrhiza, and actinomycetes (Zelles 1999). Individual PLFAs can thus be related to microbial community structure. PLFA profiles do not give information of species composition, but rather an overall picture, or fingerprint of community structure. Fatty acids make up a relatively constant proportion of the cell biomass and signature fatty acids exist that can differentiate major taxonomic groups within a community. Therefore, a change in the fatty acid profile would represent a change in the microbial population. It has been used to study microbial community composition and population changes due to chemical contaminants (Kelly *et al.* 1999) and agricultural practices (Bossio *et al.* 1998). The readers are referred to a comprehensive review of the use of fatty acid patterns of phospholipids and lipopolysaccharides to characterize microbial populations by Zelles (1999). For FAME analysis, fatty acids are extracted directly from soil, methylated and analyzed by gas chromatography (Ibekwe and Kennedy 1999). FAME profiles of different soils can be compared using multivariate analysis. This method will detect changes in the composition of the bacterial and/or fungal community, as well as enable one to follow signature fatty acids of different groups of microorganisms. Since FAME analysis directly extracts fatty acids from soils, it overcomes the limitations of culture-dependent methods. However, it is still a poor method if used as the sole analysis to study the microbial diversity. For example, a lot of material will be needed if fungal spores are used to study the potential fungal diversity. Cellular fatty acid composition can be influenced by factors such as temperature and nutrition, and the possibility exists that other organisms can confound the FAME profiles (Graham *et al.* 1995).

Bossio *et al.* (1998) used PLFA profiles to detect changes in microbial communities consistent with different farming practices. But when these researchers calculated the Shannon diversity index based on PLFA relative abundance, no difference was detected. This could be because although the community was structurally different, diversity was not, or it could represent some problems with using fatty acid profiles to measure diversity (Bossio *et al.* 1998). However, Thirup *et al.* (2003) found the PLFA method to be as good or even better at separating microbial communities in the bulk soil and rhizosphere compared to Biolog and PCR-DGGE. The method has recently been recommended for soil monitoring in Scotland and Northern Ireland (Chapman *et al.* 2000). However, individual fatty acids cannot be used to represent specific species because individuals can have numerous fatty acids and the same fatty acids can occur in more than one species (Bossio *et al.* 1998). In addition, several bacteria contain profile of FAME comprising straight chain, which do not provide sufficient variety to distinguish between species.

3.3. Functional diversity

Functional diversity is critical to ecosystem functioning because of the variety of processes for which microbes are responsible (Nannipieri *et al.* 2003). A loss of microbial taxon may not change the functioning of the system as different microbial taxon can carry out the same function, a phenomenon defined as the functional redundancy (Allison and Martiny 2008). This means that even though anthropogenic activities might affect the genetic composition of the soil microbial communities, gross microbial processes and their role in maintaining soil quality could remain unaffected (Crecchio *et al.* 2004). Functional diversity in soils includes the magnitude and capacity of soil inhabitants that are involved in key roles such as nutrient cycling, decomposition of various compounds and other transformations (Zak *et al.* 1994).

Functional diversity of microbial populations in soil may be determined by either expression of different enzymes (carbon utilization patterns, extra-cellular enzyme patterns) or diversity of nucleic acids (mRNA, rRNA) within cells, the latter also reflecting the specific enzymatic processes operating in the cells. Since changes in the activity of soil organisms may reflect changes in soil quality, indicators of microbial activity and function could be used to describe the status and trends in soil conditions due to management practices. The two most commonly used methods of measuring substrate utilization patterns are the Biolog utilization pattern and substrate induced respiration technique (SIR).

3.3.1. Community level physiological profiling (CLPP)

Carbon source utilization patterns can be measured by the Biolog assay (Garland and Mills, 1991), also named community level physiological profiling (CLPP). Over the last 20 years the soil microbiologists have been provide with an array of new and powerful tools for investigating diversity and diversity effects at all levels. These studies have shown us that the diversity in soils is vast and that microbial populations change in response to external stimuli. CLPP approach has been used to investigate the effects of management on the functional or physiological diversity of soils (Nautiyal 2009, Mishra and Nautiyal 2009, Nautiyal *et al.* 2010 a,b,c).

Biolog is a commercially available 96-well microtitre plate used to assess the potential functional diversity of the bacterial population through sole source carbon utilization (SSCU) patterns. Gram-negative (GN) and gram-positive (GP) plates are available from Biolog (Hayward, CA, USA, www.biolog.com) and each contains 95 different carbon sources and one control well without a substrate. GN and GP plates were developed originally for characterization of clinical bacterial isolates and not for community analysis. Subsequently, Biolog introduced an Eco-plate (Choi and Dobbs, 1999) containing 3 replicates of 31 different environmentally relevant carbon sources and one control well per replicate. The tetrazolium salt changes colour as the substrate is metabolized. Since many fungal species are not capable of reducing the tetrazolium salt, Biolog developed fungal specific plates SFN2 and SFP2, which have the same substrates as GN and GP plates but without the tetrazolium salt (Classen *et al.* 2003). Inoculated populations are monitored over time for their ability to utilize substrates and the speed at which these substrates are utilized. Differences in the profiles can be analyzed by multivariate statistics and experience with data interpretation is still developing (e.g. Garland and Mills 1991; Preston-Mafham *et al.* 2002). Considerable data are, however, available for a future reference database and this may facilitate data interpretation.

The Biolog assay is dependent on growth of cells under the specific conditions in the microtiter plate and thereby indicating only potential functional diversity. The number of inoculated cells are thus important for the rate and degree of color development, while the extraction method of the bacteria from soil are of less importance (Mayr *et al.* 1999; Winding and Hendriksen 1997). Optical density reading several times during color development opens the possibility of calculating various growth parameters representing the physiology of the growth responsive populations (Preston-Mafham *et al.* 2002). This method has been used successfully to assess potential metabolic diversity of microbial communities in contaminated sites, plant rhizospheres, arctic soils (Grayston *et al* 1998, Derry *et al.* 1999 Nautiyal 2009, Mishra and Nautiyal 2009, Nautiyal *et al.* 2010a, b, c).

CLPP, which can differentiate between microbial communities, is relatively easy to use, reproducible and produce a large amount of data reflecting metabolic characteristics of the communities (Zak *et al.* 1994). The technique has gained widespread use, primarily due to the ease of use and the capacity to produce comprehensive data sets. The Biolog assay has been shown to be more sensitive than microbial biomass and respiration measurements to impacts of soil management practices and of sewage sludge amendments to soil (Bending *et al.* 2000). The assay is currently implemented in the Dutch Soil Monitoring Program where it has been shown to be discriminatory to different types of soil and management practices (Schouten *et al.* 2000). The assay is also recommended for soil monitoring in Scotland and Northern Ireland (Chapman *et al.* 2000). Limitations of metabolic profiling are: the methods select for only culturable microorganisms capable of growing under the experimental conditions (Garland and Mills 1991), favors fast growing microorganisms (Yao *et al.* 2000), is sensitive to inoculum density and reflects the potential, and not the *in situ*, metabolic diversity (Garland and Mills 1991). Nonetheless, CLPP is very useful when studying the functional diversity of soils and is a valuable tool especially when used in conjunction with other methods.

3.3.2. Substrate induced respiration technique (SIR)

A logical extension to the concept of functional analysis was pioneered by Degens *et al.* (2001) who developed a catabolic community profiling technique (SIR) in which the substrate is taken to the community: catabolic responses by soils to 88 substrate were measured over the first 4 hours following addition, before any cell division is likely to have occurred and hence to detect the response of the *in situ* community. The resulting activity profiles showed a clear discrimination between different soils and soil management regimes, and recent work emphasizes this further (Degens *et al.* 2001).

3.3.3. Enzymatic activity

The enzymatic activity in soil is mainly of microbial origin being derived from intracellular, cell associated or free enzymes. Only enzymatic activity of ecto-enzymes and free enzymes is used for determination of the diversity of enzyme patterns in soil extracts. Discrimination between free and cell associated enzyme activity can be obtained by a simple filtration step to separate microbial cells from the soil extract. However, enzymes are the direct mediators for biological catabolism of soil organic and mineral components. Thus, these catalysts provide a meaningful assessment of reaction rates for important soil processes. Soil enzyme activities (i) are often closely related to soil organic matter, soil physical properties and microbial activity or biomass, (ii) change much sooner than other parameters, thus providing early indications of changes in soil health, and (iii) involve simple procedures (Dick *et al.* 1996). Disturbance of the soil microbial activity, as shown by changes in levels of metabolic enzymes, can serve as an estimate of ecosystem disturbance.

Easy, well-documented assays are available for a large number of soil enzyme activities (Dick *et al.* 1996). Measurements of soil enzyme reaction are usually based on the addition of an artificial, soluble substrate at a concentration sufficient to maintain zero-order kinetics, thus achieving a reaction rate proportional to enzyme concentration. Long incubation periods have to be omitted to avoid substrate depletion and microbial growth. Enzyme activities are usually determined by a dye reaction followed by a spectrophotometric measurement. The data are typically analyzed by multivariate statistics. If incubation times are kept short, cell growth and synthesis of new enzymes are prevented. It has been recommended that a diverse set of enzyme activities are measured, since a few dominating organisms expressing a high enzyme activity may give a biased result (Miller *et al.* 1998).

4. Advantages and Disadvantages of Different Approaches

As already discussed above, a number of culture-based and culture-independent approaches have been developed to study soil microbial diversity. Each of all methods provide valuable information about the microbial community, however, all have their own limitations. These arise not only from methodological limitations, but also from a lack of taxonomic knowledge. It is difficult to study the diversity of a group of microorganisms when it is not understood how to categorize or identify the species present. The advantages and disadvantages of different methods used to measure soil microbial diversity are presented in Table 1.

Table 1. Advantages and disadvantages of different methods to study soil microbial diversity (Kirk *et al.* 2004).

Method	Advantage	Disadvantage	Reference
Plate counts	Fast; Inexpensive	Unculturable microorganisms not detected; Bias towards fast growing individuals; Bias towards fungal species that produce large quantities of spores	Trevors 1998; Tabacchioni *et al.* 2000
CLPP	Fast; Highly reproducible; Relatively inexpensive; Differentiate between microbial communities; Option of using bacterial, fungal plates or site specific carbon sources (Biolog); Generates large amount of data	Only represents culturable fraction of community; Favors fast growing organisms; Potential metabolic diversity, not in situ diversity; Sensitive to inoculum density; Only represents those organisms capable of utilizing specific carbon source	Garland and Mills 1991; Classen *et al.* 2003; Nautiyal *et al.* 2010
FAME	No culturing of microbes; Direct extraction from soil; Follow specific organisms or Communities	If using fungal spores lot of material needed; Can be influenced by external factors; Possibility results can be confounded by other microorganisms	Graham *et al.* 1995; Siciliano and Germida 1998;
DNA microarrays and DNA hybridization	Same as nucleic acid hybridization; Thousands of genes can be analyzed; If using genes or DNA fragments, increased specificity	Only detect most abundant species; Need to be able to culture organisms; Only accurate in low diversity systems	Cho and Tiedje, 2001; Greene and Voordouw 2003
DGGE and TGGE	Large number of samples can be analyzed simultan-eously' Reliable, reproducible and rapid	PCR biases; Dependent on lysing and extraction efficiency; Sample handling influence community; One band can represent more species; Only detects dominant species	Muyzer *et al.* 1993; Duineveld *et al.* 2001
SSCP	Same as DGGE/TGGE; No GC clamp;No gradient	PCR biases; Some ssDNA can form more than onestable conformation	Lee *et al.* 1996; Tiedje *et al.* 1999
ARDRA or RFLP	Detect structural changes in microbial community	PCR biases; Banding patterns often too complex	Liu *et al.* 1997; Tiedje *et al.* 1999
RISA or ARISA	Highly reproducible community profiles; Can be automated (ARISA)	Requires large quantities of DNA; Resolution tends to be low-PCR biases	Fisher and Triplett 1999 Fisher and Triplett 1999

CLPP community level physiological profiling, *FAME* fatty acid methyl ester, *DGGE* and *TGGE* denaturing and temperature gradient gel electrophoresis, *SSCP* single strand conformation polymorphism, *ARDRA* amplified ribosomal DNA restriction analysis, *RFLP* restriction fragment length polymorphism, *RISA* ribosomal intergenic spacer analysis, *ARISA* automated ribosomal intergenic spacer analysis.

5. The Holistic Approach

The extraction of large amounts of DNA and RNA from soil and the accurate taxonomic and functional characterization of these aids can increase our knowledge of the composition of soil microflora, of pathogenicity, of trophic interactions between soil microorganisms, and of the production of bioactive molecules by soil microorganisms. These have important implications for agriculture, the environment, industry and pharmacology. However, the molecular approach does not seem sufficient for a better understanding of soil functioning (Fig. 4). In this regard it is also important to understand the link between community structure and functions in microhabitats, that is, the spatial distribution of microbial species and their spatial and temporal links with the various microbe mediated reactions *in situ*. For this reason, high-resolution techniques must be applied to soil preparations to be examined by electron microscopy.

However, it is not always rewarding to determine the composition of soil microflora and the various microbially mediated activities when assessing the functionality of the soil. Determining the composition of the soil microflora, the concentration of each metabolite, and the rate of each transformation reaction can be not only time-consuming, laborious and expensive, but also unnecessary for quantifying nutrient cycling or assessment of soil quality. Information on transformation rates of nutrients in soil has been obtained with labeled compounds, such as ^{14}C-, ^{13}C-labelled or ^{15}N-enriched compounds. In the holistic approach, the system is partitioned into pools with a functional meaning, and fluxes between these pools represent physical (such as leaching and volatilization) or abiotic or biotic transformations (Nannipieri *et al.* 1994). Then, the distribution of the isotope (reflecting the behavior of the added compound) between the various pools can be followed, and the behavior of the added compound can be discriminated with respect to that of the native carbon (C) or nitrogen (N). A better quantification of N transformations in soil can be obtained if we split the organic N pool into at least the more resistant and less resistant N organic pool, and the microbial biomass into the active and the inactive microbial pool. Current models represent both pools, but their N content is not measurable (Nannipieri and Badalucco 2003). Radajewski *et al.* (2000) have recently reported an example of this approach. They separated the ^{13}C-DNA extracted from soil treated with a ^{13}C source by density-gradient centrifugation and characterized the ^{13}C-DNA taxonomically and functionally by gene probing and sequence analysis. This technique is very promising because we might be able to calculate the active microbial C pool by multiplying the ratio between labeled and total DNA fractions by the content of microbial C in the soil. Instead of measuring the composition of soil microflora, it might be less laborious and time-consuming to monitor the behavior of key species which can function as indicators of the status of the soil microflora. However, microbial composition can be determined by PLFA (FAME) or molecular techniques (Fig. 4), depending on the degree of resolution required (Zelles 1999).

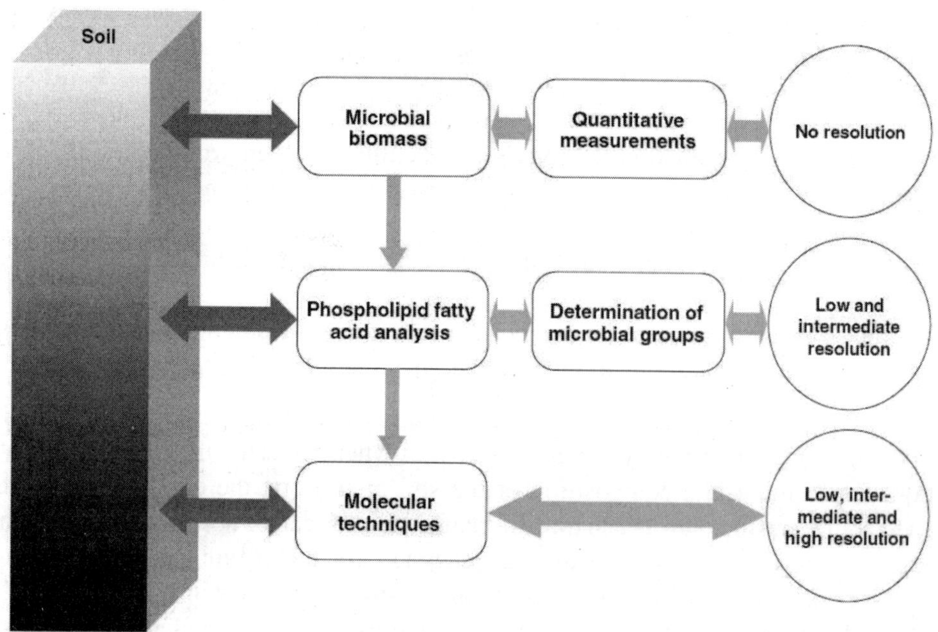

Fig. 4. Scientific insights provided by determining the composition of microbial communities or by using the holistic approach in soil (Nannipieri *et al.* 2003).

Our understanding of the links between microbial diversity and soil functions is poor because we cannot measure easily the microbial diversity, even if we can detect unculturable microorganisms by molecular techniques. In addition, the present assays for measuring microbial functions determine the overall rate of entire metabolic processes, such as respiration, or specific enzyme activities, without our identifying the active microbial species involved. According to O'Donnell *et al.* (2001), the central problem with the link between microbial diversity and soil function is to understand the relations between genetic diversity and community structure and between community structure and function. The recent advances in RNA extraction from soil might permit us to determine active species in soil (Griffiths *et al.* 2000). Further advances in understanding require us to determine the composition of microbial communities and microbial functions in microhabitats.

Griffiths *et al.* (2000, 2001) showed that the effect of microbial diversity on microbial functions depends on the function measured. Some functions increased (SIR) with decreasing microbial diversity in soil, others were not affected (NO_3 accumulation, respiratory growth response), and some declined when microbial diversity was small (short-term respiration from added grass, potential nitrification rates, Biolog) (Griffiths *et al.* 2001). No relation exists between microbial diversity and decomposition of organic matter, and a reduction in any group of species has little effect on overall soil process because the surviving microorganisms can carry out the decomposition of organic matter (Giller *et al.* 1998).

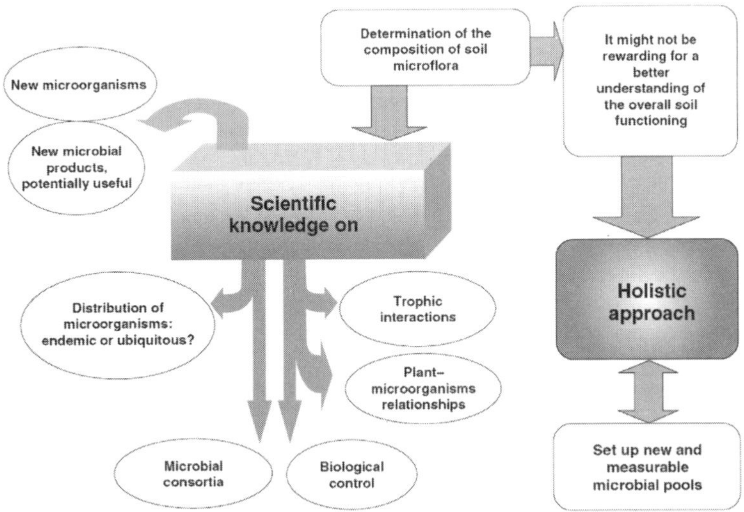

Fig. 5. Taxonomic resolution of methods used to determine microbial biomass and microbial composition in soil (Nannipieri *et al.* 2003).

The use of recent techniques has improved the determination of the composition of soil microflora (Fig. 5). Recent advances include clone libraries followed by pyrosequencing can allow the functional and taxonomic analysis of large numbers/ segments of soil metagenome with further insights into pathogenicity, competitiveness, substrate range and bioactive molecule production by soil microorganisms.

References

Amann R, Ludwig W, Schleifer KH (1995) Phylogenetic identification and *in situ* detection of individual microbial cells without cultivation. Microbiol Rev 59:143-169

Atlas RM (1984) Diversities of microbial communities. Adv Microb Ecol 7:1-47

Bending GD, Putland C, Rayns F (2000) Changes in microbial community metabolism and labile organic matter fractions as early indicators of the impact of management on soil biological quality. Biol Fertil Soils 31:78-84

Borneman J, Triplett EW (1997) Molecular microbial diversity in soils from eastern Amazonia: evidence for unusual microorganisms and microbial population shifts associated with deforestation. Appl Environ Microbiol 63:2647-2653

Bossio DA, Scow KM, Gunapala N, Graham KJ (1998) Determinants of soil microbial communities: effects of agricultural management, season, and soil type on phospholipid fatty acid profiles. Microb Ecol 36:1-12

Buckley DH, Schmidt TM (2001) The structure of microbial communities in soil and the lasting impact of cultivation. Microb Ecol 42:11-21

Chapman SJ, Campbell CD, Edwards AC and McHenery JG (2000) Assessment of the potential of new biotechnology environmental monitoring techniques. Report no SR (99) 10F, Macaulay Research and Consultancy Services, Aberdeen

Chauhan PS, Chaudhry V, Mishra S, Mishra A and Nautiyal CS (2013) Unraveling the shed of unexplored rhizosphere microbial diversity. In: Molecular Microbial Ecology of the Rhizosphere: Volume 1 & 2. de Bruijn FJ (ed.) John Wiley & Sons, Inc., Hoboken, NJ, USA, pp. 105-114

Chauhan PS, Chaudhry V, Mishra S, Nautiyal CS (2011) Uncultured bacterial diversity in tropical maize (*Zea mays* L.) rhizosphere. J Basic Microbiol 51:15-32

Cho JC, Tiedje JM (2001) Bacterial species determination from DNA–DNA hybridization by using genome fragments and DNA microarrays. Appl Environ Microbiol 67:3677-3682

Choi KH, Dobbs FC (1999) Comparison of two kinds of Biolog microplates (GN and ECO) in their ability to distinguish among aquatic microbial communities. J Microbiol Meth 36:203-213

Classen AT, Boyle SI, Haskins KE, Overby ST, Hart SC (2003) Community-level physiological profiles of bacteria and fungi: plate type and incubation temperature influences on contrasting soils. FEMS Microbiol Ecol 44:319-328

Crecchio C, Curci M, Pizzigallo MDR, Ricciuti P, Ruggiero P (2004) Effects of municipal solid waste compost amendments on soil enzyme activities and bacterial genetic diversity. Soil Biol Biochem 36:1595-1605

Cresswell N and Wellington EMH (1992) Detection of genetic exchange in the terrestrial environment. In: Genetic interactions among microorgansims in the natural environment (Wellington EMH and van Elsas JD Eds.) pp. 59–82. Pergamon Press, Oxford, UK

Degens BP, Schipper LA, Sparling GP, Duncan LC (2001) Is the microbial community in a soil with reduced catabolic diversity less resistant to stress or disturbance. Soil Biol Biochem 33:1143-1153

Derry AM, Staddon WJ, Trevors JT (1998) Functional diversity and community structure of microorganisms in uncontaminated and creosote-contaminated soils as determined by sole-carbon-source-utilization. World J Microbiol Biotechnol 14:571-578

Derry AM, Staddon WJ, Kevan PG, Trevors JT (1999) Functional diversity and community structure of micro-organisms in three arctic soils as determined by sole-carbon-source-utilization. Biodivers Conserv 8:205-221

Dick RP, Breakwell DP and Turco RF (1996) Soil enzyme activities and biodiversity measurements as integrative microbiological indicators. In: Methods for assessing soil quality. Doran JW and Jones AJ (eds). Soil Sci Soc America Inc, Madison, Wisconsin, pp 247-271

Duineveld BM, Kowalchuk GA, Keijzer A, van Elsas JD (2001) Analysis of bacterial communities in the rhizosphere of chrysanthemum via denaturing gradient gel electrophoresis of PCR-amplified 16S rRNA as well as DNA fragments coding for 16S rRNA. Appl Environ Microbiol 67:172-178

Fisher MM, Triplett EW (1999) Automated approach for ribosomal intergenic spacer analysis of microbial diversity and its application to freshwater bacterial communities. Appl Environ Microbiol 65:4630-4636

Garland JL, Mills AL (1991) Classification and characterization of heterotrophic microbial communities on the basis of patterns of community-level-sole-carbon-source utilization. Appl Environ Microbiol 5:2351-2359

Giller KE, Witter E, McGrath SP (1998) Toxicity of heavy metals to microorganisms and microbial processes in agricultural soils: a review. Soil Biol Biochem 30:1389-1414

Graham JH, Hodge NC, Morton JB (1995) Fatty acid methyl ester profiles for characterization of Glomalean fungi and their endomycorrhizae. Appl Environ Microbiol 6:58-64

Grayston SJ, Wang S, Campbell CD, Edwards AC (1998). Selective influence of plant species on microbial diversity in the rhizosphere. Soil Biol Biochem 30:369-378

Greene EA, Voordouw G (2003) Analysis of environmental microbial communities by reverse sample genome probing. J Microbiol Meth 5:211-219

Griffiths BS, Ritz K, Bardgett RD, Cook R, Christensen S, Ekelund F *et al.* (2000) Ecosystem response of pasture soil communities to fumigation-induced microbial diversity reductions: an examination of the biodiversity–ecosystem function relationship. Oikos 90:279-294

Griffiths BS, Ritz K, Wheatley RE, Kuan HL, Boag B, Christensen S *et al.* (2001) An examination of the biodiversity–ecosystem function relationship in arable soil microbial communities. Soil Biol Biochem 33:1713-1722

Gupta VVSR, Dick RP, Coleman DC (2008) Functional microbial ecology: Molecular approaches to microbial ecology and microbial habitats. Soil Biol Biochem 40:1269-1271

Hattori T (1973) Microbial Life in Soil. Marcel Dekker, New York

Ladd JN, Forster RC, Nannipieri P and Oades JM (1996) Soil structure and biological activity. In: Soil Biochemistry, Volume 9, Stotzky G and Bollag JM (eds) pp. 23-78 Marcel Dekker, New York

Heuer H and Smalla K (1997) Application of denaturing gradient gel electrophoresis and temperature gradient gel electrophoresis for studying soil microbial communities. In Modern soil microbiology. van Elsas JD, Trevors JT and Wellington EMH (eds) pp. 353-373. Marcel Dekker, Inc., New York

Hobbie JE and Fletcher MM (1988) The aquatic environment. In: Micro-organisms in action: Concepts and applications in microbial ecology. Lynch JM and Hobbie JE (Eds.) pp 132-162 Blackwell Scientific, Oxford. UK

Ibekwe AM, Kennedy AC (1999) Fatty acid methyl ester (FAME) profiles as a tool to investigate community structure of two agricultural soils. Plant Soil 206:151-161

Ingham RE, Trofymow JA, Ingham ER, Coleman DC (1985) Interactions of bacteria, fungi, and their nematode grazers: effects on nutrient cycling and plant growth. Ecol Monogr 55:119-140

Iwamoto T, Tani K, Nakamura K, Suzuki Y, Kitagawa N, Eguchi M, Nasu M (2000) Monitoring impact of *in situ* bio-stimulation treatment on groundwater bacterial community by DGGE. FEMS Microbiol Ecol 32:129-141

Kelly JJ, Haggblom M, Tate III RL (1999) Changes in soil microbial communities over time resulting from one time application of zinc: a laboratory microcosm study. Soil Biol Biochem 31:1455-1465

Lee DH, Zo YG, Kim SJ (1996) Nonradioactive method to study genetic profiles of natural bacterial communities by PCR single strand conformation polymorphism. Appl Environ Microbiol 62:3112-3120

Liu WT, Marsh TL, Cheng H, Forney LJ (1997) Characterization of microbial diversity by determining terminal restriction fragment length polymorphisms of genes encoding 16S rRNA. Appl Environ Microbiol 63:4516-4522

Mayr C, Winding A, Hendriksen NB (1999) Community level physiological profile of soil bacteria unaffected by extraction method. J Microbiol Meth 36:29-33

Miller KM, Ming TJ, Schulze AD and Withler RE (1999) Denaturing Gradient Gel Electrophoresis (DGGE): a rapid and sensitive technique to screen nucleotide sequence variation in populations. BioTechniques 27:1016-1030

Miller M, Palojarvi A, Rangger A, Reeslev M, Kjøller A (1998) The Use of Fluorogenic Substrates To Measure Fungal Presence and Activity in Soil. Appl Environ Microbiol 64:613-617

Mishra A, Nautiyal CS (2009) Functional diversity of the microbial community in the rhizosphere of chickpea grown in diesel fuel spiked soil amended with *Trichoderma ressei* using sole-carbon source utilization profiles. World J Microbiol Biotechnol 25:1175-1180

Muyzer G (1999) DGGE/TGGE a method for identifying genes from natural ecosystems. Curr Opin Microbiol 2:317-322

Muyzer G, de Waal EC, Uitterlinden AG (1993) Profiling of complex bacterial populations by denaturing gradient gel electro-phoresis analyses of polymerase chain reaction-amplified genes for 16S rRNA. Appl Environ Microbiol 59:695-700

Naeem S, Li SB (1997) Biodiversity enhances ecosystem reliability. Nature 390:507-509

Nannipieri P and Badalucco L (2003) Biological processes. In: Processes in the soil-plant system: Modelling concepts and applications. Bembi DK and Nieder R (eds) The Haworth Press, Binghamton, NY

Nannipieri P, Ascher J, Ceccherini MT, Landi L, Pietramellara G, Renella G (2003) Microbial diversity and soil functions. Eur J Soil Sci 54:655-670

Nannipieri P, Badalucco L and Landi L (1994) Holistic approaches to study of populations, nutrient pools and fluxes: limits and future research needs. In: Beyond the biomass: Compositional and functional analysis of soil microbial communities. Ritz K, Dighton J, Giller KE (eds), pp. 231-238. John Wiley and Sons, Chichester

Nautiyal CS (2009) Self-puriûcatory Ganga water facilitates death of pathogenic *Escherichia coli* 157:H7. Curr Microbiol 58:25-29

Nautiyal CS, Chauhan PS, Bhatia CR (2010) Changes in soil physico-chemical properties and microbial functional diversity due to 14 years of conversion of grassland to organic agriculture in semi-arid agroecosystem. Soil Till Res 109:55-60

Nautiyal CS, Chauhan PS, DasGupta SM, Seem K, Varma A, Staddon WJ (2010b) Tripartite interactions among *Paenibacillus lentimorbus* NRRL B-30488, *Piriformospora indica* DSM 11827, and *Cicer arietinum* L. World J Microbiol Biotechnol 26:1393-1399

Nautiyal CS, Rehman A, Chauhan PS (2010c) Environmental *Escherichia coli* occur as natural plant growth-promoting soil bacterium. Arch. Microbiol. 192:185-193

Nautiyal CS, Srivastava S and Chauhan PS (2008) Rhizosphere colonization Molecular determinants from plant-microbe coexistence perspective. In: Nautiyal CS and Dion P (eds) Molecular mechanisms of plant, microbe coexistence soil biology series, Vol. 15 Springer, Berlin, pp. 99-124

O'Donnell AG, Seasman M, Macrae A, Waite I, Davies JT (2001) Plants and fertilisers as drivers of change in microbial community structure and function in soils. Plant Soil 232:135-145.

Ogram A (2000). Soil molecular microbial ecology at age 20: methodological challenges for the future. Soil Biol Biochem 32:1499-1504

Orita M, Suzuki Y, Sekiya T, Hayashi K (1989) A rapid and sensitive detection of point mutations and genetic polymorphisms using polymerase chain reaction. Genomics 5: 874-879

Chapter – 2

Plant Virology: Diagnosis, Identification and Disease Management

S.K. Raj, Susheel Kumar and Aarti Kumari

1. Introduction

Viruses are a unique group of infectious agents. They are distinct in their simple, acellular organization; however, they are the connecting link between live and dead. Viruses are also the perfect example of obligate intracellular parasitism. Despite their simplicity, plant viruses are extremely important and drawn attention of researches because they can be used as model to study the disease establishment and progression mechanism, reproduction inside the host and trilateral host-virus-vector relationships. Their small genome also gave liberty to extensively understand the genome evolution caused by the genetic exchange or recombination. As a consequence, this will help in the development of facile methods for management and control of new emerging viruses and their strains.

This chapter is written for the beginners and students of virology subject to provide the fundamental knowledge about the origin, history of plant virology, the general features of plant viruses, their structure, shape and size, mode of spread and development of their management strategies. Besides this, the new development of non-conventional methods for developing virus-resistance in plants has also been explained for enhancing the quality and production of commercial crops in the country.

2. Virology

The science which deals with study of virus structure, its shape and size, physical, biochemical and molecular properties and possible viral disease management is called as virology.

2.1. General features of viruses

Virus is a small infectious agent that can replicate only inside the living cells of an organism. They are very minute particles, can pass through bacterial proof filters of 0.2 nm pore size. They can only be seen under electron microscope and measured in nanometers (1nm = 10^{-9} mm). They are obligate parasites and can survive only in living cells/tissues where they cause diseases. They cannot be cultured on any synthetic medium. Interestingly, they are connecting link between livings and non-livings.

Viruses can be of variable shapes and sizes e.g. spherical, geminate, rhomboid, bullet and rod shaped. Virus particles (known as *virions*) consist of two or three parts: (i) the genetic material made from either DNA or RNA; (ii) a protein coat that protects nucleic acid; and (iii) in some cases an envelope of lipid that surrounds the protein coat when they are outside the cell. Viruses can infect and cause diseases in plants, animals, fungi, bacteria and other organisms. Animal viruses are the viruses which infect animals/ human beings and cause various diseases. Some examples are: *Foot and mouth disease virus* (FMDV), *Chicken pox virus* (CPV), *Polio virus*, *Hepatitis virus* A/B/C/D/E, *Human Immunodeficiency virus* (HIV). While plant viruses are the viruses which infect plants and cause severe diseases and economic losses to the crop production. Some examples are *Tobacco mosaic virus* (TMV), *Cucumber mosaic virus* (CMV), *Papaya ring spot virus* (PRSV), *Banana bunchy top virus* (BBTV), *Tomato leaf curl virus* (ToLCV) and *Papaya leaf curl virus* (PLCV).

2.2. Discovery of plant viruses and developments in plant virology

The term virus is derived from the Latin word "*virus*" referring to "*poison*" and was used to denote a wide range of infectious agent including bacteria and microbes. Adolf Mayer (1886) reported and described a disease which was known as "*mosaic disease of tobacco*" and he further demonstrated that the disease can be transmitted by using the sap from the infected tobacco plants as the inoculums to infect healthy plants. At that time, this disease was thought to be spread by very small bacteria or toxins.

A Russian botanist (Dmitri Ivanowsky 1892) proved that the principle producing disease retained activity even after passing through bacteria filters and defined the pathogen as toxin producing entity. In 1898, Martinus Beijerinck extended Ivanowsky work by showing that mosaic agent multiplied in plant tissue and therefore could not be a toxin. He called this agent as "*contagium vivum fluidum*" and also discovered *Tobacco mosaic virus*. Later in 1926, Maurice Mulvania suggested that the mosaic disease of tobacco might be a protein of a very simple kind having characteristics of an enzyme. Boycott wrote an article entitled in 1928 on "*Transition from live to death: the nature of filterable virus*". In 1935, Wendell Meridith Stanley crystallized *Tobacco mosaic virus* and several plant viruses have been shown to consist of nucleoprotein in 1939 while Bawden in 1956, defined a virus as a obligate parasitic pathogen (http://en.wikipedia.org/wiki/ Plant_virus).

2.3. Structure of a virus and mechanism of infection

Virus consists of a single nucleic acid surrounded by a protein coat and capable of replication only within the living cells of the host like bacteria, animals or plants. During infection process, they adapts the following mechanism, say for example in plants, for its entry, uncoating, replication, assembly and release of the newly formed (virion) particles to neighbouring cells and to remote site of infection (Fig. 1).

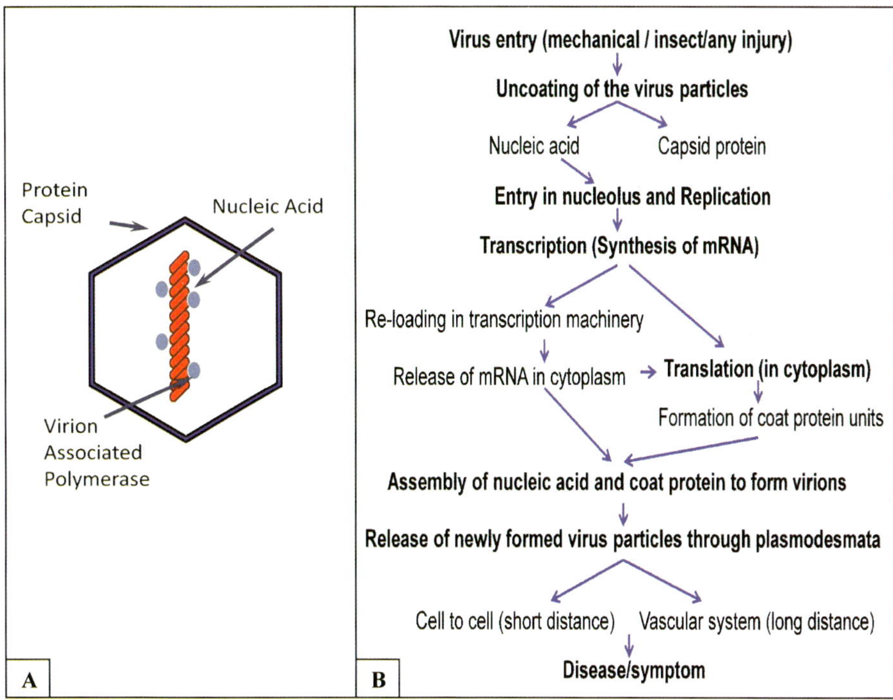

Fig. 1. (A) Generalized structure of a virus, (B) Mechanism of infection in plants by viruses.

3. Morphology of Viruses

Plant viruses are of different sizes, shapes, and DNA/RNA types. Morphological characteristics of plant viruses are generally observed as shape, size (in diameter), and length of particles under electron microscopy (EM). Morphologically, viruses are spherical or isometric (*Cucumovirus*), rod shaped (Tobauovirus) or bacilliform (Badnavirus), flexuous rod shaped (Potyvirus) and geminate or twins like (Begomovirus) as seen under electron microscope (Fig. 2).

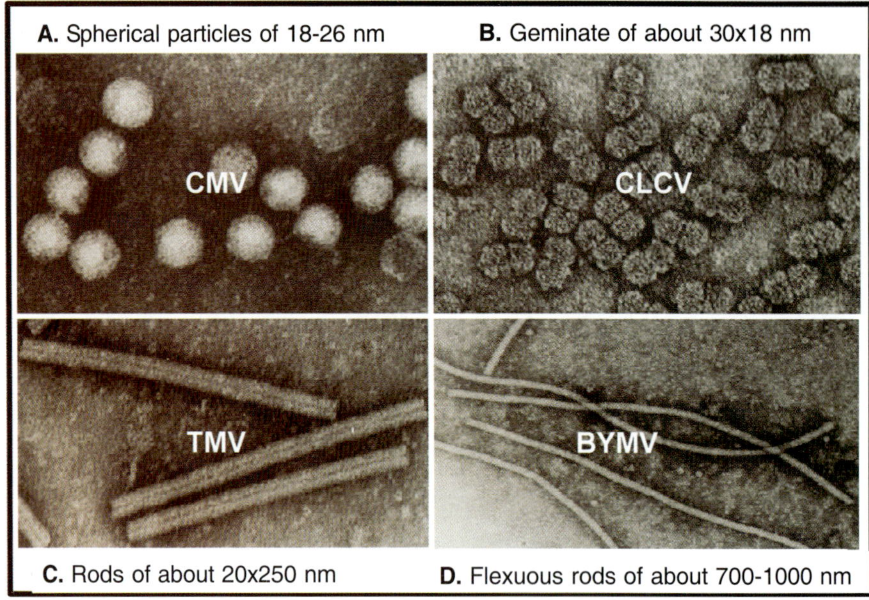

Fig. 2. Morphology of different viruses as shown under electron microscope. (A) *Cucumber mosaic virus* (CMV), (B) *Chilli leaf curl virus* (CLCV), (C) *Tobacco mosaic virus* (TMV), and (D) *Bean yellow mosaic virus* (BYMV).

3.1. Major vectors of plant virus transmission

The organisms involved in virus transmission are called virus vectors. Transmission of a virus by insects is a specific biological process.

Fig. 3. Different vectors for virus transmissions: (A) Aphids, (B) Whiteflies, (C) Thrips, (D) Mites, (E) Nematodes, (F) Fungi.

Many viruses are transmitted naturally by insect vectors and this is the most common transmission method in nature. Most of these vectors are sucking insects; e. g. aphids transmit more viruses than any other insect group, followed by leafhoppers, whiteflies, thrips, and beetles. Mites, though not actually insects, are included in this category because of their importance as vectors of some viruses (Fig. 3). Aphids and whiteflies transmit a number of RNA and DNA viruses in nature.

3.2. Other means for virus transmission

The spread of virus is termed as "transmission". Plant viruses do not penetrate the intact plant cuticle, because of this reason they are not dispersed as such by wind or water, and even when they are carried in plant sap. They generally do not cause infections unless they come in contact with the contents of wounded living cell. Viruses, however, are transmitted in a number of ways like mechanical, grafting, vegetative propagation, seed and pollen, dodder and insect vectors (Fig.4).

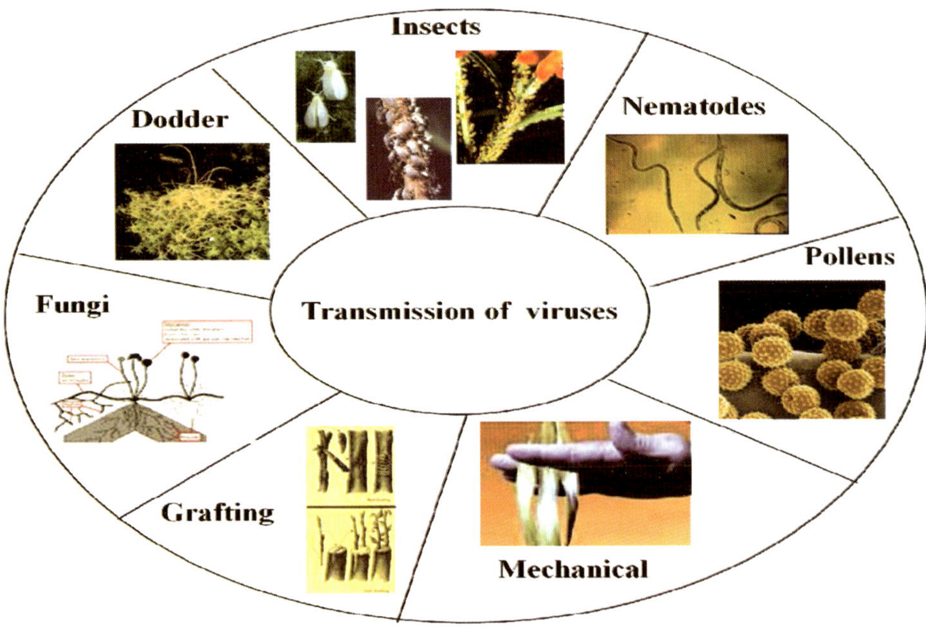

Fig. 4. Various modes of virus transmission

3.2.1. Mechanical transmission

Such transmissions may occur by the physical contact or experimental inoculation. In nature, only a few viruses are disseminated by contact e.g. *Tobacco mosaic virus* (TMV), *Potato virus* X (PVX), *Potato virus* S (PVS), and *Potato spindle tuber viroid* (PSTVd). Whereas, many viruses are transmitted by the plant extracts or sap (called as experimetal inoculum) of the diseased plants rubbed by fingers on the leaves of healthy test plants (experimetal host) like many of the tobacco species, tomato, chilli, legumes, chenopodium species, cucumber etc.

3.2.2. Vegetative propagation

Vegetative propagation is the most important and common means of virus transmission among those types of plants that propagate vegetatively by cuttings, tubers, corms, bulbs or rhizomes. The infected mother plant of this type will yield infected progeny.

3.2.3. Transmission by grafting

Grafting is considered to be a universal method for transmitting viruses, because systemic viruses can be transmitted by grafting. Grafting is particularly useful for transmission of phloem-restricted viruses that cannot be transmitted mechanically and for viruses found in low concentrations.

3.2.4. Transmission by dodder (*Cuscuta* spp.)

Many viruses can be transmitted by dodder, where transmission by grafting is impossible due to the tissue incompatibility. Two species of dodder *Cuscuta campestris* and *C. subinclusa* are frequently used that absorbs sap and viruses (if present), through its haustorium.

3.2.5. Transmission by seed

Approximately 100 of viruses are transmitted through seed say for example in case of potato. This type of transmission is known as vertical transmission.

3.2.6. Transmission by pollen

Viruses transmitted by pollen do not only infect the seed and plantlets but also they can propagate through the flower and infect the mother plant. The flower pollination with virus-infected pollen can lead to a lower fruit yield, compared to virus-free pollen. Transmission through pollen is very rare, only a few viruses are transmitted through pollen. Both PSTVd and PVT are transmitted by pollen or the ovule of infected plants.

3.2.7. Transmission by nematodes

Viruses transmitted by nematodes have a wide range of hosts and are transmitted by species of nematodes those who are feeding on the external part of the plant root. The genera *Xiphinema* and *Longidorus* are vectors of the spherical-particle viruses known as *Nepoviruses* such as tobacco and tomato ringspot viruses and *Fan leaf virus* of grapes. *Trichodorus* transmits two tubular viruses belonging to the group *Tobravirus* named as *Tobacco rattle virus* and *Green pea early browning virus*.

3.2.8. Transmission by fungi

There are four known genera of virus-transmitting fungi: *Olpidium, Synchytrium, Polymyxa* and *Spongospora.* All of these are obligate parasites consisting of one or a few cells that form thick-walled, resting in sporangia that survive in dry soil for long periods. Tobacco necrosis, lettuce big vein, tobacco stunt viruses; cucumber necrosis virus are transmitted by fungi.

4. Examples of Some Important Plant Viruses

Viruses have broad host range and can infect edible as well as ornamental plants of various economically agriculturally important families. Some are listed in Table 1.

Table 1. Examples of some important plant viruses infecting various crops

Crops	viruses	Disease	Type of virus genome
Gladiolus	*Cucumber mosaic virus*	Mosaic disease	ss RNA, linear genomess
	Bean yellow mosaic virus	Mosaic disease	RNA, linear genome
Gerbera	*Cucumber mosaic virus*	Mosaic disease	ss RNA, linear genomess
	Tomato aspermy virus	Mosaic disease	RNA, linear genomess
	Tobacco mosaic virus	Mosaic disease	RNA, linear genome
Chrysanthemum	*Cucumber mosaic virus*	Mosaic disease	ss RNA, linear genomess
	Tomato aspermy virus	Mosaic disease	RNA, linear genomess
	Chrysanthemum virus-B	Mosaic disease	RNA, linear genome
Canna	*Cucumber mosaic virus*	Mosaic disease	ss RNA, linear genomess
	Bean yellow mosaic virus	Mosaic disease	RNA, linear genomeds
	Canna yellow mottle virus	Venial streaks	DNA, circular genome
Chilli	*Chilli leaf curl virus*	Leaf curl	ds DNA, circular genome
Tomato	*Tomato leaf curl virus*	Leaf curl	ds DNA, circular genomess
	Tomato spotted wilt virus	Spotted wilt	RNA, linear genomess
	Cucumber mosaic virus	Mosaic disease	RNA, linear genome
Cotton	*Cotton leaf curl virus*	Leaf curl	ds DNA, circular genome
Papaya	*Papaya leaf curl virus*	Leaf curl	ss RNA, linear genomess
	Papaya ring spot virus	Mosaic disease	RNA, linear genome

Abbreviations: ss = single stranded; ds = double stranded.

4.1. Cucumber mosaic virus (CMV)

CMV is the type member of genus *Cucumovirus* (family *Bromoviridae*) has largest host range of any known plant virus infecting more than 1200 species. It has spherical shape (26-28 nm in diameter) causes mosaic and blistering symptoms on cucumber (Fig. 5). It is transmitted by mechanical inoculations and by aphids. The genome is single stranded made up of RNA and tripartite i.e. three RNA species in nature. The capsid protein is of 24-26 Kda.

Fig. 5. The typical symptoms of mosaic on leaves and fruits on cucumber induced by CMV.

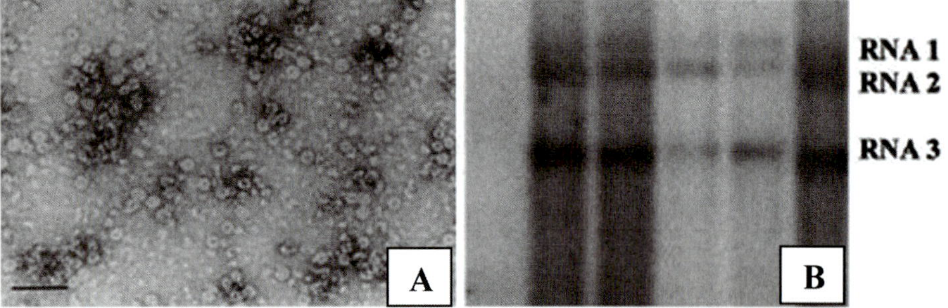

Fig. 6. (A) Spherical shape of CMV, (B) Tripartite genomic size of CMV.

CMV is tripartite plus-sense RNA virus (Fig.6) and each genomic RNAs are packaged in separate particles. CMV contains five open reading frames (ORFs). The 1a and 2a ORFs, encodes RNAs 1 and 2, respectively, are the viral components of the replicase. The 2b ORF, a gene overlapping the 2a ORF, is expressed from a sub genomic RNA, RNA 4A and encodes a suppressor of post transcriptional gene silencing. RNA 3 encodes the 3a protein, the viral movement protein, and the CP, expressed from sub genomic RNA 4 (Fig. 7) (Srivastava and Raj 2008).

Phylogeny estimations with the CP, as well as rearrangements in the 5' non translated region (NTR) of RNA 3, divided CMV strains into three subgroups: IA, IB, and II. Of which subgroup I (A and B) mostly occur in Asian countries.

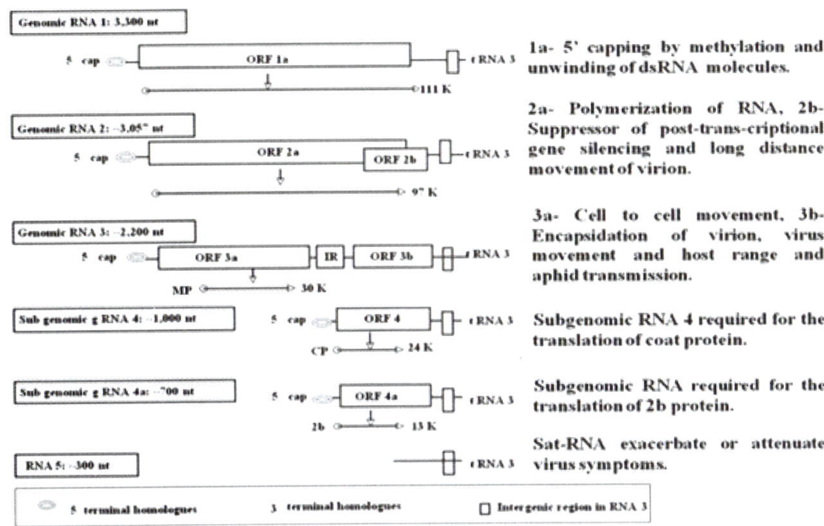

Fig. 7. Genomic organization of CMV virus

4.2. Potyviruses

Potyviruses belong to the family *Potyviridae* and named after the type virus: *Potato virus Y*. Members of this genus cause significant losses in agricultural, horticultural and ornamental crops (Fig. 8). Potyviruses spread in nature through aphids. The virus particles are of approximately 680- 900 nm long and 11-20 nm in diameter (Aminuddin *et al.*1999).

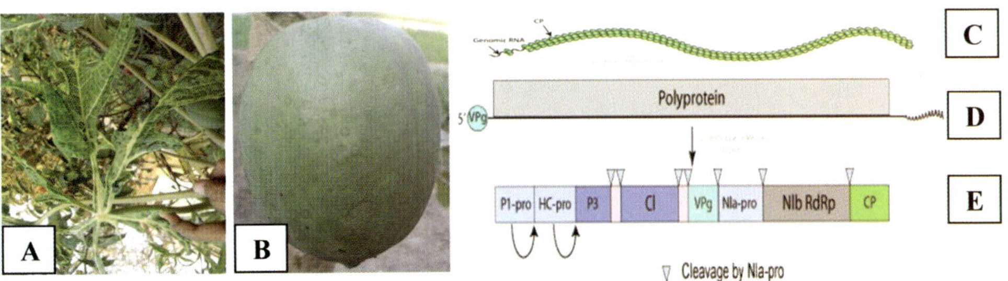

Fig. 8. (A) Ring spot symptoms on papaya leaf, (B) Ring spot symptoms on papaya fruit, (C) Morphology of Potyvirus, (D, E) Genomic organization of potyvirus.

The genome is linear positive sense ssRNA of 9-10 kb in length. The RNA genome contains one long open reading frame (ORF) expressed as a 350 kDa polyprotein precursor. This is proteolytically processed by viral and host proteases into seven smaller proteins denoted as P1, helper component (HC Pro), P3, cylindrical inclusion (CI), nuclear inclusion A (NIa), nuclear inclusion B (NIb), capsid protein (CP), as well as two small putative proteins known as 6K1 and 6K2 (Riechmann *et al.* 1992).

4.3. Geminiviruses

Geminiviruses are characterized by twined icosahedral particles (hence geminate) of approximate 18-30 nm in size (Fig. 9). They cause disease and show curling symptoms in a number of economically important crops therefore, are responsible for a significant amount of crop damage worldwide. Base on host specificity, nucleotide sequence differences, family *Geminiviridae* has been grouped into four groups: Begomovirus, Mastrivirus, Curtovirus and Topocuvirus and recently extended with four new groups: Becurtovirus, Capulavirus, Eragrovirus and Turucurtovirus. Among them, Begomoviruses have emerged globally as important agricultural pathogens.

Begomovirus have covalently closed circular single-stranded ssDNA genomes encoding six genes that diverge in both directions from a virion strand origin of replication. The genome of the begomoviruses is bipartite consisting of two DNA components: DNA-A and DNA-B or in some cases it is monopartite consists of DNA-A like molecular only (Fig.10).

Fig. 9. (A) Morphology of Geminivirus, (B) Curling and distortion symptoms on tomato, (C) Curling and distortion symptoms on papaya.

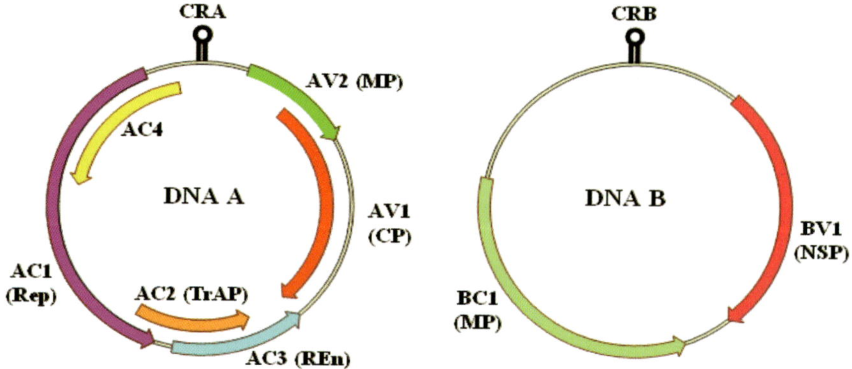

Fig. 10. Genomic Organization of Begomovirus

Genomes of bipartite begomoviruses are made up of two components, each of 2.5-2.8 kb in size. The DNA-A component typically has one gene in the virion sense and four genes in the complementary sense. AV1 (virion sense) encodes the coat protein, AC1 (complementary sense) encodes the replication-associated protein (Rep) and AC2 encodes the transcriptional activator protein (TrAP) and the. AC3 encodes the replication enhancer protein (Ren) that regulates the virus replication rate. AC4 encodes a protein which is a determinant of symptom expression in monopartite begomoviruses.

The B component has two genes, designated BC1 and BV1. BV1 encodes protein which is responsible for movement of viruses. The product of BV1 is localized in the cell nucleus and binds ssDNA, allowing the newly formed virus genome to be transported to the cytoplasm (Pascal *et al.* 1993, Pascal *et al.* 1994). The BC1 product has been extracted from cell wall and cellular membrane fractions, and its function is to increase the exclusion limit of plasmodesmata to facilitate cell to cell movement of the virus (Pascal *et al.* 1993). Both movement proteins define the viral host range but only BC1responsible for symptom severity and pathogenicity in bipartite begomovirus (Ingham *et al.* 1995, Duan *et al.* 1997b).

4.4. Badnaviruses

Badnaviruses, are the plant DNA viruses which belong to family *Caulimoviridae* (Fig. 11A, B). They are characterized by bacilliform morphology and double stranded circular genome of ~7-8 kb size (Fig.11C). The plant host usually belongs to dicotyledonous and monocotyledonous classes depending on Badna virus spp. The virus could be transmitted mainly by mealybug, and also transmitted by mechanical inoculation, by grafting, by seed.

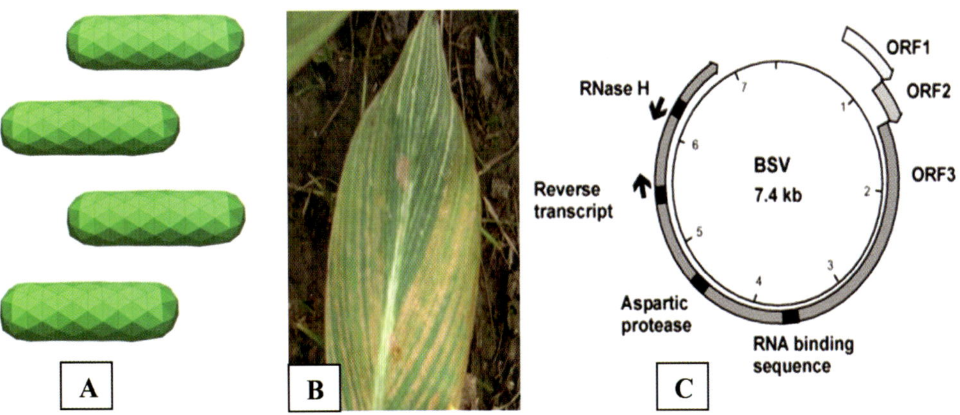

Fig. 11. (A) Morphology of Badnavirus, (B) Mosaic symptoms on canna, (C) Genomic organization of Badnavirus.

The genome of genus *Badnavirus* is one molecule of circular dsDNA (Fig. 11). The CP, aspartic protease, reverse transcriptase (RT), ribonuclease H (RNase H) and movement protein (MP) are first expressed as a large polyprotein. The polyprotein is subsequently cleaved post-translationally into functional units by the aspartic protease (Seal and Muller 2007).

5. Parameters for Characterization of Viruses

Characterization of viruses is required for identification and classification. Generally viruses are classified into families, genera, and species on the basis of following characteristics.

5.1. Biological properties

This may include type of host, host range and also the mode of transmission of the virus.

5.1.1. Symptomatology and host range

Symptoms such as mosaic, blisters, leaf deformation, ring spots, leaf curling, venial streaks, yellow stripes, yellow mosaic and stunting of whole plant are the initial step in disease diagnosis. Some symptoms indicate that the infecting virus belongs to a particular group of viruses. Some viruses have a specific host range which may greatly assist in their identification. However, they are insufficient because symptoms may result from infection of more than one virus. Different viruses may individually cause the same symptoms in the same host. Symptoms may vary with the cultivar of the host as well as influenced by the environmental conditions. Many plants are carriers of plant viruses but show no disease symptoms.

5.1.2. Types of viral symptoms

Symptoms may be local or systemic. If symptoms appeared around the site of virus inoculation are denoted local symptoms. When virus spreads from the site of inoculation and causes symptoms in other parts of the plant, this is referred to as systemic symptoms.

5.1.3. Virus transmission

Virus transmission is important parameter to characterize viruses. Transmission of plant virus takes place by several means like mechanical (rubbing the experimental sap inoculums by finger on leaves), insects, thrips, nematodes etc. For example, cucumoviruses and potyviruses can be transmitted through mechanical inoculation and by aphids. Geminiviruses may be transmitted by whiteflies while badnaviruses by mealybug etc.

5.1.4. Particle morphology by electron microscopy

Viruses cannot be seen with an optical microscope so scanning and transmission electron microscopes are used to visualize virions. To increase the contrast between viruses and the background, electron-dense "stains" e.g. negative stains are used for staining the background. The shape such as spherical, straight rods, flexuous rods, bacilliform, and size of particles as shown under the electron microscope (Fig. 2) help to understand type of virus.

5.1.5. Physical properties of the virus

Thermal inactivation point (TIP) is the highest temperature at which no symptoms appear on the inoculated test plants. Longevity *in vitro* (LIV) is defined as the length of time when virus is infective in crude sap kept at room temperature (approximately 20 to 22°C). Dilution End Point (DEP) is highest dilution (10^{-1-10}) of plant sap in which a virus is still infectious on a test host.

5.1.5.1. Serological properties and methods for virus detection

Serology explains the properties of virus proteins subunits e.g. coat protein or any other important protein. In principle, most of the serological methods are based on the precipitation produced when antibodies (aniserum) and antigens (virus) combine and make a visible line/curve of precipitation. The most commonly used serological methods are: Microprecipition test, Ouchterlony agar gel double diffusion test, Immunosorbent electron microscopy (ISEM), Enzyme-linked immunosorbent assay (ELISA), among which ELISA seems to be the most advanced and feasible technique to be used for virus-detection because it is highly sensitive technique and required less amount of antibody per reaction with capacity of detecting virus in 96 wells at once.

5.1.5.2. Molecular characterization of viruses

For virus characterization, total nucleic acid isolated from infected plant samples are used in polymerase chain reaction (PCR) with sets of primers designed from most conserved region of the virus genome for virus detection and identification. The sensitive detection methods are PCR/ RT-PCR and nucleic acid probe based hybridization. The genomic properties are also considered for molecular identification of the virus, which includes the number of genome components and sequence analysis of their associated genes and translation products. The relatedness of different sequences is often an important factor in differentiating between the species. For molecular identification, the viral components are amplified, cloned in suitable vector and sequenced. The sequence data are analyzed by various computational programmes *viz*: BLAST, Expasy translation tool, Genomatix DiAlign, MEGA, RDP etc. The BLAST tool is used for sequence identities, sequence corrections and preliminary identification. Expasy translation tool is used for ORF identification and to predict amino acid/protein translation from sequence. Genomatix DiAlign tool is used for sequence similarities and multiple sequence alignments. MEGA is used for phylogenetic relationships with other virus strains. RDP analysis is used for recombination analysis for evolution of new species strains viruses.

6. Criteria for Species Demarcation in Viruses

International Committee on Taxonomy of Viruses (ICTV) is the official body of the virology division of the International Union of Microbiological Societies responsible for naming and classifying viruses. There are different criteria for species demarcation in different family of viruses.

6.1. Species demarcation criteria in the genus Begomovirus

- Number of genomic components: presence of DNA A and B called as Bipartite begomoviruses while absence of DNA B called as Monopartite (found in Old world).
- Organization of the genome six ORFs: Presence or absence of ORF AV2 (pre-coat protein) in (New World).
- Nucleotide sequence: Identity less than 89% is generally indicative of a distinct species.
- Capsid protein characteristics: Amino acid sequence identity ~90% may be indicative of a distinct species in the first instance, but the complete sequence is necessary to confirm taxonomic status.
- Natural host range, whitefly transmission, symptom phenotype and infectivity of genomic components.

6.2. Species demarcation criteria in the Potyvirus

- The CP-gene clearly differentiate the virus species and to some extent their strains.
- On the basis of nucleotide similarity among CP genes, less than 41% similarity indicates the member of different genera of the family *Potyviridae*.
- If the similarity is less than 76%, it belongs to a different species.
- If the similarity is greater than 76%, it belongs to a same species.
- If the similarity is more than 90%, it belongs to a same strain.

6.3. Species demarcation criteria in the Cucumoviruses

- CMV strains reported from all over the world have been placed in to two subgroups I and II on the basis of: serology, nucleic acid hybridization, gene sequences, restriction fragment length polymorphism and by peptide mapping of the coat protein (CP) and nucleotide sequence identity.

- Sequence similarity: Nucleotide sequence similarity is used to distinguish subgroups within a species. Subgroups generally have at least 65% sequence similarity.
- Phylogenetic analysis of CMV strains subdivided subgroup I into IA and IB on the basis of gene sequences. Asian strains of CMV have been placed in subgroup IB.
- Protein based phylogeny (1a, 2a, 3a and 3b proteins) of CMV and its branch length of different subgroups has been demonstrated its significance to get more reliable phylogeny of the expressed genes.

7. Management of Plant Viruses

There are several strategies to manage plant viruses, which are listed as follows:

7.1. Conventional method

Conventional method is the basic method to protect plant from viruses. These methods include:

7.1.1. Pest control

Pest control refers to as use of insecticides for the control insect population of viral vectors like aphids, whiteflies, mites, nematodes etc.

7.1.2. Cultural control

Cultural control refers to those growing methods that reduce pathogen levels or reduce the rate of disease development. These include: sanitation, crop rotation, host eradication.

7.1.3. Development of virus free-plant

- Development of virus-free plant refers to the use of plant tissue culture techniques like meristem culture/ thermotherapy/ cryotherapy/ electrotherapy alone or in combination with for the elimination of viruses from the infected mother plant. The details of methods presently used are:
    - Meristem culture method: The culturing of axillary bud or apical shoot meristems is usually done.
    - Thermotherapy method: Explants are exposed to the higher temperatures (37-42° C, which are not lethal for plant cells, but they are lethal for viruses.
    - Chemotherapy method: Some chemicals which interfere with virus multiplication may be added into the culture medium for curing the virus infected explants. Virazole (ribavirin), Actinomycin D, Acyclic adenosine analogue (DHPA) etc.

- Electrotherapy method: Electric current is applied to the infected explants for the virus eradication. Exposing plant tissues to electricity increases the temperature (about 4 to 10°C) inside the cells, which is harmful to virus but not for plant metabolism.

7.1.4. Quarantine regulation

Quarantine regulation can also prevent virus transmission during import and export of plant material across the borders.

7.1.5. Breeding programme

Breeding programme deals with the development of virus tolerant and or virus resistant plant by conventional breeding. This method is cucumbersome, take much labor and money and time consuming. Therefore, development of advanced biotechnological tools seems to be essential.

7.1.6. Use of virus-free propagating materials for mass production through tissue culture

Once the virus-free mother culture developed or obtained may be used for mass multiplication and supply to the local and commercial vendors (Fig. 12). Propagation of virus-free material will help in minimizing the virus spread and load in ornamental, horticultural or other economically important industries of the country.

Fig. 12. Development of virus-free Gerbera and Gladiolus plants through tissue culture (Kaur *et al.* 2012).

7.2. Non- conventional methods

Non-conventional method based on transferring virus-derived genes, including viral coat protein, replicase, movement protein, etc. and non-viral genes including microRNAs, ribosome-inactivating proteins, protease inhibitors, dsRNAs etc. The CP gene has been successfully used for generating resistance against viruses and it is most common gene which is used to developed plant resistant against viruses.

7.2.1. Genes used for protection of viruses

AV2 (Pre-coat protein) of *Tomato yellow leaf curl virus* had been found to act as RNA silencing suppressor. AC2 (RepEn) protein of *African cassava mosaic virus* and *Tomato yellow leaf curl China virus* used for protection against homologous virus. In the similar ways, AC4 protein of the *African cassava mosaic virus* has been used for developing virus resistance in genetically transformed plants. DNA-â satellite of *Tomato yellow leaf curl China virus* has also been used successfully to suppress the expression of target genes (Xiaofeng *et al.* 2005).

In viruses of RNA world, 2b protein, used as suppressor of post-transcription gene silencing (PTGS), of CMV is being used for development of virus resistance in genetically modified plants. Satellite RNA has also been demonstrated for protection against CMV infection in China. The HC-Pro and SRB protein of *Potato virus Y* used as for suppressor study and may be used for development of virus resistance. Genetically expressed coat protein gene of CMV in host plant may be used for re-encapsidating of virus particles and will provide resistance. It has been used for protection of several RNA viruses and some DNA viruses (TYLCV, TLCV) in tomato.

7.2.2. Utilization of coat protein (CP) gene for development of virus resistance

The most common type of transgene used to develop plant virus resistance is the viral coat protein genes. This gene encapsidated the viral nucleic acid and are thought to be important in nearly every stage of viral infection including replication, movement throughout an infected plant, and transport from plant to plant as shown in the Fig. 13. and hence, checking of uncoating of virus at entry or after replication, if so occur, will help in development of transgenic plants. This is the principle for development of coat protein mediated virus-resistance.

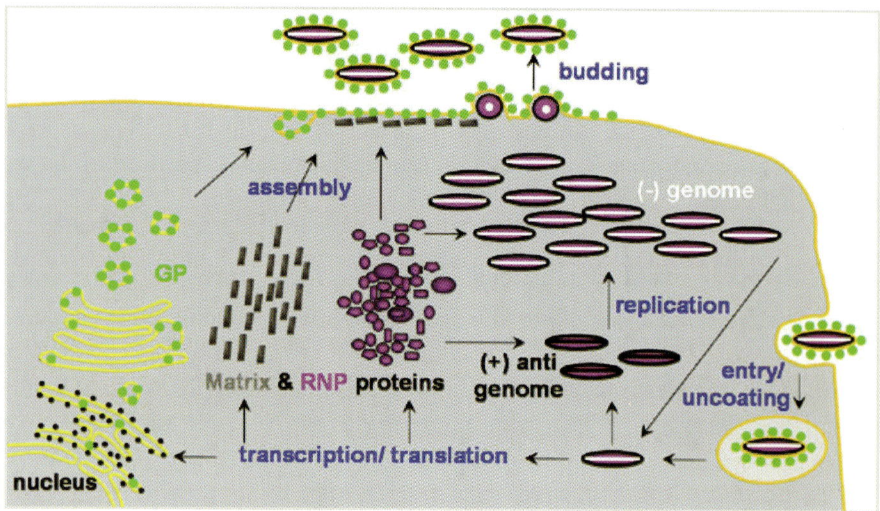

Fig. 13. Mode of action of coat protein in attenuation of life cycle of the virus

7.2.3. Development of transgenic plants of using CMV-CP construct uplifted in Agrobacterium LBA4404 strain

For the development of transgenic plants against CMV infections, the CP gene of CMV-A strain was amplified, cloned in primary and then secondary, binary, vector. The line (Fig. 14.) show diagram the arrangement of CP gene cloned in pRoK2 binary vector was used for development of transgenic plants of model plant tobacco (*Nicotiana benthamiana*). This construct is T- DNA derived in which gene of interest (CP gene) placed adjacent to CaMV35S promoter. Selection marker *npt*II is used to indicate successful plant transformation. The construct was uplifted in *Agrobacterium tumefaciens* strain LBA4404 and used for transformation of tobacco explants.

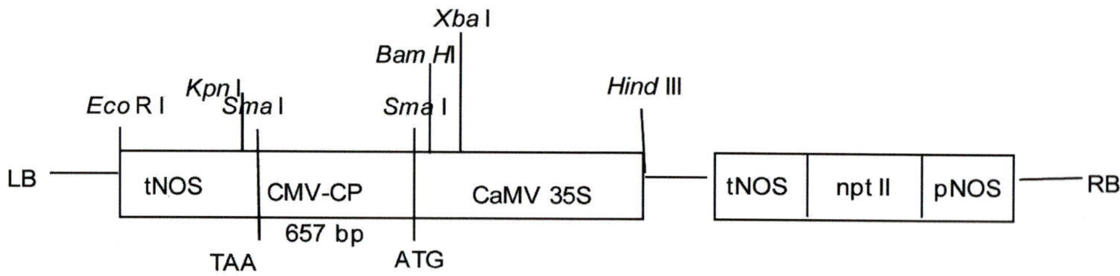

Fig. 14. Schematic diagram of binary cassette used for transformation (Srivastava and Raj 2008).

The *N. benthamiana* plants were co-cultivated and transformed with the same construct for which the steps of regeneration have been shown in the figure. The putative transgenic plants were screened for the successful integration and expression of CP gene and CP, respectively. Transgenic plants, upon virus challenge of high inoculums pressure, showed remarkable virus-resistance in tobacco lines.

Fig. 15. Developmental stages of transgenic tobacco regeneration

The success of transgenic tobacco plants, in term of virus-resistance, provoked us to use this same construct for the development of another transgenic plant of tomato, eggplants, petunia, gerbera and chrysanthemum of commercial use. The construct used again revealed significant resistance in chrysanthemum, tomato and eggplants, whereas, screening and validation of petunia and gerbera under glasshouse condition is going on.

The CP gene of a DNA virus, *Tomato leaf curl virus* (ToLCV) was also cloned in pRoK2 binary vector and used for the development if transgenic tomato plants. Few of the transgenic lines developed showed considerable resistance in tomato against the high inoculums pressure of ToLCV in glasshouse as well as field conditions. This is the first ever case where CP mediated virus resistance in tomato plants has been shown the world over.

7.2.4. Dual virus resistance

Development of virus resistance against the two plant viruses of same family or different family, RNA or DNA viruses is known as dual virus resistance and is one of the recent methodologies taking account into for the development of virus-resistant plants.

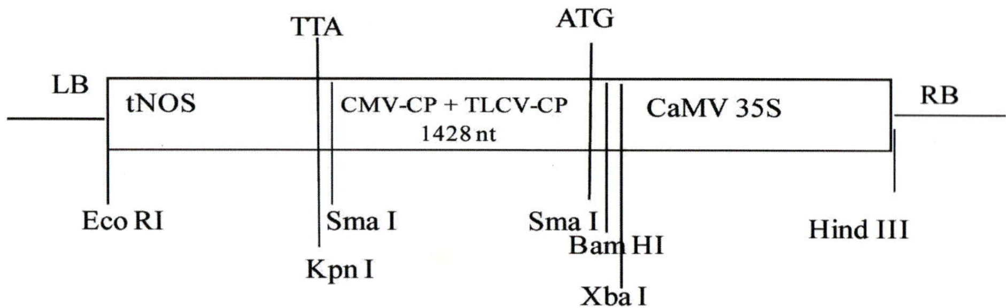

Fig. 16. Cloning map of a fusion construct of coat protein genes of TLCV and CMV

Fusion construct of coat protein genes of TLCV and CMV was prepared for development of genetic resistance against TLCV (ssDNA virus) and CMV (ssRNA virus). The coat protein genes of both the viruses were cloned in such a way that it will produced coat protein subunits of each CMV and TLCV.

Fig. 17. Development of transgenic tobacco plants and their validation for CP genes of CMV and ToLCV.

The tobacco explants were transformed using this fusion construct and regenerated plants were tested which successfully showed presence and expression of both the CP genes. The use of this fusion construct is supposed to provide the resistance against the dual infection of CMV and TLCV in transgenic plants.

7.2.5. Biological control of Aphids by use of Ladybird predators

For biological control of aphids, feeding behaviors of aphids (*Aphis gossypii*) by the ladybird (*Cocinella transversalis*) larva and its adults was studied on chrysanthemums, tobacco and dahlia plants.

Fig. 18. Biological control of aphids by ladybird on chrysanthemum, tobacco, and dahlia plants

It was observed that the larva feeds about 20-25 aphids per min. while adult feeds lesser than that. This observation may be exploited for control of aphids, the vector for virus transmission (cucumoviruses and potyviruses) in many crops.

7.2.6. Biological control of whiteflies

Whiteflies (*Bemisia tabaci*) have many natural enemies lacewings, bigeyed bugs, several small lady beetles including *Clitostethus arcuatus* and the Asian multicolored lady beetle (*Harmonia axyridis*) which feed on whiteflies. Whiteflies have a number of naturally occurring parasites that can be very important in controlling some species. *Encarsia* spp. parasites are commercially available for release in greenhouse situations; however, they are not generally recommended for outdoor use because they are not well adapted for survival in temperate zones.

References

Aminuddin, JA Khan, Raj SK (1999) Association of an unknown Potyvirus isolate with severe mosaic disease of Narcissus tazetta L. Ind J Exp Biol 37:1034-1036

Duan YP, Powell CA, Webb SE, Purcifull DE, Hiebert E (1997) Geminivirus resistance in transgenic tobacco expressing mutated BC1 protein. Mol Plant-Microbe Interact10:617-623

Ingham DJ, Pascal E, Lazarowitz SG (1995) Both geminivirus movement proteins define viral host range, but only BL1 determines viral pathogenicity. Virology 207:191-207

Kaur C, Kumar S, Raj SK, Purshottam DK and Goel AK (2012) Detection and elimination or potyvirus from Gladiolus for its better quality production. Abstract in XXI National Conference on Immunology & Management of Viral Disease in 21st Century, Indian Veterinary Research Institute, Mukteshwar, Uttarakhand (India)

Pascal E, Goodlove PE, Wu LC, Lazarowitz SG (1993) Transgenic tobacco plants expressing the geminivirus BL1 protein exhibit symptoms of viral disease. Plant Cell 5:795-807

Pascal E, Sanderfoot AA, Ward BM, Medville R, Turgeon R, Lazarowitz SG (1994). The geminivirus BR1 movement protein binds single-stranded DNA and localizes to the cell nucleus. Plant Cell 6(7):995-1006

Riechmann JL, Sonia L, Garcia JA, (1992). Highlights and prospects of potyvirus molecular biology. J Gen Virology 73:1-16

Seal S, Muller E (2007) Molecular analysis of a full-length sequence of a new yam badnavirus from *Dioscorea sansibarensis*. Arch Virol 152:819-825

Srivastava A, Raj SK (2008) Coat protein-mediated resistance against an Indian isolate of the *Cucumber mosaic virus* subgroup IB in *Nicotiana benthamiana*. J Biosci 33:249–257

Xiaofeng C, Guixin L, Daowen W, Dongwei H, and Xueping Z (2005) A Begomovirus DNA-β-Encoded Protein Binds DNA, Functions as a Suppressor of RNA Silencing, and Targets the Cell Nucleus. J Virology 79:10764-10775

Chapter – 3

Diversity, Systematics and Applications of Fungi

S.P. Singh, Shipra Pandey, Richa Shukla, Poonam C. Singh and A. Mishra

1. Introduction

Fungi are ubiquitously present in air, water, soil and organisms, or on organism surfaces. They are a group of eukaryotic system that are of great applied and scientific interest to common man, microbiologists, agriculturists and industrialists. They are a diverse group of organisms with respect to taxonomy, size, morphological appearance, mode of nutrition and reproduction. Fungi are eukaryotic spore bearing protists that lack chlorophyll. Fungi comprise the molds and yeasts. Yeasts are usually unicellular measuring only several microns in diameter, whereas molds are filamentous and multicellular which may grow up to several meters in diameter like the extremely large polypores. Fungi generally reproduce both sexually and asexually (Gow and Gadd 1995).

Although it is not easy to define fungi, biologists currently use the term "Fungus" to include nucleated, spore bearing, achlorophyllous organisms with absorptive nutrition which generally reproduce sexually and asexually, and whose filamentous branched somatic structures are typically surrounded by cell walls containing cellulose of chitin or both. A feature which all fungi have in common is their mode of nutrition. Like animals, fungi are heterotrophic organisms, which must consume preformed organic matter. They may live as saprophyte or parasites. They are of great economic importance to man and play an important role in the disintegration of organic matter and affect us directly by destroying food, fabric, leather and other commercial goods. They are responsible for a large number of diseases of plants, animals and man.

2. Cell Structure in Fungi

Fungal cells may be minute, and a single, uni-nucleate cell may constitute an entire organism. A single cell of *Olpidium* or yeast cell falls in this category. Alternatively, fungal cells may be elongated and strand – like and several may be joined to form a thread of cells – a hypha (pl. hyphae). A large number of hyphae collectively form the mycelium. The cell wall of fungi is generally made up of chitin or cellulose. The majority of the fungi have chitinous wall. Occasionally, both chitin and cellulose may occur simultaneously. One cell in the hyphae may be separated from another by a cross wall or a septum. Septa in different groups of fungi have important differences in structures. For example, the septa in the lower fungi are pseudosepta, which are perforated by so many pores that resemble like a sieve. In Ascomycota and some fungi imperfecti, the septum is perforated by a single pore. In Basidiomycota and some members of fungi imperfecti, the septum is more complex than that found in the Ascomycota. The septum here is called a dolipore septum.

Hyphae characteristically grow at the tips and can usually produce lateral branches in their older parts. The mycelium may remain microscopic and or develop easily visible organized structures, such as strand or cord like rhizomorph, compact resting bodies comprising fungal tissues made of prosenchyma or pseudoparenchyma. All growing hyphae are filled up with cytoplasm, and are frequently vacuolated and may be aseptate. Hence, they form a single multinucleate cell (coenocytes) or are divided into segments by septa. Fungal nuclei are very small. The nuclei have well- developed nuclear membranes. The nuclear envelope consists of a double nuclear membrane and is marked by large gaps. During mitosis the nuclear envelope does not disappear ordinarily, instead it consists in a dumbbell- like fashion and eventually separates into two daughter nuclei. Meiosis conforms more close to the norm in other organisms than does mitosis. In fungi ribosomes are 80S and mitochondria occur in the cytoplasm. There is also the endoplasmic reticulum. Storage products occur in the fungal cell. These may be in the form of glycogen or lipids. Golgi apparatus in fungi is similar to those in higher plants and animals have been found in some fungi. A feature which is of widespread occurrence is the presence of lysosomes. These are made up of membrane bounded tubules or vesicles. The function of lysosomes is not known in fungi.

3. Reproduction in Fungi

Fungi reproduce both asexually as well as sexually. Asexual reproduction takes place by a variety of methods such as:

- Fragmentation of the soma
- Fusion of the somatic cells
- Budding of the somatic cells

- Production of spores, such as conidia, sporangiospores- motile or non- motile.

Sexual reproduction is of widespread occurrence in fungi. Fertilization can be brought about in a variety of ways. The most common methods are:

- Planogametic copulation
- Gametangial contact
- Gametangial copulation
- Spermatization
- Somatogamy

4. Life cycle of Fungi

As in other organisms, in fungi too there is generally a cycle of haploid and diploid structures, corresponding to the gametophyte and sporophyte in the plants. The diploid phase begins with karyogamy and ends with meiosis. In the majority of fungi there is no distinct alternation of generations. Raper (1954) recognized seven basic types of life cycles. However, Burnett (1976) recognized following five basic life cycles in fungi:

4.1. ***Asexual***, in which sexual reproduction is lacking, e.g. Deuteromycetes (Fig. 1).

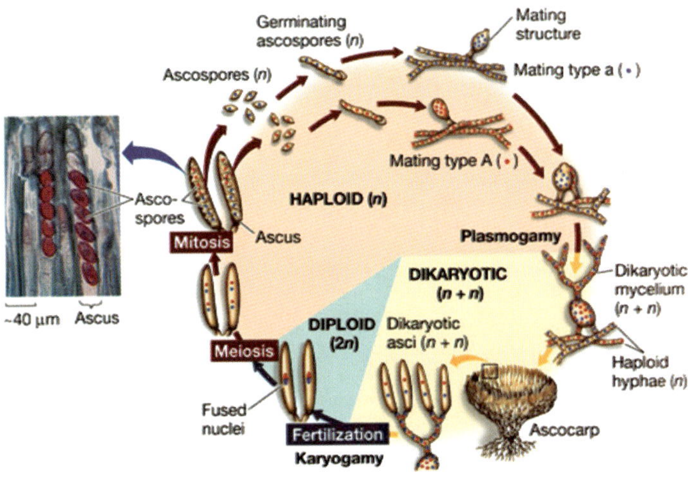

Fig. 1. Asexual life cycle

4.2. ***Haploid***, in which meiosis immediately follows nuclear fission. This is very commonly seen in fungi or some members of Ascomycota. Here the diploid phase is of minimum duration (Fig. 2).

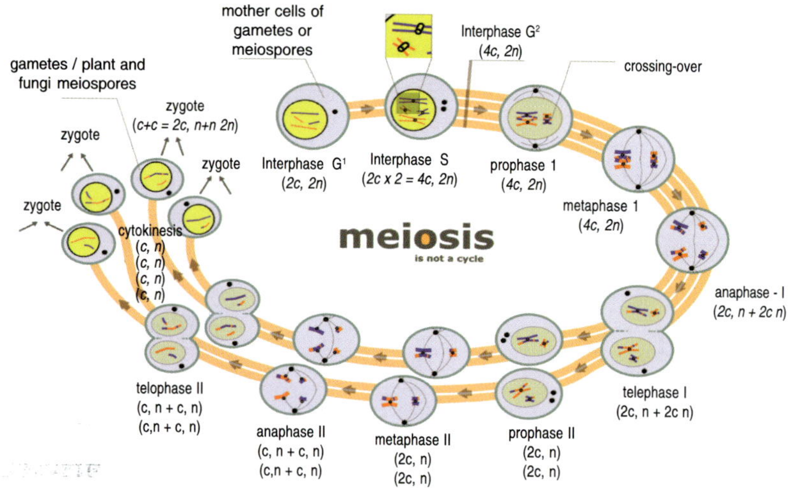

Fig. 2. Haploid life cycle

4.3. ***Haploid dikaryotic***, which is similar to the second type, except that paired, potentially conjugate nuclei persist in close physical occasion in the same hyphal segment (hence dikaryon). The examples of this are many members of Ascomycota. Binucleate ascogenous hyphae develop just prior to ascus development and such dikaryon cannot exist independently of the haploid phase. On the other hand, there may be a condition where the meiospores fuse to form dikaryon so that the fungus is dikaryotic throughout its life cycle, except for the moment of fertilization and during the subsequent meiosis. This occurs in yeast (Endomycetales) and more commonly in smuts (Ustilaginales). In the Basidiomycota, the mycelium derived from the germination of a meiospore may persist in the haploid condition as a monokaryon, but once a dikaryon is formed, it shows unrestricted and independent growth and is the 'long-lived' phase of the life cycle (Fig. 3).

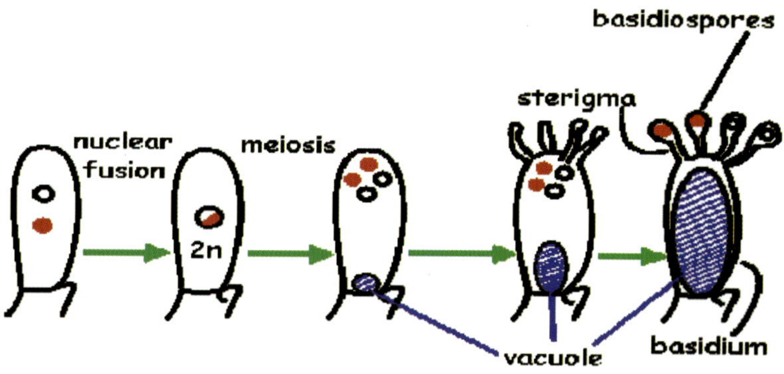

Fig. 3. Haploid dikaryotic life cycle

4.4. *Haploid-diploid*, in which the haploid and diploid phases alternate regularly. This is restricted to the order, Blastocladiales (Fig. 4).

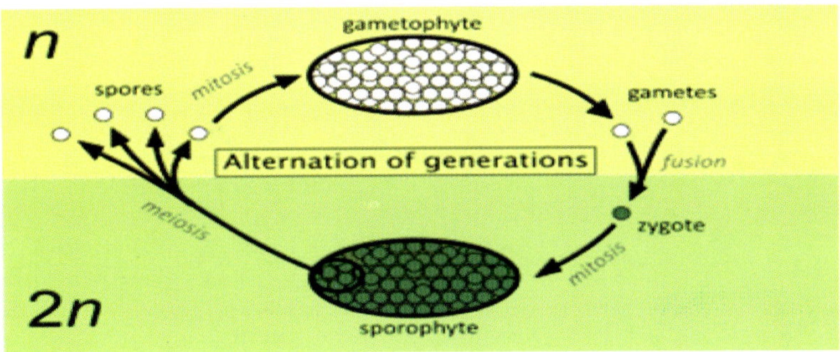

Fig. 4. Haploid-diploid alternating life cycle

4.5. *Diploid*, in which the haploid phase is restricted to the gametes. The vegetative phase is diploid. Reduction takes place at the time of gametogenesis, as in *Pythium, Phytopthora,* and *Saprolegnia* species in the Oomycetous fungi (Fig. 5).

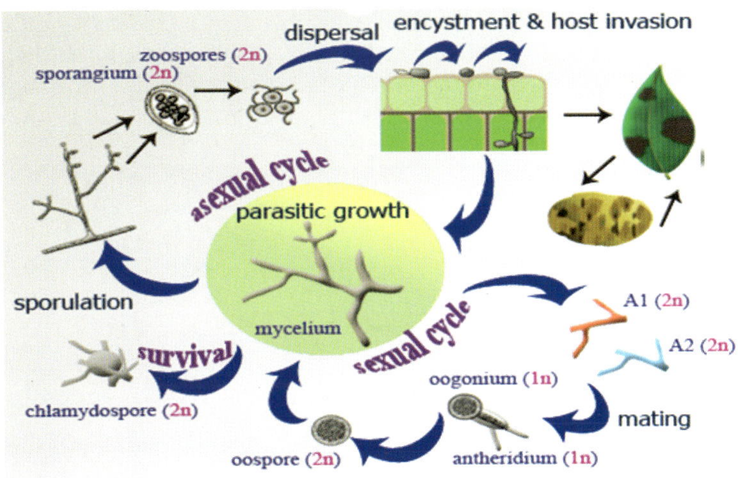

Fig. 5. Diploid life cycle

5. Classification of Fungi

Whittaker (1969) broke the tradition of a three kingdom system of classification of all living organisms. In doing so he recognized that the classification of all living organisms as Prokaryotes, Animals and Plants (including fungi) did not reflect their relationships. Whittaker added two more kingdoms in the classification *viz*. Fungi and Protista.

Table 1. Classification of fungi (Wittaker 1969)

Kingdom	Phyla
Mycota or Eumycota	1. Chytridiomycota 2. Zygomycota 3. Glomeromycota 4. Ascomycota 5. Basidiomycota 6. Mitosporic Fungi
Straminopila or Chromista	1. Oomycota 2. Hyphochytridiomycota 3. Labyrithulomycota
Slime moulds	1. Myxomycota 2. Plasmodiophoromycota 3. Dictosteliomycota 4. Acrasiomycota

6. Methods of Isolation of Fungi

Isolation of fungi from natural sources is one of the basic skills in mycology that must be mastered by almost everyone concerned with moulds. Isolation techniques are numerous and often complex but can be quite effective in yielding just the fungus one wants, while excluding all the others. Isolation techniques can be divided into two broad categories: (1) direct methods and (2) selective methods. Both are routinely used in mycology laboratories and can be further divided into a number of subtypes.

6.1. Direct isolation techniques

6.1.1. Direct transfer

The term 'direct' is applied to techniques which involves the simple transfer of a fungal culture from its natural habitat to a pure culture situation in the laboratory. At its simplest this is done by putting a piece of the habitat or substrate under a dissecting microscope so that the mould growth is easily visible. With the help of an inoculating needle some of the spores are transfered into a sterile plate of culture medium and incubated for 3-5 days. The direct transfer technique is also used to obtain pure cultures fungi.

6.1.2. Moist chambers

Direct isolation of fungi is often more effective if the natural substrate has been kept moist for one to several weeks to allow moulds to grow and sporulate. The easiest method involves a container called a moist chamber. Moist chambers can take any number of forms, but are basically containers holding a material such as cotton, paper, cloth, sterile sand or soil, or peat moss that can be kept moist for several weeks. The specimen is placed on top of the moist material and incubated at an ambient temperature

Table 2. A brief description of the fungi in different phyla with their characteristic features

Kingdom	Phyla and general characters	Subphyla Characters and examples	Figures
Mycota	**Chytridiomycota:** Smallest, oldest and simplest of the fungi. Motile cells (both zoospore and gametes) possess a single, posterior flagellum; Coenocytic thallus; Cell walls contain chitin or glucan; Nuclear divisions are intranuclear and centric.	**Spizellomycetales:** Zoospores with several lipid globules, sometimes amoeboid when in motion, rhizoids when present with blunt tips, less than 5μm in diameter predominantly from soil. Some important genera belonging to this order are: Olpidium, Rozella and Spizellomyces, Caulochytrium and Urophlyctis.	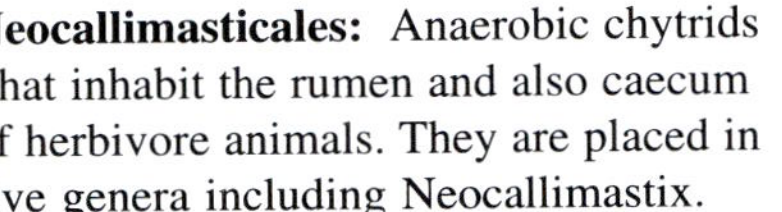 *Spizellomyces* (www.slideshare.net)
		Neocallimasticales: Anaerobic chytrids that inhabit the rumen and also caecum of herbivore animals. They are placed in five genera including Neocallimastix.	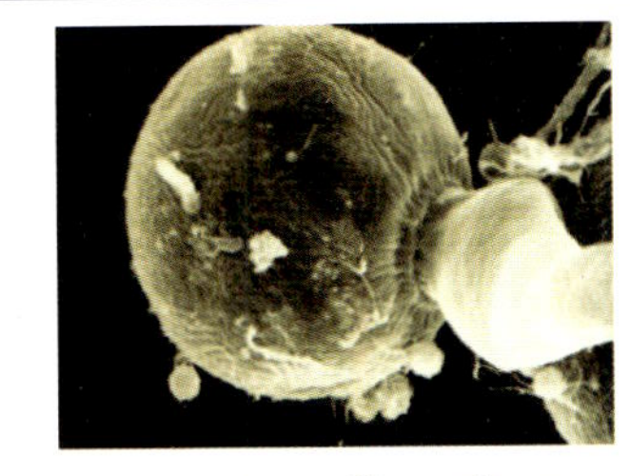 *Neocallimastix* sp. (Daniel wubah; http://www.bsu.edu)

Kingdom	Phyla and general characters	Subphyla Characters and examples	Figures
		Chytridiales: They are water or soil inhabiting fungi, many of the former parasitic on algae and water moulds, many of the latter on vascular plants. A few parasitize animal eggs and protozoa while others are saprophytic on decaying remains of dead plants. Some better known genera are Chytridium, Chytridiomyces, Polyphagus, Rhizophydium, Endochytrium, Synchytrium, Cladochytrium and Nowakowskiella. *Synchytrium endoboticum* causes the wart disease of potato.	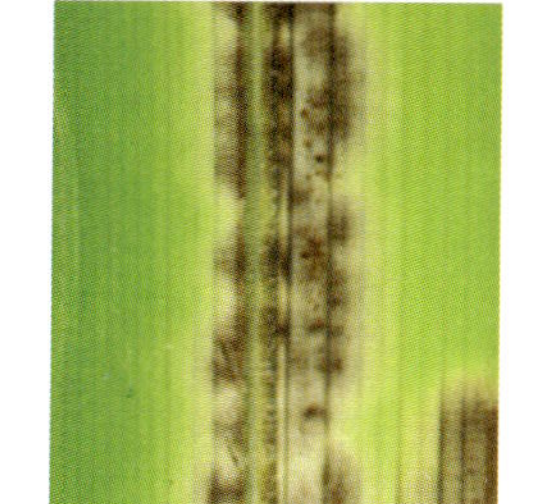 *Rhizophydium graminis* (Don Barr; http://www.bsu.edu)
		Blastocladiales: Chiefly water and soil inhabiting fungi. Blastocladiales genera are characterized by production of thick- walled resistant sporangia usually with pitted walls. Another feature linking the members of this group is the predominant nuclear cap present in the zoospore and planogametes. The order is now divided into four families, Blastocladiaceae, Coelomycetaceae, Catenariaceae and Physodermataceae.	*Physodermataceae asphodelae* (Karling, 1950; http://www.bioimages.org.uk)

Kingdom	Phyla and general characters	Subphyla Characters and examples	Figures
	Zygomycota: Moulds that live in soil or on decaying plant or animal material	Mostly terrestrial in habitat, living in soil or on decaying plant or animal material. The name comes from zygosporangia, where resistant spherical spores are formed during sexual reproduction. Zygomycetes synthesize chitosan (deacetylated homopolymer of chitin). This phylum is divided in to two groups: Zygomycetes and Trichomycetes	(e.g., *Trichoderma*) *Trichoderma koningiopsis* (DPMI, NBRI)
	Glomeromycota: Mutualistic symbiotic fungi essential for terrestrial ecosystem that benefit	Members of the Glomeromycota form arbuscular the plants they inhabit. mycorrhizas (AMs) with the roots or thalli (e.g. in bryophytes) of land plants. *Geosiphon pyriformis* forms an endocytobiotic association with *Nostoc* (Cyanobacteria). Mostly symbionts with land plants for carbon and energy, but there is recent circumstantial evidence that some species may be able to lead an independent existence.	*Geosiphon pyriformis* (Walker & Vestb.; www.agr.gc.ca)
	Ascomycota: Sac fungi that include morels, truffles and baker's yeast	**Pezizomycotina:** It is the largest subphylum of Ascomycota. It includes all filamentous, sporocarp-producing species, with the exception of *Neolecta* of Taphrinomycotina. Medically important species have been the source of both disease (e.g., *Coccidioides immitis* – Valley Fever) and life saving drugs (e.g., *Penicillium chrysogenum*, *Tolypocladium inflatum*).	*Penicillium* (http://bioweb.uwlax.edu)

Kingdom	Phyla and general characters	Subphyla Characters and examples	Figures
		Saccharomycotina: It comprises a monophyletic lineage with a single order of about 1000 known species. Fewer than 10 species accounts for most plant mycotic infections. Responsible for many biotechnological and industrial processing. e.g., *Saccharomyces cerevisiae*	*Saccharomyces cerevisiae* (M.Blackwell; http://tolweb.org)
		Taphrimycotina: They are dimorphic plant parasites (e.g., *Taphrina; Pneumocystis, Schizosaccharomyces*) with both a yeast state and a filamentous (hyphal) state in infected plants.	*Taphrina* (http://www.indexfungorum.org)

Kingdom	Phyla and general characters	Subphyla Characters and examples	Figures
	Basidiomycota: Higher fungi that include mushrooms and puffballs	**Agaricomycotina:** About 98% of the species of the Agaricomycotina are in a clade called the Agaricomycetes, which includes mushrooms, bracket fungi, puffballs, and others. The other major groups are the Tremellomycetes and Dacrymycetes. These latter groups include "jelly fungi", which have gelatinous, often translucent fruiting bodies (e.g., "witches butter" *Tremella mesenterica)*, as well as many yeast-forming species. To obtain carbon nutrition, Agaricomycotina decompose dead organic matter or enter into diverse associations (both antagonistic and benign) with plants, animals, and other fungi.	 *Tremella mesenterica* *Agaricus bisporus* (www.naturephoto-cz.com; www.svims.ca)

Kingdom	Phyla and general characters	Subphyla Characters and examples	Figures
		Pucciniomycotina: They do not form fruiting bodies. The species, parasitic on higher plants, cause leaf spots, witches'-broom (tufted growth), and galls (swellings). Particularly affected are azaleas and rhododendrons. The Pucciniomycotina include smut fungi. Few Pucciniomycotina species produce basidiocarps. They form hyphae and spore filled fruiting structure termed as "Sori". e.g., *Puccinia striiformis Puccinia striiformis*	*Puccinia striiformis* (www.naturamediterraneo.com)
		Ustilagiomycotina: They are former smut fungi and along with the Exobasidiales and one of best studied group of plant parasite. The classes of the Ustilaginomycotina are the Exobasidiomycetes, the Entorrhizomycetes, and the Ustilaginomycetes. Some examples are, *Ustilago madis; U. hordis and Ustilago tritici*	 *Ustilago madis* (slog.thestranger.com)
	Mitosporic fungi: Fungi with asexual reproduction, or by mitosis rather than meiosis. A large portion of mitotic fungi have not been correlated with meiotic states.	**Deuteromycota:** They are also known as imperfect fungi and their sexual form of reproduction has never been observed. Fungi producing the antibiotic penicillin and those that cause athlete's foot and example yeast infections are imperfect fungi. A suitable is *Aspergillus niger* (sexual cycle not known); and *Cladosporium resinae*	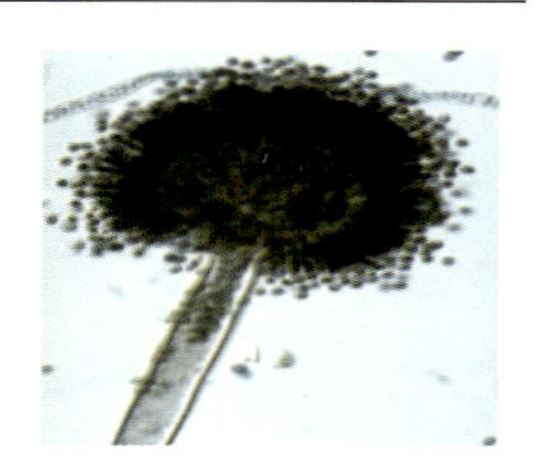 *Aspergillus niger* (www.pfdb.net)

Kingdom	Phyla and general characters	Subphyla Characters and examples	Figures
Strami-nopila or Chromista	**Oomycota:** They are filamentous protists which must absorb their food from the surrounding water or soil, or may invade the body of another organism to feed.	**Oomycota:** They are aggressive plant pathogens. They form oogonia and oospores. Mostly septa are not present and referred as coenocytic. Antheridia formation (small filamentous structures) but they not produce cell and as nuclei containing. Antheridial nuclei allow to first enter first the oogonium and then oospheres themselves. Then nuclei fuse within the oospheres to produce a diploid. Some examples are *Phytopthora* sp. *Peronospora* sp. and *Pythium* sp.	*Phytopthora sp.* (www.vn.refer.com)
	Hyphochytridiomycota: A class of microscopic fungi which form zoospores having an anteriorly directed, tinsel flagellum.	**Hyphochytridiomycota:** Flagellated fungi reproduce *asexually* by means of flagellated spores called zoospores that are produced in zoosporangia. Most primitive because of their aquatic habitat. General features are non-septate mycelium; asexual spores produced by sporamgium and lack of complex spore producing body (=sporocarp) General Names are Mastigomycotina sp., Heterokontophyta sp. and Synurid sp. etc.	*Synurid sp* (fineartamerica.com)
	Labyrithulomycota: A phylum of Stramenopila that produce networks of tubes or filaments that absorb nutrients and that the cells use as carbon source.	**Labyrithulomycota:** They are marine saprobes. These organisms cause little economic or environmental impact, but probably play an important role in nutrient cycling and detritus breakdown in marine habitats worldwide. The most important of these species is *Labyrinthula zosterae*, the causative agent of wasting disease of the eelgrass *Zostera marina.*	*Zostera marina* (www.natureblog.com)

Kingdom	Phyla and general characters	Subphyla Characters and examples	Figures
Slime moulds	**Myxomycota:** Slime molds; organisms having a noncellular and multinucleate creeping vegetative phase and a propagative spore-producing stage: comprises Myxomycetes and Acrasiomycetes.	**Myxomycota:** They produce spores that are borne in sporangia. Haploid cell which is *not* enclosed in a rigid cell wall, and ingests its food by means of phagocytosis. During this mode of ingestion, the food particles, usually bacteria, beceome surrounded by the pseudopodia of the myxamoeba. The spore surface may range from almost smooth to reticulate. Spores of *Physarum polycephalum* and *Didynium iridis* are spiny.	*P. polycephalum* (Encyclopedia of life)
	Plasmodiophoromycota: Phylum of plant parasites in the kingdom Chromista commonly referred to as endoparasitic slime molds.	**Plasmodiophoromycota:** All members are obligate parasites of algae, fungi, or plants, causing cell enlargement, especially of the roots. They are distinguished by the production of motile cells (zoospores) with two unequal anterior whiplike threads (flagella). *Plasmodiophora brassicae* causes clubroot of cabbage and related plants. *Spongospora subterranea* causes powdery scab of potato.	*Spongospora subterranea* (enfo.agt.bme.hu)

Kingdom	Phyla and general characters	Subphyla Characters and examples	Figures
	Dictosteliomycota: Important model, organisms for study of cytokinesis, signalling, chemotaxis, motility, phagocytosis etc	**Dictosteliomycota:** They are also known as the social amoebae. Their life cycle is considered among the most bizarre among microorganisms. There is no true plasmodium. Pillows of the coral-colored slime mold *Myxomycetes* grow on damp wood is the best example of this phylum	*Myxomycetes* (www.biologyreference.com)
	Acrasiomycota: No cells trapped in stalk during sporocarp formation.	**Acrasiomycota:** They have a life cycle characterized by slime like amoeboid stage and a multicellular reproductive stage. They are classified among the protists as two distinct phyla, *Dictyolsteliomycota* (the cellular slime molds) and *Myxomycota* (the plasmodial slime molds).	*Physarum polycephalum* (http://mskimrocks.tripod.com)

until moulds begin to grow on it. Moist chambers can be used for all kinds of materials such as, dung, wood, leaves, old stems, corn stalks, bark, seeds, fruits, old fungi, dead insects, and numerous other things.

6.1.3. Direct plating

Often it is most suitable to place materials that are of interest directly on a suitable agar medium. As already well known this technique encourages swiftly spreading moulds at the expense of other fungi but is nevertheless extensively used. It is a simple technique, requiring the placing of small bits of the substance on the surface of the agar or the pouring of melted but cooled agar over the fragments. After a few days' incubation fungal growth appear on the surface, and can be transferred into pure culture.

Martin's Rose Bengal Medium is a good choice for direct plating, as the rose bengal dye and antibiotics in it slow down colony growth and keep the colonies from growing together and merging with each other.

6.1.4. Dilution plating

In this technique 1 gram (dry weight) of the material to be studied is ground up (if necessary) and dispersed in 9 ml of sterile water. Subsequent dilutions are made and 1-ml (from each dilution) is pipetted to a separate Petri dish, and cooled, melted agar medium poured over it. The plate should be moved gently on the table in a figure-of-eight motion to effect proper dispersion. Alternatively, the solution can be put on the surface of solidified medium and spread evenly throughout.

After a few days' incubation, colonies will appear in varying densities, depending upon the amount of dilution from the original material. The number of spores present in the original sample can be calculated roughly by selecting the plates showing 40-100 colonies and writing down the colony count. With this information the following calculation can be performed:

Colonies per gram of original sample = Colony Count/Dilution Factor

6.1.5. Air borne fungi

Certain moulds and fungi are dispersed in air at particular time and place. These spores can serve as the infection of plant or other hosts. This technique is often helpful to know which mould or fungi species is present in the air at a particular time or place. There are two approaches to air sampling, one that yields spores for microscopic examination and one that yields cultures. Exposures of Petriplates along with solidify media in a particular place and time is enough for the spore collection but for the specific spores collection we need filters. Anderson sampler accommodates several Petri plates and will even sort spores out according to their size or mass. Such a machine is far more efficient than an open Petri plate but is bulky and is normally used only by specialists (Fig.6).

Fig. 6. Anderson sampler along with solidified media on petri plate

6.1.6. Baiting

It is based on the chemo attractive property of microbes. Some fungi and moulds have a property to utilize the specific nutrient as a carbon source for the growth and reproduction. It is specific technique used for isolation of fungi by presenting a particular substance to the environment for colonization and then later recovering it for isolation of the fungi. An example of a fungus sometimes recovered by baiting is the creosote fungus, *Amorphotheca resiniae* (Cannon and Kirk 2007). To isolate it, some scientists coat matchsticks with creosote and place them on soil in Petri dishes for two or three weeks. *Amorphotheca*, particularly able to utilize creosote, grows from spores in the soil and invades the matchsticks, where it produces abundant hyphae and conidia. From here it is transferred into pure culture. Other kinds of baits might be pieces of wood, insects, carrot chunks, plastics, hair, or anything else one can name. The bait can be submerged in a particular habitat in nature or in a moist chamber. To isolate dermatophytes, for example, it is usual to place hair on moist soil in a moist chamber and examine it periodically for sporulating moulds. The most commonly baited habitat is water, both fresh and marine. Since most moulds attracted this way are true aquatics, it is necessary to purify them by transferring them to successive changes of distilled water containing new sesame seeds or other baits. When completely isolated into pure culture they can be transferred to solid media, but they may never sporulate there. Submerged wood blocks are one of the most commonly used materials for obtaining marine ascomycetes.

6.2. Special techniques

6.2.1. Stress techniques

All moulds are capable of withstanding environmental stresses, but eventually, when the stress is great enough, they will be killed. Not all moulds have the same tolerance to stress, however, we can take advantage of this property by subjecting a material to just enough stress to kill some moulds but not others. The application of such techniques has turned up another interesting fact: that some moulds will not germinate until they have been subjected to conditions that kill most others.

To explore these interesting adaptations drastic treatment will yield a few fungal cultures that might not otherwise appear. After treatment, the substance can be handled in any of the normal ways, such as plating, or moist chambers.

6.2.2. Surface sterilization

Surface sterilization can be used as a selective technique on a slightly larger scale. A living leaf that has a dead circular spot on it if spotted directly on the agar medium will probably yield some fast-growing fungus but not the causative agent. For the fungus *inside* the leaf, not the surface adherents, a good way to obtain is to apply bleach long enough to kill the external organisms but not the internal ones. Although materials differ considerably, a sterilization time of about 1 hour in a 10% commercial bleach or 3% hydrogen peroxide or 1-10 minutes in 2% sodium hypochlorite will often work. If one is unsure, some longer and shorter times should be tried. Surface sterilization will also work for seeds, large fungi, wood, many man-made materials, dead insects etc.

6.2.3. Selective nutrient

This technique is essentially the same as the baiting methods discussed above. To select these fungi, we make a defined medium such as Czapek's medium, supplementing with the required carbon source like cellulose in place of sucrose. Another selective medium is *Trichoderma* specific medium (TSM) (Elad and Chet, 1983). TSM supplemented with benomyl is efficient for isolating *Fusarium* spp. from soil; and TSM supplemented with captan was a specific selective medium for *Trichoderma* spp., even in the presence of *Fusarium* in the soil.

6.2.4. Selective temperature

Moulds that grow easily at room temperature are said to be mesophilic. Most fungi are mesophilic and are inhibited or killed at unusually high or low temperatures. There are some, however, that actually require unusual temperatures. Moulds requiring high temperatures (40°C or more) are said to be thermophilic, and those requiring low temperatures (15°C or less) are psychrophilic. In searching for these fungi simply incubate plates or moist chambers at the appropriate temperature and isolate the moulds that develop; all isolates and subsequent transfers must be grown at their optimum temperature as well.

6.2.5. Osmophily

Many stored products, such as grain, museum specimens, and hides, undergo degradation by moulds. The phenomena is caused by fungi that can withstand unusually dry conditions. Such fungi are said to be osmophilic, a term that refers to their prevalence in environments of high osmotic potential. Osmophilic fungi grow poorly on normal culture media and often sporulate abnormally or not at all. To isolate them the water actvities of media must be drastically decreased. This means increasing the sucrose in Czapek's Agar from 20 g per litre to 200-500 g per litre, or adding 200-500 g of maltose to each litre in Leonian's Agar instead of the usual 6.25 g. For example osmophilic *Eurotium* species, mostly require them to be grown on Czapek's or malt agar with 40% sucrose or on media containing large quantities of glycerine.

6.2.6. Spore Printing

Mostly moulds, basidiomyces and ascomycetes are able to eject their spores compulsorily. This technique is fully based on this enormous property of specific fungi. For mushrooms and bracket fungi, a piece of the gill or tube tissue can be attached, perpendicular to the agar, on the lid of a petri dish. The spores will then float down on to the agar surface and later germinate. Usually petroleum jelly can be used as adhesive, but other things, such as masking tape or an agar block, may also work as well. Interesting isolations can be made from petri plates that have been opened and inverted over soil, logs, and other materials in the field, if proper care is taken to exclude air currents. Ascomycetes, such as cup fungi, can also be suspended on agar surface for spore printing.

7. Preservation

We need to preserve and maintain healthy, viable fungal cultures for many months or years for further studies. During preservation, most important factor is to stop the microbial growth or at least lower the growth rate. The following methods are used for the preservation of fungi:

7.1. Short term preservation

Short-term preservation involves maintenance of cultures for up to 1 year. Most fungal cultures can be maintained for that period by serial transfer. The method is simple, inexpensive, and widely used. Although time consuming and labor intensive, periodic transfer is a good option for small collections with cultures in constant use for short periods, less than 1 year. The method also has several disadvantages, however. Cultures must be checked frequently for contamination by mites or other microorganisms and for drying. In addition, the morphology and physiology of a cultured fungus may change over time. In particular, the ability to sporulate or to infect a host may be lost after repeated transfers. Because of these disadvantages, the technique is generally inappropriate for long-term (more than 1 year) preservation of cultures.

7.1.1. Periodic serial subcultures

Inoculum is transferred from an actively growing fungus culture to test tubes (screw cap or plugged with cotton or foam) or petri dishes (wrapped with parafilm to reduce drying) containing an agar medium of choice.

- Alternating nutrient-rich with nutrient-poor media at each transfer helps to maintain healthy cultures.
- Some fungi, such as endophytic and entomopathogenic species, have specific media requirements (Bacon 1990; Humber 1994).
- After a culture is established, it is kept at room temperature or at 4°C. Cultures must be checked periodically for contamination and desiccation.
- Fungi such as oomycetes and some basidiomycetes (e.g., *Boletus, Coprinus, Cortinarius,* and *Mycena)* should be transferred monthly if kept at 16°C (von Arx and Schipper 1978).

7.1.2. Freezing

Most filamentous fungi can survive at least 1-2 years at 4°C. Vigorous, sporulating cultures also can be sealed tightly and stored in a freezer at -20°C (Carmichael 1956; 1962) or stored at -70°C (Pasarell and McGinnis 1992) to enhance survival and increase the interval between required transfers.

7.2. Long term preservation

7.2.1. Sclerotization

Some fungi develop sclerotia or other long-term survival propagules in culture as well as in nature; preserving such structures, usually at 3°-5°C, is a good way to preserve fungal strains. Sclerotia and spherules of various myxomycetes have been germinated successfully after 1-3 years of storage. Many soil fungi, such as *Magnaporthe, Phymatotrichum,* and *Cylindrocladium* species, produce sclerotia or microsclerotia that remain viable for 2-5 years (Singleton *et al.* 1992). Instructions for inducing formation of spherules, sclerotia, and microsclerotia are available in Daniel and Baldwin (1964) and Singleton *et al.* (1992). Sometimes rice straw or toothpicks are used as substratum to promote sclerotia production in culture. Jump (1954) described a simple method for inducing sclerotium formation in *Physarum* species. A piece of sterile cellophane cut to the dimensions of a Petri dish is placed over a dish containing 1% water agar.

7.2.2. Mineral oil overlay

A low-cost and low-maintenance method for preserving cultures growing on agar slants is oil overlay. Cultures can be kept for several years or, in exceptional cases, up to 32 years at room temperature or 15°-20°C. This method is appropriate for mycelial or nonsporulating cultures that are not amenable to freezing or freezedrying. As an added benefit, oil also reduces mite infestations. Although many basidiomycetes can be maintained this way, the growth rates of the cultures slow as storage times increase (Johnson and Martin 1992; Burdsall and Dorworth 1994). The major disadvantage of the oil overlay technique is that the fungi continue to grow, and thus, selection for mutants that can grow under adverse conditions may occur. High-quality mineral oil or liquid paraffin is sterilized by autoclaving at 15-lb (6.8-kg) pressure for 2 hours. The entire agar surface and fungal culture should be submerged completely in the oil. The tubes are kept in an upright position at room temperature (15°-20°C; 12°C for *Pythium* species and *Phytophthora* species.

7.2.3. Immersion in distilled water

Most suitable for sporulating fungus. It is an inexpensive and low-maintenance method for storing fungal cultures is to immerse them in distilled water. Apparently, the water suppresses morphological changes in most fungi. Minimum storage for Basidiomycetes (2 years) at 25°C and maximum Ascomycetes, however, including their mitosporic forms, survived up to 10 years when stored at 20°C (Johnson and Martin 1992). Fungal discs are transferred into sterile screw caps test tubes. To save space, small (1.8 ml), sterile, screw-cap cryovials are filled with several discs and topped with sterile distilled water. Test tubes (loosely capped and wrapped with Parafilm) are stored at room temperature; tightly capped tubes and vials are stored at 4°C. Disks are removed aseptically and transferred to fresh agar medium to retrieve cultures.

7.2.4. Organic substrata

7.2.4.1. Wood

Wood-inhabiting fungi can be successfully stored on wood chips or toothpicks as long as the colony is growing vigorously (Nelson and Fay 1985; Singleton *et al.* 1992). Some wood-inhabiting basidiomycetes and ascomycetes can be stored on wood chips for up to 10 years. Small pieces of untreated beech wood (12-mm diameter x 6-mm thick) are added to 2% malt-extract broth (about 60 pieces of wood per 100 ml broth) and sterilized for 20 minutes at 121°C. About 15 wood chips are drained and placed on a colony of the fungus that is growing on malt extract agar in Petri dishes.The Petri dishes are sealed with Parafilm, and the fungus is allowed to colonize the wood chips. After 10-15 days, the inoculated wood chips are transferred to sterile test tubes (18 x 180 mm) containing 6-7 ml of 2%-malt agar and placed in 4°C.

7.2.4.2. Cereal grains

Fungi such as *Sclerotinia, Magnaporthe, Leptosphaeria,* and *Rhizoctonia* species have been stored for up to 10 years on seeds of oats, barley, wheat, rye, millet, and sorghum

(Singleton *et al.* 1992). To preserve isolates of *Rhizoctonia* species, barley, oat, or wheat grains (Sneh *et al.* 1991) are soaked overnight in water containing chloramphenicol (250 g/ml). The grain is autoclaved for 1 hour at 12°C over two consecutive days. Screw-cap vials are filled with the grain and autoclaved. The vials are inoculated with transfers from the margins of actively growing cultures and incubated at 23°-27°C for 7-10 days. The cultures then are dried thoroughly in a desiccation chamber. The caps are tightened and wrapped with Parafilm, and the vials are stored at -25°C.

7.2.4.3. Agar strips

Nuzum (1989) described a method of vacuum-drying fungal cultures on agar strips. *Pythium, Rhizoctonia,* and some basidiomycete species survived 18 months with this method, whereas ascomycetes and their mitosporic forms survived from 3-5 years. Fungal cultures are grown on appropriate media in Petri dishes. Strips 1-cm long are cut from the growing edge of the colony and placed in sterile Petri dishes. After 1 week at room temperature, the pieces of dried agar are transferred to sterile ampoules, vacuum-dried, and sealed.

7.2.4.4. Insect or plant tissue

The host tissue can be used as a substrate on which to maintain and store cultures of some pathogenic fungi. For example, roots of plants infected with *Pyrenochaeta* and *Thielaviopsis* can be dried and then frozen (Singleton *et al.* 1992). *Neozygites fresenii* cannot be cultured in vitro, but Steinkraus *et al.* (1993) developed a method of preserving viable conidia on frozen, infected aphid mummies. *Puccinia* spp. (Wheat rust) and *Uromyces ûcariae* (Microcyclic) can be stored on host tissues (Quilliam and Shattock 2003).

7.2.4.5. Soil or sand

Some fungi can be preserved easily and successfully for many years in dry, sterile soil or sand. This low maintenance and cost-effective method is appropriate for fungi such as *Rhizoctonia* (Sneh *et al.* 1991), *Septoria* (Shearer *et al.* 1974), and *Pseudocercosporella* (Reinecke and Fokkema 1979). Dormancy caused by dryness can take time to develop, however, and morphological changes in some fungi have been recorded.

7.2.4.6. Silica gel

The silica gel method can be used to preserve sporulating fungi if facilities for freeze-drying or for storage in liquid nitrogen are not available. It originally was developed by Perkins (1962) for *Neurospora* species. He found that sporulating fungi protected by skim milk and stored on silica gel remain viable for 4-5 years. Spores and microcysts of dictyostelids can be preserved for up to 11 years on silica gel (Raper 1984). In general, viability after storage on silica gel depends on the strain of fungus and the medium on which it was grown before storage. The advantage of silica gel is that it prevents all fungal growth and metabolism. Some researchers use glass beads instead of silica gel.

7.2.4.7. Freezing

Most fungal cultures frozen at -20° to -80°C in mechanical freezers remain viable. Cultures grown on agar slants in bottles or test tubes with screw caps can be placed directly in the freezer (Carmichael 1956). Overall failure rate for mitosporic ascomycetes, Zygomycetes, and yeasts after 5 years in storage at -20°C was 5.1% (Carmichael 1962). The cultures were preserved successfully for up to 5 years by mechanical freezing at -80°C. In general, vigorously growing and sporulating cultures survive the freezing process better than less vigorous strains.

7.2.4.8. Liquid nitrogen

Storage in liquid nitrogen is an effective way to preserve many, if not most, organisms, including those that cannot be lyophilized. Liquid-nitrogen storage is recommended for the preservation of dictyostelids (Raper 1984), amoebae, Zygomycetes including Entomophthorales (Humber *et al.* 1994), oomycetes (Nishii and Nakagiri 1991), phytopathogenic fungi (Dahmen *et al.* 1983), and yeasts (Kirsop and Doyle 1991). Ascomycetes that sporulate poorly in culture and higher basidiomycetes that generally grow only as mycelia in culture also can be stored in liquid nitrogen. Fungal cultures are kept in liquid nitrogen and store at the temperature of as same as Liquid N_2.

7.2.4.9. Lyophilization

A process used for preserving biological material by removing the water from the sample, which involves first freezing the sample and then drying it, under a vacuum, at very low temperatures. Lyophilization, or freeze-drying, a low-cost form of permanent preservation, is not appropriate for all fungi. The technique is used primarily with species that form numerous, relatively small propagules. However, Croan (2000) demonstrated that mycelial isolates of basidiomycetes can be lyophilized effectively in the presence of trehalose. Lyophilized spores of dictyostelids, with associated bacteria, were maintained successfully for up to 30 years (Raper 1984).

7.2.4.10. Cryopreservation

Cryogenic technique for long-term storage of large numbers of fungal species was introduced to ATCC in 1960 and the results have been very satisfactory. The technique was consecutively introduced to many other prominent collections, e.g., CAB International Mycological Institute. Fungal cultures are grown directly in firmly closed sterile plastic cryovials (1.8 mL) with 200 mg of perlite (Agroperlit, agricultural grade) moistened with 1 mL of wort (4° Balling) or other medium (e.g., MYA) enriched with 5% glycerol as a cryoprotectant. The grown cultures are kept into -70°C and after then placed in to liquid nitrogen.

8. Molecular Characterization

Typically, the species texture of fungal body has been resolute via combination of media culturing and identification based on macro or microscopic features. Nevertheless, these approaches have several drawbacks, including the over or underestimation of fungal species within a community based on the presence or absence of fruiting bodies, the inability of particular species to be grown in culture, and the complexity in grouping fungi to the species level based on microscopic characteristics. Rapid and simple methods for identification of the species composition of fungal communities based on sequencing a particular region of the fungal genome is a trustworthy and alternative approaches. Molecular methods are faster, more sensitive, more stable and less dependent on external factors than morphological methods (Faggi *et al.* 2001; Kim *et al.* 2001). Amplification of ribosomal genes and spacers of fungal genome using the polymerase chain reaction (PCR) is very significant, because they are comprised of highly conserved tracts with heterogeneous regions in between. These conserved regions can be amplified using universal primer (Table 3) from the heterogeneous regions.

More recently, the Internal Transcribed Spacer (ITS) regions of the ribosomal operon have been used for fungal systematics and classification (White *et al* 1990). There are 2 ITS regions in the fungal rRNA operon (Fig. 7). The first, ITS1, is found between the 18S and 5.8S rRNA genes. The second, ITS2, is located between the 5.8S and the 28S rRNA genes. Since the ITS sequences are important enough as spacer regions to be maintained by the cell, but not used for any functional purpose; they are allowed to accumulate mutations at a faster rate than the 5.8S, 18S, and 28S rRNA genes. It is this slightly increased rate of accumulated mutations which allows the ITS sequences to provide an improved level of resolution than the D2 sequence that is located with the 28S gene. It is generally accepted to sequence the entire stretch of ITS1-5.8S-ITS2 for identification and improvement of fungal classification (Fig. 8).

Table 3. Primer sequences used for amplification of ITS region

Primers	Sequence (5' b 3')
ITS1	TCCGTAGGTGAACCTGCGG
ITS2	GCTGCGTTCTTCATCGATGC
ITS3	GCATCGATGAAGAACGCAGC
ITS4	TCCTCCGCTTATTGATATGC

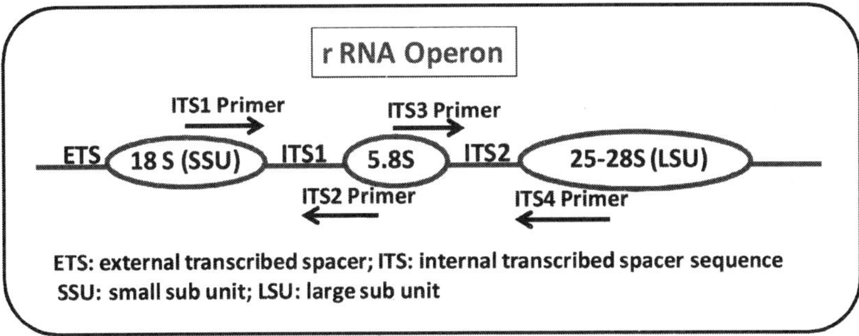

Fig. 7. Fungal rRNA operon showing ITS regions used in identification of fungi

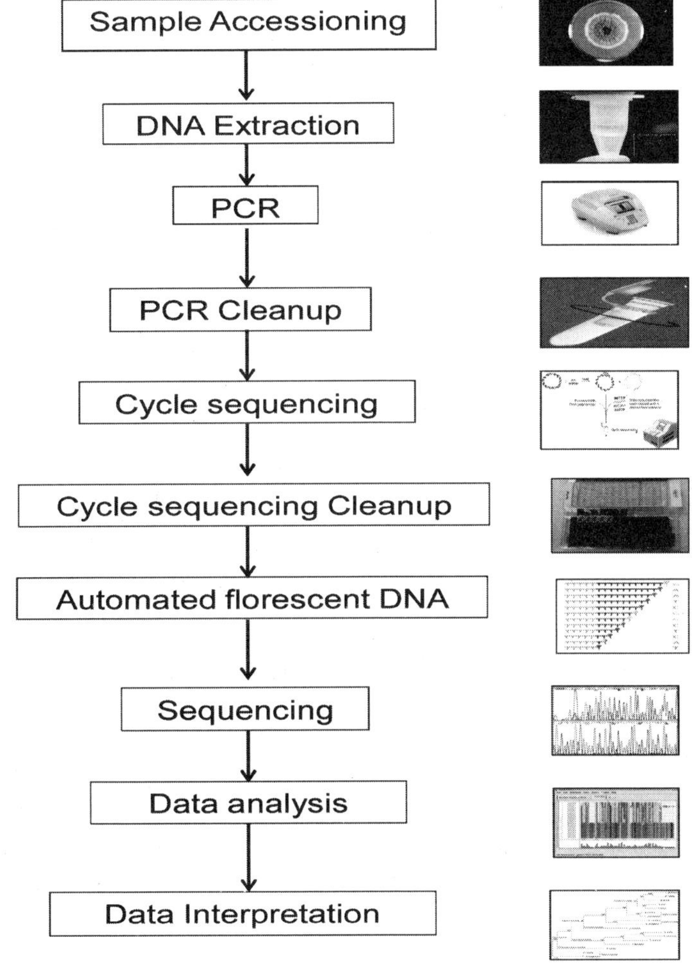

Fig. 8. Steps in molecular characterization of fungi

9. Applications of Fungi

Fungi have traditionally been the source of several useful chemical substances starting with the well known ethyl alcohol from yeast, which continues to influence human civilization all over the world. Following the antibiotics era, in the latter part of 20th century, scientists isolated many more products important in agriculture, industry and medicine from fungi.

Yeasts and filamentous fungi have been extensively used for production of industrial enzymes and human therapeutic proteins by most industries in the last few years. The ability of these organisms to grow in chemically defined medium in the absence of animal-derived growth factors (e.g., calf serum) and to secrete large amounts of recombinant protein, together with ease of scale-up and low cost of production have made these organisms the systems of choice for producing many industrial enzymes and therapeutic proteins. A list of fungi used in various industrial applications is given in Table 4.

Table 4. Different applications of fungi

Fungi	Application
Aspergillus sp., *Fusarium* sp., *Monilia* sp. *Mucor* sp., *Rhizopus* sp., *Phanerochaete chrysosporium, Trichoderma* sp.	Composting, lignin degradation industrial
Fusarium sp.	Gibberellins
Aspergillus sp., *Mucor* sp., *Rhizopus* sp.	Food, enzyme
Rhizopus sp., *Aspergillus oryzae*	Organic acids
Trichoderma sp., *Beauveria* sp., *Metharizium* sp.	Biological control, biopesticide
Amylomyces rouxii	Tape cassava, rice
Aspergillus oryzae, A.niger	Koji, food, citric acid
Rhizopus oligosporus	Tempeh, soybean, amylase, lipase
Aspergillus niger	Feed, proteins, amylase, citric acid
Pleurotus oestreatus, Lentinus edodes	Mushrooms
Penicilium notatum roquefortii	Penicillin, cheese
Aspergillus niger	Tannary
Penicillium griseofulvin, Trichoderma polysporum, Cylindrocarpon lucidum	Pharmaceuticals

9.1. Fungi in agriculture

Fungi are ubiquitous, some having beneficial effect on plants, while others may be detrimental. Mycofungicide and fungal biofertilizers have been promoted for agricultural use because of their ability to control plant disease and their ability to increase crop production in an environmental friendly manner.

9.2. Fungi as biopesticides

The use of biopesticides is growing annually while the use of traditional pesticides is on the decline because the application of chemical fungicides over a long period may result in plant pathogenic fungi developing resistance. Fungi are important enemies of many pest species and are able to control pest populations naturally. Biocontrol agents and especially antagonistic fungi and have been used to control plant diseases with 90% of applications being formulated using different strains of *Trichoderma* e.g. *T. harzianum*,*T. virens*, *T. viride* to control soil borne plant pathogens. Many species of *Chaetomium* e.g. *Chehaetomium globosum*, *C. cochlioides*, *C. cupreum* can also be antagonistic against various soil microorganisms. Entomopathogenic fungi can be used as biopesticides for they actively kill insects. There are large numbers of fungi such as *Beauveria bassiana, Metarhizium anisopliae, Paecilomyces* species and *Verticillum lecanii* that have been used widely as biopesticides.

9.3. Fungi as a biofertilizers

Plant-associated fungi fulfill important functions for plant growth and health. Direct plant growth promotion by fungi is based on improved nutrient acquisition and hormonal stimulation. Diverse mechanisms are involved in the suppression of plant pathogens, which is often indirectly connected with plant growth. *Trichoderma* are model organisms to demonstrate influence on plant health. *Trichoderma* is commonly used as a biocontrol agent, plant growth promoter and as phosphate solubilizing microorganism that improve phosphorus absorption in plants and stimulate plant growth. *Penicillium radicum* and *P. italicum* are also reported as phosphate-solubilizing taxa.

Mycorrhizal fungi are obligate symbiotic microorganisms since they cannot grow without the plant host on synthetic media. It is a symbiotic association between host plants and certain group of fungi at the root system, in which the fungal partner is benefited by obtaining its carbon requirements from the photosynthates of the host and the host in turn is benefited by improving water and nutrient uptake and provides protection from pathogen. Mycorrhizal fungi help in absorption of both water and nutrients and act as extended roots for a plant.

9.4. Fungi in food industry

Solid state fermentation performed by mainly yeast and fungi offers numerous advantages for the production of bulk chemicals and enzymes. In industries microorganism mainly fungus played a major role in the production of food and alcoholic beverages. In addition, several products of microbial fermentation are also incorporated into food as additives and supplements (antioxidants, flavours, colourants, preservatives, sweeteners). There is great interest in the development and use of natural food and additives derived from microorganisms, since they are more desirable than the synthetic ones produced by chemical processes.

The usage of fungi in food, preservation or other purposes by humans is wide-ranging and has a rich history. For example, yeasts are used to ferment beer, wine and bread; (Bekatorou *et al.* 2006), while some other species like, *Aspergillus oryzae*, is used in the production of soy sauce (Abe *et al.* 2006) and tempeh, a food product made from soya beans, with different nutritional characteristics and textural qualities. *Saccharomyces cerevisiae* (also known as baker's yeast), a single cell fungus, is used in the baking of bread and other wheat-based products such as pizza and dumplings. It is also used for the production of alcoholic beverages through fermentation. Several species, such as the *Agaricus bisporus* and the Portobello are sold as button mushrooms for consumption.

There are many species of mushroom that are harvested from the wild, and are consumed in different forma as well as sold in the market. Milk mushrooms, morels, chanterelles, truffles, black trumpets, and porcini mushrooms are all in great demand in the market and are often used in gourmet dishes.

For certain types of cheeses, it is also a common practice to inoculate milk curds with fungal spores to forment the growth of a specific species of mould that impart a unique flavour and textures to the cheese, this accounts for the blue colour in cheeses such as Stilton of Roquefort. The moulds used in cheese production are usually non-toxic and are thus safe for human comsumption however, mycotoxins may accumulate due to fungal spoilage during cheese rippening or storage.

9.5. Organic acids from fungi

There are several organic acids produced on a commercial scale from fungi (Table 5).

Table 5. Fungi used in organic acids production

Organic acid	Source
Citric acid	*Aspergillus niger*
Fumaric acid	*Rhizopus nigricans*
Gluconic acid	*Aspergillus niger*
Itaconic acid	*A. terreus*
Kojic acid	*A. oryzae*

Source: Magnuson and Lasure 2004

9.6. Application of fungi in medicine

Fungi make an extraordinarily important contribution to managing disease in humans and other animals. At the beginning of the 21st century, fungi were involved in the industrial processing of more than 10 of the 20 most profitable products used in human medicine. In 1941, penicillin from the fungus *Penicillium chrysogenum* was first used successfully to treat an infection caused by a bacterium (Dayalan *et al.* 2011).Use of penicillin revolutionized the treatment of pathogenic disease. Many formally fatal diseases caused by bacteria became treatable, and new forms of medical intervention were possible. Other useful agents from fungi that are useful in various treatments are Griseofulvin , Strobilurins , Sordarins and Echinocandins , Cyclosporin A, Calcineurin, Gliotoxins, Statins, Ergot alkaloids etc.

9.7. Application of fungi in biotechnology

Fungi are extremely useful organisms in biotechnology. Most of the yeast based therapeutic proteins are now being obtained from genetically modified baker's yeast, *Saccharmyces cerevisiae*. There are several recombinant proteins, including mammalian proteins that cannot be derived from *E. coli*, and in such cases yeasts have been found to be the answer. Yeasts have been used to express mammalian proteins at very high concentrations, as exemplified by the production of 14.8 g/l of gelatin in *Pichia pastoris* culture. This yeast has also been the source of Hepatitis-B surface antigen (HbSAg), the recombinant vaccine against the lethal virus Hepatitis-B that causes serum hepatitis. *Pichia pastoris* expression vector has been the classical example of how a fungal expression host can reduce the cost of a vaccine considerably. Earlier, a single dose of Hepatitis-B vaccine was costing about Rs. 800/- but now the cost has been reduced to Rs. 80/- per dose. This is because *P. pastoris* can be grown easily and economically in large bioreactors. Among the filamentous fungi used for expression of recombinant proteins are *Aspergillus niger, A. nidulans* and *Trichoderma reesei* have been found to be useful for the large-scale production of enzymes. *A. niger* has also been used for the development of antibodies.

9.8. Environmental applications

In the area of environmental applications, harnessing of fungi in bioremediation of industrial effluents contaminated with heavy metals and other toxic substances is a major development. Fungi have the ability to adsorb heavy metals on their surface because of the charge on their cell walls. They are also able to bind metals because of their ability to produce certain metal binding peptides and proteins. They are also able to degrade complex carbon compounds such as pesticides by virtue of an array of enzymes they normally possess.

Xenobiotics (which include substances such as pesticides and effluents) frequently pose a hazard to health, and the first step is to use expensive methods to remediate contaminated land and water. Xenobiotics are likely to persist in the environment for a long time, as most decomposers can not handle them. Their persistence is environmentally damaging, but bioremediation using fungi (and bacteria) may provide a far more environmentally-friendly solution. Fungi can bring their exoenzymes to bear on such unusual substrates called xenobiotics. Fungi have immense potential as bioremediators, and are categorize on the basis of their remediation capabilities. Saprotrophs (white rot and brown rot fungi) and mycorrhizal fungi. White rot fungi due to their non specific enzyme system and broad substrate specificity and have been implicated in the transformation and mineralization of organo pollutants and contaminants associated with soil and water. Mycorrhizal symbiosis of fungi help in degradation of xenobiotics by consortium of group of fungi. Mycorrhizae not only provide water and mineral, but also act as filter, blocking toxic compound within the mycelium. *Trichoderma* has been shown to enhance chickpea growth in the presence of Arsenic and diesel (*Tripathi et al.* 2013, Mishra and Nautiyal, 2009).

References

Abe K, Gomi K, Hasegawa F, Machida M (2006) Impact of *Aspergillus oryzae* genomics on industrial production of metabolites. Mycopathologia 162:143–153

Arx JAV, Schipper MAA (1978). The CBS fungus collection. Adv Appl Microbiol 24:215-236

Bacon CW (1990) Isolation, culture and maintenance of endophytic fungi of grasses. In: Isolation of biotechnological organisms from nature, Labeda DP (ed.) pp 259-282. New York, McGraw-Hill ISBN 0-07035-701-3

Bekatorou A, Psarianos C, Koutinas AA (2006) Production of Food Grade Yeasts. Food Technol Biotech 44(3):407-415

Berg G (2009) Plant–microbe interactions promoting plant growth and health: perspectives for controlled use of microorganisms in agriculture. Appl Microbiol Biotechnol 84:11–18

Burdsall Jr. HH, Dorworth EB (1994) Preserving cultures of wood-decaying Basidiomycotina using sterile distilled water in cryovials. Mycologia 86:275-280

Burnett JH (1976) Fundamentals of Mycology (Second Edition) Edward Arnold, London. pp 546

Cannon PF, Kirk PM (2007) Fungal families of the world. CABI International ISBN: 9780851998275

Carlile MJ, Watkinson SC, Gooday GW (2001) The Fungi (2nd Edition). Academic Press, San Diego

Carmichael JW (1956) Frozen storage for stock cultures of fungi. Mycologia 48:378–381

Carmichael JW (1962) Viability of mold cultures stored at -20°C. Mycologia 54:432-436

Croan SC (2000) Lyophilization of hypha-forming tropical wood-inhabiting Basidiomycotina. Mycologia 92:810–817

Dahmen HTS, FJ Schwinn (1983) Technique for long-term preservation of phytopathogenic fungi in liquid nitrogen. Phytopathology 73:241-246

Daniel JW and Baldwin HH (1964) Methods of culture for plasmodial myxomycetes. In: Methods in cell physiology, Vol. 1., Prescott DM (ed.), Academic Press Inc., New York pp 9-41

Dayalan SAJ, Darwin P, Prakash S (2011) Comparative study on production, purification of penicillin by *Penicillium chrysogenum* isolated from soil and citrus samples. Asian Pac J Trop Biomed 1(1):15–19

Elad Y, Chet I (1983) Improved selective media for isolation of *Trichoderma* spp. and *Fusarium* spp. Phytoparasitica 11:55-58

Faggi E, Pini G, Campisi E, Bertellini C, Difonzo E, Mancianti, F (2001) Application of PCR to distinguish common species of dermatophytes. J Clin Microbiol 39:3382–3385

Gow NAR and Gadd GM (1995). The Growing Fungus. Chapman and Hall, London

Humber RA (1994) Special consideration for operating a culture collection of fastidious fungal pathogen. J Ind Microbiol 13:195-196

Johnson GC, Martin AK (1992) Survival of wood inhabiting fungi store for 10 year in water and under oil. Can J Microbiol 38:861-864

Magnuson, JK. and Linda LL (2004) Organic Acid Production by Filamentous Fungi. In Advances in Fungal Biotechnology for Industry, Agriculture, and Medicine, Lene Lange (ed.), Kluwer Academic/ Plenum Publishers, pp 307-340

Jump JA (1954) Studies on sclerotization in *Physarum polycephalum*. Am J Bot 41:561-567

Keller S (1998) Use of Fungi for Pest Control in Sustainable Agriculture. Phytoprotection 79(4):56-60

Kim JA, Takahashi Y, Tanaka R, Fukushima K, Nishimura K, Miyaji M (2001) Identification and subtyping of *Trichophyton mentagrophytes* by random amplified polymorphic DNA. Mycoses 44:157–165

Kirsop BE, Doyle A (1991) Maintenance of microorganisms and cultured cells: A manual of laboratory methods (Second Edition) . Academic Press, San Diego, CA

Mishra A,Nautiyal CS (2009) Functional diversity of the microbial community in the rhizosphere of chickpea grown in diesel fuel-spiked soil amended with Trichoderma ressei using sole-carbon-source utilization proûles. World J Microbiol Biotechnol 25:1175-1180

Nishii T, Nakagiri A (1991) Cryopreservation of oomycetous fungi in liquid nitrogen. IFO Res Comm 15: 105–118

Nuzum C (1989) A method of vacuum drying is described for the preservation of fungal cultures which are difficult to freeze dry. Approximately 80% of cultures so far tested have survived with this technique. Australas Plant Pathol 18(4):104-105

Pasarell L and McGinnis MR (1992) Viability of Fungal Cultures Maintained at -70 C. J Clin Microbiol 30(4):1000-1004

Perkins DD (1962) Preservation of Neurospora stock cultures with anhydrous silica gel. Canad J Microbiol 8:591-594

Quilliam RS and Shattock RC (2003) Haustoria of microcyclic rust fungi *Uromyces ûcariae* and *Puccinia tumida* and other gall-forming species, *U. Dactylidis* (macrocyclic) and *P. smyrnii* demicyclic). Plant Pathol 52:104–113

Raper JR (1954) Life cycles, sexuality and sexual mechanisms in the fungi. In: Sex in Microrganisms, Wenrich DH, Lewis IF and Raper JR (eds.), Am Ass Adv Sci, Washington DC, pp 41-81

Raper KB (1984) The dictyostelids. Princeton University Press, Princeton, New Jersey, pp 453

Reinecke P and Fokkema NJ (1979) *Pseudocercosporella herpotrichoides*: storage and mass production of conidia. Trans Br Mycol Soc 72(2):329-331

Singleton LL, Mihail JD and Rush CM (1992) Methods for research on soil-borne phytopathogenic fungi. American Phytopathological Society Press, St Paul, MN, USA, pp 264

Sneh B, Burpee L and Ogoshi A (1991). Identification of *Rhizoctonia* Species. The APS Press, St. Paul, MN, USA, pp 578

Tripathi P, Singh PC, Mishra A, Chaudhry V, Mishra S, Dwivedi S, Tripathi RD, Nautiyal CS (2013) *Trichoderma* inoculation ameliorates arsenic induced phytotoxic changes in gene expression, stem anatomy and rhizospheric microbial community of chickpea (*Cicer arietinum*) Ecotox Environ Safety 89:8-14

White TJ, Bruns T, Lee S and Taylor J (1990) Amplification and direct sequencing of fungal ribosomal RNA genes for phylogenetics. In: PCR protocols: a guide to methods and applications, Innis MA, Gelfand DH, Sninsky JJ, White TJ (eds.), Academic Press, New York, USA, pp 315–322

Whittaker, RH (1969) "New Concepts of Kingdoms of Organisms". Science 163:150-160

Chapter – 4

Multidimensional Approaches in the Study of Lichens

D.K. Upreti

1. Introduction

Lichens are the most fascinating and widely distributed organisms on earth. The lichens are unique in having two microorganism in a single plant- a phycobiont (alga) and a mycobiont (fungus) forming a thallus that does not resemble either symbionts in the free living (non- lichenized) state. By virtue of the peculiar structure and physiology, lichens have high tolerance of drought and cold and are able to grow in the diverse geographical regions from icy expanses to tropical and subtropical parts and from drier, hot deserts to moist humid climate. Lichens grow on any substratum that provides a convenient foot hold to them. This may be soil (terricolous), humus (humicolous), stones, rocks, brick (saxicolous), lime plaster (calcicolous), leaves (folicolous), tree trunk (corticolous), decaying wood (lignicolous) and other man made substratum like iron pipes, asbestos sheet and glass panes. Sometimes lichens also grow on some insects and animals. Lichens which are bigger in size and shape can be easily recognized as leaf like (foliose) and thread like (fruticose) (commonly called macrolichens), while those which forming a crust over the substratum and quite smaller in size are categorized under microlichens.

Apart from altitudinal variation, the vegetation and forest types of higher plants also play an important role in determining the type of lichen flora of the region. Based on the forest types six lichen vegetation zones are known in India such as: Moist tropical evergreen forest, Cold deserts in the Himalayas, South Peninsular region, Mid Eastern Indian & Peninsular Plateau, Dry and arid regions and Indo-gangetic plains of central India, Coastal regions of India and Andaman & Nicobar Islands. The cold deserts in the Himalayas exhibit some unique group of lichens having restricted distribution only in such habitats.

The Himalayan region in India is exhaustively explored for lichen wealth in the past and the lichen flora of different Himalayan states is well worked out. Since most of the substrate exhibit dense growth of different species of lichens growing in close association, forming mixed patches, they are sometimes over looked by the collectors during collection.

Approximately 20,000 species of lichens are known from the world and India harbours more than 10% of the species. It is estimated that at present the Indian lichen flora comprises about 2319 species under 305 genera and 74 families widely distributed in tropical, subtropical, temperate and alpine regions of India (Singh and Sinha 2010). There are about 350 and 150 species of medicinal lichens in the world and India, respecitvely (Table 1). The lichen family Parmeliaceae is the largest family in India comprised of 345 species followed by Graphidiaceae, Thelotremataceae, Pyrenulaceae, Caliciaceae and Lecanoraceae represented with 279, 131, 123, 103 and 99 species, respectively. The largest lichen genus in India is *Graphis* which contains 111 species followed by genera like *Pyrenula*, *Lecanora*, *Caloplaca*, represented by 90, 83 and 65 species, respectively while *Usnea* and *Porina* are represented by 60 species each (Singh and Sinha 2010).

Table 1. Status of Medicinally Important Lichens in India vis-a-vis World

Lichen species known from the world	20,000
Lichen species in India	2,319
Medicinally Important Lichens Worldwide	350
Medicinally Important Lichens in India	150

In India, the corticolous (growing on bark) exhibit their dominance followed by terricolous (soil inhabiting) and saxicolous (rock inhabiting) lichens. Khare *et al.* (2009) recorded the occurrence of 65 lichen genera on soil from India under 22 terricolous families. The Cladoniaceae, Collemataceae, Peltigeraceae, Parmeliaceae, Sterocaulaceae, Physciaceae and Lobariaceae are the dominant families of soil lichens. The genera like *Cladonia*, *Collema*, *Peltigera*, *Leptogium*, *Lobaria* and *Stereocaulon* are the dominant soil lichens in India. Soil lichen communities best development in temperate, higher temperate and alpine regions of India where tree vegetation is lacking. The moist evergreen forests have luxuriant growth of soil lichens than the dry deciduous forests.

In temperate regions of India the corticolous lichen dominates over saxicolous and terricolous lichens. The ground flora under coniferous forest at lower temperate areas remains mostly dry and support scanty to poor growth of soil lichens. The evergreen temperate forest and coniferous forest of upper temperate regions provide a moist shady environment suitable for growth of *Lobaria*, *Peltigera*, *Stereocaulon* and *Cladonia* species to colonize on soil among mosses. The common crustose soil lichen genera of the region are *Caloplaca*, *Diploschistes*, *Diplotomma* and *Pertusaria*.

Most of the alpine region in the Himalayas exhibit dominant growth of the terricolous communities of lichens. Fruticose species of lichen genera *Cladonia* such as *Cladonia rangiferina* and *Cladonia aggregata* grow luxuriantly in moist slope in alpine regions. The cold desert in the Himalayas also exhibit good growth of terricolous lichens. Out of 81 species of lichens recorded from the cold desert of Lahul and Spiti area of Himachal Pradesh, 18 were soil inhabiting (Srivastava *et al.* 2004). *Cetraria sp.*, *Bryoria himalayana*, *Hypogymnia hypotrypa*, *Lethariella cladonioides* and *Thamnolia vermicularia* are the most common terricolous lichen species in eastern Himalayan region of India.

Despite intense effort in exploration and survey during the last five decades, our knowledge about lichens from different floristic regions of India is poor as many areas are still unexplored for their lichen wealth. The lichens are most valuable organism to biomonitors atmospheric pollution as they can be used as sensitive indicators to estimate the biological effects of pollutants by measuring changes at community or population level of an area. Lichen monitoring can be effective as an early warning system to detect environmental changes. The periodical monitoring and documentation of floristic data is necessary and useful for future study (Garty 2001).

2. Distribution of Indian Lichen Flora

Based on the altitude and prevalent climatic condition, the lichen flora of the Indian subcontinent can be grouped into four distinct vegetation zones. According to Awasthi (2000) the whole area of the subcontinent is recognized to possess (a) the tropical (upto an altitude of about 800 m), (b) the subtropical (800-1500 m), (c) the temperate (1500-3600 m) and (d) the alpine (above the altitude of 3600 m) zones. The temperate region can be further subdivided into lower temperate (1500-2500 m) and upper temperate (2500-3600 m). The region above 5000 m in the Himalayas is usually snow bound throughout the year and is thus somewhat comparable to the frigid arctic zones. Based on the altitudinal variations lichen flora of India can be summarized as below:

2.1. Tropical lichen vegetation

This region includes almost whole of the South India (leaving the crests of Nilgiri and Palni Hills) West Bengal, Andamans and part of Assam, Manipur, Arunanchal Pradesh and foot hills of the Himalayas. Evergreen moist forests have more luxuriant growth of lichens as compared to dry deciduous forests. The foliicolous lichens like *Strigula elegans*, *S. subelegans*, *Mazosia melanophthalma*, *Opergrapha* sp., *Fellhanera semicarpipholia*, *Byssoloma* sp. and *Phyllobathelium indicum* are more commonly growing lichen species in moist evergreen forests.

In the tropical region, the commonly occurring trees species of *Dipterocarpus*, *Elaeocarpus*, *Euyra*, *Mallotus*, *Artocarpus*, *Shorea*, *Terminalia* and *Schima* provide an excellent substrate for growth of many crustose lichen genera belonging to the lichen families Pyreconarpaceae, Caliciaceae, Graphidaceae, Lecideaceae, Lecanoraceae, Pertusariaceae and Arthoniaceae.

Few foliose species such as *Dirinaria applanata*, *D. consimilis*, *Bulbothrix isidiza*, *Parmotrema cristiferum*, *P. tinctorum*, *Physcia* sp., *Pyxine* sp. and *Heterodermia* sp., grow in moist open places in association with crustose species. *Diploschistes scruposus*, *Glyphis duriuscula*, *Lecanora allophana*, *Porina interstes*, *Letrouitia domingesis*, *Tylophoron moderatum* and Graphidaceous lichen are the other crustose taxa of the tropical region. *Endocarpon pusillum*, *Staurothele sp.* and *Verrucaria aethiobola* grow exclusively on stones in shady moist places together with the foliose forms species of *Leptogium cyanescens*, *Coccocarpia palmicola* and *Collema* sp., which can grow both on the ground stones and trunks of the trees. In dense close canopy forests, lichens are confined to the fringes of forests or on upper portion of the trees where enough light and wind currents are available.

Fig. 1. Common tropical lichens, (A) *Pyxine cocoes* growing on mango tree trunk, (B) *Lecanora tropica*, (C) *Dirnaria aegialita*, (D) *Lepraria lobificans*, (E) *Parmotrema tinctorum*, (F) *Heterodermia diademata*, (G) *Pyxine cocoes*

Shorea robusta tree in the tropical region is the most suitable tree species for colonization of lichens as more than 60 species are reported to grow on trunk, branches and twigs of the tree (Satya *et al.* 2005).

2.2. Subtropical lichen vegetation

The *Pinus* forest or *Pinus* growing mixed with other trees forms the major vegetation in this zone. The lichens grow on the trunks, moderately to profusely depending on the availability of moisture and sunlight. Amongst crustose genera *Anthracothecium*, *Cryptothecia*, *Pertusaria*, *Porina*, *Haematomma* and Graphidaceous taxa are abundant on the trunks of *Schima*, *Alnus*, *Quercus* and *Prunus* in association with several foliose genera such as *Bulbothrix*, *Everniastrum*, *Heterodermia*, *Hypotrachyna*, *Parmelia*, *Parmotrema* and *Pyxine cocoes*, *P. subcinaria* and *P. berteriana* belonging to Parmeliaceae and Physciaceae family. The cultivated trees of *Pyrus*, *Prunus*, *Delonix* and *Celtis* bear luxuriant growth of most of the members of Parmelioid lichens together with species of *Candelaria*, *Heterodermia*, *Dirinaria*, *Physcia* and *Pyxine*.

2.3. Temperate lichen vegetation

The rich diversity of trees in temperate region of India provides diverse bark to colonize a number of lichen taxa. Thus the lichen flora in temperate regions exhibits the greatest abundance in variety and diversity of growth. The temperate climate offers the optimum conditions for luxuriant growth of foliose and fruticose lichens. The lichens in this region prefer the bark of trees or rocks as their substratum, though they also occur on variety of other substrata. *Aesculus*, *Alnus*, *Quercus* and *Pinus* trees at lower (upto 2000 m) and coniferous trees and *Rhododendron* at higher (upto 3000 m) elevations provide excellent substratum to a large number of both macro and microlichens. Pure vegetation of *Aesculus* and *Quercus* trees near streams or moist areas at higher elevation preferred mostly by the Pyrenocarpous lichens. The trees from base to upper trunk and then twigs almost completely covered with mosses and pyrenocarpous lichens together with species of *Ochrolechia*, *Pertusaria*, *Sticta*, *Leptogium* and Parmelioid lichen genera.

Thinned out or disturbed forests of *Quercus* provide a suitable habitat for growth of light loving lichen species of *Lithothelium*, *Pertusaria* and *Lecanora*. The mixed forest of *Quercus* and *Pinus* at lower elevations are moderately dry and preferred by species of lichen genera *Phaeophyscia*, *Dirinaria* and *Heterodermia* together with Parmelioid genera. At higher elevation the coniferous trees sometime form pure strands. The dry bark is preferred by colonization of only few species of *Parmelia*, *Lecanora*, *Physcia* on trunks and *Nephromopsis*, *Usnea*, *Ramalina Bryoria*, *Sulcaria* and *Evernia* on twigs at high canopy.

Fig. 2. Common temperate lichens, (A) *Physconia detersa*, (B) *Parmotrema tinctorum*, (C) *Heterodermia diademata*, (D) *Xanthoria* elegans, (E) *Rhizoplaca chrysoleuca*, (F) *Usnea longisimma*

Apart from big trees other small trees and shrubs also act as a favourable substratum for a number of lichen species in this area. The *Berberis* and *Craetegous* bushes, small *Rhododendrons* and *Ilex* shrubs harbor luxuriant growth of crustose, foliose and fruticose lichens on their trunk and branches. Inside the dense forest, shade loving terricolous lichens species of *Leptogium*, *Collema*, *Peltigera*, *Sticta* and *Lobaria* grow abundantly on vertical slopes along with mosses. In thinned out disturbed forest, light loving saxicolous or terricolous lichen species of genera *Phaeophyscia*, *Heterodermia* and Parmelioid taxa are common on soil or/and over rock. The dry, exposed rocks are preferred by species of *Acarospora*, *Caloplaca*, *Lecanora*, *Pertusaria*, *Porpidia* and *Rhizocarpon*. The moist boulders along streams and rivers are dominated by species of *Stereocaulon*, *Dermatocarpon* and *Umbilicaria*. In disturbed forest, the under story growth of *Berberis* and *Pyracantha* bushes, also provide an excellent habitat for growth of a number of Parmeliod genera, *Cetraria*, *Ramalina* and *Heterodermia* species. *Quercus semicarpifolia* trees in the Himalayan region exhibit the maximum colonization of more than 90 species of lichens on its trunk, branches and twigs.

2.4. Alpine lichen vegetation

The alpine zone contains mainly rock inhabiting, soil inhabiting and moss inhabiting species of lichens, which form a felt-like growth over the substratum. The species of *Cladonia*, *Stereocaulon*, *Thamnolia*, *Umbilicaria* and *Diploschistes* grow abundantly on rocks or in association with mosses on the boulders and rock. The larger trees are absent in this zone, while a number of commonly growing shrubs of *Rhododendron*, *Cotoneaster*, *Rosa* and *Juniperus* provide excellent substrata for the bark inhabiting lichen species of genera like *Cetraria*, *Heterodermia*, *Parmelia*, *Ramalina*, *Usnea*, *Buellia*, *Lecanora*, *Ochrolechia* and *Pertusaria*.

Fig. 3. Common alpine lichens, (A) *Lobothallina alphoplaca*, (B) *Xanthoria elegans*, (C) *Psora himalayana*, (D) *Rhizoplaca* and *Xanthoria* sp., (E) *Umbilicaria indica*, (F) *Rhizoplaca chrysoleuca*

The boulders and stones in small rivulets coming from the glaciers harbor plenty of crustose and squamulose lichen species of genera *Acrospora*, *Rhizocarpon*, *Rhizoplaca*, *Caloplaca*, *Aspicilia*, *Lecidea* and *Rinodina*. The alpine meadows (buggyals) also exhibit occurrence of some peculiar soil inhabiting lichens. The moist habitats in the meadow exhibit growth of some members of Parmelioid lichens together with species of *Cladonia* on soil between grasses.

Apart from altitudinal variations the vegetation and forest types of higher plants also play an important role in determining the type of lichen flora of a region. Based on the forest type, Upreti (1998) has described the following six different lichen vegetation zones in India:

i. *Moist tropical evergreen forest:* This region includes north-eastern India, parts of Assam and West Bengal. The species of *Leptogium, Collema* (Cyanolichens) together with member of lichen family Physciaceae (*Heterodermia, Dirinaria* and *Phaeophyscia*) and Parmeliaceae prefer such habitats.

ii. *South Peninsular region:* The 'Shola' forest growing in patches in the south peninsular region provide a favourable habitat for good growth of both micro and macrolichen genera as all the trees found in Shola forest are equally and sufficiently exposed to rain, sunlight and wind currents.

iii. *Mid Eastern Indian and Peninsular Plateau:* The broad leaf deciduous forest trees in the mid eastern Indian and Peninsular Plateau exhibit moderate number of crustose and squamulose lichens together with few foliose forms. Hard, dry bark and bark pealing out nature of trees are common inhibitory factors responsible for poor diversity of lichens in these areas. The *Shorea robusta* trees commonly found in the region provide suitable habitat for many lichens to colonize on their base of trunk, main trunk, branches and twigs. It is a common feature in *Shorea robusta* trees that the ant and termites cover the whole tree trunk and branches with soil, which ultimately inhibit the fresh colonization of lichens and destroy the existing taxa.

iv. *Dry and arid regions and Indo-gangetic plain of central Indian regions: Mangifera indica, Anacardium, Citrus, Artocarpus, Syzygium cuminii, Murraya koenigii, Mallotus philippensis* trees are the common trees preferred by lichens. The *Mangifera indica* in mango orchard and as avenue trees provide excellent substrate for many species of lichens to colonize. The members of the family Physciaceae, which are well known for their pollution tolerant characteristics, grow on the trees. The species of foliose lichen genus *Dirinaria, Physcia, Phaeophyscia* and *Pyxine* together with crustose lichen genus like *Rinodina* exhibit their luxuriant growth on different trees cultivated or growing wild in the forest areas.

   The industrial areas and city centre are poor in lichen growth as only pollution tolerant and lime/cement loving lichens grows abundantly on walls and cement plaster of buildings and monuments. Species of lichen genera like *Endocarpon, Phylliscum, Peltula* and member of the family Lichinaceae mostly having cyanobacterium as their photobiont and having inbuilt tolerance against atmospheric pollution, which can withstand long dry periods, exhibit their rich diversity in the area.

v. The *Cocos nucifera, Areca catechu* and mangroove trees in the coastal regions of India provide suitable substrate for growth of some exclusive lichen genera such as *Roccella* and members of the family Roccellaceae. The Palm trees also exhibit luxuriant growth of Graphidaceous, Pyrenocarpous lichens. Together with species of foliose lichen genera like *Physcia, Dirinaria, Pyxine* and crust forming *Cryptothecia* and *Lepraria*.

vi. The Andaman and Nicobar Island together with eastern and western coastal region of India exhibit luxuriant growth of *Cocos nucifera, Areca catechu* and Mangroove trees. In Andaman Islands, mostly the original forest cover has been thinned out or completely removed for human habitation or for various forestry operations. Many cultivated tree species such as *Anacardium, Artocarpous, Citrus, Delonix* and *Ficus* have been introduced in many of the Islands. It is interesting to note that these tree species provide favorable habitat for different species of lichens found growing both on trunk and on leaves.

Populus another cultivated tree species introduced in the higher western Himalayan region particularly in Jammu and Kashmir state provide an excellent habitat for the members of the family Physciaceae (*Phaeophyscia* and *Heterodermia*) together with a yellow foliose lichen *Xanthoria parietina.*

3. Affinities and Endemism in Indian Lichen Flora

Indian lichen flora has about 518 endemic species (25.26%), and the crustose form exhibits high degree of endemism as compared to foliose and fruticose forms. Out of the eight lichenogeographical regions of India, the Western Ghats with 219 species exhibit the maximum endemism in lichens (Singh *et al.* 2004). The Indian lichen flora exhibits its affinities with the lichens of the following areas of the world.

i. The lichens of north western Himalayan region including Jammu and Kaskmir show affinity with lichens of Europe, Arid region of Central and Western Asia, Arid region of rock mountain in north America and Peru in South America. A large number of species of the Himalayan region are common to that of Arctic, Antarctic and Alpine region of world.

ii. Eastern Himalayan lichens show close affinities with Sino-Japanese and south-east Asian element.

iii. The cosmopolitan element of tropical India show affinity with lichens of tropical America, Africa and South East Asia.

iv. Andaman lichen flora has the predominance of endemic species, those which are not endemic show affinity with lichens of tropical insular areas of world preferably south East Asia.

Lichens exhibit their luxuriant growth on boulders and rock out crops both in exposed dry and moist areas near streams. The lichen species grow luxuriantly on exposed rocks and soil accumulated on crevices of the rocks. Most of the trees like *Eucalyptus, Salix, Myricaria* and Teak are devoid of lichens on them. Extreme cold, low precipitation and soil deficient in phosphorous and nitrogen are the conditions that impose severe limitations on the growth of many vascular plants, but mosses and lichens grow abundantly in such habitats.

The lichen species occurring in extreme environment exhibit peculiar adaptation features. Species such as *Acarospora*, *Xanthoria* and *Caloplaca* are closely appraised to grow on within the substrates. The foliose and fruticose taxa (*Umbilicaria*, *Xanrhoparmelia* and *Melanelia*) appear compact and tiny in comparison to their normal shape and size. Most of the lichens in cold deserts have strongly coloured yellow, red, brown or black pigments in their upper cortex which protect the alga present below the upper cortex from the high intensity of sun rays. Pigmentation also acts as protective layer against the irradiation and UV radiation. The dark pigments not only protects the lichens from high UV it also helps the plant to absorb heat readily. The lichens in cold desert develop a stable and elastic thalli that enable the plants to survive well on the exposed rocks and helps in anchoring against action of winds and gales. An amorphous layer of dead cells develop above the upper cortex and help in desiccation.

4. Economic Importance of Lichens

Apart from diversity, the economic use of lichens is well known in the world as well as in India from several decades. The lichens also have been well known as valuable resources and are still used as medicine, food, fodder, perfume, spices and dyes (Fig. 4). The economic importance of lichens is discussed as follows:

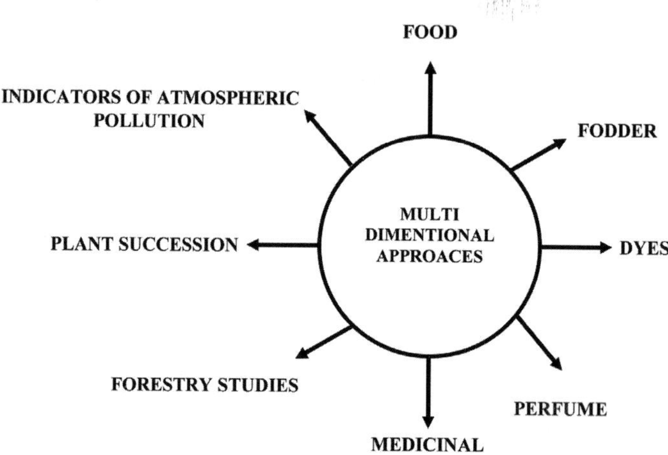

Fig. 4. Utilitarian prospects of Lichens

4.1. Lichens as food

The species of lichen genus *Umbilicaria* in Japan are eaten as salad called "Iwatake". They are rich in carbohydrates and fats. Species of *Cladonia*, *Stereocaulon, Usnea* and *Ramalina* are mixed with flour and eaten, as they are considered as good source of carbohydrates. *Lecanora esculerata* found in various parts of the world on soil, are gathered, powdered and flour is used to prepare earth-bread from it. *Cetraria islandica*, commonly known as "Iceland Moss" is used as human food. After collection and removal of certain bitter principles by allowing them to diffuse into cold water, the thallus is dried and the decoction of this dried thallus which forms a demulcent drink with milk is believed to be highly nutritious.

Parmelioid lichens available in large quantities in the market are used as food material and as condiment. In Sikkim, *Everniastrum cirrhata,* a commonly growing lichen of that area, is eaten as a vegetable after boiling and frying it in fat. *Leptogium denticulatum,* a common foliose lichen is used by 'Adi' tribe of Arunachal Pradesh as food. The local 'Adi' people collect the lichen from soil, rock and tree trunk, wash it properly and boil with water. The thallus are boiled, which becomes jelly like after boiling is used as a vegetable. In view of the high protein content and the interesting amino acids composition together with ergosterol and inorganic constituents of iron and calcium, *Dermatocarpon moulinsii* (20% crude protein), *Lobaria isidiosa* (20% crude protein), *Roccella montagneii* (14% crude protein) and *Parmotrema tinctorum* (14% crude protein) appear to have good food value (Fig. 5).

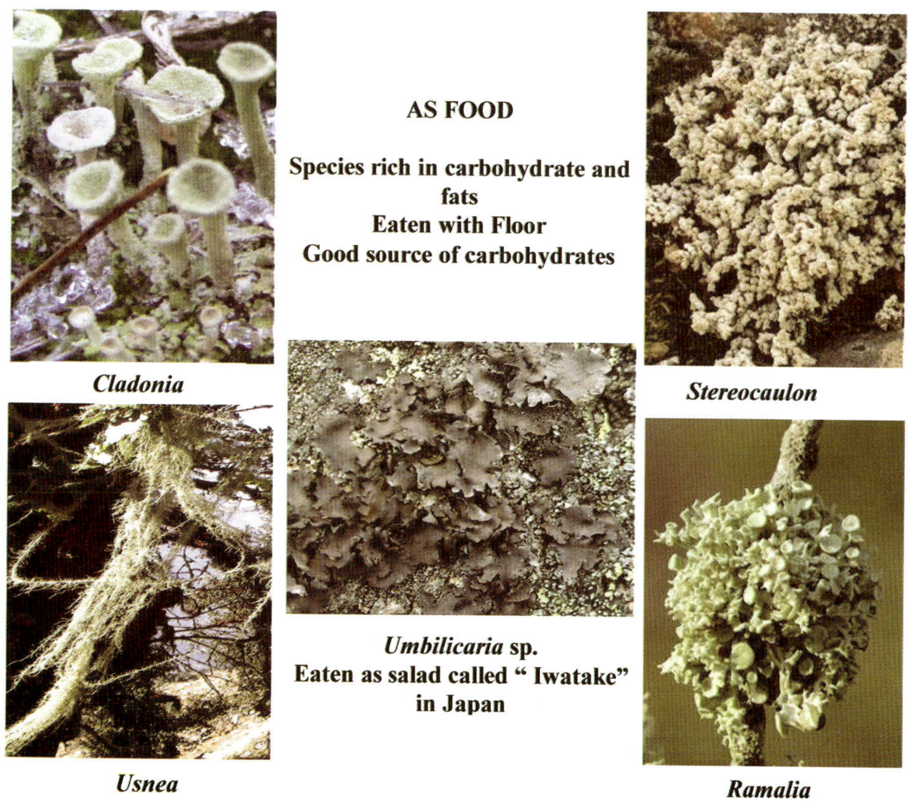

Fig. 5. Common lichens widely used as food by some ethnic groups

4.2. Lichens as spices

In Indian markets, lichens are sold by the name of 'Chharilia' are used as spices. Chharilia is a mixture of two or three species of *Evermiastrum* with other foliose Parmelioid lichen species. These lichens provide a special fragrance to meat, pulse and other important vegetables (Fig. 6).

Fig. 6. Grading of lichens for trading in various Indian markets

4.3. Lichens as fodder

Lichens are important food for animals in the arctic regions. During winter the reindeer and caribou supplement their normal diet of sedges and willow twigs with most common species of *Usnea* and *Cetraria*. Sheep in the Libyan deserts are reported to graze on the subfoliose lichen *Lecanora esculenta,* which forms a thick loose crust on soil and rocks and is easily eaten by the sheep.

In alpine meadows, the commonly growing species of lichens are common source of food for the land snails and termites. Lichens provide a protective environment for a number of invertebrates. Lichenophagous insects, such as bark lice, springtails and moth caterpillars, possess mandibulate mouth parts, with which they bite off the lichen and chew it. Species of lichen genera *Ramalina*, *Parmelia* and *Usnea* on twigs of bushes are favored by the musk deer during scarcity.

In south India, *Roccella montagnei*, which grows luxuriantly, is used as a common fodder for animals. Several new lichens species and varieties especially of *Rhizocarpon*, have been described, which are actually no more than well known species damaged by snails and mites.

4.4. Lichens used in medicine

Lichens were given importance by medicinal practitioners in medieval times, and their use has persisted to present times. In various pharmacopoeias, lichens are listed purely on the basis of their folklore medicinal use. Several species of lichens enjoyed a good position in ancient and traditional systems of Indian medicine like Ayurveda and Unani.

Table 2. Common species of lichens used in medicine

• *Cetraria* sp.	• *Peltigera polydactyla*
• *Cladonia crispate*	• *Stereocaulon himalayense*
• *Heterodermia diademata*	• *Sticta gracilis*
• *Lasallia pensylvanica*	• *Umbalicaria esculenta*
• *Lobaria orientalis*	• *Usnea longissima*
• *Parmelia sanctiangeli*	• *Usnea rubescens*

4.5. Lichens used in perfume and dyestuff

Some of the species of lichens are aromatic, of which *Evernia prunastri* (oak-moss), and *E. furfuracea* (tree-moss) are used commercially for production of aromatic resinoids. These resinoids are used extensively in perfumes, flavors and cosmetics as they have excellent odour fixatives and are universally employed in the blending of perfumes. The lichens yielding oak-moss and tree-moss resinoids occur only in Central and Southern Europe and some parts of north Africa. The main producers of oak-moss are France, Czechoslovakia, Yugoslavia and Morocco.

In India, more than thirty five species, mostly Parmelioid lichens are used for preparation of perfumes called *"Hina attar"* in Kannuj district of Uttar Pradesh.

Before the discovery of coaltar dyes, lichens had considerable economic importance as dyestuffs. In Scotland and Scandinavia woolens and tweed are still dyed with native lichen dyestuffs. The dyeing substance of lichens is an orchil substance extracted from *Ochrolechia* and *Evernia* that dyes wool and tweeds in shades of purple, red or brown for many decades. A familiar acid-base indicator (Litmus dye) in chemistry laboratories is derived from depside-containing lichen *Roccella montaignei*. It is related to orchil but represents a more complex of polymeric compounds with the 7-Oxyphenoxazon chromophore and its anion. *Buellia subsoriroides*, a crustose lichen growing commonly on rocks in Garhwal Himalayas, yield an orange dye, locally called "Maidi". The herdsmen use it for colouring their finger tips and palms like *"Henna"*- a well known red-orange dye obtained from the leaves of *Lawsonia inermis*. The herdsmen when visit the high temperate region spit saliva on the rock over lichen and start rubbing it with the help of small pieces of rough stone until a small amount of paste accumulates. The paste thus obtained is applied in the form of drops on the finger tips and palms to make designs. More than sixty species of macro lichens are proposed to have dyeing properties for fibers particularly silk (Fig. 7).

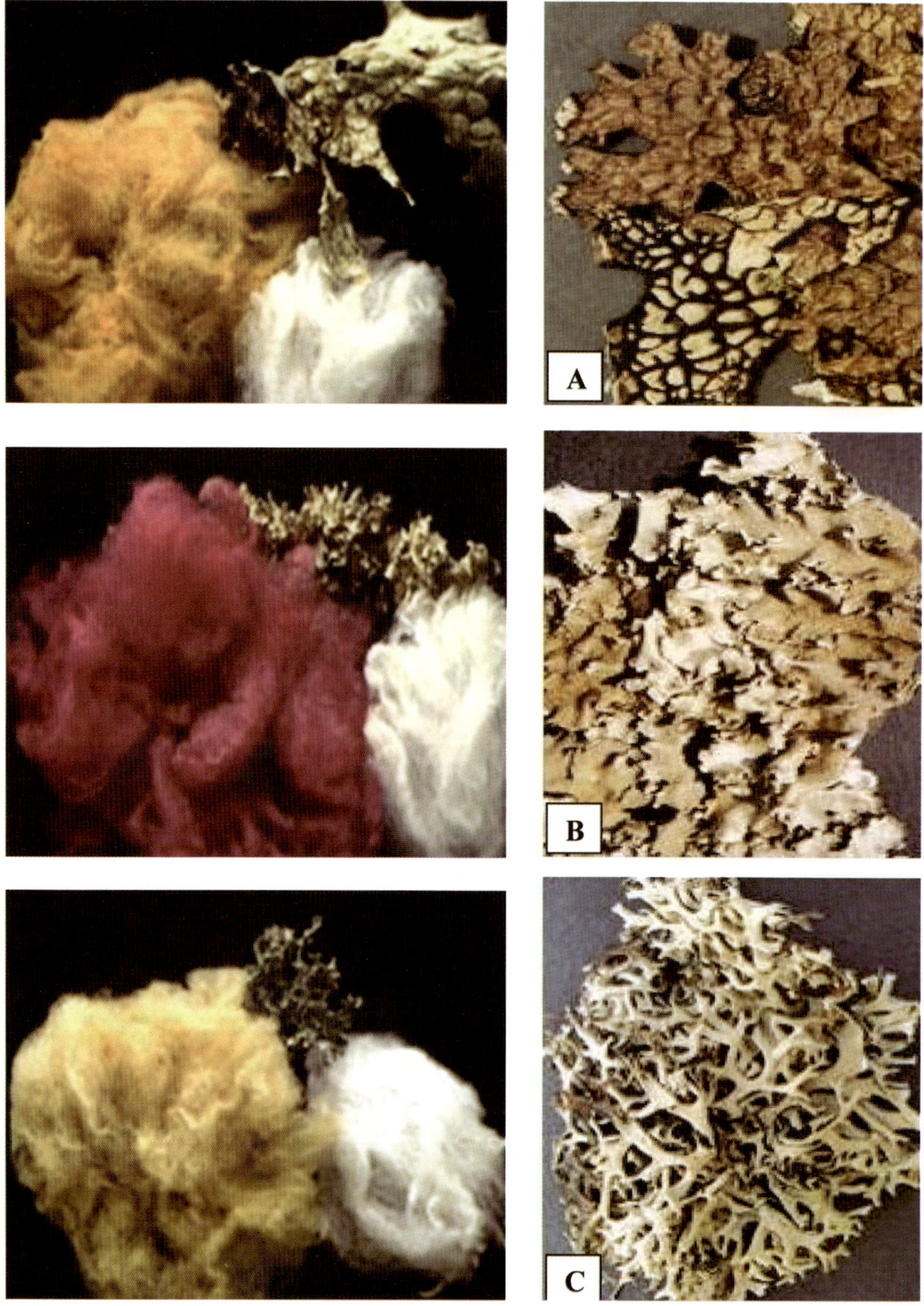

Fig. 7. Dye yielding lichens, (A) Golden yellow dye form *Lobaria pulmonaria*, (B) Purple dye from *Parmotrema tinctorum*, (C) Brown or light yellow dye from *Everniastrum cirrhatum*.

4.6. Lichens in geochemical studies

Lichenometric technique is useful in dating moraine ridges on recent glacier forelands in alpine regions. The method was originally developed and used by Beschel (1961). Glaciers and their retreat are recognized as being among the most sensitive indicators of climate change, advancing substantially during climate cooling and retreating during climate warming. Lichens due to their slow growth rate and uniform growth size, help in dating the exposure time of the sequences of the rock forming glacier moraines due to retreat of the glacier thus providing the approximate time of glacier retreat. The study is based on lichen size/age correlation and population distribution that involves the measurement of large specimens growing on large boulders that are supposed to be unaffected by the prevailing climatic conditions as well as human and animal interferences. Field studies of climate change impact in India can be conducted by initiating lichenometric studies in relation to climate change.

4.7. Pedogenic significance

Lichens are important agents in the biogeochemical and biochemical weathering of minerals and rocks and in certain situation play an important role in plant succession. The substances execrated by the thalli of lichens are obviously too weak to alter the rocks by hydrogen ion exchange, but chelation could be an important mechanism in mineral breakdown. The lichen acids bind metal atoms of the substrates between their own molecules and form a metal complex. The metal complex is an unstable attachment and the metal is easily released in free form and the complex relents to its original form.

4.8. Plant succession, soil development

The crustose forms of lichens are the world's greatest pioneers. No organism other than a crustose lichen can maintain itself on a perfectly plane, clean rock surface. Thus, after colonizing on a substratum, lichens accumulate several elements, frequently in large amounts. The accumulated N, P and sulphur elements can be used by mosses and higher plants which may replace lichens during soil development. The mixing of organic matter from the decay of the thallus mineral particles detached from the substratum and atmospheric dust trapped by the lichens thallus (due to their spongy nature) may produce a primitive soil.

4.9. Lichens in pollution monitoring (Biomonitoring)

Lichens are one of the most valuable biomonitors of atmospheric pollution and can be used as sensitive indicators to estimate the biological effects of pollutants by measuring changes at community or population level and as accumulative monitors of persistent pollutants. The high capability of lichens to accumulate air pollutants, resistance to environmental stress and longevity are the major features that make them most suitable organisms for biomonitoring studies. Free diffusibility of lichen thalli due to lack of cuticle enables quick penetration of toxic compounds from the atmosphere to the photobiont layer. Thus, the response of lichens to the environmental pollution is more sensitive than the higher plants. Owing to their dependence on atmosphere for nutrient supply and capacity to biomagnify accumulated environmental contaminants, lichens can provide details on the presence of persistent pollutants in the atmosphere and their biological effects (Garty 2001).

Table 3. Methods for pollution monitoring utilizing lichens

Transplant Technique	Transplanting healthy lichens into a polluted area and measuring thallus deterioration
Lichen zone mapping	To indicate the diversity of pollution with reference to distance from source, as reflected by the number of species absent
Sampling of individual lichen species	Measuring contaminants accumulated within the thallus

Lichens are excellent bioindiactors of air pollution due to their sensitivity to acidic gases, exhibit distinctly the incited damage in relation to morphological and/or physiological symptoms and are also excellent accumulators of pollutants. Broad geographical distribution which allows documentation of the wide spread pattern, perennial, slow growth rate uniform morphology and not shedding parts as in higher plants provide ability to cumulatively accumulate pollutants are the feature which make lichens more suitable organism for pollution monitoring.

All lichens are not equally sensitive to air pollutants. Rather, different lichen species exhibit differential sensitivity to specific air pollutants. As a consequence, lichens are well suited as biological indicators for monitoring environmental quality. During the last three decades a number of studies devoted to assessing the effect of air pollution on lichens were carried out throughout the world (Garty *et al.* 2002). The mapping of lichen communities is one of the major areas of research to study the variation in lichen communities. The frequency of occurrence of certain species is related to specific air pollutants and in some cases to their concentrations. Apart from distribution map the morphological and anatomical changes in responses to air pollutants further provide an assessment of the effect of environmental pollutants on living organism. The physiological reactions and changes in the lichen thallus due to air pollutants can also be measured and predict the environmental conditions of particular area (Table 4).

Table 4. Common bio-indicator species of lichens

Phytogeographical (Altitudinal) zones	Bioindicator species
Tropical areas	*Dirinaria consimilis, Rinodina sophodes, Pyxine cocoes, Lepraria lobificans, Cryptothecia punctulata*
Subtropical areas	*Phaeophyscia hispidula, Pyxine subcinerea, Parmotrema praesorediosum, Parmelinella wallichiana*
Temperate areas	*Cladonia praetermissa, Heterodermia diademata, Candelaria concolor, Dermatocarpon vellereum, Usnea longissema*
Alpine areas	*Rhizocarpon geographicum, Aspicilia* sp., *Xanthoria elegans, X. fallax*

The level of airborne pollutants arising from anthropogenic (point and line) sources such as power plants, smelters, automobiles, industry and agriculture can be easily monitored through lichens. The degradation of chlorophyll in the symbiotic photobiont is one of the most obvious signs of the damage that occurs in sensitive lichens. Heavy metals are known to interfere with chlorophyll synthesis either through the direct inhibition of an enzymatic step or through the induced deficiency of an essential nutrient. The species accumulate relatively high amounts of heavy metals and contain less chlorophyll, which clearly indicates that lichen chlorophyll contents interfere with the thallus metal contents (Loppi *et al.* 2000).

The lichen species express particular symptoms or response to indicate the changing environment (bioindicator); distribution or population which is studied over time and compared to some standard or baseline survey (biomonitor); species accumulating particular environmental substance within their fruiting bodies, thallus and rhizine (bioaccumulator); physiological and biochemical changes in sensitive species caused by environmental pollutant (biomarkers).

The biomonitoring with lichens compared to instrumental methods offers other advantages such as low cost, independence of power supply, easier sample handling and trace elements determination methods. The perennial nature of lichens, absence of root or other special organs for uptake of nutrients and lack of cuticle enable them to absorb metals directly from the atmosphere and these characteristics make them ideal bio-monitoring organisms (Seaward and Richardson 1989). Lichens are used as bio-indicators and /or bio-monitors in two ways: *in situ* i.e. passive monitoring, and the active monitoring i.e. transplant of lichens from one place to other.

4.9.1. Lichens as passive biomonitors

Passive biomonitoring is the use of organisms, organism associations, and parts of organisms which are a natural component of the ecosystem and appear there spontaneously. Lichen communities are currently used as indicator of forest ecosystem function in several context. Studies of lichens of particular forest type often have goals of monitoring effects of forest management practices and landscape context, including a variety of indirect human impacts on forest environment (Dettki and Essen 1998). Lichens have also been identified to indicate forest stand temporal continuity and condition and age structure of forest. The main groups of lichen bioindicator species known from India are described below:

i. **Calcioid group:** The calcioid group or the "pin-head" lichens are indicators of old growth forests. Many of these species are dependent on snags and old trees with stable rough bark (Fig. 8A).

ii. **Alectoroid and Usnioid group:** The tufted and pendulous fruticose lichens including genera such as *Sulcaria*, *Bryoria*, *Ramalina* and *Usnea* have been found to be useful as indicators of old forest with better air quality. The *Usnea* species indicate older growth forest and vanish from the forest onset of habitat destruction, (Fig. 8B).

iii. **Cyanophycean group:** The variation in diversity and abundance of epiphytic cyanolichens appears useful as an indicator of forest ecosystem function. Cyanophycean lichens play an important role in forest nutrient cycle and indicate forest age and continuity (Mc Cune 1993), (Fig. 8C).

iv. **Lobarian group:** The Lobarian group comprises of *Lobaria*, *Pseudocephallaria*, *Peltigera* and *Sticta* are sensitive to air quality and are reliable indicators of species rich old forest with long forest continuity (Fig. 8D).

v. **Xanthoparmelioid group:** According to Eldridge and Koen (1998), the yellow foliose morphological group comprising of foliose lichen species of *Xanthoparmelia* is consistently correlated with stable productive landscape i.e. landscapes with no accelerated erosion (Fig. 8E).

vi. **Graphidioid and Pyrenuloid group:** The growth of graphidaceous (*Graphis*, *Opergrapha*, *Scareographya*, *Phaeographis*) and pyrenocarpous (*Anthracothecium*, *Pyrenula*, *Lithothelium*, *Porina*) lichens is influenced by the nature of bark. Both groups mostly prefer to grow on a smooth bark tree in evergreen forests (Fig. 8F).

vii. **Lecanorioid group:** The group comprises *Lecanora*, *Lecidella* and *Biatora*. The group indicates well illuminated environmental condition of the forest with considerable exposure of light and wind. They prefer to grow on trees in thinned out, regenerated or disturbed forest with more open area to receive more light and wind (Fig. 8G).

viii. **Parmelioid group:** The group comprises mostly the species of lichen genera such as *Bulbothrix*, *Flavoparmelia*, *Parmotrema*, *Parmelia*, *Punctelia* and other genera of Parmeliaceae. The forest with closed canopy and less sunlight support few species of Parmelioid genera to grow while the open thinned out forest with more sunlight exhibit dominance of Parmelioid lichens (Fig. 8H, I).

ix. **Pertusorioid group:** The group includes species of lichen genus *Pertusaria* and indicates old tree forest with rough-barked trees. The *Shorea robusta* tree in the dry deciduous forest appears to be an excellent host for this group of lichens to colonize (Fig. 8J).

x. **Lecideoid group:** The member of the group such as *Lecidea*, *Protoblastedia*, *Haematomma, Bacidia*, *Buellia* and *Schadonia* colonized mostly on bark of deciduous trees in sheltered and well lit exposed sides (Fig. 8.K).

xi. **Leprarioid group:** The species of *Chrysothrix*, *Cryptothecia* and *Lepararia* are the common lichens of the Leprarioid group, which forms powdery thallus on the substrates indicates moist and dry vertical slopes, rough-barked trees of moist and dry habitats. The species of *Chrysothrix* appears first after forest fire (Fig. 8L).

xii. **Physcioid group:** The species of *Physcia*, *Pyxine*, *Dirinaria*, *Heterodermia*, *Phaeophyscia* and *Rinodina* belong to this group. The Physcioid lichens are considered pollution tolerant and have ability to grow on varied substrates in both moist and dry habitats (Fig. 8M).

xiii. **Teloschistacean group:** The species of *Caloplaca*, *Letroutia*, *Brigantiaea* and *Xanthoria* having yellow thallus and apothecia belongs to this group. The members of this group have an ability to grow both on exposed and sheltered rocks. The dark orange pigment present on the upper cortex of the thallus act as a filter and to protects the lichens from high UV radiation (Fig. 8N).

xiv. **Lichinioid group:** The genera of the lichen family Lichninaceae mostly having cyanobacteria as their photobiont belong to this group. The members of the group prefer dry rocks and barks having higher concentration of calcium and indicate presence of calcareous substrates in the habitats (Fig. 8O).

xv. **Peltuloid group:** The species of lichen genus *Peltula* belong to this group of lichens. The presence of the species of this group indicates a stable rock substratum (Fig. 8P).

Fig. 8. Different Indian lichen bioindicator communities, (A) Calcioid, (B) Usnioid, (C) Cyanophycean, (D) Lobarian, (E) Xanthoparmelioid, (F) Graphidioid, (G) Lecanorioid, (H-I) Parmelioid, (J) Pertusorioid, (K) Lecideoid, (L) Leprarioid, (M) Physcioid, (N) Teloschistacean, (O) Lichinioid, (P) Peltuloid

Lichen zone mapping

Sernander (1926) recognized the disappearance of lichens from cities and conducted the systematic mapping and proposed three distinct zonations. The first zone is the 'Lichen desert', which is found in the city centre, where the tree trunks were bare of lichens. The second zone is called 'Struggle zone', which comprised of areas outside the city centre with tree trunks poorly colonized with lichens, followed by the 'normal zone' where lichen communities on the tree trunks were well established. Subsequently, the large number of similar city maps showed that these zonations were well correlated with the degree of pollution, the size of the urbanization area and the prevailing winds.

A case study on lichen zone mapping

The lichen zone mapping in India with reference to Lucknow city, Capital of Uttar Pradesh was initiated in the year 2004 after dividing the city into four major areas i.e. North, East, South and West in a 1 x 1 sq Km grid.

The distribution data of lichens collected from all the four area of the district provide an idea about the overall picture of the lichen distribution in the district and the detailed distribution to segregate the area into four different zones. Zone A - no lichen growth, an area in the centre of the city up to 5 Km all around; Zone B- presence of some calcicolous lichens, mostly the areas with old historical buildings; Zone C- scarce growth of few crustose and foliose lichens, areas with scattered mango trees; Zone D- normal growth of different epiphytic lichen taxa together with some follicolous lichens. Among the four zones of Lucknow district, north-east zone has the highest concentration of the metals such as Fe, Ni, Zn and Hg, while south-west zone of the district exhibits higher accumulation of Pb. The key sources for the metal accumulation in lichens are heavy motor traffic, frequent use of generator sets for electricity, burning of fuel wood and use of pesticides in the Mango orchards (Saxena *et al.* 2007).

The comparison of the lichen diversity with an earlier study carried out during 1960-80s showed a distinct change in the Lucknow city. In and around the city of Lucknow, out of the 18 species recorded in the past, 14 species are common to the present study. It seems that the remaining 4 species (*Julella* sp., *Opergrapha herpetica*, *Peltula euploca* and *Phylliscum macrosporum*) of the former study might have become totally extinct from the area. The change in lichen communities in the district is mainly due to change in the environmental condition during the last 25 years. This indicates the replacement of the sensitive species of lichens with tolerant ones in the district.

Similarly in Indian Botanical Garden (IBG) Howrah and Kolkata out of the 53 species earlier reported in 1865 (140 year before) only 5 species are common in the past and present study. The IBG exhibited dominance of crustose lichens (21 out of 25 species), which are more tolerant to air pollution. The pollution tolerant crustose lichen act as pioneer colonizer in a new environmental and replace sensitive species (Upreti *et al.* 2005).

The lichen flora of Lalbagh Garden, Bangalore was also compared with an earlier enumeration and it was interesting to note that in the last 18 years, the lichen flora of the area has been changed significantly as only four species were common between the two studies. The fast pace of urbanization together with air pollution may probably be the reason for the change in the lichen flora of the different Indian cities. The Pune city also exhibit poor growth of lichens within the city centre however the lichens are growing luxuriantly in area having dense tree canopy in the out skirts of the city.

4.9.2. Lichens as active biomonitors

Active biomonitoring includes all methods which include organisms under controlled conditions into the site to be monitored. Both active and passive bio-monitoring are able to find out the indicator species of area with respect to pollutants. The bioindicators are commonly grouped into accumulation indicators and response indicators. Accumulation indicators store pollutants without any evident changes in their metabolisms. Response indicators react with cell changes or visible symptoms of damage when taking up even small amounts of harmful substances.

In the active monitoring, such areas where lichens are in scarcity, the transplant technique (active) were employed for determining the levels pollutants in the area. The same size of thallus along with substratum glued on cardboard of 20 x 20 cm and fixed vertically on exposed pole boundary wall of same height at the selected (absence of lichen) sites. After that samples taking from the transplant site to the laboratory for the further analysis.

4.9.2.1. Inorganic pollutants accumulation in lichens

The accumulation of metal (Al, Cd, Cu, Cr, Fe, Pb, Ni, Zn) pollutants in lichen thallus by passive as well as active principals are well known from different cities of the country such as Uttar Pradesh (Faizabad, Lucknow, Kanpur and Raebareli districts); Madhya Pradesh (Dhar, Katni and Rewa districts) and West Bengal (Hooghly and Nadia district); Maharashtra (Pune and Satara district); Uttrakhand (Dehradun, Pauri districts). The accumulation level of different metals decrease with increasing distance from the city centre. The metals Cr, Cu and Pb were more at the higher vertical position (20-25 feet) where as other metals (Zn, Fe) accumulated maximum at lower vertical position (4-5 feet) (Bajpai *et al.* 2004).

The damage caused by the metallic pollutants in the lichen *Pyxine subcinerea* Stirton, by measurements of Chl a, Chl b, total Chl, carotenoid and protein and OD 435/415 ratio significantly exhibits the changes in physiology. It was observed that the Cu, Pb and Zn significantly affect the physiology of the lichen. Multiple correlation analysis revealed significant correlation (<0.001) among the Fe, Ni, Cu, Zn and Pb metals analyzed. Cd did not correlate with any other metals except Fe ($P<0.05$). Cu, Pb and Zn are the main constituents of the vehicular emissions had significant positive correlation ($P<0.001$) with protein content while the OD 435/415 ratio values decreased statistically ($P<0.001$) with increase in amount of Cu, Pb and Zn (Shukla and Upreti 2008, 2009).

Pyxine cocoes, a foliose lichen commonly growing on mango trees in tropical regions of India, is an excellent organism for determining the pollutants emitted from coal-based thermal power plant and accumulated in lichens after prolonged exposure. The diversity and distribution of lichens in and around such power plant provide a useful tool to measure the extent of pollution in the area. The distributions of heavy metals from power plant showed positive correlation with distance for all directions. The speed of wind and direction plays a major role in dispersion of the metals. The accumulation of Al, Cr, Fe, Pb and Zn in the thallus suppressed the concentration of pigments (chlorophyll a, chlorophyll b and total chlorophyll) however, enhanced the level of protein. Further the concentration of chlorophyll content in *P. cocoes* increased with decreasing the distance from the power plant, while protein carotenoid and phaeophytisation exhibit significant decrease.

The morphology, chemistry and anatomy of lichens play important role in accumulation of metals. Another common tropical lichen species *Phaeophyscia hispidula* belongs to the same lichen family (Physciaceae) as of *Pyxine* has distinct morphology and chemistry. A thick tuft of rhizinae (hair like structure) on the lower surface of the thallus in *Phaeophyscia hispidula* acts as a metal reservoir and thus exhibit higher accumulation of most of the metals than *Pyxine*. The crust forming lichens attached tightly to the substrates through their whole lower surface have the highest accumulation of Al in the metal sequence while the squamulose, foliose form show Fe in the higher concentrations. The lichens have special affinity with iron and they accumulate iron in greater amount than other metals.

4.9.2.2. Organic pollutants accumulation in lichens

Apart from inorganic metals, lichens are excellent indicators of Polycyclic Aromatic Hydrocarbon compounds (PAHs) too. The PAHs accumulation studies in Indian lichens are initiated recently in the Himalayan region of Uttrakhand. The PAHs accumulation in lichens of different localities of Dehra Dun city, and on way to Badrinath are estimated recently. The first baseline data on the distribution and origin of polycyclic aromatic hydrocarbons (PAHs) in *Phaeophyscia hispidula* collected from nine different road crossings of Dehra Dun city and enroute to Badrinath of Uttarakhand exhibit the presence of 13 types of PAHs (Naphthalene (0.14-5.65 ppm), Acenaphthylene (0.89-22.13 ppm), Fluorene+acenaphthylene (0.07-3.38 ppm), Phenanthrene (0.06-6.47 ppm), Anthracene (0.01-0.38 ppm), Fluoranthene (0.01-3.58 ppm), Pyrene (0.13-14.46 ppm), Benzo (a) anthracene + chrysene (0.01-0.13 ppm), Benzo (k) fluoranthene (0.01-0.03 ppm), Benzo (b) fluoranthene (0.02-0.09 ppm), Benzo (a) pyrene (0.00-0.03 ppm), Dibenzo (a,h) anthracene (0.17-0.31 ppm), Indino (1,2,3-cd ppm) pyrene + benzo (ghi) perylene (0.00-0.20 ppm) (Shukla and Upreti 2009).

The PAHs were of mixed origin, a major characteristic of urban environment. Significantly higher concentration of phenanthrene, Pyrene and acenapthalene indicates road traffic as major source of PAH pollution. The acetyl polymalonyl patway in lichens results in biosynthesis of secondary metabolites of depsides and depsidones containing highly reactive –OH radicals (due to ortho effect). The depsides and depsidones easily provide their hydroxyl group for adduct formation. The higher accumulation of 2 and 3 ring PAH in lichens may be because most of the species contains depsides and depsidones with active –OH sites, which readily combine with PAH to from an adduct. *Phaeophyscia* and *Pyxine* have skyrin triterpine and lichenoxanthone (having hydroxyl group) which readily combine with most of the PAHs.

The growth from of lichens may also play a significant role in the accumulation of PAHs. The saxicolous, crustose and squamulose species growing on rocks mostly accumulated uniform concentration of low molecular weight 2 and 3 ringed PAHs compounds. The higher vehicular activities or excess uses of wood and coal in a particular area are responsible for higher concentration of PAHs. The study establishes the utility of *Phaeophyscia hispidula* as an excellent biomonitoring organism in monitoring of PAHs from foot hill to sub-temperate area of Garhwal Himalayas (Shukla and Upreti 2007).

4.9.2.3. Metalloid (As) pollutants accumulation in lichens

The ability of lichens to uptake As, translocate, metabolize and accumulate is one of the determinating factors for phytotoxicity of this element. The arsenic contamination is of particular interest because of its high toxicity to plants, and artificially created arsenic rain could provide information regarding their impact on biological system.

Among the different growth forms, the leafy form (foliose) accumulates higher amounts of arsenic followed by the powdery (leprose) form. The squamulose (crust to leafy) and crustose (crust forming) form accumulate lower concentration of arsenic that ranged between 0.46 ± 0.03 and 20.99 ± 0.58 µg g^{-1}DW, while the foliose and leprose lichens accumulate arsenic in the ranges of 10.98–51.95 and 28.63–51.20 µg g^{-1}DW respectively.

The cyanolichens (with blue green photobiont) exhibit higher concentration of arsenic than the green photobiont- containing squamulose form. The active monitoring (transplant) of the same metalloid also adopted to investigate the toxicity of excess arsenic on physiochemical process of foliose lichen *P. cocoes*. The arsenic solution with concentrations 10, 25, 50, 75, 100 and 200 µM were sprayed on lichen thallus for 45 days. The arsenic content in the thalli was then correlated with the pigments degradation, total protein concentration and the activities of antioxidant enzymes focusing on superoxide dismutase, catalase and ascorbate peroxidase. The resultant information was utilized to assess the suitability of *P. cocoes* as biomarker against arsenic pollution in tropical environment (Fig. 9).

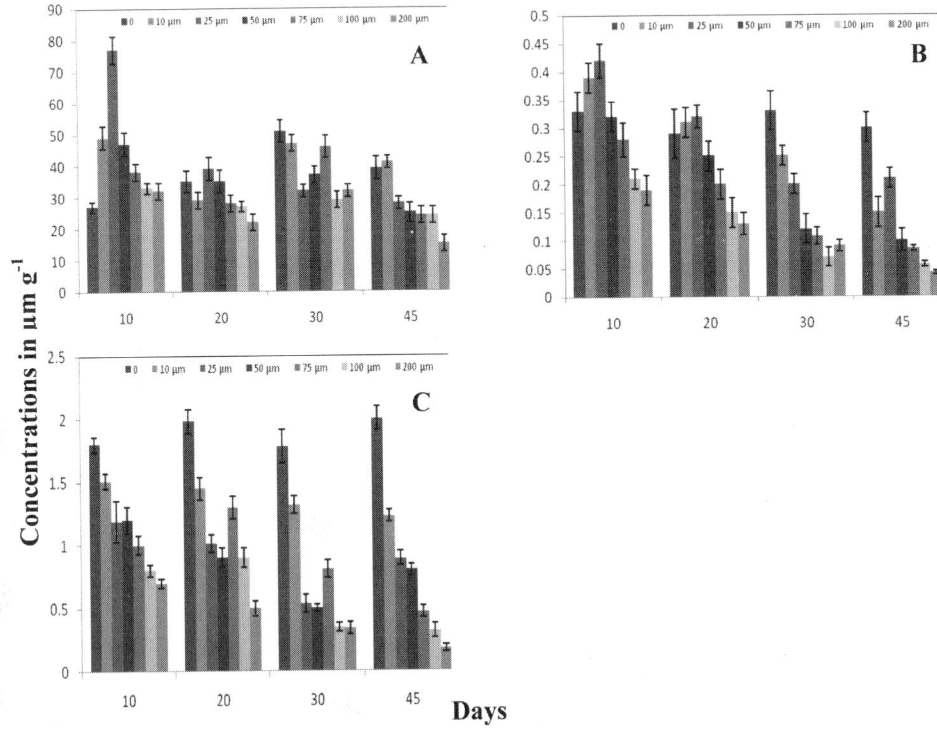

Fig. 9. Comparison between days and various concentration of arsenic on enzymatic activities of *P. cocoes*; A. superoxide dismutase; B. ascorbate peroxidase; C. catalase.

4.9.2.4. Biomarkers

Pollutants cause damage to living organisms in different ways. Damage can occur at all levels of biological organization, from the components of individual cells to ecosystems. Traditionally, the rate of accumulation of contaminants, geographical distribution or morphological modifications has been studied in "indicator" species. However, it is now realized that the impact of pollutants can be measured more quickly by testing their effects on certain physiological processes termed as "biomarkers". Lagadic *et al.* (1997) define a biomarker as "an observable and/or measurable change at a molecular, biochemical, cellular, physiological or behavioral level, which revals the present or past exposure of an individual to a chemical polluting substance."

An ideal biomarker should be easy to measure and produce distinctive symptoms that are not confused with those caused by other environmental stresses. When properly used, biomarkers can "forecast" impending harmful effects. Ideally, an environmental survey based on biomarkers can be used as a warning by early detection of the effects of pollutants, and by detecting pollution below the dose that causes irreversible damage. Results from such survey can be used to argue for a more intensive survey of the particular ecosystem. Several parameters are best used in lichens for suitability as biomarkers are well known.

For a long time lichens have been known as valuable resources for various purposes and still they are most important sources of dyes, medicines, food, fodder and perfumes. In recent years their use as a means of monitoring gaseous pollutants and extraction of HIV-antigen product from these plants will definitely attract the attention of research workers to conduct more intensive researches in this branch i. e. lichenology. It can thus be inferred that there is a vast treasure of medicines hidden in this small group of plants and proper study may unfold a vast fund of new information leading to discovery of some potential lichen species.

References

Awasthi DD (2000) Lichenology in Indian subcontinent- A suppliment to a hand book of lichens. Bishen Singh and Mahender Pal Singh, Dehra Dun, India pp.9-14

Bajpai R, Upreti DK, Mishra SK (2004) Pollution monitoring with the help of lichen transplant technique at some residential sites of Lucknow. J Environ Bio 25(2):191-195

Beschel R (1961) Dating rock surface by lichen growth its application to glaciology and physiography, In: Geology of the Arctic (eds. Rassch G.O.) University of Toranto Press, Toranto 1:1044-1062

Dettki H, Esseen PA (1998) Epiphytic macrolichens in managed and natural forest landscapes: a comparison at two spatial scales. Ecography 21:613-624

Eldridge DJ, Koen TB (1998) Cover and floristic of microphytic soil crusts in relation to indices of landscape health. Plant Eco 137:101–114

Garty J (2001) Biomonitoring atmospheric heavy metals with lichens: theory and application. Crit Rev Plant Sci 20 (4):309-371

Garty J, Tomer S, Levin T, Lehr H (2002) Lichens as biomonitors around coal fired power station in Israel. Environ Res 91:186-198

Khare R, Upreti DK, Nayaka S, Gupta RK (2009) Diversity of soil lichens in India. In: Gupta RK, Kumar M and Vyas D (Eds.) *Soil Microflora*, Daya Publishing House, New Delhi pp 64-75

Lagadic L, Caquet T, Amiard JC, Ramade F (1997) Biomarqueurs en écotoxicologie: Aspects fonda-mentaux., Edition Masson. 33:53-97

Loppi S, Putorti E, Pirintsos SA, Dominicis VD (2000) Accumulation of heavy metals in epiphytic lichens near municipal waste incinerator (central Italy). Environ Moni Ass 61:361-371

McCune B (1993) Gradients in epiphyte biomass in three *Pseudotsuga-Tsuga* forests of different ages in western Oregon and Washington. Bryologist 96:405-411

Satya, Upreti DK, Nayaka S (2005) *Shorea robusta*, an excellent host tree for lichen growth. Curr Sci 89 (4):594-595

Saxena S, Upreti DK., Sharma N (2007) Heavy metal accumulation in lichens growing in north side of Lucknow city India. J Environ Bio 28(1):49-51

Seaward MRD and Richardson DHS (1989) Atmospheric sources of metal pollution and effect on vegetation. In: Heavy metal tolerance in plants: Evolutionary Aspects. Pp. 75-92. Shaw, A.J., (ed.). CRC Press, Boca Raton, FL

Sernander R (1926) Stockholms Natur. Almguist and Wiksella, Uppsala, Sweden

Shukla V, Upreti DK (2007) Heavy metal accumulation in *Phaeophyscia hispidula* en route to Badrinath, Uttaranchal, India. Environ Monit Ass 131:365-369

Shukla V, Upreti DK (2008) Effect of metallic pollutants on the physiology of lichen P*yxine subcinerea* in Garhwal Himalayas. Environ Monit Ass 141:237-243

Shukla V, Upreti DK (2009) Polycyclic aromatic hydrocarbon accumulation in lichen *Phaophyscia hispidula* of DehraDun city, Garhwal Himalayas. Environ Monit Ass 149:1-7

Singh KP and Sinha GP (2010) Indian Lichens: An annotated checklist. Botanical Survey of India (MoEF), Kolkata

Singh KP, Sinha GP, Bujarbarua P (2004) Endemic lichens of India. Geophytology 33(1/2):1-16

Srivastava R, Yadav V, Upreti DK, Nayaka S (2004) An emumeration of lichens from Shimla disctict, Himachal Pradesh. Geophytology 33:29-34

Upreti DK (1998) Diversity of lichens in India. In: Perspectives in Environemnt. Agarwal SK, Kaushik JP, Koul KK and Jain AK (eds.), APH Publishing corporation, New Delhi pp 71-79

Upreti DK, Nayaka S, Bajpai A (2005) Do lichens still grow in Kolkata City?. Curr Sci 88 (3):338-339

Chapter – 5

Methods and Techniques in Collection, Preservation and Identification of Lichens

Sanjeeva Nayaka

1. Introduction

The word 'lichen' has a Greek origin, which was referred to the superficial growth of fungus like organism on the bark of olive trees. Theophrastus, the Father of Botany coined the term 'lichen' during 300 BC and introduced this group of plants to the scientific world.

Lichen is a combination of two organisms, an alga and a fungus, living together in symbiotic association. Sometimes instead of an alga a cyanobacterium (blue-green alga) may be present in the lichen thallus. The algal component in the lichen is generally called as 'photobiont' as it contains photosynthetic pigment. The fungal part is called as 'mycobiont' (myco = fungus). The photobiont and the mycobiont loose their original identity in symbiotic association and the resulting organism is the lichen which behaves as a single organism, both morphologically and physiologically. Hence, the lichen is called as a composite organism. The lichen thallus is made 90% of mycobiont which provides shape, structure and colour to the lichen. In a lichen thallus whatever visible from outside is fungal part, which holds algae inside its body (Fig. 1) and therefore, the lichens are placed within the Kingdom - Fungi. The fungi present in lichens are called as 'lichenized fungi'. Among the 20,000 lichen species present in the world, 98.9% of them belongs to Ascomycetes group while Bacidiomycetes and Deuteriomycetes groups are represented by only 0.1% and 1%, respectively.

Fig. 1. Vertical section of a lichen thallus

2. Nature of Symbiosis

Schwendener (1867), a Swiss Botanist demonstrated for the first time the dual nature (presence of both fungus and alga) of lichen thallus. There after a great debate started discussing the nature of relation between the fungus and the algae. It was proposed that the relationship is 'parasitism', where fungus is a parasite on alga. It is observed that the haustorium of fungal hypae enter in to algal cells for drawing nutrients. However, this theory is discarded, because in parasitism the host would ultimately die and finally the association will come to an end, whereas in case of lichen the algal fungal association is permanent. Sometimes the relation is called 'controlled or balanced parasitism', because the fungus does not kill the alga, but at the same time it will not allow alga to flourish. It can also be called as 'helotism' or master and slave relationship, where fungus is the master and the alga is a slave. Some lichenologists called the relation an 'endosaprophytism'. The fungi are usually saprophytes and thrive on dead organic matter. Inside some lichens dead algal cells were observed and hence it was thought that fungus may be feeding on them. However, most of the lichenologists believe it is 'mutualism'. The alga by having photosynthetic pigments prepares food and supplies it to fungus; the fungus in turn provides shelter to the alga and supplies water and nutrients. Due to the difference of opinion regarding the nature of relationship between fungus and alga, it is suggested to call the relation as 'symbiosis', which simply means 'living together'.

3. Lichen Growth and Growth Forms

Lichens are very slow growing organisms. They grow just few millimeter or centimeter in a year. Lichens can grow in diverse climatic conditions and on various substrates. The lichens growing on bark or tree trunk are called corticolous, on twigs as ramicolous, on dead wood - legnicolous, on rocks and boulders - saxicolous, on moss - muscicolous, on soil - terricolous and on evergreen leaves – foliicolous (Fig. 2A-G). The lichens can grow under water rocks, but not purely in water or ice. The lichens are widely distributed in almost all the phytogeographical regions of the world. Adequate moisture, light, altitude, unpolluted air and undisturbed substratum favours luxuriant growth of lichens. By their appearance the lichens can be grouped into three main growth forms.

Fig. 2. (A) Lichen growing on tree bark (corticolous), (B) On twig (ramicolous), (C) On dead wood (legnicolous), (D) On rock (saxicolous), (E) On moss (muscicolous), (F) On soil (terricolous), and (G) On leaf (foliicolous).

3.1. Crustose lichens

The lichen thallus is closely attached to the substratum without leaving any free margin. The thallus usually lacks lower cortex and rhizine (root like structure). Such lichens have to be collected along with their substratum for the study (Fig. 3A).

3.2. Foliose lichens

Foliose lichens are also known as leafy lichens. The thallus in this case is loosely attached at least at the margin. Such lichens can be collected by scraping them from the substratum (Fig. 3B).

3.3. Fruticose lichens

Here the lichen thallus is attached to the substratum at one point and remaining major portion is either growing erect or hanging. The lichen usually appears as small shrub or bush (Fig. 3C).

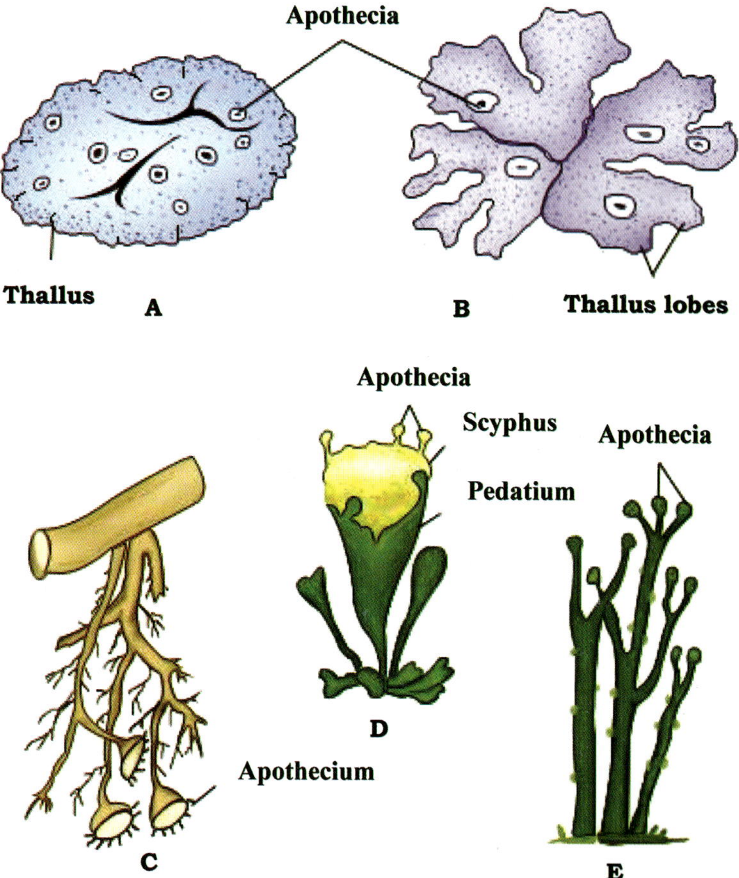

Fig. 3 (A) Crustose lichen, (B) Foliose lichen, (C) Fruticose lichens (D-E) Dimorphic lichen

There are few intermediate categories of growth forms such as:

(a) **Leprose lichens.** Here the lichen thallus is powdery or granular, does not form perfect smooth thallus.

(b) **Placodioid lichens.** In this case the lichen thallus is closely attached to the substratum at centre and lobate or free at the margin, but lacking rhizine.

(c) **Squamulose lichens.** Here the lichen thallus is in the forms of minute lobes, having dorsiventral differentiation.

(d) **Dimorphic lichens.** In this case single lichen thallus has the characters of both foliose/ squamulose and fruiticose lichens. Here the leaf like structures are called 'phyllocladia' and erect, stem like structures are called as 'podetia' (Fig. 3D-E).

All the above mentioned growth forms of lichens can be arranged according to their imaginary evolution as; Leprose (pioneer) → Crustose → Placodioid → Squamulose → Foliose → Dimorphic → Fruticose (latest). The leprose, crustose, some placodioid and squamulose lichens are generally called as 'microlichens', because of their smaller size and mostly require microscope for their identification. The foliose, dimorphic and fruticose lichens on the other hand are called as 'macrolichens'. The macrolichens have comparatively larger thallus and a hand lens or dissection or stereozoom microscope or is sufficient for identification.

The crustose, foliose, squamulose, placodioid and sometimes dimorphic forms of lichens usually grow in a circular and centrifugal manner. The rough and uneven surface of the substratum may change the shapes of the thallus. The leprose lichens forms irregular patches of thallus on the substraum. The fruticose lichens of smaller size usually grow erect while larger ones hang from the substratum with their growing point located at the tips.

Differentiating lichens from other groups of plants

The non-lichenized fungi, algae, moss, liverworts (bryophytes) are the plants, which grow on rocks, bark and soil, and may confuse with lichens at least for the beginners in the field. However, lichens can be easily differentiated from these plants. The lichens are never greener as algae, liverworts and mosses. Foliose lichens in the moist places or in wet condition may look greener, but have thick, leathery thallus while liverworts have non-leathery and slimy thallus. The dimorphic forms of lichens such as *Cladonia* may confuse with the leafy liverworts and mosses. However, leafy liverworts and mosses have dense small leaf like structures throughout the central axis of the plant, while in case of dimorphic lichens the squamules of semicircular shape are usually present at the base of the central axis or sparce throughout. Algal mat are usually found in water-flooded habitat. The beginners may confuse the dried algal mat on rocks and bark for lichens. By spraying some water on these mat one can make out whether it is algal mat or lichen.

The non-lichenized fungi are the most confusing ones with crustose lichens in the field. Such fungus usually forms patches with loosely woven hyphae, which will be evident under lens. The lichens on the other hand form smooth, perfect thallus. The fungus are usually whitish in colour and lichens are usually grayish, off white, yellowish, yellowish- green or sometimes bright yellow or yellow orange in colour. The lichens by having algal cells inside exhibit greenish tinge in colouration. The lichen thallus usually bears cup like structures called apothecia, or bulged, globular structures called perithecia or finger like projections called isidia or granular, powder like structure called soredia. Some crustose lichens belonging to family Graphidaceae bear worm like structures, which are nothing but modified apothecia. While collecting lichens it is necessary to look for such structures with the help of hand lens. When a lichen thallus does not have any such structures, it becomes difficult to differentiate it from fungus.

In any case it is observed that a beginner usually collects fungus and other plants in place of lichens. Usually fungus of various colour (mostly appearing like mushroom) are confused for lichens and collected by the beginners. Such specimens can be identified by taking a thin section of the thallus and studying them under a microscope. If the section contains both fungal tissue and algal cells, then the specimen is lichen, otherwise it is something else. In India basidiolichens (looking like mushroom) are rare or absent.

4. Methods

4.1. Collection and preservation of lichens

The micro or macrolichens are visible to the naked eye in the field. However, a hand lens, preferably of 10×, is necessary to examine the structure of the thallus and confirm while collecting the lichens. A sharp, flat edged chisel (1 to 2 inch) and a hammer (1 to 2 kg weight) are the tools required for collecting lichens. Carpenter use such flat chisels and it can have either wooden or plastic handle. Sometimes sharp, hard knife can also be used. The pointed, long chisel and heavy hammer are recommended for collecting lichens growing on rocks. Polythene packets (small (6 × 12 inch) and bigger sizes), rubber bands, labeling stickers, Global Positioning System (GPS), a field notebook, pen, pencil, plant press, old news papers or blotters, nylon ropes, collection bags, herbarium packets are the other necessary items needed during lichen collection trip.

The lichens are usually collected along with their substratum irrespectively of their growth form. Only the lichens that are very loosely attached to substratum are scraped out and collected. In case of saxicolous lichens smaller pieces of the rocks are collected. Collection of lichen samples in 'sufficient' amount (at least 2 thallus) is necessary, as the material will be used for detailed microscopic as well as chemical studies. Care should be taken to collect intact thallus, at least the margins should be clearly visible in the specimens.

In case of corticolous lichens one should try to collect superficial bark to avoid damage to the trees. The collected lichen samples are transferred to the polythene packets, labeled and closed with the help of rubber bands. Several such packets are then transferred to larger polythene or collection bags. One can also keep the collected material in newspaper or blotter packets. The lichen specimens should not be kept in polythene packets for longer duration as they spoil due to fungal attack and colour of the thallus will also change. After returning from field all the specimens should be transferred to newspaper or blotter packets. The lichen specimens on wet barks should be kept in plant press and tied tightly. Otherwise the bark gets curled up as it dries, makes uncomfortable to preserve in herbarium packets and will give a shabby look. No poisoning methods are available for lichen preservation. Lichen samples are thoroughly dried and preserved in the herbarium packets.

The lichen herbarium packets should be of thick, white or brown hand made paper. The hand made paper sheet of dimension 13.5 × 11.5 inches is folded lengthwise twice and then side ways to produce the packets of dimension 7 × 5 inches with upper flap of 3.5 inches to stick the label (Fig. 4). The herbarium label should contain the information on name and family of the lichen (which can be written after the identification), details of locality, altitude, longitude, latitude, date of collection, a reference number, collectors name and notes on its substratum and any other interesting observations. After the identification name of the person who identified (determined) the specimen along with date can also be mentioned.

The dried lichen specimen should be pasted on to a thick, hard paperboard of dimension 6.5 x 4.5 inches (little less than the total packet size) and then placed inside the herbarium packet. The board also should have the same reference number as on the label. A herbarium packet should have specimen belonging to single species and mixtures should be avoided. Once the packets are ready they can be stacked in rectangular boxes (like shoe box) and such boxes can be kept inside the almerahs or wooden cupboards.

The method for collecting lichens depends on the objective of the study. For a simple floristic study the method mentioned above is sufficient. However, for air pollution, ecological or phytosociological study the data gathering method may be different. Hence, the objective and the methodology should be clear before starting the field work. Only representative few samples should be collected and bulk collection should be avoided.

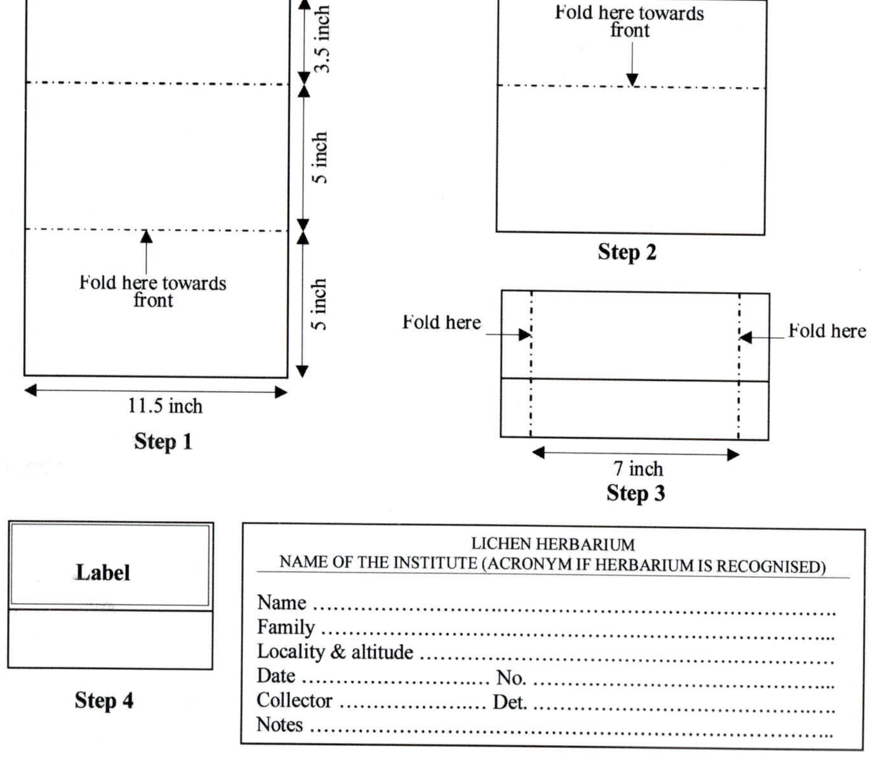

Fig. 4. Method of preparing lichen herbarium packet

4.2. Identification of lichens

The collected lichen specimens are initially segregated according to their growth forms. Within the growth forms the specimens can be further grouped according to the type of fruiting bodies (apothecia, perithecia, sterile) (Fig. 5).

Fig. 5 (A) Apothecia, (B) Perithecia, (C) Streched apothecia, lirellae

The lichens are identified by studying their morphology, anatomy and chemistry. The micro and macrolichen keys of Awasthi (1991, 2007) are the important literature referred for identification of Indian lichens. The beginner should have a glossary of technical terms while identifying the lichens. Illustrated glossary given in 'Lichen flora of Great Britain and Ireland' (Purvis *et al.* 1992) would be very useful. A botany student or one with mycological background can better follow the terminology and identification keys.

The morphological and anatomical characters to be observed in a lichen specimen differ from genus to genus or group to group. However, some common characters to be noted are given in Fig. 6.

4.2.1. Morphology

The morphological characters of a lichens specimen are studied under dissection or stereo microscope. Type of thallus or growth form (leprose, crustose, foliose, squamulose, dimorphic, fruticose), its shape (irregular, circular) and size should be recorded.

(a) Upper surface

The colour of the thallus, texture (smooth, rough, warty), presence of finger like projections (isidia), granular structures (soredia), fine powder (pruina), black dots (pycnidia) and whitish decorticated areas (pseudocyphellae) have to be noted. The branching pattern, length and breadth of marginal lobes, presence of hair like structures (cilia) in case of foliose lichens has to be noted. In case of fruticose lichens length of the thallus, branching pattern, flatness or cylindricalness of the thallus has to be noted down (Fig. 7).

The morphology of fruiting bodies have to be studied separately. In case of apothecia, shape [rounded or stretched (Fig. 5)], size, attachment (stalked or not), colour and texture of the margin and disc, presence or absence of powder (pruina) on the disc, shape of the disc (convex or concave) are essential characters. In case of perithecia, its colour, shape, size and the position of its opening (ostiole, apical or lateral), single or grouped has to be noted.

Some lichens thallus emits florescence (yellowish, bluish) when observed under UV light due to the presence of lichen substance called lichexanthone. Such lichens are examined by keeping them in a closed UV lamp chamber and exposing UV light with wavelengths 254 and 365 nm.

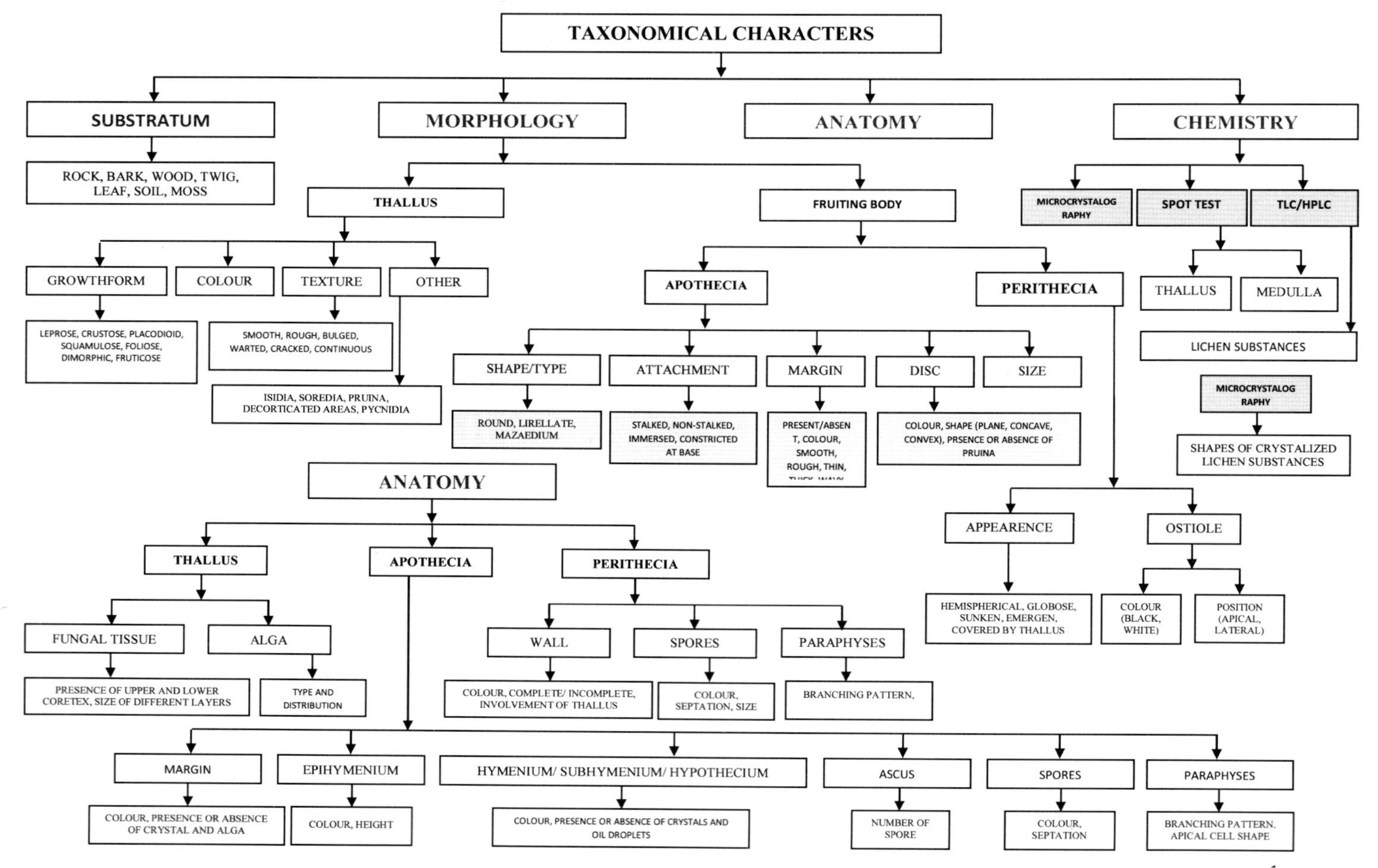

Fig. 6. The important characters to be observed for identification of lichens

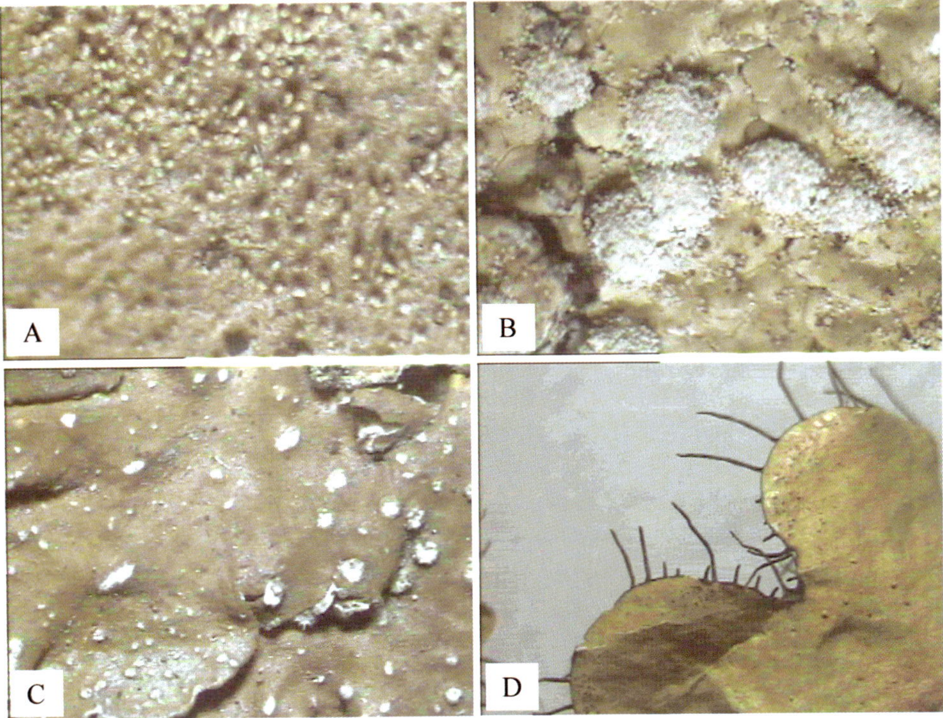

Fig. 7. (A) Isidia, (B) Soredia, (C) Pseudocyphellae, (D) Marginal cilia

(b) Lower surface

The lower surface of only foliose lichens can be seen as it is absent in crustose lichens while dimorpohic and fruticose lichens do not show dorsiventral differentiation. The colour of lower surface, presence of any pores (cyphaellae, pseudocyphellae), presence or absence of rihizines (root like structures), their colour, distribution, branching, abundance are to be noted.

4.2.2. Anatomy

The anatomy of lichen thallus and fruiting bodies is examined under compound microscope with minimum magnification of 40X. The anatomy of the thallus is occasionally studied to see the thickness of various layers (upper cortex, algal layer, medulla, lower cortex), type of algae and their distribution (stratified – heteromerous or uniform – homeomerous) and arrangement of fungal hyphae (vertical or horizontal) within the thallus. The section of thallus can be cut with snapper or razor blade by keeping the fragment of thallus in potato or papaya pith. Microtome sections are very helpful but time consuming. Just to check the type of algae present in the thallus one need not cut a section. By the colour of the thallus one can make out the type of alga (at least group) present within. Lichen with blackish, bluish, slate grey thallus usually has blue green alga, while grayish, yellowish, brownish, greenish thallus has green alga. However, it is better to confirm the type of alga present by the following easy procedure. The algal layer of the lichen thallus is exposed by scraping the upper cortex with blade and alagal part (which appears dark green, blue green, black) is picked up with blade or needle, transferred to the slide and examined under microscope.

The anatomical character of fruiting bodies (ascocarp) is very important identification aids especially in case of crustose lichens. The type of spore (simple, septate), colour (hyaline, brown), their shape, size, number of spores in a spore-sac (ascus), colour of ascocarp wall (exciple), presence or absence of crystals and algal cells in the wall, colour and height of different layers (hymenium, epi and subhymenium, hypothecium) within in the ascocarps are to be noted (Fig. 8). The branching pattern and arrangement of paraphyses, shape and colour of apical cell are important character to be noted.

The thin, hand section of ascocarp is taken with the help of blade while it is still attached to the thallus or substratum and by viewing through dissection or stereo microscope. The ascocarp is made wet with a drop of water before cutting the section. Few sections are enough to observe the character.

The sections of thallus and ascocarps are mounted with plain water to observe basic characters. The lactophenol cotton blue and other stains can be used to colour different tissue as per requirement. Semi-permanent slides can be prepared by adding a drop of glycerol water mixture (1:1) to the section slide and sealing the cover slip with quick fix.

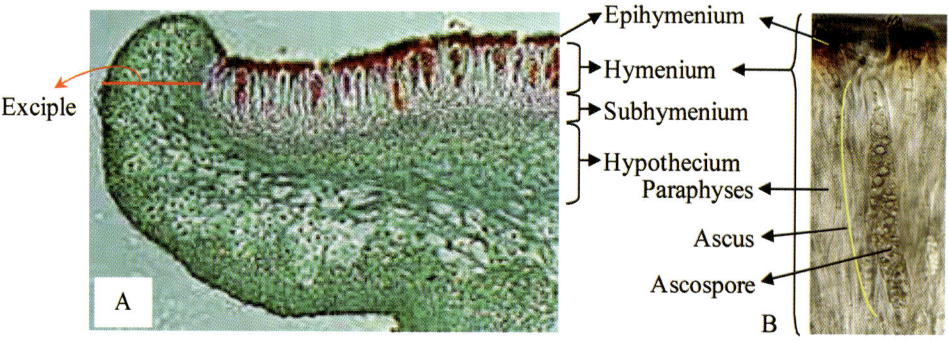

Fig 8. (A) Vertical section of apothecia showing anatomical structures, (B) Part of hymenium showing ascus, ascospores & paraphyses

4.2.3. Chemistry

Lichens produce more than 1000 secondary metabolites that are popularly known as lichen substances. Out of the 1000 lichens substances around 950 are unique to the lichens and are not available in any other groups of plants, only small portion of about 50 – 60 occur in other fungi or higher plants (Elix and Ernst-Russel 1993). For example, the anthraquinone parietin, the orange pigment that is common in most 'Teloschistales' occur in non – lichenized fungal genera *Achaetomium, Alternaria, Aspergillus, Dermocybe, Penicillium* as well as in the vascular plants *Rheum, Rumex* and *Ventilago.* The lecanoric acid also occurs in fungus *Pyricularia,* while the sterol, brassicasterol of higher plant is also available in the lichens. Most of these lichen substances act as an important character for identification of lichens (chemotaxonomy). The lichen substances are identified by performing colour spot test, microcrystalography, thin layer chromatography (TLC) or by high performance liquid chromatography (HPLC).

4.2.3.1. Colour spot test

Three chemical reagents commonly used for colour spot test are aqueous potassium hydroxide (K), bleaching powder or aqueous solution of calcium hypochlorite (C) and aqueous solution of paraphyenldiamine (Pd). K-test is performed either on upper surface of thallus (cortex) or on the medulla by exposing it with blade, or on both. A drop of K solution is placed on the cortex or medulla and colour reaction is noted. Usually C and Pd test are performed on the medulla and colour changes are recorded. KC-test is performed by applying K solution first and then immediately C soloution over earlier K solution drop. The colour of the cortex or medulla changes due to presence of particular lichen substances in lichen thallus (Fig. 9). Sometimes Iodine (I) test is also performed on the cortex to check the presence or absence of polysaccharides. William Nylander in 1860s introduced I, K, C and KC test. Later Asahina *et al.* introduced Pd test in 1930s (Culberson 1969). The composition of the reagents is given below and the possible mode of action is given in the box 1.

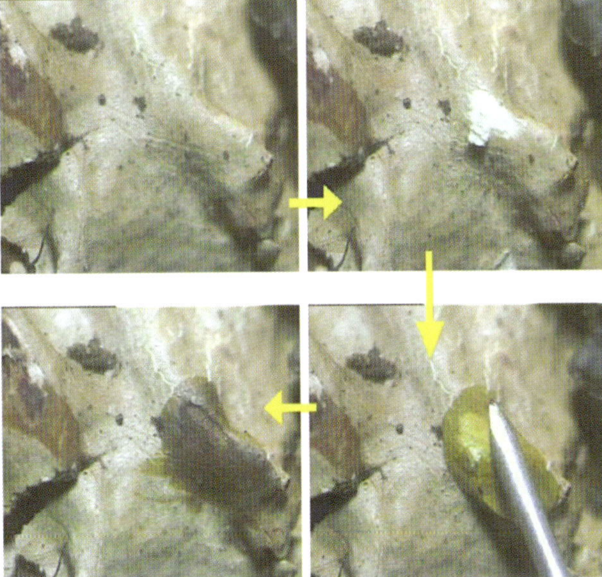

Fig. 9. Performing colour test on lichen thallus

K = 10% aqueous KOH solution

a. Turns yellow then red with most *o*-hydroxy aromatic aldehydes.
b. Turns bright red to deep purple with anthraquinone pigments.

C = saturated aqueous $Ca(OCl)_2$ or common bleach (NaOCl) solution

a. Turns red with *m*-dihydroxy phenols, except for those substituted between the hydroxy groups with a –CHO or $-CO_2H$.
b. Turns green with dihydroxy dibenzofurans.

KC = 10% aqueous KOH solution followed by saturated aqueous $Ca(OCl)_2$ or common bleach (NaOCl) solution

a. Turns yellow with usnic acid.
b. Turns blue with dihydroxy dibenzofurans.
c. Turns red with C- depsides and depsidones which undergo rapid hydrolysis to yield a *m*-dihydroxy phenolic moiety.

PD = 5% alcoholic *p*-phenylenediamine solution

a. Turns yellow, orange or red with aromatic aldehydes.

Box 1. The colour spot test reagents and their possible reactions

(a) Reagent 10% K

Potassium hydroxide pellets 10 g dissolved in 100 ml of distilled water. The reagent should be prepared fresh, it absorbs carbon dioxide from the air and gradually becomes ineffective, it should be replaced when it becomes cloudy. Confirmation test can be performed on the medulla of *Parmelinella wallichiana* or any other lichen which gives K+ red colour.

(b) Reagent C

One part of calcium hypochlorite should be added to double the volume of distilled water (1:2 ratio) and shaken well. The reagent may be allowed settle down and supernatant solution can be used for the spot test. The reagent should be prepared fresh and it should be discarded when it stops emitting chlorine smell. The confirmation test can be performed on the medulla of *Puctelia borreri* or any other lichen that gives C+ pink colour.

(c) Reagent Pd (Steiner's solution)

The *para*-phenylenediamine 1 g, sodium sulphite 10 g, detergent liquid 0.5 ml are dissolved in 100 ml of distilled water. The reagent should be prepared fresh; oxidation of the reagent gives false result and should be discarded. Confirmation test can be performed on the medulla of *Parmelinella wallichiana* or on the lichen which yield Pd+ orange colour. *Para*-phenylenediamine is a suspected carcinogen which should be handled with great care.

4.2.3.2. Microcrystalography

The method for crystalization of lichen substances was introduced by Asahina and Shibata (1954). The experiment is usually carried out on the glass slide. Small pieces of lichen are placed at the centre of the thallus and chemical substances present in it are extracted on to the slide by dropping acetone over it. The acetone drops should be smaller and each drop should be added after the evaporation of the earlier drop. It gives concentric white rings of crystals around the lichen pieces. Then the material should be removed gently and a drop of crystallizing reagent is added over the concentric rings. After placing a cover-slip on the crystallizing agent the slide should be gently warmed over the spirit lamp and then allowed to cool down and to form crystals (Fig. 10). Then the slide is observed under compound microscope and the lichen substances are identified based on the shape of the crystals. The compositions of various crystallizing reagents are given in the box 2.

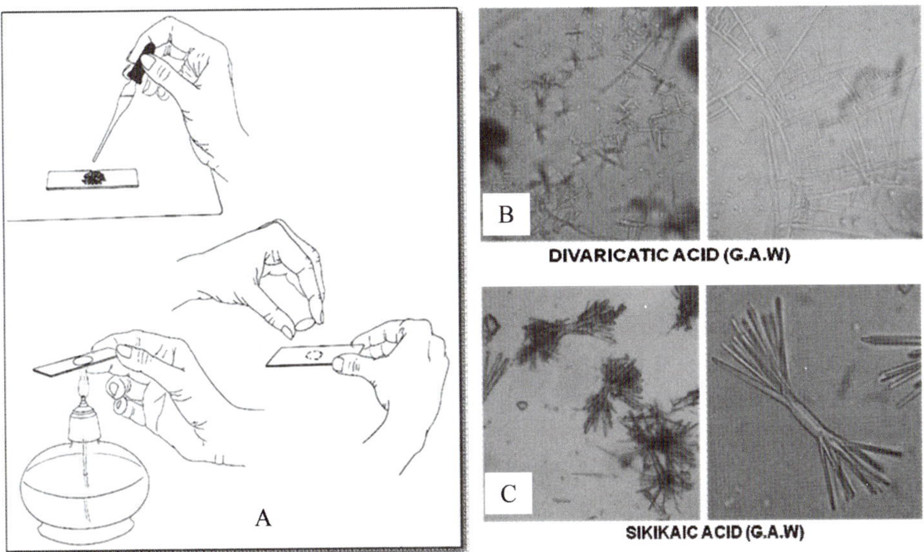

Fig. 10. (A) Procedure involved in microcrystallography, (B) Crystals of Divaricatic acid in GAW, (C) Crystals of Sikikaic acid in GAW.

G.E.: Glycerine-acetic acid, 1:3.
G.A.W.: Glycerine-alcohol-water, 1:1:1.
G.A.*o*-T.: Glycerine-alcohol-*o*-toluidine, 2:2:1.
G.A.An.: Glycerine-alcohol-aniline, 2:2:1.
G.A.Q.: Glycerine-alcohol-quinoline, 2:2:1.

Box 2. Various crystallizing reagent used in microcrystallography

Microcrystalography is believed to be more accurate than spot tests. However, may be difficult to identify mixture of substances with this method, and also minor substances may be undetectable. This method is superseded by more accurate Thin Layer Chromatography (TLC) method.

4.2.3.3. Thin layer chromatography

Many lichen substances are undetectable in colour spot test or the colour spot test may not give proper result. In such cases TLC have to be performed. In TLC various lichen substances present in a lichen thallus get separated as spots on TLC plates. With the help of lichen TLC manuals, these spots can be identified. The TLC method was standardized by Culberson in 1950s and it is now a widely used technique (Culberson 1969). It is a simple, relatively inexpensive, speedy method and helps accurate recognition of secondary metabolites. The steps involved in performing TLC are given below.

i. Extraction of lichen substances

Fragments of the thallus are placed in small test tubes and few drops of Acetone are added to it. The lichen substances in the thallus get extracted in to the acetone. The test tubes should be numbered serially including the control sample.

ii. Preparation of TLC plate

Silica gel pre-coated thin aluminium plates are used for the TLC and 20 × 20 cm Merck 60F is highly recommended. A line is drawn at 2 cm from base of the plate (loading line) and another at 15 cm (finishing line). On the 2 cm line several spots are marked at equal distances. The number of spots should be corresponding to the number of test tubes of samples which are needed to be checked for chemistry. The minimum distance between two spots should be 0.8 cm. Farer the spots on the loading line lesser would be the overlapping of developed spots. The width of the plate can be cut in to various sizes as per requirement and number of samples.

iii. Loading

The acetone extract in the test tube is spotted on the TLC plate with the help of capillary tube. The number on the test tube with the extract should be corresponding to the number on the TLC plate. The capillary tube should be thin and separate tube should be used for each extract to avoid contamination. The spot on the TLC plate should be concentrated enough and it can be achieved by repeated loading of the extract on the same spot.

iv. Preparation of solvent system and TLC tank

Different organic solvent can be used to separate the compounds present in the lichens (Box 3). Solvent should be freshly prepared and older solvent give false result. Rectangle specimen jar can be used as TLC tank and it should be covered with glass lid. The tank is made air tight by applying grease or Vaseline at the rim of the jar where the glass lid touches. In side the TLC tank towards back side filter paper sheet can be placed. The wet filter paper provides uniform vapourous atmosphere inside the tank and help separation better of lichen compounds. The quantity of the solvent should be sufficient enough but should be just below 2 cm loading line of TLC plate. If the bottom of the TLC tank is not flat a broader glass slide can be placed at the base to provide flat 'flat form' for placing TLC plate.

Solvent A = Toluene–dioxane–acetic acid (180:45:5) is reputed to owe its distinctive characteristics to the ability of dioxane to associate with phenolic hydroxy groups.

Solvent B = Hexane–methyl *tert.*-butyl ether–formic acid (140:72:18) gives good separation of compounds that differ only slightly due to the length of side chains or the number of *C*-methyl substituents.

Solvent C = Toluene–acetic acid (170:30) is an excellent general solvent for a wide variety of different compounds.

Solvent E = Cyclohexane–ethyl acetate (75:25) is recommended for less acidic compounds, that have high R_f values in solvents A, B, and C (e.g. many pigments, esters, triterpenes: (Elix *et al.* 1988).

Solvent G = Toluene–ethyl acetate–formic acid (139:83:8) is particularly useful in separating compounds with relatively low R_f values in solvents A, B and C (e.g. β-orcinol depsidones, secalonic acids).

Box 3. Different solvent systems and their composition

v. Running

The spotted TLC plate is placed inside the TLC tank. The solvent rises up on the TLC plate passing through the loaded spots. The heavier lichen substance within the spot settles down near to the base of the plate while the lighter substances are carried away as the solvent rises upwards; hence the lichens substances get separated. The solvent is allowed to touch the finishing line drawn at 15 cm and then removed out of the TLC tank. The process takes about 40 – 50 minutes.

vi. Colouring the spots and charring

The lichen substances separated on the TLC plate are usually invisible or paler in colour. They are made more visible by spraying colouring reagent and heating. 10% sulphuric acid solution is sprayed over the TLC plate. The spray particles of solution should be very fine and it should just wet the plate, overflow of the solution should be avoided. Good quality glass spraying gun should be used for this purpose. After spraying the TLC plate is kept in the hot air oven for few minutes. The oven should be pre-heated 110° C before keeping the plate. The TLC plate can be taken out of the oven once the spots are developed properly after 3 – 5 minutes.

vii. Identification of the spots

The spots appeared on the TLC plate are identified as per their colour and the distance they travelled from the loading point. The distance travelled by a lichen substance (spot) is either referred as Rf value or Rf class and are calculated as follows,

Rf value = (distance travelled by substance (spot) ÷ distance travelled by solvent) × 100

Rf class = Divide TLC plates into approx. 7 equal parts from the loading line to the last spot (atranorin), each division is a Rf class. If *Parmelinella wallichiana* is used as control, it gives 2 spots, salazinic acid and atranorin. When the TLC plate is divided into 7 parts from loading line to atranorin, the salazinic acid spot appears at Rf class 2. Hence, Rf classes of salazinic acid and atranorin can be used for referring other spots. The lichen sample containg norstictic (Rf class 4) acid and atranorin can also be used as control. TLC manuals are available for identification of the lichen substances (Fig. 11).

Observation of TLC plate under shortwave UV lamp is sometimes necessary before and after heating. The spots appear differently while some spots emit fluorescence and such spots are marked with pencil. These are added characters for identification of lichen substances.

viii. Identification of fatty acids

After removing the TLC plate from the tank it is sprayed with distilled water. The fatty acids appear as oily, grey spots as the water dries up. They are circled with pencil as dotted lines. The suphuric acid solution is sprayed only after this step and then the plate is heated. The fatty acid spot does not give any colour after heating. The fatty acids are also identified based on their Rf class or Rf value.

4.2.3.4. High performance liquid chromatography

This technique provides a powerful complement to the established TLC methods. The bonded reverse phase columns are used here and all the aromatic lichen products are suitable for analysis with this method. Samples are dissolved in methanol and injected in to the appropriate portion column, through which an appropriate solvent or sequence of solvents is passed under high pressure. The substances separated and detected using UV detector. The retention time (Rt or time of passage) and peak intensity are recorded by a chart recorder. HPLC is also used to measure either absolute or relative concentrations of lichen compounds, because the peak intensity (area under curve) is proportional to the concentration. Most workers use HPLC to detect lichen compounds combine this technique with TLC and/or mass spectrometry to verify the identification of the peaks.

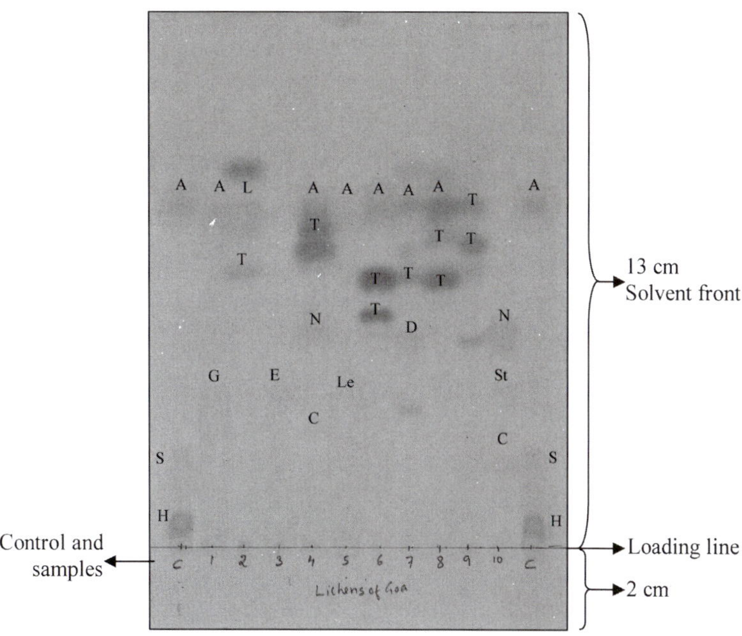

Fig. 11 A. Thin Layer Chromatogram of some lichens. C. Control (*Parmelinella wallichiana*), 1. *Parmotrema sancti-angelii*, 2. *Pyxine cocoes*, 3. *Roccella montagnei*, 4. *Pyxine cylindrica*, 5. *Parmotrema tinctorum*, 6. *Heterodermia diademata*, 7. *Dirinaria aegialita*, 8. *Lecanora cenisia*, 9. *Buellia disciformis*, and 10. *Graphis capillacea.* A – atranorin, S – salazinic acid, H – hypostictic acid, G – gyrophoric acid, L – Lichenoxanthone, T – triterpen, E – erythrine, N – Norstictic, C – constictic acid, Le – Lecanoric acid, D – divericastic acid, St – stictic acid.

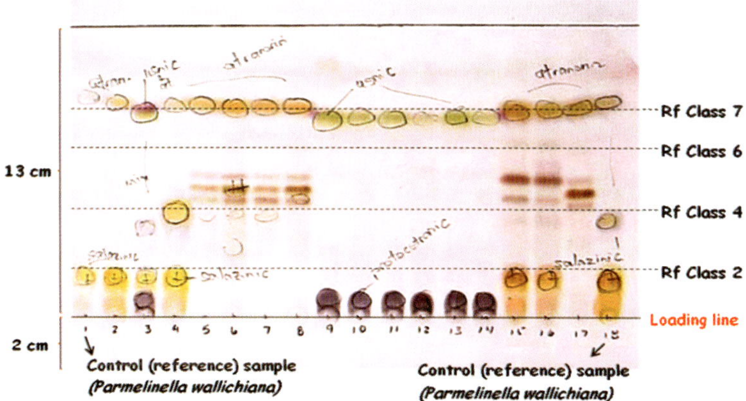

Fig. 11 B. Another example for developed TLC plate

There are several other methods used for identification of lichen compounds, but later discontinued because of their inefficiency. These techniques involve Paper chromatography, High Performance Thin-layer Chromatography (HPTLC), Gas Chromatography and Lichen Mass Spectormetry (GCLMS).

The identification of lichens involves combination techniques. Many macrolichens specimens can be identified with their external morphology and colour spot test, and rarely anatomy of thallus or fruiting bodies. Whereas in case of microlichens, identification mostly involves observation of anatomical details of fruiting bodies along with colour spot test and TLC. While observing the anatomical details of fruiting bodies sometimes the sections are also treated with reagents K, P, and rarely with C, and colour reaction is noted. As discussed earlier, characters to be observed and techniques to be applied for identification of lichens differ from group to group or genus to genus. As chemistry plays a very important role in identification of lichens, it also helps in segregating complex group of lichens. Hence, taxonomic significance of lichen chemistry is discussed here with a few exmples.

4.2.3.5. Taxonomic significance of lichen chemistry

The chemical constituents of lichen can be categorized in to two major classes: 1. Primary metabolites, and 2. Secondary metabolites. Primary metabolites are intracellular products, which are directly involved in the metabolic activities of the lichens such as growth, development and reproduction. They include proteins, amino acids, polyols, carotenoids, polysaccharides and vitamins, which are bound to the cell walls and protoplasts. They are often water soluble and can be extracted with boiling water. The primary metabolites are either of fungal or algal origin or both. They are also non-specific and present in free living alga, fungus, higher plants and other organisms. The secondary metabolites in lichens are of fungal origin. They are produced by utilizing the primary metabolites through three major pathways: 1. Acetyl-polymalonyl pathway, 2. Mevalonic acid pathway, and 3. Shikimic acid pathway. These secondary metabolites are not involved in the direct metabolism of lichen. They are the byproduct of primary metabolism and biosynthetic pathways. They act as storage substances, few have an important ecological role, but role of several secondary metabolites is poorly known. The secondary metabolites are deposited on the surface of the hyphae rather than within the cells, hence they are called as extra-cellular compounds. Popularly they are known as lichen acids or lichen substances. Secondary metabolites are insoluble in water, but can be extracted using organic solvents.

5. Biochemical Systematics

The first chemical test conducted on lichen thalli for taxonomic purpose was carried out by Nylander in 1860 (Hale 1983). He detected the presence of various colourless lichen substances by spotting chemical reagents on lichen thallus to produce characteristics colour change (spot test). In recent times spots test is much standardized and one of the important steps in identification of lichens. Nylander separated C+ red *Cetrelia olivetorum* from the identical *C. cetrarioides* (C-); however he was condemned at the time by most of the lichenologists (Fig. 12A,B). Today species discrimination on the basis of chemistry has become inseparable procedure in lichen taxonomy. Lichenologists have employed "chemical characteristics" in the taxonomy of lichen-forming fungi for nearly 120 years! They have employed a wide variety of terms/ classifications method to these chemical population variations including; SPECIES, SUBSPECIES, VARIETIES, FORMS, "Chemovars", "Chemical Strain No.__", "Chemotypes" and "Chemical races". However, before jumping into any conclusion Lichenologists must be satisfied with answers to following questions; Do differences in chemical composition mean two individuals (which may look the same) belong to different species? Are there ecological, morphological, geographical correlations with the observed chemical differences? What taxonomic weight (if any) should be applied to these differences?

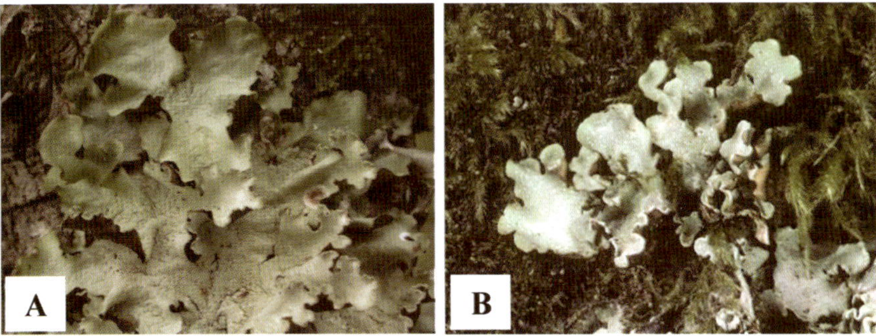

Fig. 12. (A) *Cetrelia olivetorum* (C+ red) and (B) *C. cetrarioides* (C⁻), separated by W. Nylander in 1860 based on the colour test.

The secondary metabolites are stored either in the cortex or in medulla of lichen. The cortical substances usually have an ecological role and most of them act as light screens (e.g., usnic acid, atranorin, chloroatranorin, anthraquinones, pulvinic acid derivatives, xanthones). They may absorb sunlight to warm up quickly in colder regions so as to initiate active metabolism of the thallus. They act as a filter and regulate the amount of sunlight reaching to sensitive algal layer. Given their apparent physiological importance it seems likely that their formation would have evolutionary significance. Some of the cortical substances are correlated with higher taxonomic ranks, example – at generic level presence of vulpinic acid is the characteristic feature of genus *Lethraria* (Fig. 13). Morphologically similar looking genera *Physcia* and *Phaeophyscia* can be identified on the basis of presence or absence of atranorin, which is present in former species and absent in later species. Similarly, *Flavopunctelia* (usnic acid present) and

Punctelia (usnic acid absent) (Fig. 14 A,B) are separated. Superficially, the usnic acid containing lichen would have yellowish tinge and atranorin yield yellow colour for K spot test. At family level presence of anthraquinones and particularly parietin is characteristics of Teloschistaceae.

Fig. 13 *Lethraria columbiana*

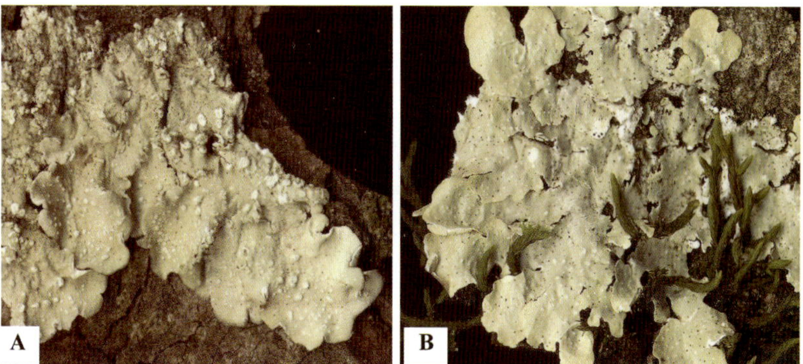

Fig. 14. Example for role of cortical chemistry in biochemical systematics. (A) *Flavopunctelia* (usnic acid present), (B) *Punctelia* (usnic acid absent).

The medullary chemistry of lichen is usually complicated. The secondary metabolites present in the medulla of lichen are used primarily as discriminators at the species level but also occasionally at generic or suborder level. For example, *Cetralia* (with orcinol derivatives) and *Platismatia* (with fatty acids or beta-orcinol derivatives) are separated on the basis of their chemistry). The discovery of chemical differences often led to an appreciation of the importance of previously overlooked morphological features as in *Punctelia subrudecta* with lecanoric acid and a pale tan lower surface, and *P. borreri* with the related tridepside gyrophoric acid and a black lower surface (Fig.15 A,B). Fortunately, most morphologically defined species have a constant chemistry, irrespective of their geographical origin, substrate or ecology, and this justifies the use of chemistry in lichen taxonomy. Within a complex of morphologically similar species, three common patterns of chemical variation are observed; replacement of compounds, chemosyndromic variation, and accessory type compounds.

Fig. 15. Example for role of medullary chemistry in biochemical systematics. (A) *Punctelia subrudecta* (Lecanoric acid present, (B) *Punctelia borreri* (Gyrophoric acid present).

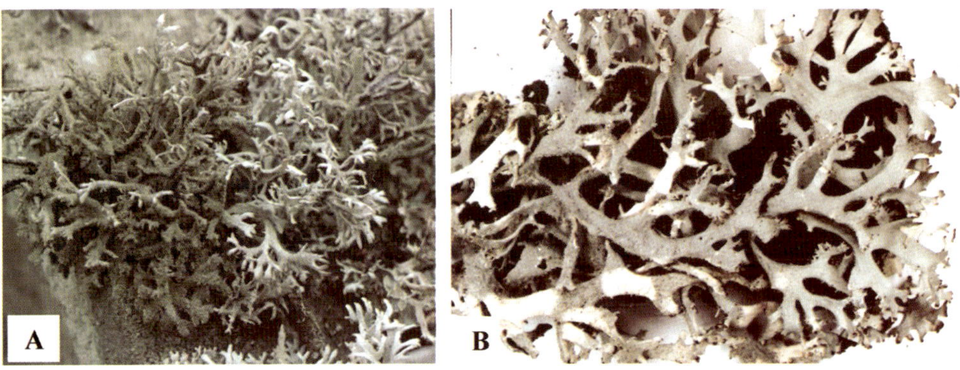

Fig. 16 (A) *Pseudoevernia furfuracea,* (B) *P. conssocians*

In case of replacement, simple replacement of one compound with another takes place and the replaced compound would be from same family of chemicals or from same biosynthetic sequence (biosequential). Morphologically these lichen populations are sometimes indistinguishable, but they have well defined, constant variations in chemical composition. This can be explained with the help *Pseudoevernia furfuracea* (Fig. 16A), which has three chemical races; an olivetoric acid race from northern Europe; a physodic acid containing race from southern Europe and north Africa; and a lecanoric acid containing race from North America. Biogenetically the first two races appear closely related, the metabolites can be considered biosequential, because one can be derived from the other by a single biosynthetic step. But the third race is not related as the lecanoric acid is biosynthetically remote from the other two compounds. It is now generally accepted that, when there is a biogenetic demarcation allied with a biogeographical separation, such taxa should be recognized as species and the North American taxon is distinguished as *P. conssocians* (Fig. 16 B). Further studies on first two races of *P. furfuracea* revealed that though they possess distinctive, but overlapping chemistries do not show significant habitat ecology or morphological differences, hence considered as single species. It is suggested that chemical variation among lichen species should be genetic rather than being environmentally determined. The occurrence of intermediates in an area indicates that such races are either in the process of speciation or hybridization is occurring between the races.

The chemosyndrome refers to a group of biosynthetically related metabolites and in this pattern of chemical variation the major metabolite (or metabolites) in any one taxon is invariably accompanied by minor quantities of several bio-sequentially related substances. Further, the major constituents of one species become minor in related taxa and vice versa. Hence a true chemical intermediate cannot simply be defined as containing both of two replacement compounds, but would have to contain both chemical constellations in comparable concentration. Several examples are available among lichens for the occurrence of chemosyndrome. In *Cladonia chlorophea* group about 14 chemotypes are available, while in *Usnea longissima* (Fig. 17A) seven chemical strains are identified. Sometimes it is used for discrimination of species, which is logically wrong. Example, chemosyndromic variation leading to misidentification of species can be seen in case of *Relicina samoensis* complex (Box 4).

Species (distribution)	Echinocarpic	Conechinocarpic	Hirtifructic	Gyrophoric	Fatty acids	Distribution
R. samoensis	major	minor	—	—	—	Pan-Pacific
R. amphithrix	major	minor	—	—	—	Australia/Indonesia
R. terricrocodila	major	minor	trace	—	—	Australia
R. fijiensis	—	—	major	—	—	Fiji
R. niuginiensis	—	—	major	trace	minor	Papua New Guinea
R. relicinula	—	—	—	—	major	Indonesia

Box 4. Chemosyndromic variation in *Reclicina samoensis* complex

Fig. 17. (A) *Usnea longissima,* (B) *Ramalina siliquosa*

Ecology plays an important role in differentiation of chemical races. Different chemical races were found to be ecologically sorted into distinct habitats in their range of sympatry. For example, different species within the *Ramalina siliquosa* (Fig. 17B) complex are recognized due to the difference in chemistry and habitat preferences. *R. siliquosa* complex occur on maritime rocks on the coast of Europe (microhabitat preferences); the one which occupies lowest zone is identified as *R. cuspidata* and it has stictic acid; *R. crassa* is most sheltered species that grows in land-facing habitats and contains hypoprotocetraric acid; while *R. stenoclada* occupies the region in between and contain norstictic acid. Here all the three species do not have distinct morphological difference and the lichen substances are biogenetically related. They could have been chemical races rather than species. Culberson (1986) of the opinion that the ecological and biological characteristics of the major chemotype should be better considered as sibling species rather than as components of traditional morphological species. The population of sibling species would have reproductive isolation along with ecological but may not have morphological differentiation.

Accessory metabolites occur sporadically in a species, in addition to constant constituents but usually have no correlation with any morphological or distributional variations. Hence have less or no taxonomic significance. Such compounds commonly occur as accessory compounds in more than one species and often vary in quantity from deficiency to abundance. On the contrary in North Carolina the edaphic preferences of *Cladonia* along with difference in chemistry separates them into two distinct species. *C. polycarpoides* prefers "clay-type" soils and contains norstictic acid, while *C. polycarpia* prefers "sandy" soils and contain an additional substance, atranorin along with norstictic acid (Fig. 18 A,B). Here the chemical characteristics are correlated with ecological, but not morphological. Hence, it could have been a variety rather than a different species.

Fig. 18 (A) *Cladonia polycarpia* (atranorin and norstictic acid) grows on "sandy" soils; (B) *C. polycarpoides* (norstictic acid) occur on "clay-type" soils

Cell wall polysaccharides

These are primary metabolites and are polymorphic storage products of lichens require different techniques for their detection. Some of the well known polysaccharides in lichen are lichenan, isolichenan, and galactomannan. Some polysaccharides are taxonomically significant at the highest levels of classification. For example presence chitin, chitosan, or cellulose in the cell wall is a feature that helps define the classes of fungi. This indicates the conservative features of the taxa in evolution and helps tracing the phylogeny. In lichen taxonomy utility of cell wall polysaccharides can be seen in family Parmeliaceae where four types of lichenan can be identified in four distinct groups; isolichenan, *Xanthoparmelia*-type lichenan, *Cetraria*-type lichenan, and an intermediate-type lichenan. Chemically the polysaccharides differ in stereochemistry of the glycosidic bonds. These lichenan differ mostly in their staining properties with iodine and hence can be easily identified (Box 5). The morphologically similar lichen genus *Hypogymnia* (Fig. 19A) contains *Cetraria*-type lichenan, while *Menegazzia* (Fig. 19B) contains isolichenan. The polysaccharides are also used as one of the primary discriminators to differentiate yellow parmelioid genera, *Psiloparmelia* and *Flavoparmelia* (containing isolichenan) from *Arctoparmelia* (*Cetraria*-type lichenan) and *Xanthoparmelia* (*Xanthoparmelia*-type lichenan).

Fig. 19. (A) *Hypogymnia* with *Cetraria*-type lichenan, (B) *Menegazzia* with Isolichenan

To resolve taxonomic dilemma due to varying chemical, morphological, ecological and geographical characteristics four patterns can be illustrated.

Pattern 1. Replacement of one substance by one or more biogenetically distinct substances.

A. Correlated with morphological and (or strong) ecological differences = SPECIES

B. Correlated with major geographical differences = SPECIES

C. Not correlated with morphological or geographical differences = NONE

Polysaccharide	20–0.15% IKI	0.15% LPIKI	1.5% IKI	CaIKI	ZnIKI	SIKI	Meltzers
Isolichenan	blue	pale blue	bluish	bluish	bluish	—	bluish
Cetraria-type lichenan	—	—	red	deep red	—	red ppt.	orange
Xanthoparmelia-type lichenan	intense blue	—	red	deep red	purple	red ppt.	deep red
Intermediate-type lichenan	pale blue	—	red	deep red	—	red ppt.	red

IKI, iodine, potassium iodide solution; LP, lactophenol; S, 10% sulfuric acid; Ca, calcium chloride; Zn, zinc chloride; Meltzers Reagent, chloral hydrate + iodine potassium iodide.

Summarized from Common (1991).

Box 5. Four types of lichenan present in Parmeliaceae and their reaction with reagents.

Pattern 2. Replacement of one substance by one or more biogenetically closely related substances

A. Correlated with morphological and distributional differences = SPECIES

B. Correlated with local geographical differences or tendencies = VARIETY

C. Correlated with ecological or microhabitat differences = VARIETY

D. Not correlated with morphological, ecological or geographical differences = NONE

- Can be called as chemotype, chemical strain or race

Pattern 3. Presence of one or more unreplaced (additional) substances

A. Correlated with major geographical differences = SPECIES

B. Correlated with differences in ecological amplitude = VARIETY

C. Correlated with local distributional differences or tendencies = VARIETY

D. Not correlated with morphological, ecological or geographical differences = NONE

Pattern 4. Variations in concentration of particular substances = NONE

A. Correlated with light intensity

B. Correlated with heavy metal contents of substrate

C. Correlated with any other ecological factor

In biochemical systematics or chemotaxonomy of lichens considerable controversy still remains. American, Australian and Japanese workers generally recognize chemical variation at the species level while European lichenologists prefer to "lump" but are more likely to recognize subtle morphological traits. However, most morphologically defined species have a constant chemistry, irrespective of their geographic origin, substrate or ecology. This clearly justifies the use of chemistry in lichen taxonomy. So far only about 6000 lichens are screened for metabolites, approx. 30% of known species are studied, remaining 70% are yet to be explored. The structures of several metabolites are unknown, while biosynthetic pathway has to be revisited with molecular approaches. Apart from taxonomy the commercially important metabolites should be identified and the genes responsible for their production should be recognized for mass production and commercialization and hence "Lichen products" should be popularized.

References

Asahina Y, Shibata S (1954) Chemistry of Lichen Substances. Japan Society for the Promotion of Science, Tokyo

Awasthi DD (1991) A Key to the Microlichens of India, Nepal and Sri Lanka. Bibliotheca Lichnenologica 40. J Cramer, Berlin-Stuttgart

Awasthi DD (2007) A Compendium of the Macrolichens from India, Nepal and Sri Lanka. Bishen Singh Mahendra Pal Singh, Dehra Dun

Culberson CF (1969) Chemical and Botanical Guide to Lichen Products. The University of North Carolina Press, Chapel Hill

Culberson CF (1986) Biogenetic relationships of the lichen substances in the framework of systematics. Bryologist 89:91-98

Elix JA and Ernst-Russell KD (1993) A Catalogue of Standardized Thin Layer Chromatographic Data and Biosynthetic Relationships for Lichen Substances, 2nd Edition. Australian National University, Canbera

Hale ME Jr. (1983) The Biology of Lichens (3rd Edition.). Edward Arnold Ltd., London

Purvis OW, Coppins BJ, Hawksworth DL, James PW and Moore DM (Eds.) (1992). The Lichen Flora of Great Britain and Ireland. Natural History Museum, London

Schwendener S (1867) Ueber die wahre Natur der Flechten. Verh Schweiz Naturf Ges 1867:88-90

Chapter – 6

Diversity and Systematics of Algae

M.R. Suseela, Kiran Toppo and M.A. Usmani

1. Introduction

India, as a party to the Convention on Biological Diversity (CBD), has obliged to document a whole range of organism diversity within her territorial boundaries, make all attempts to conserve these bio-resources and monitor the efficacy of the conservation measures adopted. With its diverse ecological conditions and unique geographical location, Indian landmass supports rich diversity of algal flora. However, this algal wealth needs to be systematically surveyed and documented from all possible habitats. Ecological patterns emerged from recent studies within India are assessed with the identification of knowledge gaps and conservation implications in algal systematics.

During last one decade, great advances have been made in the field of Algology/ Phycology throughout the world. Over the years algae has attained a greater reputation and popularity because of its economic importance, particularly as food, for commercial products, water contamination, toxin production, sewage oxidation, municipal water purification in fisheries and above all for the study of fundamental scientific information. But before the algae put in to such important use, it is imperative to have a baseline data about the material in question. Without knowing the status of algal resources of any area one cannot have projections for their utilization or gain knowledge of their ecological role and functions. The taxonomy of algae is, therefore, a prerequisite for all applied fields of algae apart from academic interest. Recent global surveys have revealed that there is lack of knowledge about fresh water algal diversity of tropical countries including India.

There are about 6,500 species of algae belonging to 780 genera are known to occur in India. Local and regional Floras are available for some groups of freshwater algae but information on algal flora is scanty or has not been compiled for most part of India. Floras usually provide ranges but distribution of many species may be discontinuous due to various reasons. Filling the gapsin our knowledge on algal flora of India will require consolidated efforts.

What are Algae?

Algae are simple photosynthetic organisms not included among the mosses, liverworts (and other bryophytes) or the vascular plants because of their unique structural variation. This fascinating group of organisms forms the basis for the science of Phycology/ Algology - the study of algae.

The eukaryotic algae are those with compartmentalized cells that are part of a diverse group of organisms called protists. The kingdom Protista (once included partly in the animal kingdom and partly in the plant kingdom) is now divided into a number of as yet in adequately circumscribed kingdoms. The prokaryotic algae (the cyanobacteria) are included with bacteria in the kingdom Monera. Algae are ubiquitous in nature and grow in almost every habitat in every part of the world. The following are examples of 'freshwater' habitats.

2. Habitat

All major bodies of water have algae in abundance, including lakes, small streams, large rivers, and even waterfalls. Algae occur in fresh water, to the saline water of the sea, and even in saltpans. There are also algae that thrive in the hot springs. In the sea they may occur below the range of tidal exposure in the sub tidal zone, as well as in the harsh intertidal environment of the seashore, where they may be beaten by waves. In some parts of the world, intertidal algae are even scoured by sea ice, yet they persist in living in this environment. Those algae, which live attached to the bottom of a water body, are called benthic algae, and the ecosystems of which they are part are referred to as benthos. Small, microscopic algae that drift about in bodies of water, such as lakes and oceans, are called phytoplankton. Phytoplankton algae are the most important in freshwater and marine food webs, and are probably responsible for producing much of the oxygen that we breathe. Some forms of algae are even able to grow in Arctic and Antarctic sea ice, where they can be quite productive and support a whole associated food web. Some algae can even grow on the seabed, beneath a thick blanket of Arctic or Antarctic sea ice, even though they are in total darkness for a considerable part of the year.

In some parts of the world, blooms of snow algae may paint the snow beds red in spring. Algae even occur in the driest deserts. Algae are also found in the air, for there are many algae that colonize new bodies of water by simply drifting about through the air. There is even a unicellular green alga called *Prototheca*, which causes disease in humans. It produces skin lesions, mainly in patients whose immune systems have been damaged by other serious diseases. There are the algae that enter into symbiosis with other organisms, for example, the symbiotic organism lichens, formed from the association of an alga and a fungus. The stony corals, which construct coral reefs in warm tropical seas, are only able to build up these massive and beautiful structures because they have formed a symbiotic partnership with tiny single-celled algae called *Zooxanthellae*. The *Zooxanthellae*, which live in the tissues of the coral, share with it the organic products of their photosynthesis, as well as helping the coral with the construction of its limestone skeleton. Even the chloroplast of land plants had its origin as a blue-green alga that lived within the cells of the ancestral organism. Such a special symbiotic relationship, where one organism lives inside the cells of another, is called endosymbiosis. Some of the substrate habitats of algae are given below.

2.1. Streams and Rivers

Acidic streams and river support a diverse macroalgal flora and are highly susceptible to eutrophication. Alkaline streams and river are to be species poor, but due to restricted occurrence they may support rare taxa. Eutrophic streams and river are species poor, but sometimes native taxa remain and, due to shortage of oligotrophic streams in some areas, they may represent the last fragments of a previously wide distribution (Fig. 1).

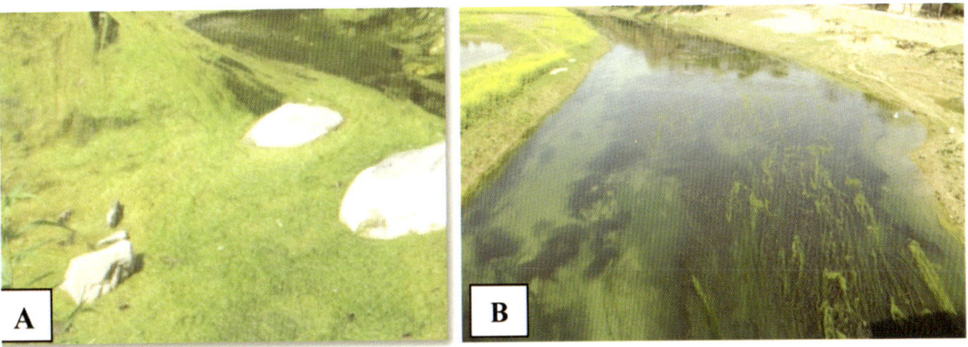

Fig. 1. (A) Algae in a flowing stream, (B) Algae in a flowing river

2.2. Saline and Hypersaline lagoons

These coastal habitats have an interesting microalgal flora, but are very sensitive to changes in the water table.

2.3. Saline lakes and marshes

These habitats are genrally species poor, but support a distinctive algal flora.

2.4. Salt marshes and salt lakes

This is not a strictly 'fresh' water habitat, but a distinctive non-marine one. *Enteromorpha* and *Rhizoclonium* also grow in brackish to saline waters.

2.5. Hot springs

Little is known about hot springs algae, but as elsewhere the blue-green algae (Cyanobacteria) dominate except where sulfide concentrations are high. (Fig. 2)

Fig. 2 Hot springs (Tatwani, H.P.)

2.6. Snow

'Red snow' is usually coloured by *Chlamydomonas*, but other algae occur in this habitat as well.

2.7. Aquatic plants

Algae grow on and inside water plants (including other algae), but are generally not restricted to a single host. *Anabaena* grows inside the aquatic fern *Azolla*, *Nostoc* within members of the Anthocerotales. (Fig. 3)

Fig. 3. Algae on aquatic plants

2.8. Billabongs and lagoons

These are microalgal rich habitats, particularly for desmids. Their ecology is poorly understood and changes in flooding re*gime may* alter the algal flora.

2.9. Bogs, marshes and swamps

These are known for their rich desmid population in aquatic plants root system.

2.10. Farm dams

These artificial water-bodies may allow many algal species to extend their natural range (Fig. 4).

Fig. 4. Algae in Reservoir

2.11 Mud and sand

The surface of submerged soil in shallow waters supports many species of algae.

2.12. Rock

Rock inhabiting algae are called lithophytes. Some algae live within rocks and on rock surface (Fig. 5A,B,C).

Fig. 5. (A, B and C) Algae on rocks

2.13. Animals

Reported substrates include turtles, snails, rotifers, worms, crustacea and many other animals (Fig. 6A).

2.14. Soil

Cryptogam mats in dry soil surface and many cyanobacteria, green algae grow luxuriantly on the moist soils (Fig. 6B).

Fig. 6. (A) Algae on Snail, (B) Algae on soil

2.15. Symbiotic in higher plants

Anabaena grows in nodules formed on the surface roots of cycads.

2.16. Terrestrial plants

Several algae are terrestrial also called soil algae. *Trentepohlia* and related species, which grow on tree trunks and branches have a habit very similar to lichenised fungi. *Apatococcus* (earlier *Protococcus*) is universally common as a green powdery covering on the shady side of trees. A variety of blue-green algae readily colonize damp walls and other moist surfaces. Algae also live on the surface of, or penetrate into, leaves; it is possible that such species are extremely diverse in rainforests (Fig. 7A).

Fig. 7. (A) *Trentepohlia* on tree trunks, (B) Algae on artificial substrate surface

Table 1. Some of the important classes and characters of algae

S. No.	Class	Order	Typical colour	Typical morphology	Salient features	Examples
1.	Chlorophyceae (Green Algae)	1. Volvocales 2. Chlorococcales 3. Ulotrichales 4. Cladophorales 5. Chaetophorales 6. Oedogoniales 7. Conjugales 8. Siphonales 9. Charales	Grass-green	Microscopic or visible unicellular, colonial filamentous and branching filaments.	Main pigments are Chlorophyll a, b, Carotenoids and Xanthophylls. The reserve food material is starch.	*Chlamydomonas*, *vovox*, *Spirogyra*, *Pithophora*
2.	Bacillariophyceae (Diatom, Golden brown algae)	1. Centrales 2. Pennales	Yellow or Golden brown	Microscopic unicellular or filamentous colonies	Main pigments are Fucoxanthin, Diatoxanthin and diadinoxanthin. The reserve food materials are fat and volutin	*Epithemia*, *Fragillaria*
3.	Rhodophyceae (Red algae)	1. Bangiales 2. Nemalionales 3. Gelidiales	Red	Microscopic or visible unicellular or colonial	The main pigments are two type of phycobilins (phycoerythrin and phycocyanin), Chlorophyll a,b. The reserve food material is solid polysaccharide and floridean starch.	*Batrachospermum*, *Bangia*

(Contd.)

S. No.	Class	Order	Typical colour	Typical morphology	Salient features	Examples
4.	Euglenophyceae (Eugnenoid)	1. Euglenales 2. Euglenamorphales 3. Eutreptiales 4. Heteronematales 5. Rhabdomonadales 6. Sphenomonadales	Various colour	Microscopic unicellular	The main pigment is Chlorophyll, each cell has many chromatophores which are pure green. The reserve food material is polysaccharide paramylon.	*Euglena Phacus*
5.	Cyanophyceae (Blue-green algae)	1. Chroococcales 2. Chamaesiphonales 3. Pleurocapsales 4. Nostocales 5. Stigonematales	Blue-green	Microscopic or visible unicellular or colonial or filamentous and false branching filament.	The main pigments are c-phycocyanin and c-phycoerythrin. The reserve food material is cyan ophycean starch.	*Synechocystis, Microcystis, Oscillatoria*

2.17. Artificial substrates

Wooden posts and fences, cans and bottles, etc., all provide algal habitats (Fig. 7B).

3. Algal Classification

Regarding the algal classification, there are two schools of thoughts. According to one school (Pascher 1914; Smith 1933, 1955; Papenfuss 1946; Prescott 1969) the algae should be divided first in to several "phyta"(i.e. divisions, such as Chlorophyta and Phaeophyta) and then in each "phyta" there should be different "phyceae" (i.e. classes, such as Chlorophyceae and Phaeophyceae). Phycologists supporting the second view (Fritsch 1935, 1945 and his followers) have been of the opinion that algae are equivalent to a division, and therefore it can only be further divided into Classes ('phyceae') and not into many divisions ('phyta'). So there is no word like 'phyta' in classification proposed by Fritsch and his supporters of the second view. Some other important suggestions of his classification are: (i) Class Conjugatae of Pascher's classification should be treated only as an order (Conjgales) of class Chlorophyceae. (ii) Division Charophyta of Pascher's (1914) class Charophyceae as Smith's (1933, 1955) classification should be treated only as an order (Charels) in class Chlorophyceae. They show similarity in pigmentation and product of assimilation. (iii) Fritsch does not recognize Desmokontae (Sharma1986).

Fritsch's classification (1935, 1945) published in his book entitled "The Structure and Reproduction of the Algae is now widely accepted. According to Fritsch (1935) algae broadly classified into eleven classes: Chlorophyceae, Xanthophyceae, Chrysophyceae, Bacillariophyceae, Cryptophyceae, Dinophyceae, Chloromonadineae, Euglinophyceae, Phaeophyceae, Rhodophyceae, Cyanophyceae and Myxophyceae.

Based on color, food storage substances, and the composition of cell walls etc. members of each class have distinctive colors, depending on the photosynthetic pigments in their cells. These pigments absorb light. All algae contain the pigment chlorophyll *a*. However, different divisions of algae also contain other forms of chlorophyll, such as b, c or d, each of which absorbs a different wavelength of light. Algae also vary in methods of reproduction. Recent classification given by Lee (2008) as mentioned in the flowchart (Fig. 8).

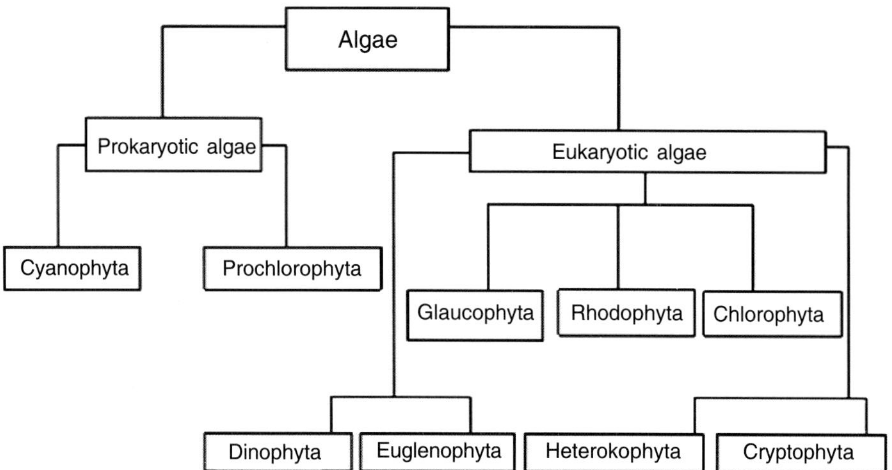

Fig. 8. Classification of algae (Lee 2008)

3.1. Class: Chlorophyceae (Green algae)

Main pigments are Chlorophyll *a*, Chlorophyll *b*, Carotenoids and Xanthophylls; Reserve food material is starch; Chloroplasts have pyrenoids; Starch grains are usually aggregated around the pyrenoids, Flagella, if present, are of equal length, whiplash type and inserted at the anterior end.

Class Chlorophyceae has been divided into following nine orders:

1. Volvocales 2. Chlorococcales 3. Ulotrichales 4. Cladophorales 5. Chaetophorales 6. Oedogoniales 7. Conjugales 8. Siphonales 9. Charales

Members of this class, also called green algae, are a diverse group of organisms of over 7,000 species. They can be unicellular, colonial, filamentous, or thalloid. Most green algae are aquatic, although many inhabit moist terrestrial environments such as soil, rock surfaces and tree trunks.

The following three observations led to conclude that green algae are the ancestors of plants. First, both green algae and plants have chloroplasts that contain chlorophylls a and b. Second, both green algae and plants store food as starch. Finally, both green algae and plants have cell walls made of cellulose (Prescott, 1951). The diversity of unicellular algae is extreme. *Chlamydomonas*, unicellular algae, is common in soils and in freshwater ponds and streams. It has a single-cup shaped chloroplast. Each chloroplast contains a pyrenoid, where starch is made. Two interior flagella enable the organism to swim. An eyespot, which is an area sensitive to light, enables the algae to move either toward or away from light. More than 500 genera and 8000 species of green algae have been identified. Some familiar green algae include the genus *Spirogyra*, known for its spiral-shaped chloroplasts (Fig. 9) and the desmids, recognized by their characteristic shape—two symmetrical halves, joined by a small bridge.

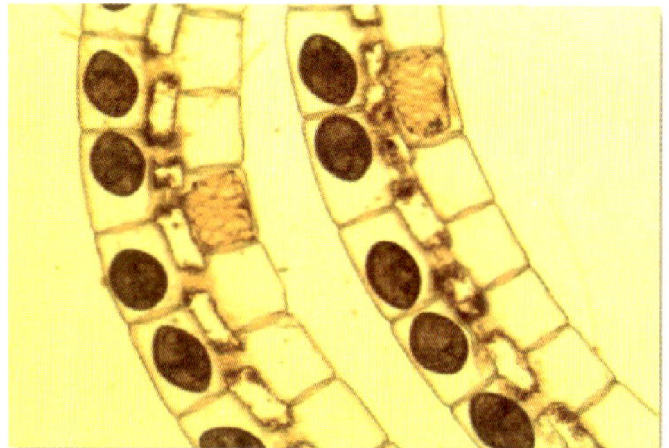

Fig. 9. Conjugation in *Spirogyra*

Unlike unicellular algae, colonial algae have some characteristics of multicelluar organisms. *Gonium* is perhaps the simplest colonial algae, consisting of a colony that is one cell thick and shaped like a rectangle. A *Volvox* colony, on the other hand, is round and much larger, containing up to 60,000 cells and exhibiting a remarkable division of labor. Intercellular communication allows the coordination of the many cells in the colony. Volvox cells are connected by fine cytoplasmic strands that enable adjacent cells to chemically communicate with each other.

Spirogyra is a filamentous green algae with unusual spiral chloroplasts that stretch from one end of the cell to the other. *Oedogonium* is another common freshwater filamentous green algae. members of this genus have net like chloroplasts. The green algae known as *Stoneworts* often grow several feet in length. Their name comes from calcium crusts that make them feel like stone. Most green algae reproduce both sexually and asexually. Alternation of generations, where the algae alternates between gametophyte and sporophyte generations, is common among the multicellular green algae.

Desmids

Desmids are unusual unicellular algae that essentially free floating and frequently occur in great abundance in small ponds, lake and in other localities where the water is not alkaline. In fact, the presence of desmids often indicates degree of water pollution. Each cell is divided into half and these halves are called semi cells. Desmids like *Micrasterias*, *Cosmarium*, *Staurastrum* etc. are commonly known as 'Pond Star' (Fig. 10 ABC) These are the some desmids:

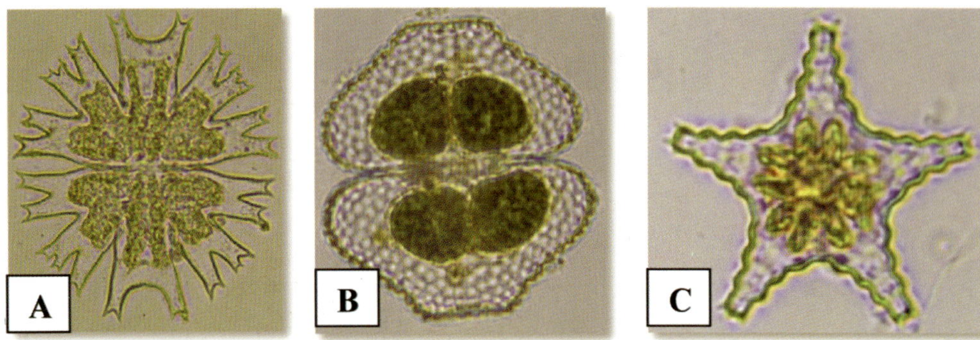

Fig. 10. (A) *Micrasterias*, (B) *Cosmarium*, (C) *Staurastrum*

3.2. Class : Bacillariophyceae (Diatoms, yellow or golden-brown algae)

The main pigments are fucoxanthin, diatoxanthin and diadinoxanthin. The reserve food materials are fat and volutin, Pyrenoids are present.

Class Bacillariophyceae has been divided into following two orders:

1. Centrales
2. Pennales

There are over 10,000 species of golden-brown algae, the majority of which are commonly called diatoms. These organisms contain chlorophylls a and c and the accessory pigment fucoxanthin. Diatoms store food in the form of oil, not starch. Diatoms form the large and complex organisms ranging in size from a fraction of a millimeter to more than 100 m (300 ft) long. They have carotenoid secondary pigments that tend to mask the green of the primary chlorophyll pigment, giving them a golden or golden-brown appearance.

Diatoms are unicellular or colonial, non flagellated, photosynthetic algae with silica-impregnated shells like *Cymbella, Rhopaldia,Gomphonema,Caloneis* and *Epithemia* (Fig. 11A-F) They inhibit both freshwater and marine environments and are an essential component of phytoplankton. Diatoms are so abundant in some marine environments that they are responsible for the bulk of worldwide photosynthesis. These algae are mostly unicellular or colonial, swimming or floating in lakes and oceans as phytoplankton. In shallow ponds that dry up in summer or freeze completely in winter, golden-brown algae survive by forming protective cysts that can withstand the harsh conditions.

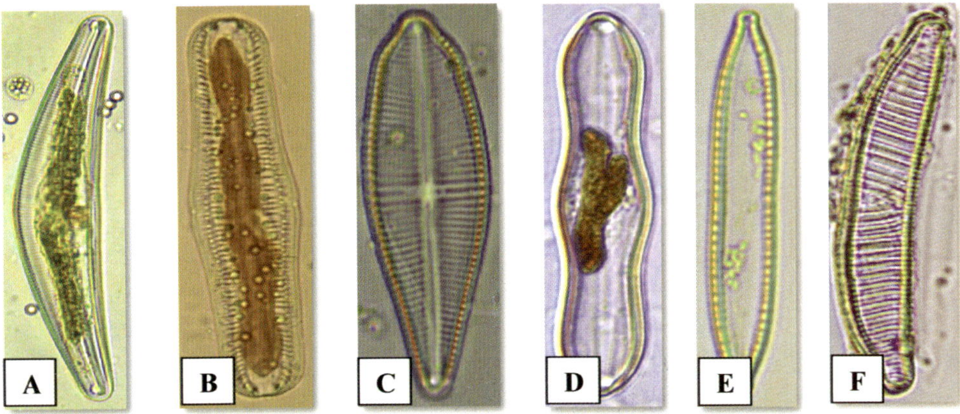

Fig. 11. (A) *Cymbella,* (B) *Rhopalodi,* (C) *Gamphonema,* (D), *Caloneis* (E), *Nitzschia* and (F) *Epithemia*

Diatoms

When favorable conditions return, the algae emerge from the cysts. Like so many other algae, the unicellular algae tend to reproduce through fission, while the multicellular and colonial forms reproduce either through fragmentation or through spore production. Diatoms are best known for their glasslike cell wall made of silica. The cell wall has ornate ridge patterns. A diatom consists of two overlapping halves that fit together like a shoebox or a petri dish, with the lid slightly larger and fitting over the base.

Diatoms have highly ornamented double walls containing silicon dioxide. The two halves of the wall fit together like the two parts of a box. Each half is called a valve. Centric diatoms have circular or triangular valves and are most abundant in marine waters. Pennate diatoms have rectangular valves and are most abundant in freshwater ponds and lakes. Some pennate diatoms move by secreting threads that attach to the surface of water.

Because their cell walls contain silicon dioxide, diatom shells do not decompose. Rather the rigid shells of dead diatoms sink and eventually form a layer of material called diatomaceous earth. Diatomaceous earth is slightly abrasive; it is an ingredient of many commercial products such as detergents, paint removers, fertilizers, and insulators.

3.3. Class: Rhodophyceae (Red algae)

The main pigments are two types of phycobilins (phycoerythrin and phycocyanin) besides chlorophyll a and d, the reserve food material is solid polysaccharide and floridean starch. Some lower forms have pyrenoid like bodies and motile stages are totally absent.

Class Rhodophyceae includes following seven orders:

1.Bangiales 2.Nemalionales 3.Gelidiales 4.Cryptonemiales 5.Gigartinales 6. Rhodymeniales 7. Ceramiales

Members of the Rhodophcae are called red algae. Red algae form with approximately 500 genera and 6000 species. Found in warm coastal waters and in water as deep as 260 m (850 ft), red algae species adapt to varied water depths by having different proportions of pigments. Their red color is due to a red pigment, phycoerythrin, which is well suited to absorb the blue light that penetrates deeper into water than the other colors of light. Red algae found in deep water may be almost black due to a high concentration of phycoerythrin. At moderate depths red algae appear red, while in shallow water they may appear green because a smaller proportion of phycoerythrin is unable to mask the green of chlorophyll. Red algae can survive at greater depths than any other algae can. Red algae contain chlorophyll's a and d as well as accessory pigments called phycobilins. Phycobilins absorb violet, blue and green light that penetrates the depths at which these algae grow.

Most red algae are multicellular and come in a variety of shapes, including filaments, which are shaped like a blade of grass, and seaweed shapes (Fig. 12A-D). Unlike most other eukaryotic algae, red algae have no flagella. Red algae use diverse strategies to reproduce, including fragmentation and spore production.

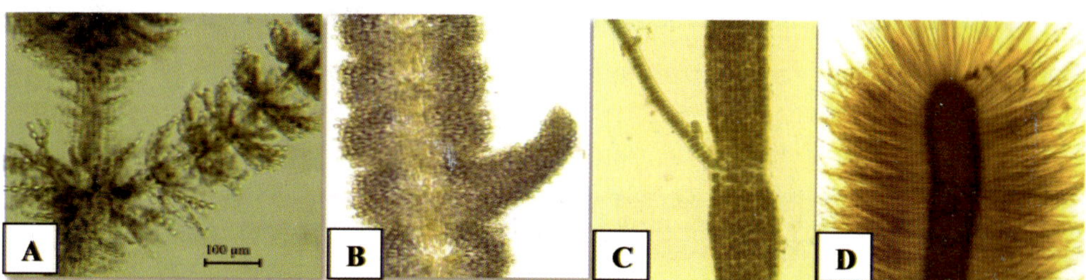

Fig. 12. (A,B) *Batrachospermum*, (C) *Compsopogan*, (D) *Thorea*

3.4. Class: Euglinophyceae (Euglenoid)

The main pigment is chlorophyll, each cell has many chromatophores which are pure green, the reserve food material is polysaccharide paramylon, pyrenoid like bodies are found in some forms, the plants are motile flagellates, flagella may be one or two arising from the base of canal-like invagination at the front end.

Class Euglenophyceae includes six orders:

1. Euglenales 2. Euglenamorphales 3. Eutreptiales 4. Heteronematales 5. Rhabdomonadales 6. Sphenomonadales

The approximately 1,000 species of the division Euglinineae are unicellular algae with many features in common with green algae. However, they also have many characteristics similar to those of protozoa. Euglenoids contain chlorophyll's a and b and store food as starch, but they do not have cell walls. Unlike other algae, euglenoids are not completely autotrophic. If a euglenoid is placed in the dark, it will become heterotrophic. For these reasons some scientists classify euglenoids as protozoa.

The most familiar genus of euglenoids is *Phacus* and *Euglena* (Fig. 13 A,B,C) members of which are abundant in freshwater ponds and lakes.

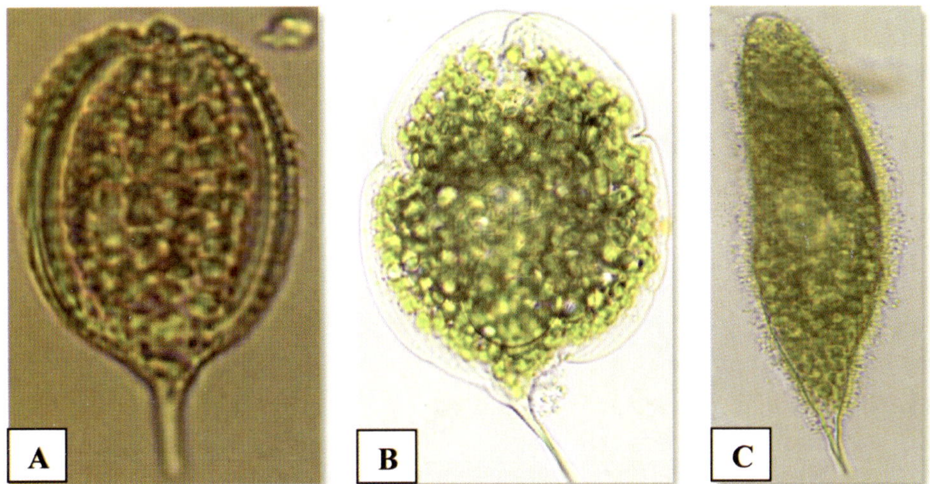

Fig. 13. (A,B) *Phacus,* (C) *Euglina*

Euglena gracillis is generally elongate and fairly flexible. Like all species of Euglena. it is able to change shape because of the presence of a pellicle, a flexible proteinaceous covering. *Euglena gracilis* contains a structure called a reservoir with small openings that lead to the outside. A contractile vacuole is located in the reservoir. The contractile vacuole functions to rid the organism of excess water. Growing out of the reservoir and extending far beyond the cell is the long flagellum. A second flagellum is contained within the reservoir. Only the emergent flagellum moves the euglenoid about. A red-orange eyespot functions as a light detector and guides the photosynthetic algae toward bright areas.

3.5. Class: Cyanophyceae (Blue green algae)

Cell is prokaryotic, main pigments are c-phycocyanin and c-phycoerythrin, reserve food is stored in the form of cyanophycean starch, cell wall is made up of cellulose, fucinic and alginic acids, sexual reproduction is absent; asexual reproduction takes place by hormogonia, fragmentation and akinetes, motile cells are absent (Desikachary, 1959).

Cyanophyta includes only five orders:

1. Chroococcales 2. Chamaesiphonales 3. Pleurocapsales 4. Nostocales 5. Stigonematales

Unlike other algae, the cyanobacteria are prokaryotes-single-celled organisms with characteristics that cause biologists to debate whether they are really algae or bacteria. Cyanobacteria are found nearly everywhere, occurring in typical aquatic and terrestrial habitats as well as in such extreme sites as hot springs with temperatures as high as 71° C (160° F) and crevices of desert rocks. Cyanobacteria make up the phylum Cyanophyta, and this phylum contains about 150 genera and 2000 species worldwide. Like other bacteria, cyanobacteria do not have organelles such as nuclei, mitochondria, or chloroplasts. Cyanobacteria are distinguished from bacteria by the presence of internal membranes, called thylakoids, that contain chlorophyll and other structures involved in photosynthesis. While higher plants have two kinds of chlorophyll, called a and b, cyanobacteria contain only chlorophyll a. Cyanobacteria color varies from blue-green to red or purple and is determined by the proportions of two secondary pigments, c-phycocyanin (blue) and c-phycoerythrin (red), which tend to mask the green chlorophyll present in the thylakoids. Cyanobacteria reproduce asexually by binary fission, spore production, or fragmentation, forming singular cells, colonies, filaments, or gelatinous masses.

Although most lack flagella and are non-motile, filamentous forms such as *Oscillatoria rotate* in a screw like manner, and the gelatinous forms glide along their slimy mucus Cyanobacteria may be both beneficial and harmful to humans. Some act as natural fertilizers in some habitats, especially rice paddies, whereas others produce toxins. Mild toxins of toxic cyanobacteria in lakes and oceans cause a rash known as swimmer's itch, while powerful neuromuscular toxins released by other cyanobacteria can kill fish living in the water or the animals that drink the water. In certain conditions, cyanobacteria may form dense blooms, which may produce toxins that make seafood poisonous to humans. Even if the cyanobacteria do not produce toxins, blooms can cause water to have an unpleasant taste and odor.

4. Potential Biotechnological Applications of Algae

Algae are extremely diverse group of photosynthetic organisms that range from single celled organisms to complex thalli, cells are unicellular like *Aphanothece* and *Microcystis* (Fig. 14A) Colonial form *Nostoc* and *Chroococcus*, (Fig. 14 D,F) Filamentus form *Anabaena*, *Oscillatoria Tolypothrix*, (Fig. 14, B,C,E) These organisms are important as primary producers in contributing to the fertility of soil, in providing substrate for other organisms and in defining aquatic environments. Blue-green algae, green algae and diatoms are the major components of aquatic ecosystems and occur as planktonic, epiphytic, epizoic and endozoic forms. Sub-aerial algae were the inhabitants of moist tree barks, walls of buildings and dropping rocks.

Algal biotechnology has come a long way since the early days of agronomy/mariculture for food, agars, carrageenans, alginates etc. Some current biotechnology applications of algae, for example, include single cell protein, carotenoids, pharmaceuticals, neutraceuticals, cosmetics etc. are already commercially exploited. Microalgae play a key role as a food source, cycling nutrients in aquatic systems and balancing CO_2 between the oceans and atmosphere. For last several decades considerable attention has been devoted to study these micro-organisms for obvious basic and applied reasons. In fact some algal species provide a simple model system for the study of cellular processes at molecular and functional level. Microalgal biotechnology involves the commercial exploitation of these organisms via mass cultivation and conversion of the harvested biomass into a range of value added products. This is not only limited to their use as a source of biologically useful compounds but also antimicrobial, antiviral, anti-cancerous compounds and removal of toxic heavy metals from contaminated sites. Increasing knowledge of algal physiology and biochemistry, combined with genetic engineering is opening up new vistas. The cultivation of microalgae can also be linked to several co-processes or services, such as capture and conversion of waste CO_2 flu gases, remediation of agricultural and industrial waste streams and energy generation from residual waste materials, biofertilizers etc. The utilization of waste in this manner, coupled with re-cycling of by-products, makes the production of microalgae increasingly more viable.

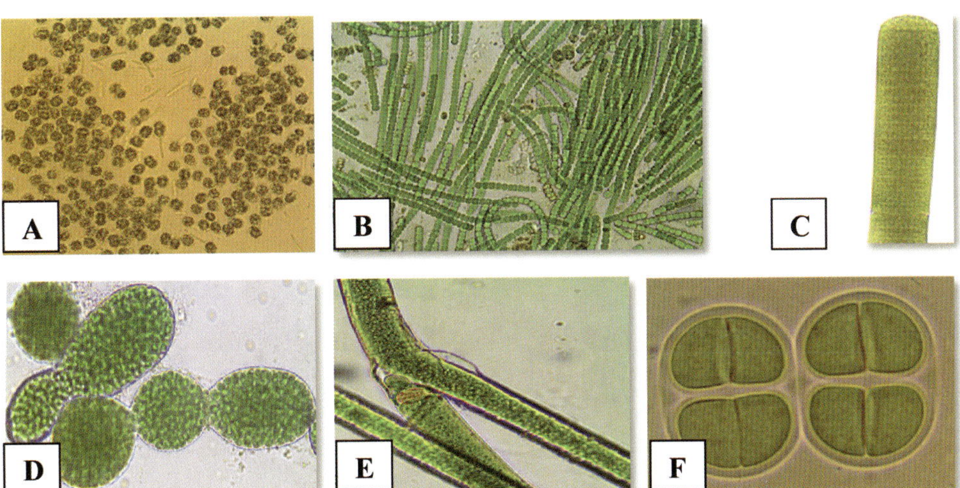

Fig. 14. (A) *Microcystis*, (B) *Anabaena*, (C) *Oscillatoria*, (D) *Nostoc*, (E) *Tolypothrix*, (F) *Chroococcus*

The application of the ability to grow microalgae to very high yields in bioreactors, to some percentage of a theoretical maximum, as opposed to ponds, holds great promise for the future of algal biotechnology. A growing proportion of today's promising pharmaceutical and nutraceutical research focuses on the production of promising compounds from algae. Thus, the untapped potential of algae in the field of pharmaceuticals and nutraceuticals has to be still explored to grow and capitalize on tremendous global marketing opportunities

4.1. Algal products and applications in medicines, nutrition and cosmetics

Fresh or dried microalgae are being increasingly used for animal and human nutrition. Algae have high protein content, essential amino acids, *β-Carotene*, astaxanthin and significant quantities of omega 3 and 6 fatty oils with proven health benefits. Annual production of biomass for human consumption has seen dramatic increase over the past decades, especially in Asian regions where it is sold either as capsules, tablets or health drinks. Recent studies suggest that inclusion of microalgae in animal diets can have several health promoting promising effects and could offer a cheaper and more sustainable replacement for fish meal. With the aquaculture industry seeing a huge increase in output, there is now a substantial market for live and dried algae to supplement the diets of juvenile fish and shellfish. Some algal species which are commercialized and their important benefits are described below.

4.1.1. Spirulina

Spirulina is one of the most concentrated natural sources of nutrition known. It contains proteins, all the essential amino acids, rich in chlorophyll, beta-carotene and its co-factors, and other natural phytochemicals. *Spirulina* is the only green food rich in GLA (gamma-Linolenic acid) essential fatty acid. GLA stimulates growth in some animal and makes skin and hair shiny and soft yet more durable. GLA also acts as anti-inflammatory, sometimes alleviating symptoms of arthritic conditions. *Spirulina* acts as a functional food, feeding beneficial intestinal flora, especially *Lactobacillus bifidus*. Maintaining a healthy population of these bacteria in the intestine reduces potential problems from opportunistic pathogens like *E. coli* and *Candida albicans*. (Fig. 15)

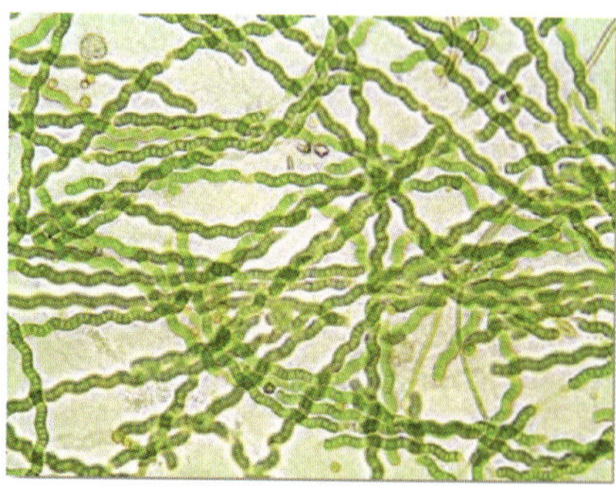

Fig. 15. Shows micrograph of *Spirulina* sp.

Spirulina is also proved to be a powerful tonic for the immune system by regulating some important key cells and organs of the body and improve their ability to function in spite of stresses from environmental toxins and infectious agents.

4.1.2. *Chlorella*

Chlorella is a green alga that grows in fresh water and has the highest content of chlorophyll of any known plant on earth. It is extremely high in enzymes, vitamins and minerals, including the full vitamin-B Complex. There are fossils from the pre-Cambrian period that clearly indicate the presence of *Chlorella* (Fig. 16).

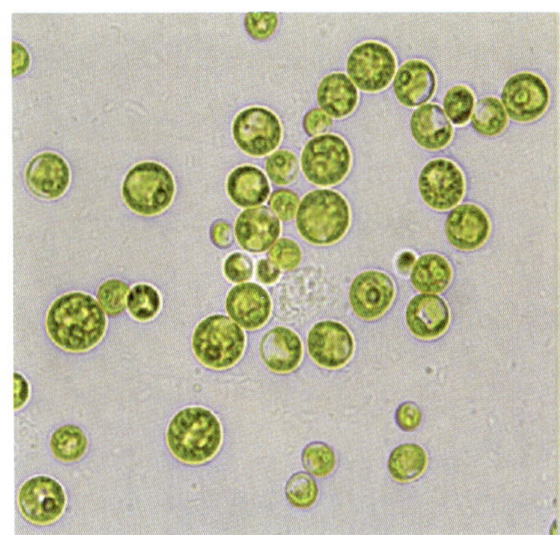

Fig. 16. Shows micrograph of *Chlorella* sp.

The following have been associated with the consumption of *Chlorella*: increased production of interferon; cleansing the blood stream, liver, kidneys, and bowel; stimulates production of red blood cells; increases oxygen to your body's cells and brain; aides digestion; promotes proper growth in children; stimulates tissue repair; helps raise the pH of your body to a more alkaline state; helps keep the heart functioning normally; and helps promote the production of friendly flora in your gastrointestinal tract. *Chlorella* can be safely used as health food. It reduces fatigue, high or low blood pressure, cardiovascular problems, memory loss, high cholesterol, digestive problems, obesity, headaches, infections, aged skin, toxemia, poor circulation, joint stiffness and pain, sleep disorders, allergies, injuries, and improves overall health. The protein found in *Chlorella* contains all the amino acids known to be essential for the nutrition of animals and human beings. There are also vitamins found in *Chlorella* including Vitamin C, pro-vitamin A (*β-Carotene*), thiamine (B_1), riboflavin (B_2), pyridoxine (B_6), niacin, pantothenic acid, folic acid, Vitamin B_{12}, biotin, choline, Vitamin K, lipoic acid, and inositol. Minerals in *Chlorella* include phosphorus, calcium, zinc, iodine, magnesium, iron, and copper (http://sunchlorillausa.com/sites).

4.1.3. *Haematococcus*

Astaxanthin is the most stable antioxidant and has action up to 500 times that of vitamin E, which has led some experts to term it as "Super Vitamin E". Typically, an antioxidant becomes unstable after quenching a free radical, and in the process becomes a free radical itself. However, astaxanthin due to its unique molecular structure that allows it to span the cell's lipid bi-layer and also because of its long double-bonded chain, can stabilize free radicals by adding them to its structure, rather than sacrificing an atom or electron. This makes it a better antioxidant than others. Astaxanthin is derived from the microalga *Haematococcus pluvialis*.

Haematococcus is a freshwater species of Chlorophyceae from the family Haematococcaceae. This species is well known for its high content of the strong antioxidant astaxanthin, that is important in aquaculture, various pharmaceuticals and cosmetics. The high amount of astaxanthin is present in the resting cells, which are produced and rapidly accumulated when the environmental conditions becomes unfavorable for normal cell growth. Examples of such conditions are i.e. high light, high salinity or low nutrients. *Haematococcus pluvialis* is usually found in temperate regions around the world. Their resting cysts are often responsible for the blood-red colour seen in the bottom of dried out rock pools and bird baths (Fig. 17). This colour is caused by astaxanthin which is believed to protect the resting cysts from the detrimental effect of UV-radiation, when exposed to direct sunlight. The microalga is cultivated in a carefully monitored process to maintain Carotenoids stability. (Guerin *et al.* 2003)

Fig. 17. *Haematococcus* pond (Sueela and Toppo, 2006)

4.1.4. Dunaliella

β-Carotene, the major carotenoid accumulated by *Dunaliella sp.*, is a valuable chemical, in high demand as a natural food coloring agent, as pro-vitamin A (retinol), as additive to cosmetics, and as a health food. Some *Dunaliella* strains can accumulate very large amounts of this carotenoid.

The biotechnological potential of *Dunaliella* as a source of β-carotene was investigated relatively early (Aharon 2005). The first pilot plant for *Dunaliella* cultivation for *â*-carotene production was established in the USSR in 1966. The commercial cultivation of *Dunaliella* for the production of β-carotene throughout the world is now one of the success stories of halophile biotechnology. Different technologies are used, from low-tech extensive cultivation in lagoons to intensive cultivation at high cell densities under carefully controlled conditions. One of the methods used in such biotechnological operations to induce massive carotenoid accumulation is reduction of the growth rate by deprivation of nutrients. That high carotenoid content of the cells may be caused by nutrient limitation as well as by high light intensities.

As the red coloration occurred especially in old cultures, it was reasonable to assume a correlation with the nutritional conditions and in particular with the lack of one or more compounds.

4.2. Polyunsaturated fatty acids (PUFA)

Polyunsaturated fatty acids (PUFAs) are essential part of our diets and key for child development. Only plants are able to synthesize PUFAs, and microalgae supply aquatic food chains with these vital components. Fish which feed on algae provide oils that are the common sources of PUFAs for human nutrition, but with increasing toxic blooms and accumulation of their toxins are affecting the beneficial algal resources. Microalgal PUFAs therefore have a promising market for food and feed. As algae could be relatively easily cultivated at different stress conditions, they offer the prospect of a good source of PUFA for the nutraceutical market. Recently, attention has focused on n-3 PUFAs from algae, especially eicosapentaenoic acid (EPA) and docosahexaenoic acid (DHA), due to their association with the prevention and treatment of several diseases (atherosclerosis, thrombosis, arthritis, cancers, etc.). Long-chain polyunsaturated fatty acids of the omega 3 and omega 6 series are straight chain carboxylic acids of 20 or more carbon atoms that contain 3 or more double bonds. The conventional source of EPA and DHA is marine fish oil, but higher amount of EPA and some DHA can be produced by the use of algae

Microalgae have been an attractive source of PUFA due to their inherently high PUFA content. Cyanobacterium *Spirulina* is rich in γ-linolenic acid (GLA) (and poor in the α-isomer) and thus is a good source for the purification of this PUFA. *Chlorella minutissima* is a eukaryotic species with a fast growth rate and high PUFA content and could be another important source of a PUFA-rich nutraceutical supplement. The red algae, *Porphyridium* contain more than 30% of total fatty acids as AA (Arachidonic Acid) and equally high concentrations of EPA. Fatty acids (FA) from marine macroalgae are generally richer in polyunsaturated fatty acids (PUFA) and a higher degree of total unsaturation. Both Phaeophyta and Rhodophyta species of macroalgae were rich in arachadonic acid (AA) and eicosopentaenoic acid (EPA) (http:www.bioline.org.).

4.3. Antimicrobial, antiviral and antifungal compounds from algae

Both microalgae and macroalgae exhibit antimicrobial activity which finds use in various pharmaceutical industries. Microalgae, such as *Ochromonas* sp., *Prymnesium* and a number of blue green algae (*Microcystis, Nostoc, Gloeotrichia, Oscillatoria* etc.) produce toxins that may have potential pharmaceutical applications. Various strains of cyanobacteria are known to produce intracellular and extracellular metabolites with diverse biological activities such as antibacterial, antifungal and antiviral activity. The biological activities of the algae may be attributed to the presence of volatile compounds, some phenols, free fatty acids and their oxidized derivatives. There are numerous reports of macroalgae derived compounds that have a broad range of biological activities, such as antibiotic, antiviral, anti- neoplastic, antifouling, anti-inflammatory, cytotoxic and antimitotic. In the past few decades, macroalgae have been widely recognised as producers of a broad range of bioactive metabolites. Such antimicrobial properties enable macroalgae to be used as natural preservatives in the cosmetic industry. The highest percentage of antimicrobial activity was found in Phaeophyceae (84%), followed by Rhodophyceae (67%) and Chlorophyceae (44%). Red and brown macroalgae extracts show significant potential as anti-pathogenic agents for use in fish aquaculture (Wefky and Ghobrial 2008).

4.4. Anti-cancer effects

Several studies show that *Spirulina* or its extracts can prevent or inhibit cancers in humans and animals. Some common forms of cancer are thought to be a result of damaged cell DNA running amok, causing uncontrolled cell growth. Cellular biologists have defined a system of special enzyme called endonuclease which repair damaged DNA to keep cells alive and healthy. When these enzymes are deactivated by radiation or toxins, errors in DNA go unrepaired and cancer may develop. In vitro studies suggest the unique polysaccharides of *Spirulina* enhance cell nucleus enzyme activity and DNA repair synthesis (Hayashi *et al.*1996).

4.5. Bio-fertilizers

Algae especially blue green algae/cyanobacteria possess the twin abilities of photosynthesis as well as biological nitrogen fixation. Besides fixing nitrogen, they also excrete Vitamin $B_{12,}$ auxins, ascorbic acid etc, which acts as growth promoting substances. Nitrogen fixation takes place in specialized cells called 'heterocysts' on the algal filament. However, unicellular and filamentous, non-heterocystous forms like *Plectonema, Lyngbya* and *Oscillatoria* exhibit abilities to fix molecular nitrogen. Heterocysts form like *Tolypothrix*, *Anabaena* (Fig. 18 A,B) are large, thick walled, cells grow in between pigmented cells on the algal filament. Vegetative cells and heterocysts are interdependent when active nitrogen fixation takes place (Rao 1981). Under rice field conditions, nitrogen fixing blue green algae form the most efficient biological system contributing to the fertility of the soil. In addition to adding about 30kgN/ha/season they also help the crop in utilizing more of the applied nitrogen by providing growth regulators. They also been shown to reduce the oxidizable matter

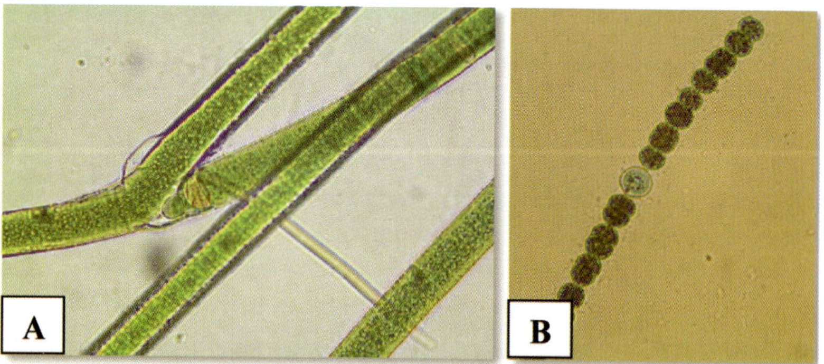

Fig. 18. (A) *Tolypothrix,* (B) *Anabaena*

and total sulfide and ferrous ion content of the soil, mobilize the insoluble phosphate and help in the formation of soil aggregates.

4.6. Remediation of waste water resources

Wastewater generated from agricultural, industrial and domestic sources often contains high concentrations of organic matter, nitrogen, phosphorous, heavy metals and is in plentiful supply. Processing of these wastes represents a serious environmental challenge and requires costly treatments to remove organic compounds and inorganic chemicals to prevent eutrophication in water bodies.

The use of algae in wastewater treatment has long been appreciated, and their ability to efficiently accumulate nutrients and toxic metals when grown in nutrient-rich environments represents a low cost and more sustainable strategy compared to other treatment processes. For these reasons, the use of such wastewaters for the cultivation of microalgae is been suggested as a plausible route to economical biomass production, off-setting the cost of biomass production in particular for biofuels, offering savings on costly waste remediation, a reduction in the use of fresh water for alga culture and a cheap source of nutrients for growth.

4.7. CO_2 and flue gas sequestration

The process of capture and storage of carbon dioxide being produced by the combustion process by microalgae offers the possibility of reducing the emission of anthropogenic greenhouse gases as well as saving money on carbon dioxide emission certificates. Microalgae, the major natural nitrogen fixing organisms on the planet can be successful solution for that issue. They are able to transform CO_2 from flue gases directly into organic material such as valuable proteins, lipids and carbohydrates by the photosynthetic process that is driven by sunlight. This process requires land, water, sunlight as well as nitrates and phosphates as nutrients. (www.algae-biotech.com)

The application of vertical tubular photobioreactors in industrial scale can achieve high cell densities that are directly linked to CO_2 fixation rates. At the same time the vertical orientation will reduce the necessary land space that is needed for microalgal culturing. One of the key advantages of using algae is that they are able to withstand high concentrations of CO_2 during culture. Discrete, intensive microalgal growth platforms could be used for efficient carbon capture systems high-CO_2 content flue gases from power stations and other industry. It has also been found that many algae are able to remove nitrogen oxide and sulphur dioxide from gas streams, which are common causes of acid rain. These approaches could help to reduce carbon emissions and associated taxation, and helps to reduce costs of algal product manufacture. Additionally, residual biomass, rich in carbon can be converted to a form used to enhance agricultural soil, a further CO_2 mitigation opportunity.

4.8. Bioenergy

Bioenergy or biofuel is a term used to describe renewable energy generated from biological materials. Microalgae are increasingly being recognized as a sustainable feedstock for biofuel production due to their rates of production and the significant fraction of the biomass that is made up of useful lipids and carbohydrates. Algal biofuels have seen considerable research effort and investment from both academic and industrial sectors.

Algae are well suited to biodiesel production as many species have high lipid contents that can readily extracted and converted to biodiesel by transesterification on an industrial scale. Microalgae are also an ideal candidate for bioethanol production due to their high content of fermentable sugars, lack of the structural component lignin (http://www.oilgae.com).

However, algae can also generate a whole suite of bioenergy products. Methane can be generated from microalgal biomass by gasification, pyrolysis or anaerobic digestion. This can then either be combusted directly for generation electricity and heat or alternatively as biogas as a liquid fuel replacement or for domestic purposes. Microaglae can also be utilized for photo-biological hydrogen production for use in fuel cells, essentially generating a sustainable, emission-free hydrogen gas from sunlight and water. A number of other technologies exist for the production of range of biofuels from microalgae, including bio-butanol and bio-gasoline for use as liquid fuels. However, significant technological challenges remain to be answered.

Transgenesis in algae is a complex and fast-growing technology. Selectable marker genes, promoters, reporter genes, transformation techniques, and other genetic tools and methods are already available for various species and currently ~25 species are accessible to genetic transformation. Fortunately, large-scale sequencing projects are also planned, in progress, or completed for several of these species; the most advanced genome projects are those for the red alga *Cyanidioschyzon merolae*, the diatom *Thalassiosira pseudonana*, and the three green algae *Chlamydomonas reinhardtii, Volvox carteri* and *Ostreococcus tauri* (Hallmann 2007).

The main focus of algal biotechnology continues to be on high value fine chemicals. Molecular markers have been used to study algal photosynthesis, respiration, storage granules and reproduction. Genetic engineering techniques can make it possible to enhance the desired molecules in the cell system. So there is great scope for molecular biologists to innovate novel methods to engineer the algal genome for enhanced desirable products. Given the existing technology and algal biotechnology, high value products from microalgae and algal biomass for antimicrobial compounds and algal biofuels have the greatest commercial potential in the future.

References

Aharon O (2005) A hundred years of *Dunaliella* research: 1905–2005. Saline Systems, 1:2

Desikachary TV (1959) Cyanophyta, I.C.A.R. Monograph on blue green algae, New Delhi, pp 686

Fritsch FE (1935) The structure and reproduction of the Algae, Vol. I, Cambridge Univ Press, London, pp 791

Fritsch FE (1945) The structure and reproduction of the Algae, Vol. II, Cambridge Univ Press, London, pp 939

Guerin M, Huntley M E, Olaizola M. 2003. Haematococcus astaxanthin: application for human health and nutrition. Trends Biotechnol. 21(5):2010-216

Hallmann A (2007) Algal transgenics and biotechnology. Transgenic Plant J 1:81-98

Hayashi K, Hayashi T and Kojima I (1996) A natural sulphated polysaccharide, calcium spirulan, isolated from Spirulina platensis: *in vitro* and *ex vivo* evaluation of anti-herpes simplex virus and anti-human immunodeficiency virus activities. Aids Res Hum Retroviruses. 10:12(15):1463-1471

http://en.wikipedia.org/wiki/Spirulina_(dietary_supplement)

http://sunchlorillausa.com/siti(UniqueclinicalBenefitsofsc)

Lee RE (2008) "Phycology" (4th Edition). Cambridge university Press, Colorado State University, USA

Merchant RE, Andre CA (2001) A review of recent clinical trials of the nutritional supplement chlorella pyrenoidosa in the treatment of fibromyalgia, hypertension, and ulcerative colitis. Alternative Therapies 7(3):79

Papenfuss GF (1946) Proposed names for the phyla of algae. Bull Torrey Bot Club 73:217-218

Pascher GF (1914) Ueber Flagellaten und Algen, Ber. Deutsch. Bot Ges 32:136-160

Prescott GW (1951) Monograph algae of the Western great lakes area. Cranbrook Institute of Science, Michigan Pp 944

Prescott GW (1969) The Algae: A review. Thomas Nelson and Sons, London

Sharma OP (1986) Diversity of microbes and cryptogams Algae. Tata McGraw-Hill Education, Jan 1, pp 396

Smith GM (1933) Fresh water algae of the United State, McGraw-Hill Book Co., New York

Smith GM (1955) Cryptogamic Botany, Vol. I, McGraw-Hill Book Co., New York

Subba Rao NS (1981) Biofertilizer in Agriculture, II edition, Oxford and IBH Publishing Co. Pvt. Ltd. New Delhi, Bombay, Calcutta

Suseela MR, Toppo K (2006) *Haematococcus pulvialis*- A green algae, richest natural source of astaxanthin. Curr Sci 90(12):1602-1603

Wefky S, Ghobrial M (2008) Studies on the Bioactivity of Different Solvents Extracts of Selected Marine Macroalgae Against Fish Pathogens. Res J Microbiol 3(12):673-682

Chapter – 7

Methods and Approaches Used in Collection, Preservation and Identification of Algal Resources

Kiran Toppo, M.A. Usmani and M.R. Suseela

1. Introduction

The objective of algal taxonomic collections is to represent the natural population, flora and diversity in size and form. Therefore, it is important to collect the representative samples seasonally from various habitats. Collection of information about the habitats and other details are also very important. For algal sampling the first thing to consider is how to collect the specimens. The freshwater algae show an ability to tolerate a wide range of environmental condition. Under natural conditions, they usually grow in mixed communities which may include many species and genera. Field study provides information about the site, environment, season and size and shape of the living algal thallus in its natural surroundings. Many algae can be tentatively identified by their appearance, shape and colour of thallus in a particular habitat. Algal population can be sampled in a number of different ways, depending on the nature of the habitat and the location of the algae within the water body. The specimens must be prepared carefully so that important morphological characters are displayed as fully and completely for further researches.

2. Methods

2.1. Collection of sample materials

Techniques for sampling enumeration can be conveniently separated in relation to these two main groups of algae-nonplanktonic and planktonic (substrate-associated) organisms (Fig. 1A, B). Macroalgae and the attached microalgae can be collected by hand or with a knife, including part or all of the substrate (rock, plant, wood etc.), if possible. It is recommended to explore all habitats in the water body including the edge of stones in fast-flowing water, aquatic plants, dam walls, and any floating debris. In running or slightly turbid waters, a simple viewing box made from transparent perspex can be used to find algae more easily. A hand lens is often useful to determine if material is in reproductive stage (essential for species determination in some genera and helpful for generic placement).

Fig. 1. (A) Collection with forcep (Non Planktonic), (B) Collection with spoon (Planktonic)

Microscopic floating algae (the phytoplankton) (Fig. 2) can be collected with a mesh net (e.g. with 25-30 μm pores) or, if in sufficient quantity (i.e. colouring the water), by simply scooping a jar through the water. Water samples can be left overnight allowing the algae to settle and concentrate on the bottom of the container. Squeezing *Sphagnum* and other mosses, or some aquatic flowering plants is a good way to collect a large number of desmids and diatoms.

Fig. 2. Floating algae

Microscope slides suspended in a water body for 2-4 weeks will reveal many algal species colonies. The slides should be kept submerged until algal colonization are ready to examine under the microscope. One side can be wiped clean and a cover slip placed over the other. Algae growing on soil are difficult to collect and study, but many culturing techniques are already available for isolation, multiplication and identification (Edward and David 2010).

2.2. Storage and preservation of fresh water algae

Algae can be stored initially in a bucket, jar, bottle or plastic bags, with some water from the collecting site. The container should be left open or only half filled with liquid and wide shallow containers are better than narrow deep jars. Glass/plastic vials are commonly used to collect algae. If refrigerated or kept on ice soon after collecting, most algae can be kept alive for short periods (a day or two). If relatively sparse in the sample, some algae can continue to grow in an open dish stored in a cool place with reduced light. For long-term storage, specimens can be preserved in liquid, dried, or made into a permanent microscope mount (preferably all three). Even with ideal preservation, examination of fresh material is sometimes essential for an accurate determination. Motile algae particularly must be examined while flagella and other delicate structures remain intact.

2.2.1. Liquid preservation

Commercial formalin (which is a solution of 40% formaldehyde), diluted between 1/10 and 1/20 with the collecting solution, is the most commonly used fixative. (*Note: Formaldehyde is thought to be carcinogenic and all contact with skin, eyes and air passages should be avoided.*) FAA (by volume, 40% formaldehyde 1: glacial acetic acid 1: 95% alcohol 8: water 10) or 6-3-1 (by volume, water 6: 90% alcohol 3: 40% formaldehyde 1) solutions give better preservation results for some of the most fragile algae, whereas the standard alcohol and water mix (e.g. 70% ethyl alcohol or industrial methylated spirit) will ruin all but the larger algae.

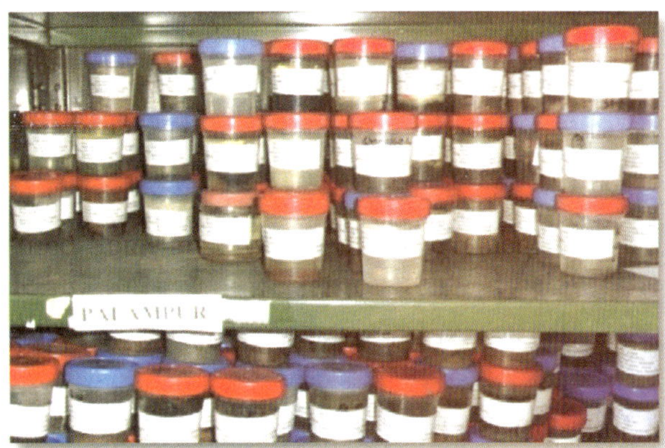

Fig. 3. Preserved algal samples

Algae can be kept in diluted formalin for a number of years (Fig. 3), but the solution is usually replaced by 70% ethyl alcohol with 5% glycerin (the latter to prevent accidental drying out). Lugol's solution is commonly used for short-term (e.g. a few months, but possibly a year or more) storage of microalgae. Dissolve one gram of iodine crystals and two grams of potassium iodide in 300 ml of water. Use three drops of this solution in a 100 ml sample (it should look like very weak tea).

2.2.2. Herbarium specimens

Dried herbarium specimens can be prepared by 'floating out' similar to aquatic flowering plants. Ideally, fresh specimens should be fixed prior to drying. Most algae will adhere to absorbent herbarium paper. Smaller, more fragile specimens or tangled, mat-forming algae may be dried onto mica or cellophane. After 'floating out', most freshwater algae should not be pressed but simply left to air dry in a warm dry room. If pressed, they should be covered with pieces of waxed paper, plastic or muslin cloth so that the specimen does not stick to the drying paper in the press (Fig. 4). To examine a herbarium specimen add a few drops of water to the specimen. After a minute or so, when the specimen swells up, lift a part slightly from the paper. Carefully remove a small portion of the specimen with forceps or a razor-blade.

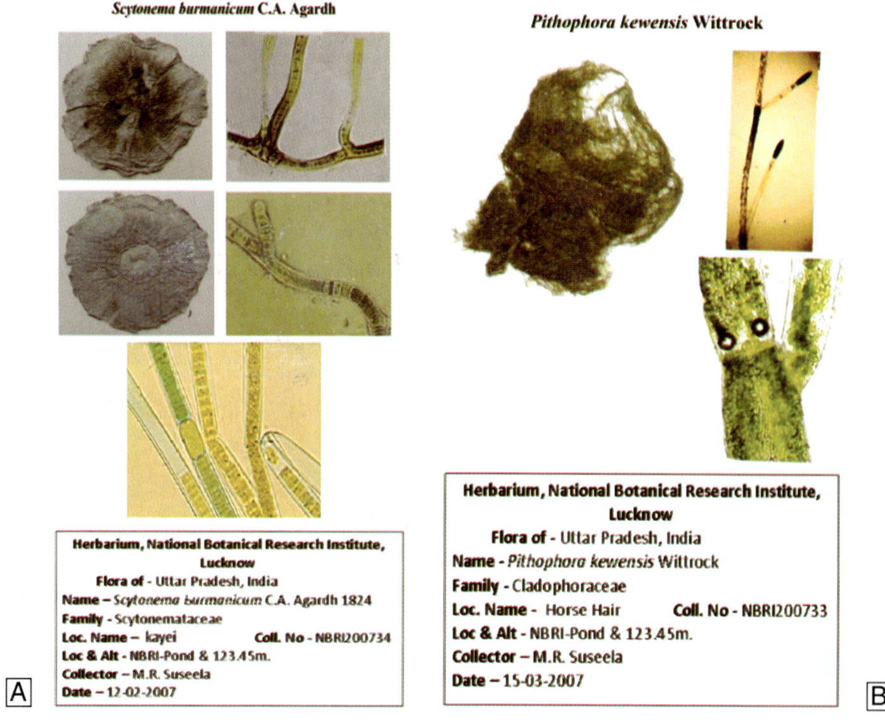

Fig. 4. (A) *Scytonema* Herbarium Sheet, (B) *Pithophora* Herbarium Sheet

2.3. Collection of field information

The collector's name, date of collection, brief description of habitat and collecting number should be penciled onto the herbarium sheet (and later replaced with a full label) or onto a collecting tag inserted into the solution. The accompanying notes should include standard information such as the locality, date of collection, and the collector's name and collection number, coordinate data whether the water is saline, brackish or fresh; whether the collection site is terrestrial, or a river, stream or lake; whether the alga is submerged during water level fluctuations or floods; whether the water is muddy or polluted; whether the alga is free floating or attached, and if the latter, the type of substrate to which it is attached; and the colour, texture and size of the alga.

3. Identification of Algae

3.1. Microscopic examination of fresh material

Observations (preferably including drawings or photographs) based on living material are essential for the identification of some genera, and a valuable adjunct to more leisurely observations on preserved material for others. The simplest method is to place of drop of the water including the alga onto a microscope slide and carefully lower a coverslip onto it. It is always tempting to put a large amount of the alga onto the slide but smaller fragments are much easier to view under the microscope. Start by observing the algae at lower magnification (10X, 20X, 40X) and move sequentially up if necessary. Microalgae may be better observed using the 'hanging drop method': place a few drops of the sample liquid on a cover slip and turn it over onto a ring of paraffin wax, liquid paraffin or a 'slide ring'.

3.2. Permanent slides

A permanent slide is a valuable addition to wet and dry herbarium specimens. Aniline blue (1% aqueous solution with 4% molar HCl), Toluidine blue O (0.05% aqueous solution) and Potassium permanganate (2% aqueous $KMnO_4$) are useful stains for macroalgae (different stains suit different species) and Indian Ink is a good stain for highlighting mucilage and some flagella-like structures. After staining for 30 seconds to five minutes (depending on the material), rinse in water, then add a drop or two of 10% corn syrup solution to a small piece of the algae placed on a microscope slide then carefully lower the coverslip, the sides of the coverslip can be readily sealed with nail polish.

Glycerine solution (75% glycerine, 25 % water) is another useful mounting agent and starting with a very dilute solution and building up to 100% glycerine. Sealing with nail polish is essential. These two mountants are unsuitable for most unicellular algae which should be examined fresh or in temporary mounts of liquid-preserved material. Magnifications of between 40 and 1000 times are required for the identification of all but a few algal genera. A compound microscope is therefore an essential piece of equipment for anyone wishing to discover the world of algal diversity. Microscopes with 10X eyepiece and 4X, 10X and 40X objectives (Fig. 5) would be suitable for identifying almost all algae (Fig. 6). An oil immersion objective 100X, would be a useful addition, particularly when identifying to species level. A camera lucida attachment is helpful for producing accurate drawings while an eyepiece

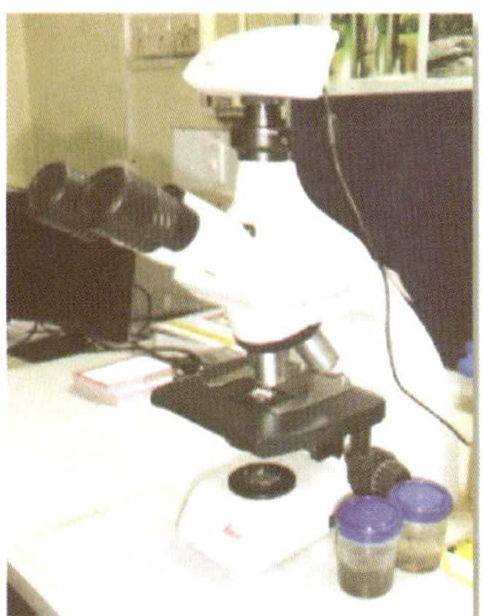

Fig. 5. Microscope

micrometer is important for any species level identifications. Phase-contrast or interference microscopy can improve the contrast for bleached or small specimens. A dissecting microscope provides magnifications up to 40 or 50 times is a useful aid but is secondary to a compound microscope. Scanning and transmission electron microscopes are boon for identifying desmids and diatoms.

Fig. 6. (A) *Hydrodictyon* bloom, (B, C) culture and (D) microphotograph.

4. Isolation of Microalgae

There are many recipes for media preparation for cultivation of algae under laboratory condition. The main consideration in developing a nutrient media for isolating and culturing of algal isolates are; total salt concentration of major ionic compounds, nitrogen source, carbon source, pH, trace elements and vitamins. A number of culture media for growing microalgae are available. Sometime organism specific media composition may also be used.

4.1. Preparation of the media

Media are generally composed of three components: Macro and micronutrients, trace elements, and vitamins. All three are often prepared as stock solutions. Adjustment of pH level depends on the constituents and requirement of culture conditions. The common algal culture media are:

- Blue Green Algae Agar (BG-11) (Stainer *et al.*1971)
- Bold Basal Media (BBM) (Nichols and Bold 1965)
- Chu 10 (Gerloff *et al.*1950)

The detailed composition of two media are following:

4.1.1. Bold basal media (BBM)

(a) 940 ml of distilled water add a volume of 10ml of the following stock solution (g/ 400 ml of H_2O)

i.	$NaNO_3$	10.0g
ii.	$CaCl_{2.}2H_2O$	1.0g
iii.	$MgSO_4.7H_2O$	3.0g
iv.	K_2HPO_4	3.0g
v.	KH_2PO_4	7.0g
vi.	NaCl	1.0g

(b) To this add 1ml volume of each of the following solution (g/100ml of H_2O)

i.	Thiamine	0.1g
ii.	Biotin	$25.0x10^{-6}$g
iii.	Vitamin B_{12}	$15.0x10^{-6}$g

4.1.2. BG-11 (Blue Green Algae Agar)

Dissolve the following in 850 ml of distilled water:

i.	$NaNO_3$	1.5g
ii.	K_2HPO_4	0.04g
iii.	$MgSO_4$-$7H_2O$	0.075g
iv.	$CaCl_2$-$2H_2O$	0.036g
v.	Citric acid	0.006g
vi.	Ferric ammonium citrate	0.006g
vii.	EDTA	0.001g
viii.	Na_2CO_3	0.02g
ix.	Trace metal mix	1.0ml
x.	Agar	10g

(a) Adjust pH to 7.1.

(b) Bring to 1000ml with distilled water

(c) Autoclave the medium

(d) To prepare trace metal mix dissolve the following in 850ml of distilled water:

i.	H_3BO_3	2.86g
ii.	$MnCl_2$-$4H_2O$	1.81g
iii.	$ZnSO_4$-$7H_2O$	0.222g
iv.	Na_2MoO_4-$2H_2O$	0.39g
v.	$CuSO_4$-$5H_2O$	0.079g
vi.	$Co(NO_3)_2$-$6H_2O$	0.0494g

4.2. Isolation methods

4.2.1. Streaking plate method

In a streak plate, from the loop containing algae, varying numbers of algae adhere to the surface of the medium and towards the end of the streak the number gets so much reduced to from separate colonies. Prepare agar plates containing appropriate medium. Flame the inoculation needle and cool it dry. With the drop on the edge away from the body, streak the culture back and forth, edge in parallel lines, moving towards the body. (Fig. 7A).

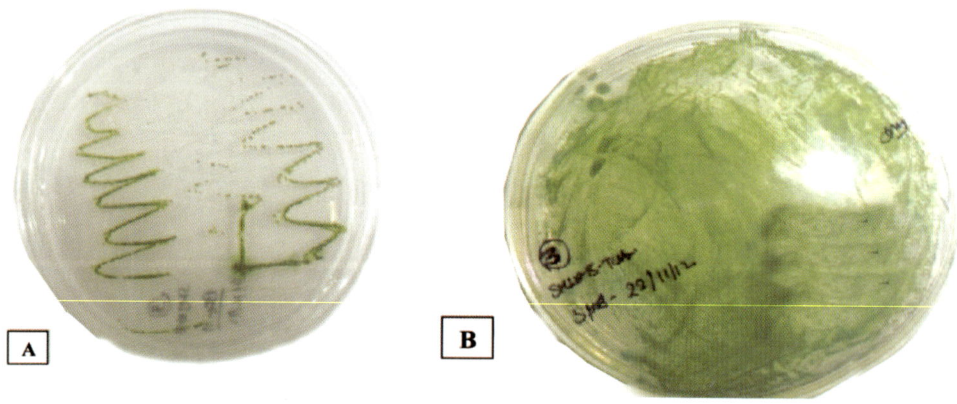

Fig. 7. (A) Streak plate, (B) Spread plate

4.2.2. Spread plate method

In spread plate, the algae in liquid medium are directly spreaded over the entire surface of the solid medium resulting in separation at many places. Pour the sterilized medium into petri-dishes and allow them to solidify. Inoculate 0.1 ml dilute algal sample mixture on to surface of agar and spread into surface of agar and spread it thoroughly with the 'L' glass rod or sterile disposable spreader (Fig. 7B).

4.2.3. Pouring method

In pour plate, there is direct dilution of cyanobacteria while being suspended in pour agar resulting in separation at the time of plating. The addition of solidifying agent to liquid medium containing algal cell trap the individual cells in place in the agar medium instead of floating around when they multiply, as in the liquid medium, they produce a fixed colony of the cells or filaments and grow to form separate colonies.

4.3. Incubation conditions

The cultures are normally incubated at 28± 2°C under continuous illumination (4-5 Klux). Because of the high concentration of water in agar, condensations of water may form in petriplate during incubation and moisture is likely to drip from the cover into surface of the agar and spread out, resulting in a confluent mass growth and running individual colony formation. To avoid this, plates are incubated bottom side up. The slant culture should be placed in a slanting position, with the streak towards light source. Always label all petri dishes, tubes, flasks and culture bottles with name, data and identification of the content.

4.4. Purification

4.4.1. Repeated liquid subculture

This technique has been successfully used when a natural collection is particularly rich in specific algae (Gerloff *et al.*1950) (Fig. A-D).

4.4.2. Fragmentation

Homogenization of filaments with a glass homogenizer for 5-10 minutes revealed short filaments of 4-8 cells long to be obtained. Individual colonies can be obtained when suspension is streaked on agar plates containing suitable medium.

4.4.3. Antibiotics

It may be difficult to remove certain contaminants by repeated and/or ultrasonic treatment. In such instance the use of chemical method rather than a physical method is preferred. One such chemical method uses antibiotics singly or in combination to kill or inhibit the growth of contaminants.

4.4.4. UV irradiation

This method has been widely used to obtain some strains including those frequently used culture of Blue green algae (Gerloff *et al.*1950; Bowyer and Skerman 1968; Koch 1964).

4.4.5. Higher temperature incubation

Thermophilic forms can be isolated by enrichment at about 40°C.

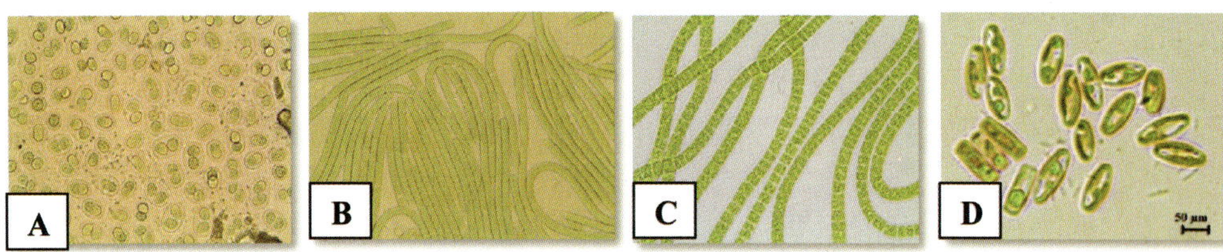

Fig. 8. After purification some algal cultures (A) *Aphanothece* (B) *Lyngbya* (C) Diatom (D) *Ulothrix*

5. Maintenance

All strains are maintained by sub culturing in liquid media and agar slants. The maintenance of algal cultures usually requires propagating working culture by subculture and keeping alive and growing in the course of investigations. (Fig. 9).

Fig. 9. Algal culture in controlled conditions

For a long term preservation, cultures are usually stored in a lyophilize or deep freezern (cryopreservation) to prolong their viability and to reduce changes due to the occurrence of spontaneous mutations. But, the ability of algal cells to remain viable under lyophilized or cryopreserved condition can vary and cannot be taken for granted.

Compared to continuous sub culturing in liquid media, growing them in slants can prove useful for all strains, and mutation, although cannot be avoided, can be minimized. Usually, maintenance of working culture requires the periodic transfer of stain to fresh minimal agar media. The culture should be maintained in very low light (1.5 Klux).They can be stored in to the refrigerator in dark at 5-8°C to reduce metabolic activity while maintaining viability; some strains cannot be stored in refrigerator as they would lose all the properties.

6. Preservation

The primary purpose of preserving cultures is to maintain algal population in a viable state for considerably longer period. During preservation, all physiological processes of an organism are considerably slowed down, without affecting viability. For preservation, most algae do at well at less temperature (15-20°C), with few exceptions where algae prefer higher temperature for survival.

6.1. Dry culture slants

Algal cultures can be preserved in dried conditions in agar slants and can be revived as and when required by supplying nutrient to the cultures (Fig. 10).

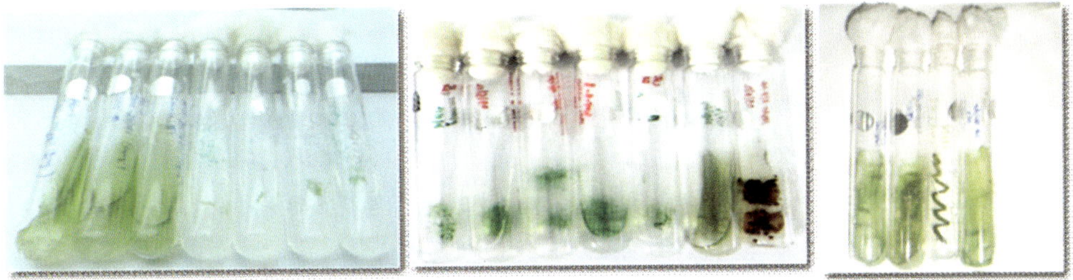

Fig. 10. Algal Culture Slants

6.2. Lyophilization

Freeze drying or Lyophilization is a technique where drying is avoiding the liquid state, through sublimation. In this process, the ice crystals forms in cells are directly converted in to water vapors through evacuation at low temperature. Since the lyophilized are hygroscopic and show loss in viability when exposed to molecular oxygen, they should be preserved in evacuated sealed glass ampoules. Such sample show very little metabolic activity and remain viable for many years.

References

Bowyer JW, Skerman VBD (1968) Production of axenic cultures of soil-borne and endophytic blue-green algae. J Gen Microbiol 54:299-306

Edward GB and David CS (2010) Freshwater algae identification and use as bioindicators,Wiley-Blackwell, A John Wiley and Sons, Ltd, Publication, pp. 271

Gerloff GC, Fizgereld GP, Skoog F (1950) The isolation, purification and culture of Blue-green algae. Am J Bot 37:216-218

Koch W (1964) Cyanophyceenkulturen. Anreicherungs una Isolierverfahren. Zentbl Bakt Parasitenk I Suppl, 1 , pp. 415-431

Nichols HW and Bold HC (1965) *Trichosarcina polymorpha* gen.*et*. sp. nova. J Phycol. 1:34-38.

Stainer RY, Kunisawa R, Mandel M, and Cohen-Bazire G (1971) Applied Environmental Microbiological Purification and properties of unicellular blue-green algae (Order Chroococcales). Bacteriol Rev 35:171-205

Chapter – 8

An Introduction to Bryophyte Diversity and Systematics

A.K. Asthana

1. Introduction

Bryophytes are atracheate, archegoniate non-flowering group of highly diversified plants. Among plants they form the second largest group comprising of 15000 – 25,000 species (Gradstein *et al.* 2001, Crum 2001). They occur in every continent on globe, in every location habitable by photosynthetic plants and grow luxuriantly between altitude of 1000-3500m. In India, nearly 850 species of liverworts, 41 species of hornworts and 2000 species of mosses, such as are known so far (Singh 1997, Vohra and Aziz 1997). These plants exhibit a variable range of size as some of them are only 0.5-2 mm in size, while on the other end mosses *Polytrichum commune* and *Dawsonia superba* can attain a height of 50 and 70 cm, respectively. These plants play a vital role in soil conservation and formation of fertile substrata for other plants in forest ecosystem. On account of unique moisture retaining capacity, *Sphagnum* and other pleurocarpous mosses are being used as mulching substances, moss grass, moss sticks and bags in horticulture industry. These plants are very sensitive to environmental pollution due to simple structure and devoid of cuticle, and accumulate heavy metals too, hence serve as good bio-monitor of pollution. Some novel compounds viz., marchantin, riccardin, lunularic acid, terpenes and other phenolic compounds have been detected in several thalloid as well as leafy liverworts.

Bryophytes exhibit heteromorphic alternation of generations in which the gametophytic phase (haploid phase) is a prominent and nutritionally self- sufficient independent phase in the life cycle while the sporophyte (diploid phase) is attached and dependent on the gametophyte. The main plant body (Gametophyte) is either thalloid or leafy (differentiated into axis and leaves). They are attached to substratum by simple hair like structures called as rhizoids. The rhizoids are usually unicellular, simple (smooth walled) in hornworts; simple, tuberculate or sinuate in liverworts or multicellular, oblique septate in mosses. Sex organs, antheridia and archegonia produce antherozoids and eggs respectively. After fertilization the zygote is formed which develops into the sporophyte. The sporophyte consists of foot, seta and capsule. The foot is embedded in the gametophytic tissue and draws nourishment for the development of sporophyte. Seta is generally stalk like may be of variable length which holds the capsule. The capsule is the main fertile portion of the sporophytic generation. The spore mother cells, after meiotic division, form spore tetrads having haploid spores. The elater mother cells form elaters, which are sterile and help in the dispersal of spores. Bryophytes are homosporous. Spores germinate to form young gametophyte in liverworts and hornworts, while in mosses, spores first give rise to protonema then buds develop on the protonema to form new gametophytes.

2. Classification and Characteristic Features of Different Classes

Bryophytes have been classified into three major Classes: 1. Hepaticopsida (Hepaticae – liverworts). 2. Anthocerotopsida (Anthocerotae – Hornworts). 3. Bryopsida (Musci – Mosses).

Table 1. Classification of Bryophytes

Class Hepaticopsida (Schuster 1984)	**Class Anthocerotopsida (Duff *et al.* 2007)**	**Class Bryopsida (Vitt 1984)**
Subclass Jungermannidae	**Order Leiosporocerotales**	**Subclass Sphagnidae**
Order Calobryales	**Family Leiosporcerotaceae**	(1 family, 1 genus)
Order Metzgeriales	(*Leiosporoceros*)	**1. Order Sphagnales,**
Suborders	**Order Anthocerotales**	Sphagnaceae – *Sphagnum* L.
1. Fossombroniineae	Family Anthocerotaceae	**Subclass Andreaeidae**
2. Pelliineae	(*Anthoceros*, *Folioceros*,	(2 families, 2 genera)
3. Phyllothalliineae	*Sphaerosporoceros*)	**2. Order Andreaeales**
4. Pallaviciniineae	**Order Notothyladales**	(1 family, 1 genus)
5. Blasiineae	Family Notothylaceae	**Subclass Takakiideae**
6. Metzgeriineae	(*Notothylas, Hattorioceros,*	**3. Order Takakiales**
7. Hymenophytineae	*Mesoceros*, *Phaeoceros,*	(1 family, 1 genus)
Order Jungermanniales	*Paraphymatoceros*)	**4. Order Andreaeobryales**
Suborders	**Order Phymatocerotales**	(1 family, 1 genus)
1. Herbertineae	Family Phymatocerotaceae	**Subclass Bryidae**
2. Antheliineae	(*Phymatoceros*)	(88 families, 800 genera)
3. Lepicoleineae	**Order Dendrocerotales**	**5. Order Polytrichales**
4. Jungermanniineae	Family Dedrocerotaceae	**6. Order Tetraphidales**
5. Geocalycineae	(*Dendroceros* , *Megaceros,*	(1 family, 1 genus)
6. Brevianthineae	*Nothoceros, Phaeomegaceros*)	**7. Order- Bryales**
7. Perssonillineae		(85 families, 765 genera)
8. Balantiopsidineae		**Suborders**

(*Contd.*)

Class Hepaticopsida (Schuster 1984)	Class Anthocerotopsida (Duff *et al.* 2007)	Class Bryopsida (Vitt 1984)
9. Lepidoziineae		1. Archidiineae
10. Cephaloziineae		2. Funariineae
11. Ptilidiineae		3. Splachnineae
12. Lepidolaenineae		4. Orthotrichineae
13. Porellineae		5. Bryineae (Eubryales)
14. Radulineae		6. Hypnineae
15. Pleuroziineae		7. Leucodontineae
Subclass Marchantiidae		8. Hookeriineae
Order Sphaerocarpales		9. Buxbaumiineae
Suborders		10. Encalyptineae
1. Riellineae		11. Pottiineae
2. Sphaerocarpineae		12. Dicranineae
Order Monocleales		13. Fissidentineae
Order Marchantiales		14. Seligeriineae
Suborders		15. Grimmiineae
1. Marchantiineae (families- Cleveaceae, Aytoniaceae, Lunulariaceae, Conocephalaceae, Exormothecaceae, Marchantiaceae, Monoseleniaceae, Cyathodiaceae		
2. Targioniineae (Targioniaceae, Cyathodiaceae,		
3. Carrpineae (Carrpaceae)		
4. Corsiniineae (Corsiniaceae)		
5. Ricciineae (Ricciaceae, Oxymitriaceae)		

2.1. Class - Hepaticopsida

Word 'Hepatics' (Latin word 'Hepatica' means 'liver') was coined for these plants since ancient time. They are commonly also called as 'liverworts'. The study on these plants was actually started during prelinnean period. Class Hepaticopsida is chiefly characterized by dorsiventrally flattened gametophytic plants, which may be thalloid or leafy (differentiated into axis and leaves). The position of sex organs is always dorsal and they are formed by superficial cell of dorsal surface of thallus (exogenous in origin). In some cases they are terminal in certain leafy forms. The sporophyte possesses distinct foot, seta and capsule. Foot is embedded, seta is massive and elongates after the maturation of the capsule. Sporogenous tissue has endogenous origin and Sporophyte exhibits determinate growth.

2.1.1. Calobryales

Plants of this order are characterized by the heterotrichous habit (differentiated into prostrate rhizomatous axis and erect aerial axis). The aerial branches are radially symmetrical having three vertical rows of spirally arranged leaves, which are transversely inserted on the axis. The leaves are usually unistratose, but sometimes bistratose at the base lacking midrib. The rhizoids are absent. The plant body is densely covered with enormous mucilage produced by number of mucilage papillae (2-3 celled) on the axis. Growth of the axis takes place by a pyramid shaped apical cell with three cutting faces. The stem is internally differentiated into outer cortical and inner medullary regions. The cortical cells are large parenchymatous filled with starch grains. The medullary cells are comparatively smaller. The sex organs are present on somewhat flattened apical portion of the axis where the leaves are closely arranged and they are larger in size also. The plants are dioecious. They may be acrogynous (*Calobryum,* Fig. 1) or anacrogynous (*Haplomitrium*). Antheridia have short stalk and somewhat spherical to ovoid antheridial body. Archegonia have long, twisted neck (Fig. 6L) with four rows of cells in the jacket enclosing 16-20 neck canal cells. The sporophyte has distinct foot, long massive seta and cylindrical capsule (Fig. 7A). The capsule has unistratose jacket except the apical portion of the capsule. The cells have longitudinally oriented ring like thickening. The spores are small and elaters are long tapered. The capsule dehisces into 2 (-4) valves.

The order has single family Haplomitriaceae and two genera (*Calobryum* and *Haplomitrium*) in India.

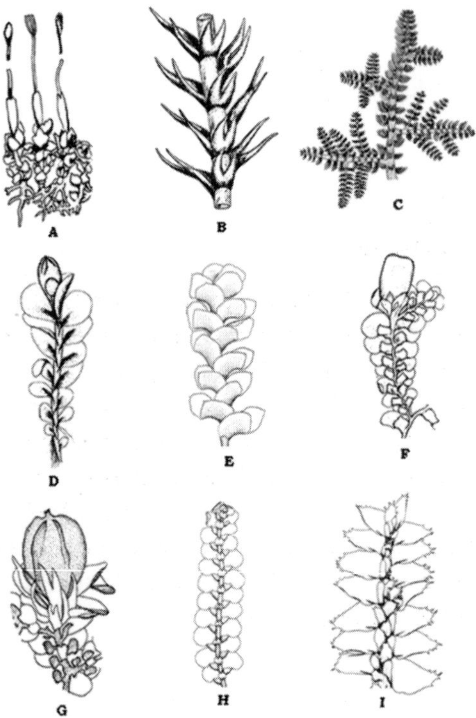

Fig. 1. (A-I). Diversity in Leafy liverworts. (A-C) Three rowed leaves, (A) *Calobryum*, (B) *Herbertus*, (C) *Trichocolea*, (D) Two rowed leaves in *Jungermannia*, (E-F) Bilobed condition, (E) *Scapania*, (F) *Radula*, (G-I) Three rowed complicate bilobed condition, (G) *Frullania,* (H) *Cheilolejeunea*, (I) *Porella*

2.1.2. Metzgeriales

Order Metzgeriales includes the plants which are characteristically very simple in their morphology and anatomy and have anacrogynous condition. It includes both the morphoforms, thallose (being undifferentiated) and leafy (differentiated into axis and leaf like lateral lobes). Rhizoids are simple and scattered on the ventral surface of the plant. In leafy forms, the leaves are longitudinally inserted. The conducting strand is distinct in some plants viz., *Pallavicinia*, while other members are devoid of conducting strand like *Pellia* (Fig. 2), *Riccardia* and *Fossombronia*. Plants may be monoecous or dioecous. Antheridia are usually globose and arehegonia have 5 rows of cells in neck portion. The sporophyte has distinct foot, long seta and spherical or elongated capsule (Fig.7Ia). Capsule has two - layered capsule wall, small spores and elongated elaters, apart from this sometimes short stumpy elaters and fixed sterile cells (elaterophores) are also present. Elaterophores may be basal in case of *Pellia* (Fig. 7Ib), while it is apical in *Metzgeria* and *Riccardia* (Fig. 7Jb). Dehiscence of capsule may be two valved (*Pallavicinia*) or four valved (*Pellia,* Fig. 7Jb).

Metzgeriales include 13 families (including Treubiaceae) and 30 genera (including *Treubia* and *Apotreubia*). In India, the order is represented by 7 families, 12 genera and 58 taxa (Srivastava 1998).

2.1.3. Jungermanniales

The order Jungermanniales includes a large number of leafy liverworts, which exhibit enormous morphological and habitat diversity (corticolous, foliicolous, rupicolous, saxicolous and terricolous (Fig. 9). It includes about 85% of the total Hepatics (Schuster 1984). Plants are differentiated into axis and the leaves. They are usually bilaterally symmetrical or sometimes radially symmetrical in some primitive forms. Leaf arrangement may be in two or three rows showing isophylly (three rows of leaves alike) e.g. *Herbertus* (Fig. 1B) or anisophylly (two rows of dorsolateral leaves are large and alike, the third ventral row has smaller leaves, which are generally called as underleaf/ventral leaf or amphigastrium) e.g. *Frullania, Cheilolejeunea* and *Porella* (Fig. 1G). The leaves are subtransverse-obliquely attached to the axis showing overlapping each other. Leaf arrangement on the axis may be succubous (*Jungermannia*, Fig. 1D) or incubous (*Radula*, Fig. 1F). The leaves are always unistratose without mid rib. The leaf-cells may be thin walled or thick walled forming diverse trigones at the corners. The rhizoids are simple, smooth walled and they are present all along the ventral surface of the axis or restricted to the base of under leaf forming a group or fascicle. The axis has simple anatomy, exhibiting differentiation into cortical and medullary cells. Plants may be dioecious or monoecious. Antheridia are usually globose, axillary in position and are protected by male bracts. Archegonia are always terminal (acrogynous) occurring singly or in group, protected by a specialized structure perianth (Fig. 7Da), formed by the fusion of perichaetial leaves. Archegonial neck has five vertical rows in the jacket. Sporophyte is differentiated into foot, seta and capsule. Capsule has multilayered [2-8(10)] capsule wall, large number of small spores (Fig. 8C,Ca,D,Da,E,Ea) and elongated elaters (Fig. 8a,b,c). Dehiscence of capsule takes place into 4 distinct valves (Fig. 7Da).

The order has 39 families and 263 genera (Schuster 1984). In India, the group is represented by 13 families, 65 genera and 520 species (Srivastava 1998).

2.1.4. Monocleales

The order Monocleales is monotypic with the single family Monocleaceae and single genus *Monoclea* (Fig. 2). It is characterized by exclusively large, pure green, translucent thalloid gametophyte, resembling in many features with order Metzgeriales. Internally the thallus is undifferentiated. Rhizoids are simple and smooth walled. The sex organs are dorsal in position on the thallus. Antheridia are embedded in the thallus tissue, while archegonium is protected by a flap like involucre and is anacrogynous in position. Sporophyte is differentiated into foot, elongated seta and cylindrical or elongated capsule. Dehiscence of Capsule takes place through single lateral slit. In India the order is not represented.

2.1.5. Sphaerocarpales

The order Sphaerocarpales is remarkable in having very delicate, translucent, green gametophytes (generally short-lived). Plant body has central thick midrib portion and lateral wing portion (unistratose).

Plants are dioecious. Sex organs are borne on dorsal surface of thallus in acropetal manner and protected inside bottle shaped involucre. Archegonial neck has six rows of cells. Sporophyte has small bulbous foot and short (abbreviated) seta, which never elongates even after the maturity keeping sporophyte inside the involucre. Capsule is small, spherical. Spores are present in tetrad form. Elaters are absent, however some sterile nurse cells are present. Capsule is cleistocarpous and the spores are dispersed after the decaying of capsule wall and the surrounding tissue.

Order includes two families, three genera and 12-14 species (Schuster 1984). The family Riellaceae is monotypic with the genus *Riella* Mont. (Fig. 2) and Sphaerocarpaceae with two genera: *Sphaerocarpos* (Mich.) Boehm. (Fig. 2) and *Geothallus* Campb. In India the order is represented by single genus - *Riella*.

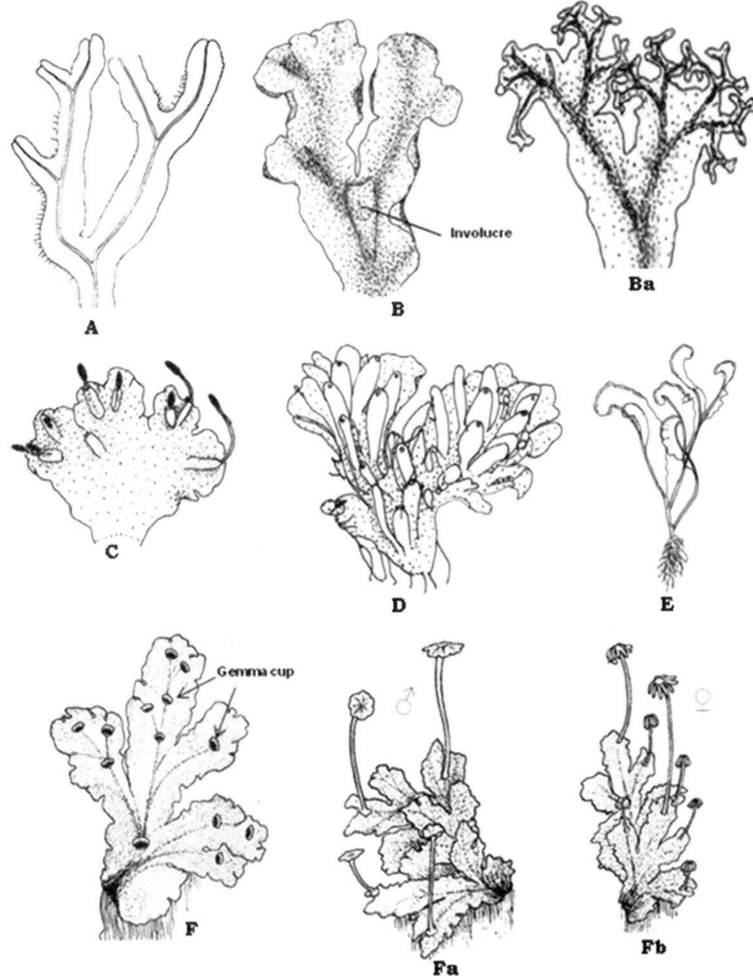

Fig. 2. (A-Fb) Diversity in thalloid liverworts, (A) *Metzgeria*, (B) *Pellia*, (Ba) Thallus with apical dichotomous branching, (C) *Monoclea*, (D) *Sphaerocarpos*, (E) *Riella*, (F) *Marchantia* (thallus with gemma cups), (Fa) Thallus with antheridiophores (male receptacles), (Fb) Thallus with archegoniophores (female receptacles)

2.1.6. Marchantiales

Marchantiales is the largest order which include only thalloid forms. Some of them exhibit drought resistant features and have evolved many anatomical devices to check the water loss and survive in stress conditions. Plants of this group are basically adapted to grow in exposed sites with intense light conditions. The plant body is thalloid, dorsiventrally flattened and dichotomously branched, which is attached to the substratum by means of simple and tuberculate rhizoids. The presence of ventral scales in the thalloid plant body is a unique feature of this order. Internally thallus plant body is differentiated into upper assimilatory zone and lower storage zone. Sex organs are present on a specialized stalked (Fig. 5F, Fig. 6O) or sessile receptacle. Archegonial neck has six rows of cells. Sporophyte is determinate in growth with well differentiated foot, short seta and capsule (Fig. 7K), while in *Riccia* foot and seta is absent (Fig. 7L). Capsule has a unistratose capsule wall, number of small to large spores and elongated elaters. Sporoderm patterns are prominent (Fig. 8J,K,L,M,N) and of high taxonomic value.

Marchantiales has been divided into 13 families and 29 genera (Schuster 1984). In India the order is represented by 10 families and 21 genera.

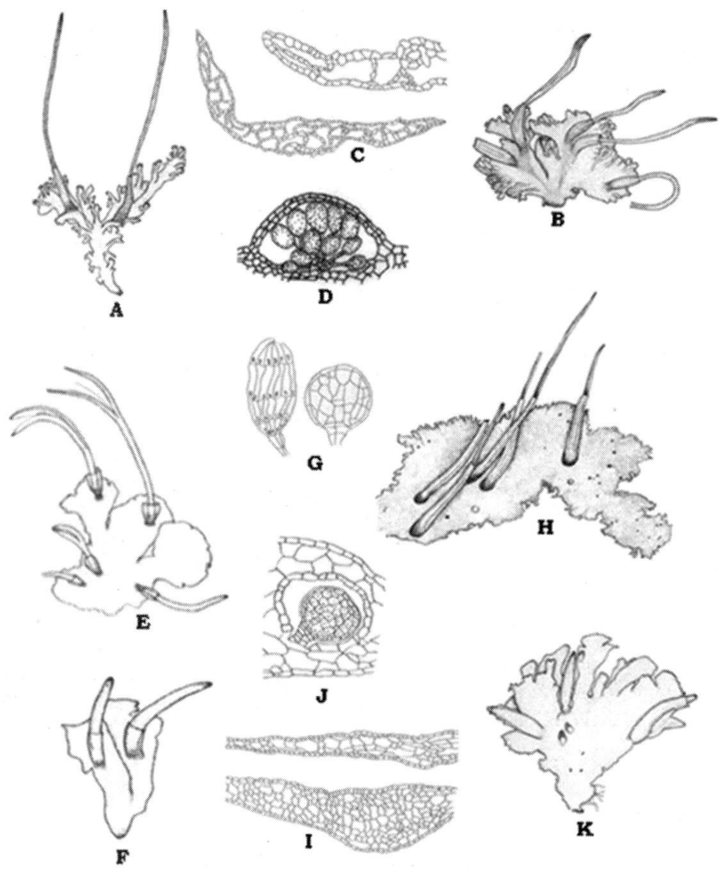

Fig. 3. (A-K) Diversity in thallus morphology and anatomy in Hornworts, (A) Female thallus of *Folioceros*, (B) Female thallus of *Anthoceros*, (C) Cross section of spongy thallus, (D) Cross section through androecial chamber in *Anthoceros*, (E,F) Female thallus of *Phaeoceros*, (G) Antheridial body with stalk and antherial body, (H) *Megaceros*, (I) Cross section of solid thallus, (J) Cross section of thallus through androecial chamber with Single antheridium, (K) Thallus of *Notothylas*

2.2. Class – anthocerotopsida

Class Anthocerotopsida is a small group of plants in which gametophyte is simple and thalloid, while sporophyte is comparatively complex and horn like or needle shaped (*Anthoceros*, *Folioceros,* Fig. 3). Hence members of this group are commonly called as 'Hornworts'. The thalloid plants are dorsiventral, lobed without any internal tissue differentiation and may be compact or spongy. Air chambers and pores are absent. Rhizoids are smooth walled. The ventral scales are totally absent. The epidermal cells of thallus usually have single, large, plate like chloroplast with conspicuous pyrenoid bodies. Distinct *Nostoc* chambers are present in thallus. Sex organs are embedded in the gametophytic tissue. Antheridia are endogenous in origin and occur in groups or single within the androecial chamber (Fig. 3). Archegonia are exogenous in origin and embedded in thallus tissue. Sporophyte is differentiated into capsule and foot only Seta is absent. Capsule is cylindrical 'horn' like and indeterminate in growth due to presence of meristematic tissue at base. Capsule wall is 4-6 layered, chlorophyllous with or without stomata. The archesporium is amphithecial in origin which forms spores (Fig. 8.O, P, Q, R) and 1-4 celled elaters (pseudoelaters, Fig. 8. I, J, K). Columella is endothecial in origin.

In India the group is represented by 5 genera: *Anthoceros, Folioceros, Phaeoceros, Megaceros* and *Notothylas* (Fig. 3), (Asthana and Srivastava 1991; Singh 2003).

2.3. Class - bryopsida

Class Bryopsida (Musci) is mainly characterized by exclusively leafy gametophytes. They are differentiated into axis and spirally arranged leaves. They may be erect growing (acrocarpous e.g. *Pogonatum*, Fig. 4D) or prostrate growing (pleurocarpous e.g. *Hypnum* and *Entodon*, Fig. 4. G, H). Some epiphytic forms may be hanging or pendulous. Leaves are usually with distinct midrib being multistratose in midrib portion and unistratose in wing portion. Rhizoids are multicellular and obliquely septate. Sex organs are always terminal in position and present in groups, exogenous in origin and develop by means of an apical cell. They grow in association with multicelullar, uniseriate paraphysis (Fig. 5P) and are protected by perigonial (male bracts) or perichaetial (female bracts) leaves. Antheridia have stalk and clavate to elongated cylindrical antheridial body (Fig. 5.Q,R,S). Archegonia are also stalked with long neck (Fig. 6. I,J). The sporophyte in mosses is of determinate growth and partially independent due to presence of chloroplasts and stomata in the capsule wall. Sporophyte consists of foot, seta and capsule. The capsule is generally cylindrical-elongated with outermost multilayered capsule wall, the central sterile column-columella, and in between the spore sac, which has spores only. Elaters are totally absent. The sporogenous tissue is endothecial in origin. The capsule has a distinct operculum, which gets removed at maturity. There is a fringe of peristome teeth at the mouth of capsule, it may be Arthrodontous (*Funaria*, Fig. 4) or Nematodontous type (*Pogonatum*, Fig. 4), which help in the dispersal of spores. The spores after germination first form a protonema. On the protonema, buds are produced which develop into adult gametophytes. Thus a single spore gives rise to number of gametophytes.

Bryopsida includes a large number of mosses, which are variously classified.

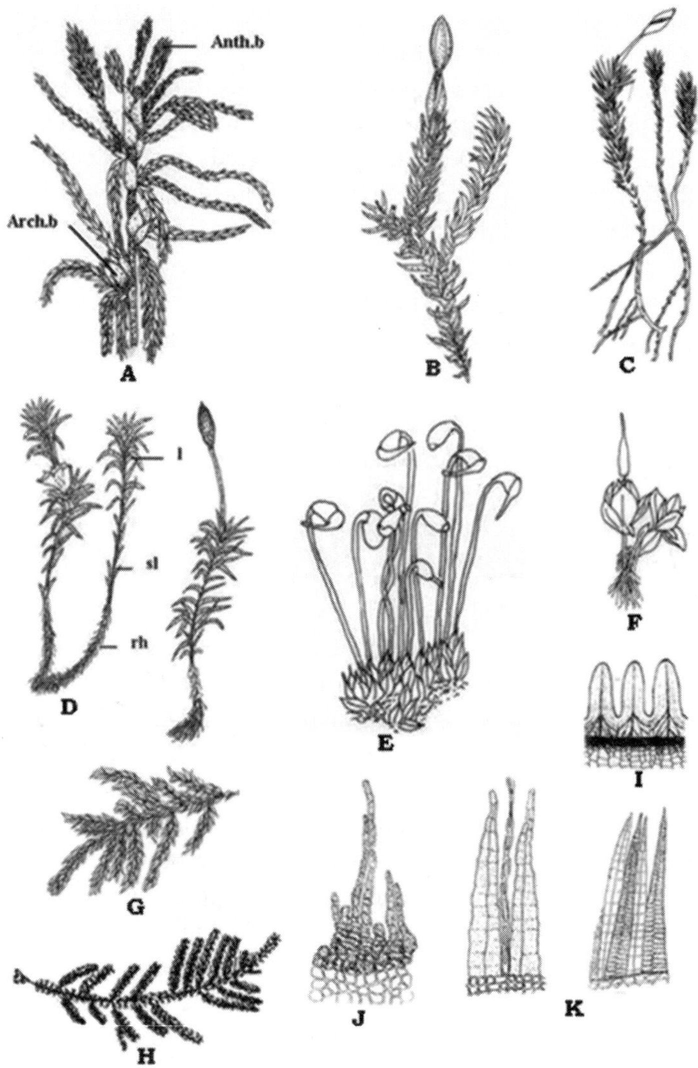

Fig. 4. (A-H) Diversity in Habit and morphoforms of mosses. (A) *Sphagnum*, (B) *Andreaea*, (C) *Takakia*, (D) *Pogonatum* (acrocarpous habit), (E) Plant of *Pogonatum* with capsule, (F) *Funaria*, (G) *Hypum*, (H) *Entodon* (pleurocarpous habit). (I-K) Peristome types, (I) Nematodontous. (J-K) Arthrodontous type, (J) Haplolepidous, (K) Diplolepidous

2.3.1. Sphagnales

The order is monotypic with a single subclass – Sphagnidae, order Sphagnales, family Sphagnaceae and single genus *Sphagnum* (Fig. 4).

Sphagnum is commonly called as 'Bog Moss' / 'Peat Moss'/ 'Moss Cotton' and it is an aquatic moss. The plants are differentiated into axis and spirally arranged leaves. There is a cluster of short, stout 'comal' branches forming a compact head at apex. Lateral branches are typically axillary and develop at certain intervals being pendent or divergent. The leaves are unistratose, without costa having dimorphic leaf-cells (hyaline and chlorophyllous). The chlorophyllous cells are elongated with numerous discoid chloroplasts, while hyaline cells are large, polygonal, without chloroplasts having distinct pores and thickening bands. Rhizoids are totally lacking in adult gametophyte while present in protonemal phase. The axis in cross section shows differentiation into cortex and medulla. In some species e.g. *S. molluscum*, few enlarged, flask shaped cells are present in the cortex, called as 'Retort Cells'. Vegetative reproduction in plants takes place by death and decay of older portion of plants, resulting into separation of lateral branches, which develop into new plants. Sex organs are present on special branches. Antheridial branch is short, spindle shaped. Antheridia occur in the axil of male bracts /perigonial leaves (Fig. 5H). The archegonial branch is short bud like having large, bright green conspicuous perichaetial leaves with 1-3 archegonia. The sporophyte possesses bulbous foot and spherical capsule only (Fig. 7Sb). Seta is lacking but pseudopodium is present, which is a leaf less extension of the gametophytic axis to raise the capsule. The capsule wall is multilayered with rudimentary stomata. Archesporium is dome shaped which arches over the central columella (Fig. 7Sb). The archesporium is amphithecial in origin, while columella is endothecial in origin – a feature common to hornworts. Annulus is present and Peristome is lacking. The dehiscence of capsule takes place through special 'Gun-shot Mechanism' or 'Air Gun Mechanism' in which operculum is thrown off with jerk. Spores are tetrahedral in shape with distinct triradiate mark. Spores germinate and form exosporic thalloid protonema, which further may develop into secondary protonema. Buds develop on protonema to form new gametophyte.

2.3.2. Andreaeales

This order is also a monotypic group with single family Andreaeaceae and genus *Andreaea* (Fig. 4). *Andreaea* usually occurs at high altitude, growing on fairly exposed, dry, silicious rocks and is commonly known as 'Granite Moss'. The plants of this genus are dark brown in colour and differentiated into axis and spirally arranged leaves, which may be with or without costa (midrib). The midrib portion is multistratose while wings are unistratose. Rhizoids are characteristically multicellular, cylindrical or flat, plate like which help the plants for anchorage by penetrating into the rock crevices or by sticking to the rock surface. The stem in a cross section exhibits outer, more or less thick walled cortex and comparatively thin walled medulla with slightly larger cells. Plants are mostly monoecious. Antheridia occur in groups at terminal position having a short stalk and elongated club shaped antheridial body, along with multicellular, uniseriate paraphysis. Archegonia are typically moss like with a short stalk, swollen venter and long neck. Both of the sex organs originate through the apical cell.

Fig. 5. Diversity in androecia and antheridia in bryophytes (A) *Calobryum indicum*, VLS through antheridia. (B) *Frullania tamirisci*, male branch and male bract with two antheridia. (C) *Aphanolejeunea cornutissima*, bract with solitary antheridium. (D) *Pellia epiphylla,* VTS thallus through antheridia. (E) *Metzgeria pubescens*,ventral branch shows antheridia on costa. (F) *Marchantia polymorpha*, antheridiophore and a cross section of the same. (G) *Anthoceros gemmulosus*, male thallus and androecial chamber showing two antheridia. (H) Antheridial branch in *Sphagnum.* (I) Male branch in *Funaria hygrometrica*. (J) Antheridial head in *Mnium*. (K) Antheridial head in *Pogonatum*. (L-V) Antheridia. (L) *Calobryum indicum*. (M) *Lejeunea* sp. (N) *Lophozia* sp. (O) *Marchantia* sp. (P) *Pogonatum* sp. (Q) *Sphagnum* sp. (R) *Funaria hygrometrica*. (S) *Polytrichum* sp. (T) *Adreaea* sp. (U) *Phaeoceros himalayensis*. (V) *Anthoceros erectus.* (A after Udar and Chandra 1965, B after Cavers, C after Schuster 1966, D,F after Parihar 1959, E After Schuster 1966, G after Asthana and Srivastava 1991, H after Parihar 1959, I-K, P-T after Udar 1976, L after Udar and Chandra 1965, M after Muller 1948, N after Schuster 1966,O after Parihar 1959, P,Q,R,S,T after Udar 1976, U,V after Asthana and Srivastava 1991).

The sporophyte possesses almost ovoid capsule and swollen foot. Seta is lacking but pseudopodium is present. The capsule wall is multilayered. Archesporium is dome shaped which arches over columella.

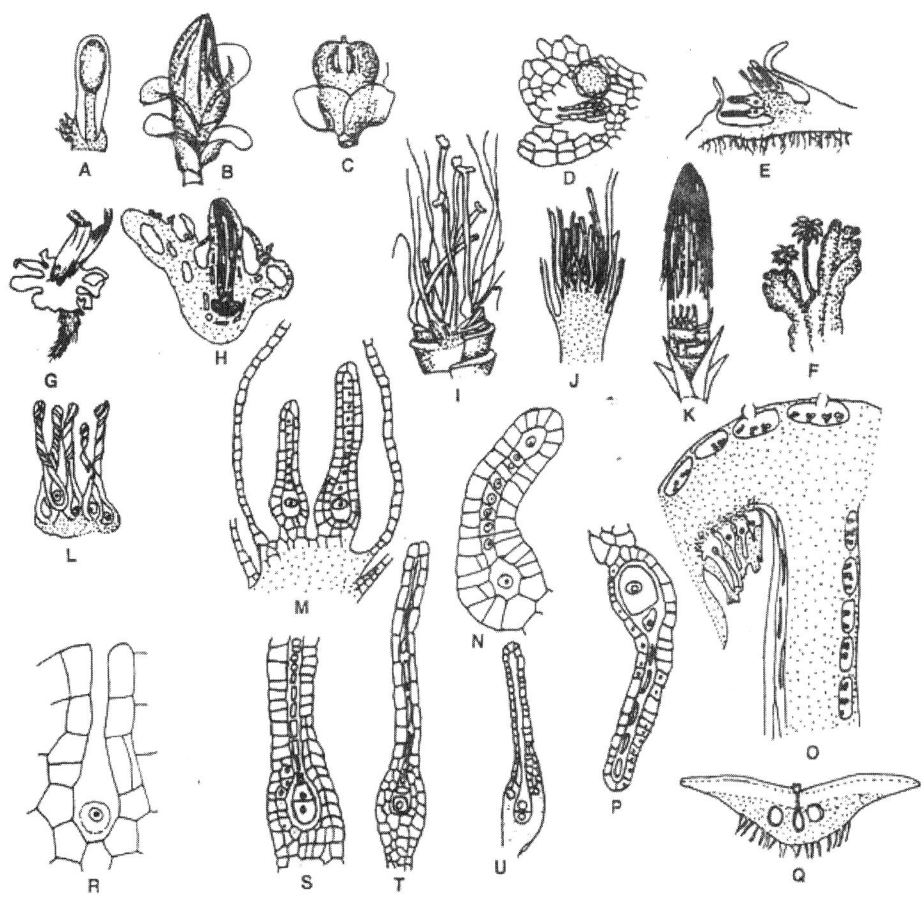

Fig. 6. Diversity in archegonial structure and their protection devices (A) Shoot calyptra in H*aplomitrium hookeri*. (B) Plicate perianth in *Solenostoma* sp. (C) Perianth in *Rectolejeunea* sp. (D) Synoecia in *Riccardia multifida*. (E) Cross section of involucre in *Pellia endaevifolia,* (F) Archegoniophore in *Marchantia polymorpha*. (G) Involucre in *Anthoceros erectus*. (H) A section through young sporophyte. (I) Apex of female gametophores in *Polytrichum* sp. (J) Vertical section at the apex of female gametophore in *Funaria*. (K) Apex of the same with hairy calyptra. (L-P) Structure of arhegonia. (L) *Calobryum indicum*. (M) *Frullania* sp. (N) *Pellia epiphylla*. (O) *Marchantia polymorpha* (Vertical section of archegoniophore). (P) Archegonia of the same. (Q) Cross section of thallus through archegonium in *Riccia billardieri*. (R) *Anthoceros*. (S) *Pogonatum*, (T) *Funaria*. (U) *Sphagnum* sp. (A-D after Schuster 1966, E,F,G,H,I, J,K,M,N,U after Udar 1976, L after Udar and Chandra 1965, O,P,R after Parihar 1959, Q after Udar 1957).

Fig. 7. Diversity in sporophyte and their dehiscence patterns in bryophytes (A) *Calobryum indicum*, V.L.S. through young sporophyte. (CAL calyptra, CAP capsule, S seta, F foot, SA shoot apex). (B) *Calobryum dentatum*, sporophyte in shoot calyptra. (C) *Haplomitrium hookeri*, dehiscence of capsule. (Ca) Through single longitudinal slit, (Cb) Completely dehisced capsule with double spoon head, (Cc) Fixed elaters at apex and base of capsule. (Da) *Jungermannia tetragona*, perianth and mature capsule, (Db) *Jungermannia* (Plectocolea) *tetragona,* dehisced capsule. (E) *Cololejeunea madathecoides*, perianth and four-valved dehisced capsule. (Fa) *Jackiella javanica*, marsupium. (Fb) Two-valved dehisced capsule. (G) *Geocalyx graveolense*, marsupium with four-valved dehisced capsule. (H) *Frullania*, perianth and dehisced capsule and fixed elaters. (Ia) *Pellia epiphylla*, involucre and mature sporophyte on thallus. (Ib) Dehisced capsule with basal elatophore. (Ja) *Riccardia levieri*, involucre, calypta and mature sporophyte. (Jb) Dehisced four-valved capsule. (K) *Marchantia* sp., Calyptra, perigynium and dehisced capsule. (L) *Riccia,* embedded sporophyte. (M) *Notothylas indica*, dehisced capsule with columella. (N) *Anthoceros erectus*, thallus with involucre, dehisced capsule and columella. (Oa) *Pogonatum,* female plant with sporophytes. (Ob) *Pogonatum*, Capsule with calyptra lifted. (Oc) Capsule with detached operculum showing peristome rim. (Pa) *Funaria* sp., female branch with sporophyte. (Pb,c) Capsule with operculum removed to show peristome teeth. (Q) *Andreaea rupestris,* portion of plant with sporogonium. (Ra) Elongated pseudopodium and mature sporogonium. (Rb) Mature dehisced sporophyte with columella. (Rc) Sporophyte with calyptra. (Sa) *Sphagnum,* plant apex with mature sporogonium. (Sb) Cross section of capsule. (Sc) Dehisced capsule.

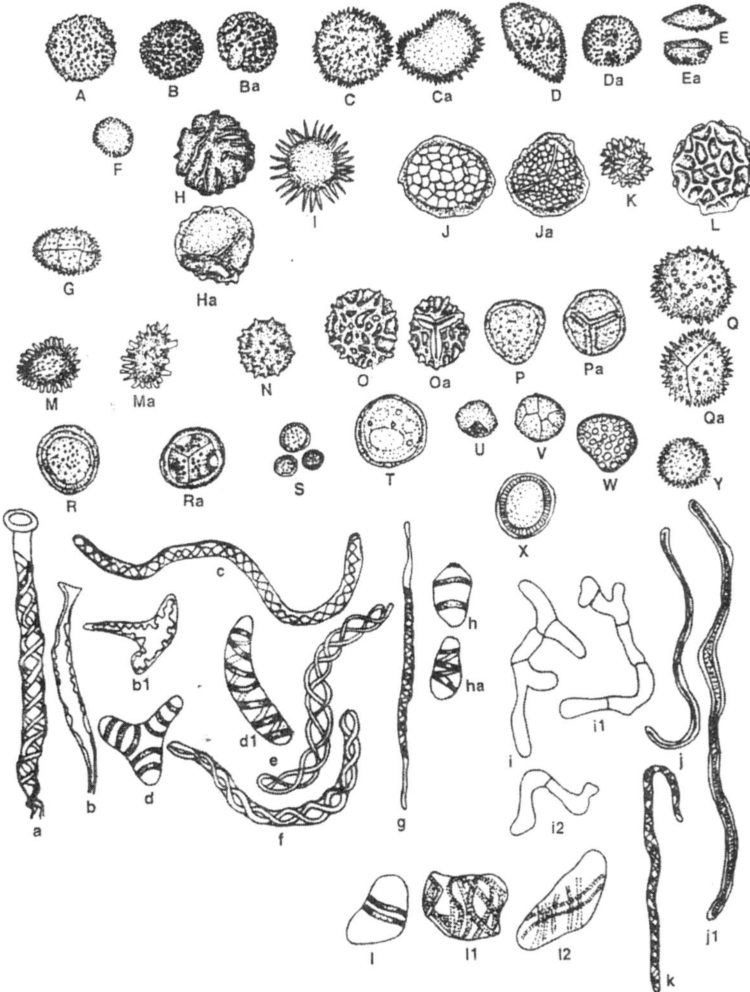

Fig. 8. Diversity in Spores and elaters in bryophytes. (A-Y) Spores. (A) *Calobryum dentatum*. (B, Ba) *Haplomitrium hookeri*. (C,Ca) *Radula tabularis*. (D,Da) *Schiffenriolejeunea indica*. (E,Ea) *Lejeunea indica*. (F) *Metzgeria indica*. (G) *Pellia epiphylla*. (H,Ha) *Fossombronia himalayensis*. (I) *Calycularia crispula*. (J,Ja) *Riccia pandei*. (K) *Athalamia pinguis*. (L) *Asterella* sp., (M,Ma) *Cyathodium indicum*. (N) *Cyathodium tuberculatum*. (O,Oa) *Anthoceros erectus*. (P,Pa) *Phaeoceros laevis*. (Q,Qa) *Folioceros udarii*. (R,Ra) *Notothylas dissecta*. (S) *Pogonatum himalayanum*. (T) *Oligotrichum* sp., (U) *Encalypta* sp. (V-Y) *Fissidens*, range of sporoderm pattern. (a – l) Elaters. (a) *Schiffineriolejeunia indica*. (b b1) *Lejeunea indica*. (c) *Radula tabularis*. (d-f) *Fossombronia* sp. (g) *Riccardia sikkimensis*. (h,ha) *Targionia indica*. (i,i1).*Anthoceros* sp .(j,j1) *Folioceros* sp. (k) *Megaceros* sp. (l, l1, l2) In *Notothylas* spp. (A after Gupta and Udar 1986, B after Udar and Srivastava 1981, C after Udar and Kumar 1982, D, E after Udar and Awasthi 1981,1982, F after Udar and Srivastava 1970, G after Schuster 1966, H, Ha after Srivastava and Udar 1975, I after Pande and Udar 1956, J, Ja after Udar 1959, K after Udar 1960, L after Gupta and Udar 1986, M N after Srivastava and Dixit, 1996, O,P,Q R, after Asthana and Srivastava 1991, S-Y after Chopra and Kumar 1981, a,b,b1 after Udar and Awasthi 1981,1982, c after Udar and Kumar 1982, d,e,f after Srivastava and Dixit, 1996, g after Srivastava and Udar1976, h after Udar and Gupta 1983, i- l after Asthana and Srivastava 1991).

Archesporium as well as columella both are endothecial in origin. Annulus, operculum and peristome are lacking. The capsule dehisces along four longitudinal lines forming four distinct valves but these valves remain attached at apex and base forming four slits (Fig. 7Q, Ra, Rb, Rc). The spores germinate to form multicellular globose sporeling within the stretched spore coat (endosporic thalloid protonema).

2.3.3. Takakiales

It is also a monotypic order with single family Takakiaceae and single genus *Takakia* (Fig. 4 C). Plants are characterized by heterotrichous nature, differentiated into prostrate rhizomatous axis and erect growing aerial axis or shoot. Shoots are densely covered with mucilage papillae. The leaves are present in three rows, which are spirally arranged. Leaves are cylindrical, with 2-4 finger-like lobes. Rhizoids are absent. Antheridia are in groups and are terminal in position. Archegonia are stalked with long neck, terminal or on stem surface. Sporophyte has distinct foot, long seta and ovate-elongated capsule with multilayered capsule wall. Archesporium is dome shaped over central columella. Operculum, annulus and peristome are lacking. The outermost layer of the capsule wall is thickened causing cell lumen appearance more or less flask shaped. Capsule dehisces through single longitudinal line (slit). Spores are tetrahedral in shape with distinct tri-radiate mark.

2.3.4. Polytrichales

Plants of order Polytrichales are characterized by erect growing, robust (giant sized) gametophytes, differentiated into axis, spirally arranged leaves and multicellular, branched, rhizoids having oblique septa. Leaves have sheathing base and upper limb portion with distinct midrib and lateral wings. Vertical photosynthetic lamellae are usually present on the Leaf surface. Stem is highly differentiated internally with specialized hydroid and leptoids cells. The sporophyte is differentiated into distinct foot, long and rigid seta and erect, elongated capsule. The capsule has a distinct operculum and nematodontous type of peristome (Fig. 4) with 16-32-64 teeth. Calyptra is hairy (Fig. 7Oa,Ob). The sporogenous tissue is present all around the columella and is endothecial in origin. The protonema is filamentous.

The order has a single family Polytrichaceae consisting of 21 genera. Among these, *Pogonatum* (Fig. 4), *Polytrichum, Dawsonia* are large sized mosses. In India 5 genera: *Atrichum, Oligotrichum, Lyellia, Pogonatum* and *Polytrichum* are reported.

2.3.5. Tetraphidales

This order has a single family Tetraphidaceae and the genus *Tetraphis* (*Georgia*). The genus is characterized by the presence of four nematodontous type of peristome teeth in capsule. It is not represented in India.

2.3.6. Bryales

This order consists of large number of true mosses which are characterized by small to large sized, erect to prostrate growing plants, differentiated into axis and two - many rows of spirally arranged leaves with mid rib. The rhizoids are multi-cellular and oblique septate. The sporophyte has a distinct foot, elongated seta and variously shaped capsule with a definite operculum (Fig. 7Pb,Pc) and arthodontous type of peristome, present in one ring (haplolepidous) or two rings (diplolepidous) - Fig. 4. The sporogenous tissue is present all around the columella and is endothecial in origin. The protonema is filamentous. The group has 15 suborders, 85 sub familes and 765 genera (Vitt 1984).

3. Methods for Collection, Preservation and Identification

A systematic approach is required for the taxonomical studies at each and every step starting from the collection of plants, their preservation in herbarium and proper investigation for correct identification of plants.

3.1. Collection of bryophytes

The collection of Bryophytes requires careful handling because these are very delicate and tiny plants. The ideal way to learn and collect the plants for the first time is, to join field excursion trip led by experts. A systematic collection provides the base for taxonomic research. During collection a careful screening of limited area is more fruitful than visiting different places at one time. Some important points to be remembered and the materials required during collection of bryophytes are mentioned below. The collection kit should contain the following items:

- Herbarium packets
- Polythene Bags
- Specimen Bottles/Tubes
- Preservative (90 % Ethyl alcohal)
- Field/Hand lens
- Sharp edged knife/Scalpel
- Small saw/twig cutter
- Blotting paper/ News paper
- Pen/Pencil / Marker
- Slip pad/ Diary
- Altimeter
- Thermohygrometer

- GPS
- Field camera
- Field Book

Bryophytes usually grow on soil, rock, soil covered rock, stem bark, twigs and leaf surfaces (Fig. 9). To collect them, plants should be gently scraped with a sharp edged knife from the surface without damaging the plant body and natural population. A sufficient patch of plant should be kept in the herbarium packets or polythene bags after removing the extra substrate. Not more than required quantity for the study should be collected. In case of corticolous and epiphyllous plants, a small portion of twigs and leaves should be kept in packets. While collecting the plants the major portion of the populations should be kept in paper packets, but a small portion of the plants with fruiting bodies may be fixed in 70% alcohol in small specimen bottles or tubes.

Fig. 9. Bryophytes in different habitats. (A) on rock surface (rupicolous), (B) On soil (terricolous), (C, D) On stem bark (Corticolous), (E, F) On leaf surface (epiphyllous)

All the field data (including GPS Data) must be recorded in the field book/diary and on the packet itself. Field data should have the following parameters:

- Locality (Collection site)
- Date of collection
- Habitat details
- Associates
- Altitude, Longitude, Latitude, etc.
- Name of Collector

The name of plants (genus/species), their associates and the name of the determinator can be added after investigation and identification of taxon. If the plants are required to be investigated in living condition they should be kept in polythene bags.

3.2. Preservation of bryophytes in herbarium

Preservation of Bryophytes in Herbarium does not require any kind of poisoning or special kind of treatment as these plants naturally possess some antimicrobial compounds which protect them from damage caused by Fungi and other micro-organisms. These can be preserved safely for several years. The collected plants should not be left moist for longer period. They should be dried without pressure in between blotting paper. For the study of oil bodies in liverworts, the plants can be kept fresh for a month under refrigeration.

To maintain a proper herbarium of Bryophytes we require:

- Herbarium packets
- Labels
- Herbarium sheets
- Boxes
- Herbarium Almirah

The herbarium packets should be of standard size (5"x 4"). These can be made by folding a thick paper (bamboo paper) of good quality. These herbarium packets can be glued on Herbarium sheets which may be placed in folders and the folders inside Almirah. Herbarium packets can also be arranged directly in the boxes inside the Almirah. The arrangement may be serial wise according to years, or alphabetically genera-wise or family-wise. For floristic studies they may be arranged area-wise.

3.3. Investigation of bryophyte specimens

The following (Fig. 10) are needed for the investigation of bryophyte specimens:

- Petridishes
- Forceps
- Needles
- Brush
- Scissors
- Sharp edged Blade
- Paper
- Pencil
- Notebook or Register
- Camera-lucida
- Compound light microscope
- Dissecting binoculor microscope

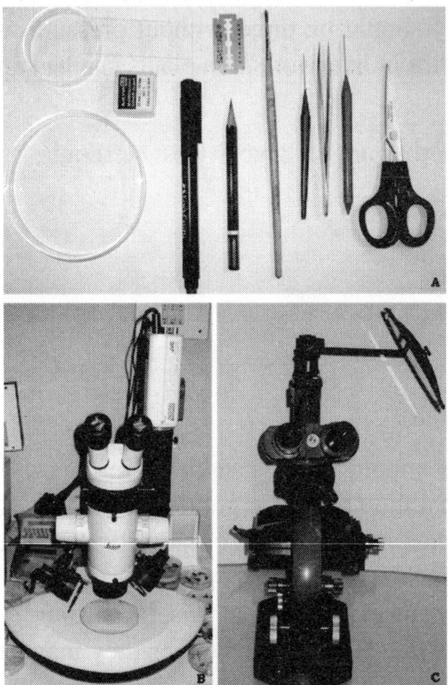

Fig. 10. (A) Articles required for investigation of Bryophytes. (B) Stereoscopic dissecting Zoom Binoculor Microscope. (C) Compound Research Microscope attached with Camera Lucida

3.4. Making of Herbarium packets to keep the Bryophyte plants

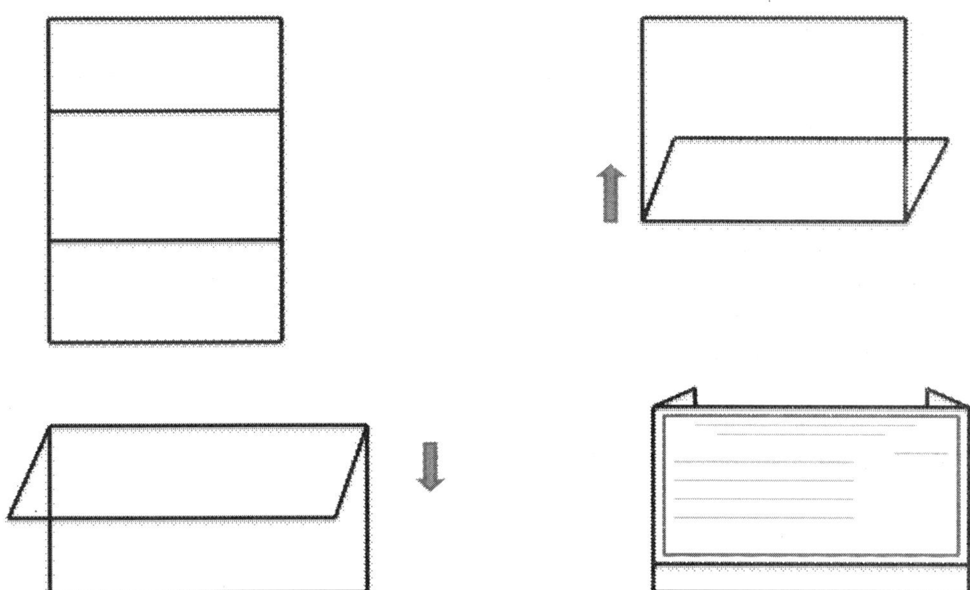

Sample of Herbarium Label

HERBARIUM
NATIONAL BOTANICAL RESEARCH INSTITUTE
LUCKNOW

No.:LWG202147 **Date:** 19.04.1965
Name: A. *Ptychanthus striatus* (Lehm. & Lindenb.) Nees
Family: Lejeuneaceae
Locality: Tiger Hill, Distt. Darjeeling (West Bengal), E. Himalaya
Altitude: ca 8,250 ft (2500 m)
Habitat: Epiphytic
Associates: B. *Plagiochila elegans*, C. *Frullania pariharii*
Collected by: S. Chandra
Determined by: V. Nath, A.K. Asthana & V. Sahu

4. Investigation

In order to investigate the plants, the entire collected population in a specimen packet should be soaked in water at normal temperature for stretching the plants so as to regain their original shape and size. The leafy plants take few seconds to few minutes and thalloid plants take few minutes to few hours to get stretched. Normally plants regain their exact natural shape but sometimes they fail to do so up to some extent in older collection, hence it is ideal to fix few plants in 70% or 90% alcohol at the time of collection.

The stretched population should be labelled with same number of original herbarium packet by marking the number on petridishes or by putting a paper slip. The whole population should be scanned properly under the dissecting binoculor microscope to ascertain that how many taxa are present in the population. If there is more than one taxon, all the different plant types should be separated and kept in small petridishes. After thoroughly cleaning the individual plant, it is examined under Stereo Zoom Binoculor microscope for morphological study. For anatomical study glycerine mounts of cross section of thallus or stem, dissected leaves, underleaves and other parts are prepared on glass microslides and examined under Compound Research Microscope. Illustrations/line drawings are prepared by using Camera Lucida attached with the Microscope.

4.1. Key for identification of bryophytes

1. Plants leafy (Foliose) .... 2.
2. Midrib present in leaves, rhizoids multicellular and branched with oblique septa, elaters absent .... **Mosses**
2. Midrib absent in leaves, rhizoids unicellular and unbranched, elaters present .... **Liverworts**
1. Plants thalloid ....3.
3. Sporophyte with foot, seta and capsule, columella absent in the capsule, capsule globose/elliptical .... **Liverworts**
3. Sporophyte with foot and capsule only, columella usually present in the capsule, capsule horn/needle shaped .... **Hornworts**

(i) Key for Identification of Bryopsida (Mosses)

1. Peristome absent 2
2. Capsule dehiscence through operculum (Gun-shot mechanism) .... I. Subclass – **Sphagnidae** (Peat Moss)
2. Capsule dehiscence through longitudinal slit 3
3. Capsule dehiscence through four longitudinal slits .... II. Subclass - **Andreaeidae** (Granite Moss)
3. Capsule dehiscence through single longitudinal slit .... III. Subclass – **Takakiidae**

1. Peristome present .... IV. Subclass - **Bryidae** (True Moss). 4
4. Nematodontous type of Peristome 5
5. Peristome present in a ring of 16, 32 or 64 teeth .... Order – **Polytrichales**

5. Peristome present in a ring of 4 teeth .... Order – **Tetraphidales**
4. Arthodontous type of Peristome .... Order – **Bryales.** 6
6. Peristome present in a single ring .... **Haplolepideae (Pottiineae, Dicranineae, Fissidentineae, Seligeriineae, Grimmiineae)**
6. Peristome present in two rings .... **Diplolepideae (Archidiineae, Funariineae, Splachnineae, Orthotrichineae, Bryineae, Hypnineae, Leucodontineae, Hookeriineae, Buxbaumiineae, Encalyptineae)**

(ii) Key for Identification of Liverworts

1. Plants may be leafy or thalloid ....**Jungermannidae** 2
2. Plants with leafy organization.. 3
3. Leaves present in 3 rows, transversly inserted on the axis .... I. Order – **Calobryales**
3. Leaves present in 2-3 rows, obliquely inserted on the axis (Succubous or Incubous)...II. Order – **Jungermanniales**
2. Plants thalloid (sometimes leafy) .... III. Order - **Metzgeriales**
1. Plants thalloid only .... **Marchantiidae** 4
4. Thallus internally undifferentiated into assimilatory and storage zones, ventral scales absent.. 5
5. Thallus with bottle shaped involucre, seta reduced .... IV. Order – **Sphaerocarpales**
5. Thallus with pouch shaped involucre, seta elongated................V. Order – **Monocleales**
4. Thallus internally differentiated into assimilatory and storage zones, ventral scales present .... VI. Order **Marchantiales**

(iii) Key for Identification of Hornworts

1. Thallus spongy with mucilage chambers 2

2. Elaters vermiform with dark central lumen.. I- ***Folioceros***

2. Elaters short without lumen .... II - ***Anthoceros***

1. Thallus compact without mucilage chambers 3

3. Capsule short inside the involucre or slightly exserted .... III - ***Notothylas***

3. Capsule long, exserted 4

4. Elaters with spiral thickening bands .... IV - ***Megaceros***

4. Elaters without spiral thickening bands ....V-***Phaeoceros***

A

B

C

D

E

F

Fig. 11. (A-D). Thalloid liverworts. (A) *Plagiochasma*, (B) *Reboulia*, (C) *Asterella*, (D) *Marchantia*, (E,F) Leafy liverworts. (E) *Plagiochila*, (F) *Frullania*

Fig. 12. Different taxa of Mosses. (A) Acrocarpous moss, (B) Pleurocarpous moss, (C) *Polytrichastrum*, (D) *Rhodobryum*, (E) *Sphagnum*

4.2. Taxonomic parameters for the study of bryophytes

The following Characters are generally used for the systematic study of liverworts, hornworts and mosses.

Liverworts (Hepaticopsida)

The gametophytes of Liverworts show both leafy and thalloid organization. They may be erect growing (primitive leafy liverworts) or prostrate growing (thalloid as well as leafy liverworts). The gametophytes are attached to the substratum by means of unicellular and unbranched rhizoids.

Leafy liverworts

The leafy liverworts have axis and 2-3 rows of leaves.

Rhizoids

The rhizoids are smooth walled present either on the axis throughout (*Jungermannia*), or at the base of underleaf (*Frullania*) or at the leaf lobule (*Radula*).

Branching

The branching may be terminal or intercalary, acroscopic or basioscopic, Frullania-type, Microlepidozia-type, Acromastigum- type, Zoopsis - type, Radula - type, Lejeunea - type, Bryopteris -type, thecal or athecal -type.

Stem anatomy

Cross section of stem clearly exhibits the cortical and medullary regions. These may be undifferentiated or differentiated. If differentiated, then thickness, size and number of cortical cells signify the generic characteristics (Lepidoziineae, Lejeuneaceae).

Leaves

The leaves may be present in 2-3 rows. Sometimes all the leaves are identical (isophyllous) or they may differ in size having two rows of large dorsolateral leaves and one row of small underleaf. They are unistratose and without midrib. The leaf may be simple, unlobed or bi-tetra lobed. Sometimes they are complicate bilobed having large leaf- lobe and ventral leaf-lobule.

Leaf shape

The shape of the leaves is very important parameter. They may be ovate, obovate, oblong, lanceolate, linear or spathulate.

Leaf margin

The leaf margin may be entire, dentate, serrated or crenate. Sometimes the leaves have specialized cells: ocelli or vitta cells which are present at the base of leaf either singly or in rows (*Frullania*). Sometimes, leaves are bordered by hyaline cells (*Cololejeunea*).

Leaf arrangement

The leaves may be distant or closely arranged either succubously or incubously.

Sporophyte

The sporophyte has distinct foot, seta and capsule.

Spores

The spores show various sporoderm patterns: papillate, spinate, reticulate, lamellate or echinulate.

Elaters

Elaters elongated with tapering ends having spiral thickening bands.

Thalloid liverworts

The plants are dorsiventral and usually dichotomously branched.

Rhizoids

The rhizoids are smooth walled, tuberculate and sinuate. They are present on the ventral surface of the thallus.

Ventral scales

Ventral scales are present in 1-2 (3) rows on the ventral side of the thallus in Order Marchantiales.

Epidermal pores

Epidermal pores are present on the dorsal side of the thallus in Order Marchantiales. They may be simple, compound or barrel shaped (e.g. *Marchantia*) depending upon the cells arrangement.

Thallus anatomy

The thallus may be compact, internally undifferentiated (e.g. Order Metzgeriales) or may be internally differentiated in to upper assimilatory zone & lower storage zone (e.g. Order Marchantiales).

Sporophyte

The sporophyte has distinct foot, seta and capsule.

Spores

The spores show various sporoderm patterns: papillate, spinate, reticulate, lamellate or echinulate (Fig. 8).

Hornworts (Anthocerotopsida)

Gametophyte

The hornwort gametophytes show thalloid organization. They are prostrate growing and sometimes form distinct rossettes. The thallus cells are thin walled having chloroplasts with distinct pyrenoid bodies. The number of chloroplasts is usually one per cell of epidermal layer in *Anthoceros, Folioceros, Phaeoceros* and *Notothylas*, while more than one in *Megaceros.* The shape of chloroplast and nature of pyrenoid bodies play important role to delimit some taxa.

Thallus anatomy

Thallus anatomy is an important parameter for generic segregation. The thallus may be spongy having chambers filled with mucilage (*Anthoceros, Folioceros*) or compact, without chambers (*Phaeoceros, Megaceros* and *Notothylas*).

Rhizoids

The thalli are attached to the substratum by means of unicellular, unbranched and smooth walled rhizoids.

Sporophyte

The sporophyte in hornworts is needle or horn shaped having foot and capsule only. The seta is absent however a compressed meristematic portion is there. The capsule is elongated and cylindrical having multilayered capsule wall with or without stomata, archesporium (spores & pseudoelaters) and central columella, which is lacking in some species of *Notothylas*. The spore production is continuous due to presence of meristematic tissue at the base of the capsule. The capsule dehisces in to two valves releasing the spores and pseudoelaters.

Spores

The spores show various ornamentation patterns ranging from papillate to lamellate with or without equatorial crossitudo or hump in *Phaeoceros, Megaceros and Notothylas*, while spinulate, baculate, reticulate, punctate and reticuloid in *Anthoceros* and *Folioceros*. The colour of the spores and sporoderm patterns are very important for generic and species delimitation.

Pseudoelaters

Pseudoelaters are short and stumpy in *Anthoceros* and *Notothylas* with spiral and annular thickenings in latter, while vermiform longish with central dark lumen in *Folioceros* and with spiral thickening bands in *Megaceros*.

Mosses (Bryopsida)

Gametophyte

The moss gametophytes show leafy organization having axis and leaves. They may be erect growing (Acrocarpous) or prostrate growing (Pleurocarpous). The growth habit is one of the important taxonomic parameter used for the classification of mosses (Acrocarpi & Pleurocarpi). The plant (axis) may have limited growth in acrocarpous mosses with apical, terminal archegonia / sporophyte on the main axis or unlimited growth in pleurocarpous mosses with apical, terminal archegonia / sporophyte on lateral branches.

Rhizoids

The gametophytes are attached to the substratum by means of multicellular, branched, oblique septate rhizoids which are the characteristic feature of the mosses as in liverworts and hornworts the rhizoids are unicellular and unbranched. They perform the functions of anchorage and absorption. Sometimes they are present in such quantities as to form tomentum.

Axis

The branching may be regular or irregular pinnate (bipinnate – tripinnate).

Leaves

The leaves may be present in 2- many rows in spiral manner. The number of rows of leaves is determined by the nature of apical cell: bifacial (2 rows in *Fissidens*) or tetrahedral and their derivative segments (3-many rows).

Leaf shape

The shape of the leaves is an important parameter. They may be ovate, obovate, oblong, lanceolate, linear or spathulate.

Leaf margin

The leaf margin may be entire (smooth), wavy (flexuose), serrated (saw like toothed edge). The teeth or dentations may be small and indistinct to large and prominent. They may be present throughout the margin or confined to apical portion only.

Leaf apices

The leaf apices are remarkably variable and taxonomically important. It may be stout abrupt point (mucronate), long stiff point (cuspidate), gradually tapered narrow point (acuminate), short and sharply pointed (acute), turned to one side (falcate) and turned to form circle (circinate).

Leaf costa

The leaves are usually unistratose with distinct, multistratose midrib / nerve /costa (costate). The midrib may be absent in some mosses like *Sphagnum*, *Andreaea* (ecostate). The costa may be single, double or forked. The costa when fails to reach the leaf tip (vanishing), when reaches to the tips (percurrent) and when extends beyond the tip, often forming a hair like awn (excurrent).

Leaf arrangement

The leaf arrangement is variable and taxonomically important. The leaves may be arranged in two rows (complanate), close to stem (erect), with 45^0 or less divergence (patent), with 45^0 or over divergence (spreading), turned back (squarrose), turned to one side (secund).

Leaf-cells

The leaf cells may vary in shape and size. They may be rectangulate, elongated flexuose, sometimes with or without papillosity. Basal alar cells are very important for generic and species segregation.

Sporophyte

A typical moss sporophyte has distinct foot, seta and capsule.

Foot

The foot is an absorbing organ embedded in the gametophytic tissue.

Seta

The seta is the stalk of the sporophyte which holds and carries the capsule above the perichaetial leaves. It may be short, long, straight, curved (arcuate) or twisted.

Capsule

A typical moss capsule has apophysis (basal sterile portion), theca or urn (middle fertile portion) and operculum (apical lid). They may be erect, pendent, horizontal or curved. They may be immersed (perichaetial leaves reaching up to the apex), emergent (perichaetial leaves reaching up to the base) or exerted (high above the perichaetial leaves). The shape of the capsule may be globose, ovate, oblong, elongated, cylindrical, pyriform, asymmetrical, arcuate.

Calyptra

The calyptra is a protective covering over the moss capsule derived from archegonial venter. It has taxonomic importance to some extent. It covers the capsule partly or completely. It may be deciduous or persistant, mitrate or mitriform (splitting on two or more sides), cuculate – hood shaped splitting on one side only. It is very characteristic and hairy in the members of Polytrichales (*Pogonatum* & *Polytrichum*) covering the capsule up to the base.

Operculum

The operculum is the lid or cap of the capsule with less taxonomic significance. It may be convex, conic, conic-apiculate, beaked – long rostrate or short rostrate.

Peristome

Peristome is an important taxonomic tool. The presence (peristomate) and absence (gymnostomous) of teeth, number of teeth (4, 16, 32, 64), type of peristome (Nematodontous and Arthodontous) and the number of rings in peristome (one ring in haplolepidous and two rings in diplolepidous) are important taxonomic parameters.

Dehiscence

The moss capsule may be cleistocarpous (lacking lid and peristome) or Stegocarpous (having lid and peristome).

Important Herbaria with Bryophyte Collection

- Conservatoire et Jardin Botaniques, Geneve, Switzerland (G)
- British Museum (Natural History), London, England (BM)
- New York Botanical Garden, Bronx, New York, U.S.A. (NY)
- Royal Botanical Garden, Kew, London, U.K. (K)
- Herbarium Haussknecht, Jena, Germany (JE)
- Hattori Botanical Laboratory, Miyazaki ken, Japan (NICH)

- Farlow herbarium of Cryptogamic Botany, Cambridge, Massachusettes, U.S.A. (FH)
- University of California , Berkley Herbarium, California, U.S.A. (UC)
- Lucknow University Herbarium, Lucknow, India (LWU)
- Herbarium National Botanical Research Institute, Lucknow, India (LWG)
- Central National Herbarium, Kolkata, India (CNH)

Important Bryological Journals

- Journal of Bryology, UK
- The Bryologist, USA
- Cryptogamie Bryologie, France
- Tropical Bryology, Germany
- Evansia, USA
- Journal of Japanese Botany, Japan
- Taiwania, Taiwan
- Yushania, Taiwan
- Lindbergia, The Netherlands
- Journal of Hattori Botanical Laboratory, Japan

References

Asthana AK, Srivastava SC (1991) Indian Hornworts (A taxonomic study). Bryophyt Biblioth 42:1-230

Chopra RS, Kumar SS (1981) Mosses of western Himalayas and adjacent plains. Chronica Botanica, New Delhi, pp. 1-142

Crum HA (2001) Structural Diversity of Bryophytes. The University of Michigan Herbarium, Ann Arbor, MI pp. 379

Duff RJ, Villarel JC, Cargill DC, Renzaglia KS (2007) Progress and challenges toward a phylogeny and classification of the hornworts. The Bryologist 110 (2):214-243

Gradstein SR, Churchill SP, Salazan AN (2001) Guide to the Bryophytes of Tropical America. Mem N Y Bot Garden 86:1-577

Gupta A, Udar R (1986) Palynotaxonomy of selected Indian liverworts. Bryophyt Biblioth 29:1-202

Kamimura, M (1961) Monograph of Japanese Frullaniaceae. J Hattori Bot Lab 24:1-109

Muller, K (1948) Morphologische und anatomische Untersunchungen und antheridien Jungermannien. Bot Not 1948:71-80

Pande SK, Udar R (1956) Studies in Indian Metzgerineae-III. *Calycularia crispula* Mitt. Phytomorphology 6:331-346

Parihar NS (1959) An Introduction to the Embryophyta.I. Bryophyta. Central Book Depot, Allahabad

Schuster RM (1966) The Hepaticae and Anthocerotae of North America. East of the hundredth meridian. Vol.I. the Columbia University Press, New York

Schuster RM (1984) Evolution, Phylogeny and Classification of the Hepaticae. In: New Manual of Bryology, Schuster RM (ed.), The Hattori Botanical Lab, Japan, pp. 892-1070

Singh DK (1997) Liverworts In: Floristic Diversity and Conservation strategies in India, Mudgal, V and Hajra, PK (eds.), Botanical survey of India, Kolkata, pp. 235-299

Srivastava SC (996) Distribution of Hepaticae and Anthocerotae in India In: Topics in Bryology, Chopra RN (ed.), Allied Publishers Ltd. New Delhi, pp. 53-88

Srivastava SC, Dixit R (1996) The genus *Cyathodium* Kunze. J Hattori Bot Lab 80:149-215

Srivastava SC, Udar R (1975) The genus *Fossombronia* Raddi in India with a note on the Indian taxa of family Fossombroniaceae. Nova Hedwigia 36:799-845

Srivastava SC, Udar R (1976) Indian Aneuraceae- A monographic study. Biol. Mem. 1:121-154

Udar R (1957) Culture studies in the genus *Riccia* (Mich.) L.I. spore germination in *Riccia billardieri* Mont. et Nees. J Indian Bot Soc 36:46-50

Udar R (1959) Genus *Riccia* in India-IV. A new *Riccia, R. pandei* Udar from Garhwal with a note on the species of the genus from the western Himalayan territory. J Indian Bot Soc 38:146-159

Udar R (1960) Studies in Indian Sauteriaceae-II. On the morphology of *Athalamia pinguis* Falc. J Indian Bot Soc 39:56-57

Udar R, Chandra V (1965) On a new species of *Calobryum, C. indicum* Udar *et* Chandra from Darjeeling, India. Rev Bryol et Lichenol 33:556-559

Udar R, Gupta A (1983) Differentiation of the genus *Targionia* L. in India-II. The East Himalayan and South Indian complex and description of a new species of *Targionia*. Geophytology 13:83-87

Udar R, Kumar D (1982) The genus *Radula* Dumort. in India-I. J Indian Bot Soc 61: 177-182

Udar R, Srivastava SC (1970) A new species of *Metzgeria* from India. Rev Bryol *et* Lichenol 37:361-365

Udar R, Srivastava SC (1981) On some noteworthy features in the sporophyte of *Haplomitrium hookeri* from the western Himalayas. J Indian Bot Soc 60:179-181

Udar R, Srivastava SC, Kumar D (1982) *Geocalyx* Nees: a rare marsupial genus from India. Proc Indian Acad Sci (Plant Sci) 91:139-143

Udar R (1976) An Introduction to Bryophyta. Shashi Dhar Malviya Prakashan, Lucknow

Udar R, Awasthi US (1981) A new species of *Lejeunea* from India. Cryptogamie Bryol. Lichenol. 8:55-59

Udar R, Awasthi, US (1982) The genus *Schiffneriolejeunea* Verd. (Hepaticae) in India. Lindbergia 8:55-59

Vitt DH (1984) Classification of the Bryopsida In: New Manual of Bryology, Schuster RM (ed.), The Hattori Botanical Lab, Japan, pp. 696-759

Vohra JN, Aziz, MN (1997) Mosses In: Floristic Diversity and Conservation strategies in India, Mudgal, V and Hajra, PK (eds.), Botanical Survey of India, Kolkata, pp. 300-372

Chapter – 9

Pteridophytes: Collection, Identification and Preservation

A.P. Singh and P.B. Khare

1. Introduction

Pteridophytes consist of ferns and their allies and are one of the oldest groups of plants on the earth, and they are commonly known as vascular cryptogams. Pteridophytes have a long history of evolution and were the dominant group of plants during the carboniferous period. The ferns, which constitute major element of the pteridophytes, preferentially grow in shade, moist habitats with moderate temperature but also occur throughout from high altitude in arctic-alpine to rain forests of tropic regions of the world. The most favoured habitat of these plants, however, is tropical mesic environment, where 65% of the living species of fernsthrive vigorously and grow luxuriantly (Page, 1979). Ferns are distinct in having rhizomes, stipes, fronds and are homosporous, whereas the fern allies exhibit scale like leaves and are heterosporous (mega and micro). In ferns, the spore matures into gametophyte that bears both male and female gametangia in a definite pattern. In fern allies, both the mega and microspore mature into female and male gametophyte that bears female and male gametangia, respectively. Pteridophytes are unique in having two different phases-the gametophyte and sporophyte. There has been a wide range of variation in spatial distribution, habitat, growth pattern and morphology of pteridophytes. Fern and fern allies, are comprised of six distantly related classes *viz.* Lycopodiopsida, Selaginellopsida, Isoetopsida, Equisetopsida, Psiloptopsida and Polypodiopsida (Copeland 1947; Pichi Sermoli 1987).

There are about 12000 species of the pteridophytes belonging to 300 genera in the world, whereas India is represented by about 1200 species (Chandra, 2000). Beddome (1892) during the first quarter of the twentieth century provided a comprehensive account on ferns of British India, and subsequently, Clarke (1880) and Hope (1899, 1900, 1902,

1903, 1904) made significant contributions in taxonomic studies of pteridophytes mainly the ferns of north western India. Several workers (Nayar and Kaur 1974; Dixit 1984; Chandra and Kaur 1987, 1994; Manickam *et al.* 1992; Khullar 1994, 2000; Chandra 2000; Singh 2003; Singh and Panigrahi 2005a, 2005b) have significantly contributed towards the taxonomy of Indian pteridophytes.

2. Methods

2.1. Collection and preservation

The collection and preservation methods of pteridophytes are the same as followed in general for other vascular plants. Readers may refer to chapter 12 for details about the collection and preservation methods. Some important tips for fern collection are that the ferns should be collected with because the type and shape of rhizome, hairs and scales on rhizome are important taxonomic characters. If the fronds are too big, a portion of frond with basal portion must be collected and the exact size should be recorded in the field note book, along with other details. Fronds without sori are of little value in taxonomy.

2.2. Identification

2.2.1. Plants as a whole (Sporophyte)

Pteridophytes complete their life cycle in two different phases. These phases are sporophyte and gametophyte. The sporophyte is the predominant phase in their life span, where it begins from the very first mitotic division in zygote. The zygote after subsequent transverse and vertical division give rise to dichotomously branched young sporophyte, which later metamorphose in to the whole plant. The easiest way to distinguish and identify the ferns with their allies is by their leaves. The fern allies except the *Isoetes,* have small scale like structures known as microphylls and the true ferns have large leaves (the familiar fern fronds) known as megaphylls. The orientation of plants in nature and their habitat also discriminate the plant group in to different category. The example is prostrate growth of *Lycopodium cerenuum*, *Selaginella chrysocaulos;* climbing behaviour of *Lygodium flexuosuum,* erect habit of most of the ferns viz. *Dryopteris wallichiana* including tree fern (*Cyathea* sp). Sporophyte of ferns comprised of root, rhizome, scale, stipe, frond, rachis, pinnae, venation and sori. There is huge variation in all part of the plants, and these variations are considered as taxonomic parameter to categorize the ferns and their allies in to different taxonomic rank (genus and species).

2.2.2. Rhizome

It is modified form of stem in pteridophytes which may be short, long, creeping, procumbent or tufted. They may be short creeping, medium creeping or long creeping in which cases the growing point will be situated at one end (*Polypodium*); or they may be erect and tufted, in which cases the growing point will be situated in the centre of the stem at apex (*Dryopteris*). It may be small, erect, tufted, caudex as in tree fern *Cyathea*. The rhizome also exhibits different branching pattern. It may be dichotomous or forking (*Gleichenia*), lateral branching (*Lastreopsis*) or stolons (*Nephrolepis*). Rhizomes are an important feature for fern identification (Fig. 1).

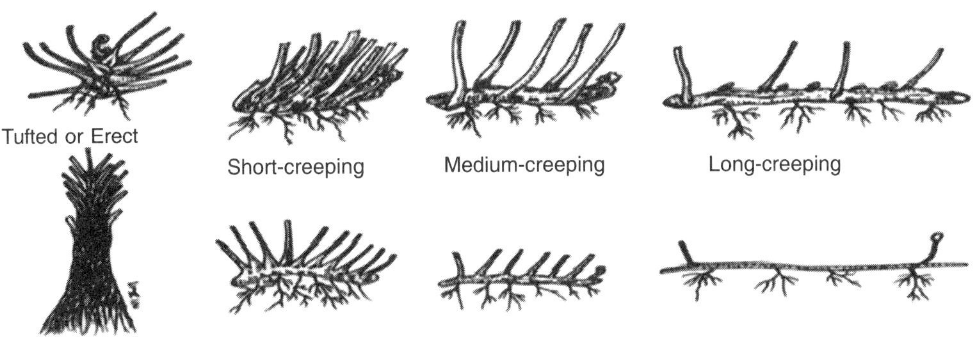

Fig. 1. Rhizome (Jones 1987)

2.2.3. Scales

The young parts of rhizome or even the older part of fronds are protected by yellow to brown coloured, papery, dry structures which may be called scales, bristles or hairs. The shape size and colour of these structures vary but will be useful as diagnostic character, and can serve as a good taxonomic parameter for taxa delimitations at various ranks. These scales are papery in texture and are one celled thick. They consist of an arrangement of flat cells, and are attached to the fern by a point on their edge (base) or by some point on their flat surface (when peltate). If the cell wall of the scales is thickened to form a lattice work pattern on the surface, they are termed clatherate (*Asplenium*). The edge of scales may be entire, lacerate, toothed or ciliate. Scales may be lanceolate (*Gleichenia*), hyaline (*Dryopteris*), peltate irregular (*Pyrrosia*), linear setiferous (*Cyathea procera*), bullate (*Cyathea carrii*), flabellate (*Cyathea oinops*), irregular stellate (*Elaphoglossum*), twisted (*Gleichenia*), oblong irregular (*Elaphoglossum*), linear setiferous (*Blechnum*), circular peltate (*Pyrrosia lingua*). In some cases golden and densely packed (*Hypodematium crenatum*) and forms a covering around the rhizome (Fig. 2).

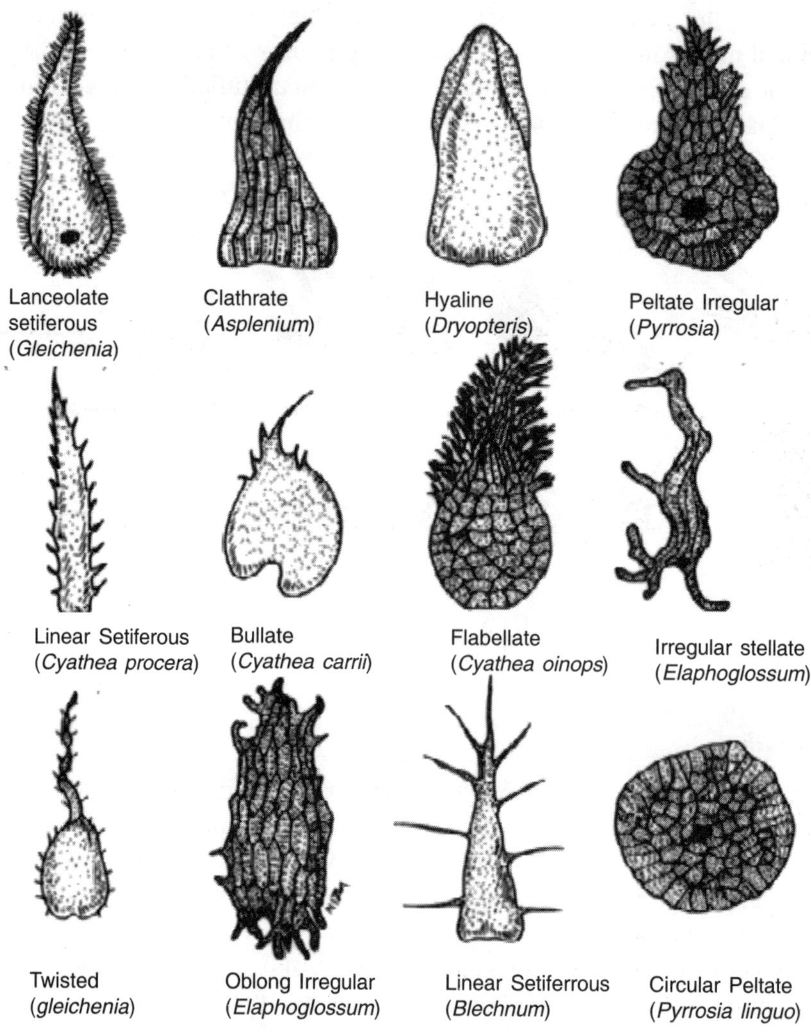

Fig 2. Scales (Jones 1987)

2.2.4. Hairs

Pteridophytes exhibit hairs on the aerial part of rhizome, stipe, rachis, and more abundantly on fronds (*Christella, Pyrrosia*). Hairs are elongated structures which consist of one drawn-out cell or a single row of cells. Their length, color, thickness, rigidity and abundance may be useful diagnostically. Hairs vary from species to species and include the following types: simple or spreading, septate, hairs on raised protuberance, bristles, glandular hairs, tangled hairs, appressed hairs, hooked hairs, stellate hairs and branched hairs (*Gleichenia*). A bristle is similar to a hair but tends to be very rigid and has an expanded base which is more than one cell thick (Fig. 3).

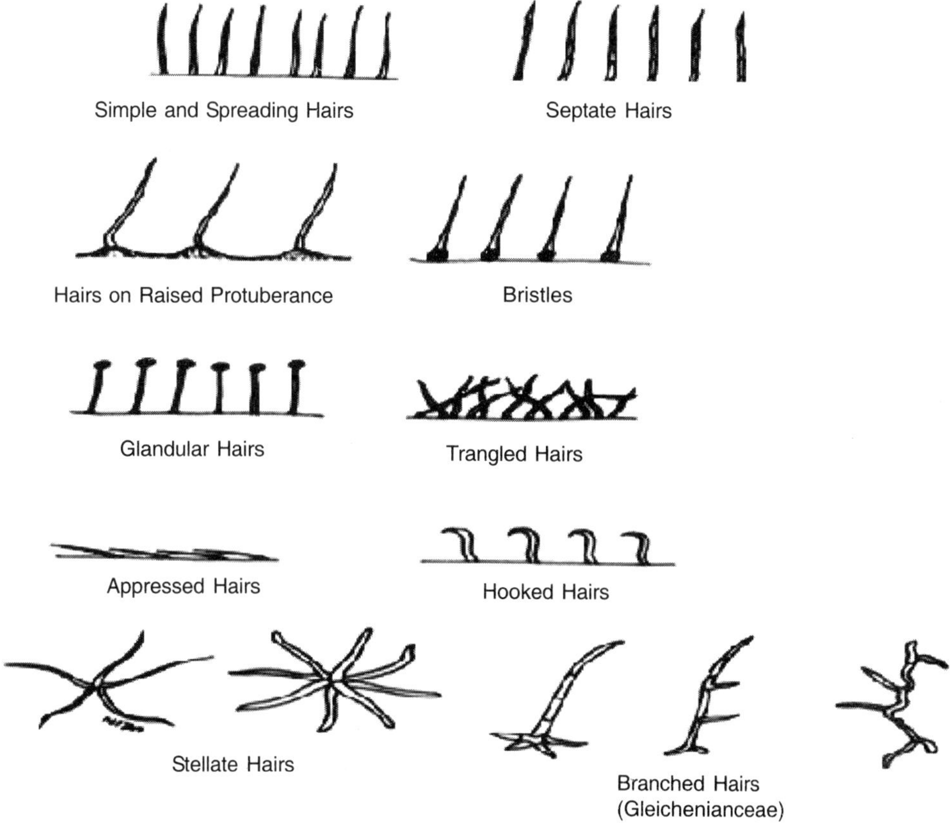

Fig. 3. Hairs (Jones 1987)

2.2.5. Fronds

The leaves in pteridophytes (fern) are usually called fronds. In most of the ferns, the fronds are crowded or scattered along the rhizome. But in ferns with an upright growth habit they are arranged in a spreading or arching crown. Young fern fronds are coiled and are termed croziers. The types of fronds most easily recognized are those which are simple and entire. These have a uniform lamina with no divisions or lobes whatsoever (*Microsorum membranaceum*). Simple entire fronds may divide once or many times, resulting variable shape (pinnate to multi-pinnate) of fronds.

(i) **Shape of Fronds:** In simple fronds, their shape may vary from lanceolate (*Paraleptochillus*), paddle-shaped or elliptic (*Pyrrosia*), cordate or heart shaped (*Ophioglossum*), ovate, reniform or kidney shaped (*Anemia rotundifolia*), linear, saggitate (*Doryopteris ludens*) to rounded or orbicular (*Adiantum pedestum*). The shape and division of frond also varies from genera to genera. It may be didymous

(*Dipteris*), pseudodichotomous and tiered (*Sticherus*), trilobed (*Microsorum*), quadrifoliate (*Marsilea*), palmatifid and pedate (*Doryopteris*), palmatisect and pedate (*Doryopteris*), pedate (*Adiantum*). Simple fronds, which are divided but with the division reaching or not reaching to the rachis can be arranged into following groups (Fig. 4 and Fig. 5).

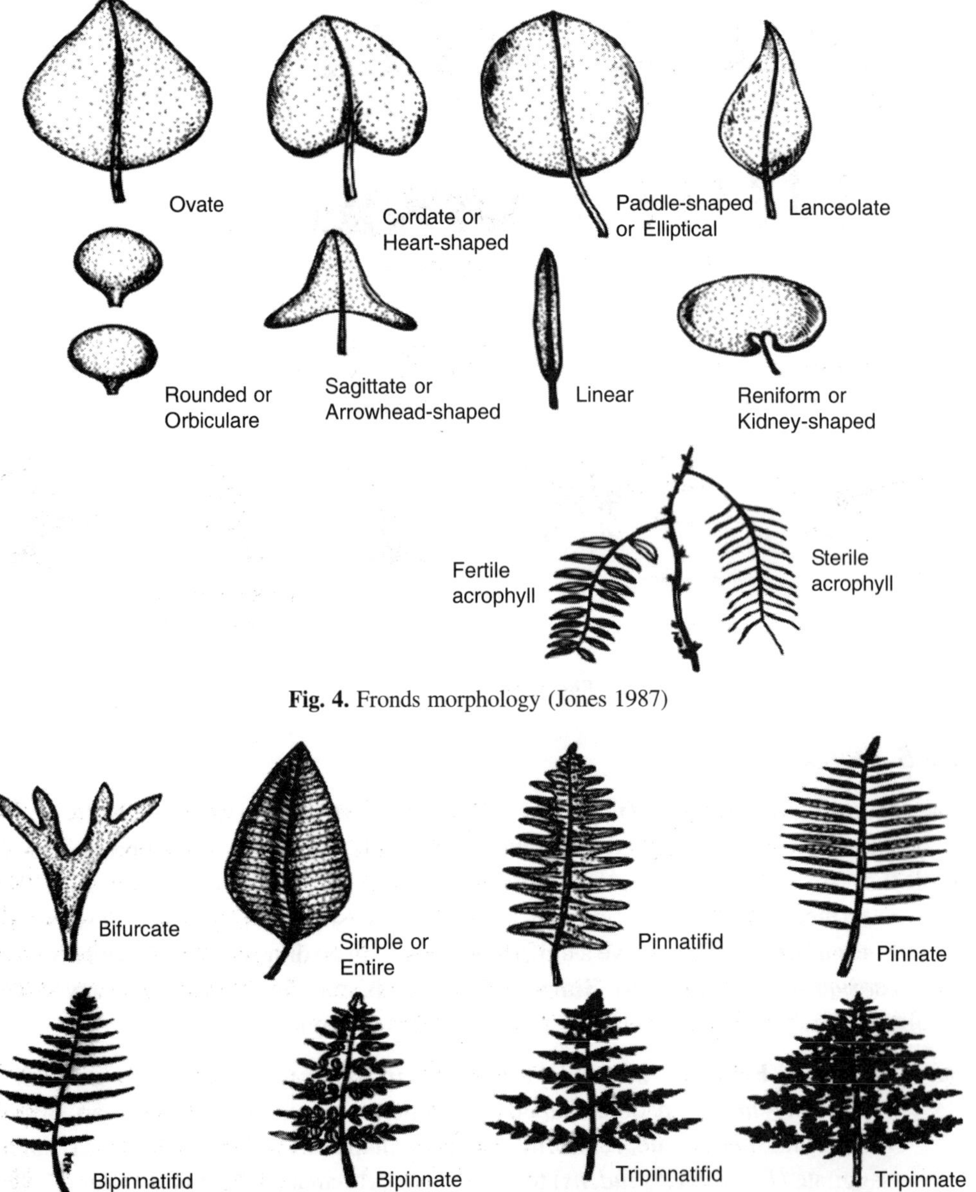

Fig. 4. Fronds morphology (Jones 1987)

Fig. 5. Frond division (Jones 1987)

(ii) **Pinnatifid:** The simple frond divides in such a way that the division extends about one quarter to half way to the rachis. Thus, if a frond is divided once, and when the division does not reach the rachis, the frond is called pinnatifid.

(iii) **Pinnatipartite:** There are many ferns wherein division of simple frond does not reach up to the rachis. Simple frond divides in such a way that the division extends half to two thirds of the way to rachis (*Tectaria*).

(iv) **Pinnatisect:** In many of the pteridophytes, simple fronds divide in such a way that the division reaches more than three quarters of the way to rachis (*Pteris excelsa*).

(v) **Pinnate:** If the frond is divided once, with the divisions extending to the midrib. In such cases the frond is called pinnate and the lobes as pinnae (*Blechnum*).

(vi) **Bipinnatifid:** The condition in which pinnae are lobed but the lobes do not reach the secondary rachis. When the pinnae divides in such a way that the division extending about one quarter to half way to the secondary rachis, it formed, is called bipinnatifid (*Christella dentata*).

(vii) **Bipinnatipartite:** There are many ferns wherein division of pinnae does not reach up to the rachis. Pinnae divide in such a way that the division extending half to two thirds of the way to secondary rachis is called bipinnatipartite.

(viii) **Bipinnatisect:** In certain ferns, the pinnae divide in such a way that the division reaches more than three quarters of the way to secondary rachis. Such fronds are is called bipinnatisect.

(ix) **Bipinnate:** If the pinnae are divided all the way to secondary rachis, the frond is twice pinnate or bipinnate, and the segments are called pinnules. Division of this nature can increase up to five times in pinnate fronds, depending on the depth of the division. Thus, the fronds may be simple, bifurcate, pinnatifid, pinnate, bipinnatifid, bipinnatipartite, bipinnatisect, bipinnate, tripinnatifid, tripinnatipartite, tripinnatisect, tripinnate, and pentapinnate.

(x) **Margin of Lamina:** Similarly, the margin of lamina is a significant attribute for identification of ferns. The margin may be entire, crenate, crenulate, undulate, ciliate, denticulate, dentate, incised, lacerate or lobed. The apices of fronds are also quite remarkable to differentiate the taxa. Apex may be acute, acuminate, mucronate, obtuse, truncate, retuse or emarginate, where as the pinnae attachment on the rachis is a remarkable feature for identifying the taxa. The pinnae may be attached by vein and leaflets bases, it may be decurrent, attached by vein only or surcurrent (Fig. 6).

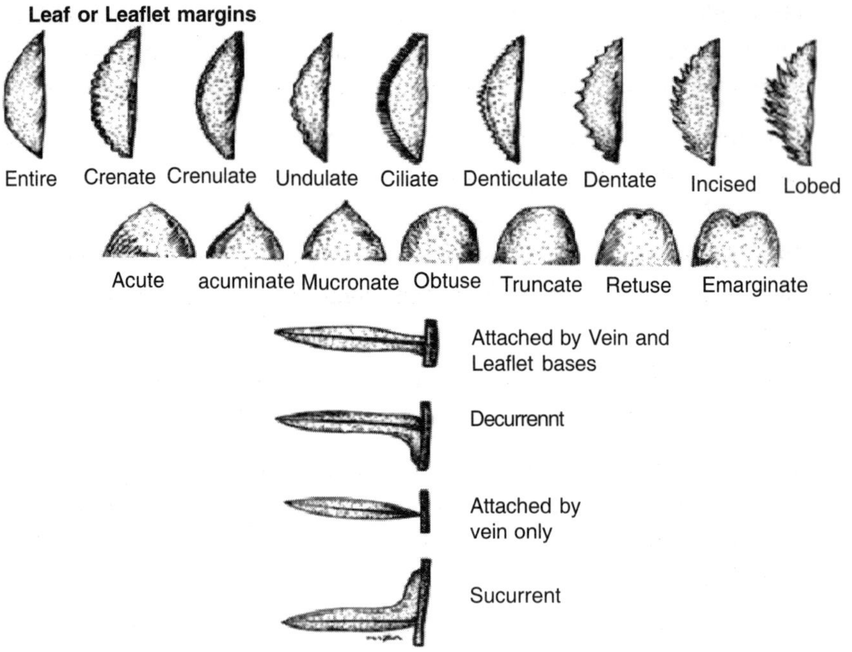

Fig. 6. Lamina margin (Jones 1987)

2.2.6. Stipes

The basal woody structure of a fern frond, which extends from rhizome to the first leaflet or segment, (analogous to the petiole of a leaf) is known as the stipe. It may be entire and rounded in section or grooved on the upper surface. The surface is smooth or roughened, or bears distinct thorns or spines. Most stipe bears scales or hairs; these are usually smaller than those on the rhizome but are identical in all other respects. The stipes may be stramineous, grooved, circular or ridged.

2.2.7. Venation pattern

A vein is actually a trace of vascular bundles and is usually quite prominent on the surface of a frond. In species with flesh fronds the veins may be hidden. The pattern made by the veins is known as the venation and this arrangement is an important taxonomic feature of the members of Polypodiopsida. The venation may be simple, when the veins fork one or more times but the branches do not join again. The vein commonly ends at the edge of the lamina, frequently in a lobe or marginal sorus. All of these veins are free. In some ferns the grouping of free veins may be pinnate and resembles that of a fish's backbone. The main vein of a leaflet is termed as midrib or costa and the smaller side veins are known as costules. In many ferns the veins rejoin after branching or join on to neighbouring veins. This results in a reticulate network and this type of venation is called anastomosing. The small areas of tissues enclosed by each network of veins are known as an areole. The veins may be free, forked, forked and free, pinnate vein group, anastomosing and free, anastomosing with simple included veinlets, anastomosing with branched included veinlets (Fig. 7).

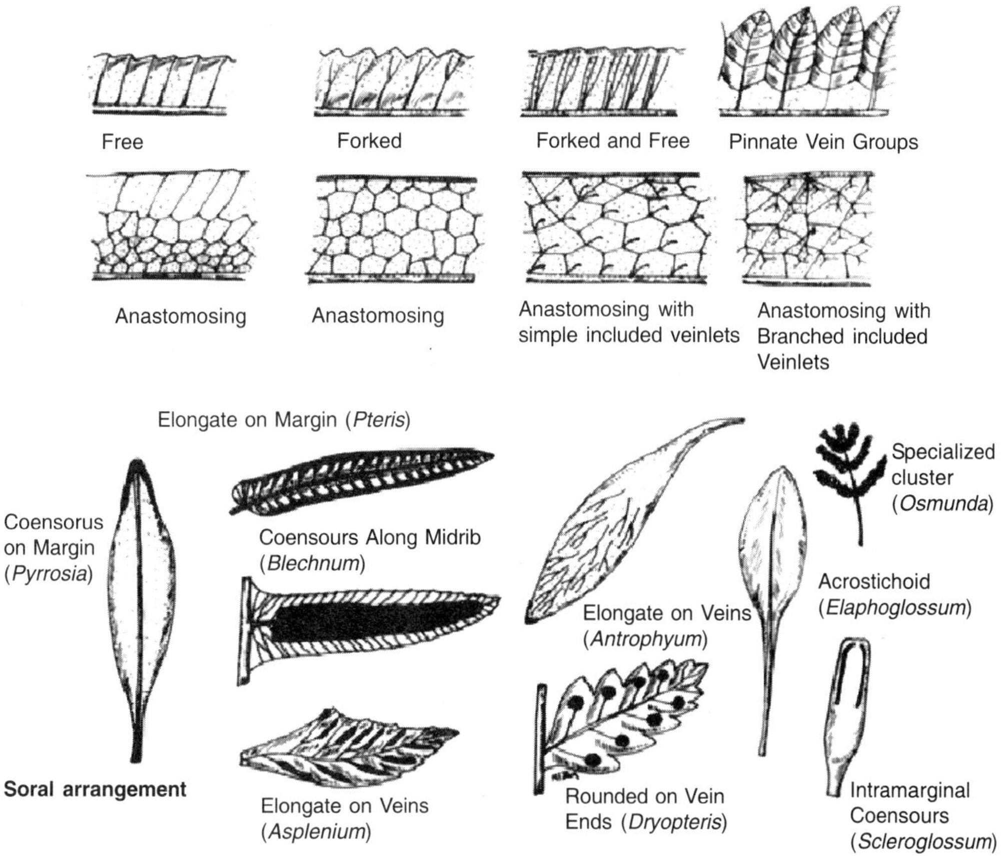

Fig. 7. Venation and sori arrangement (Jones 1987)

2.2.8. Sori

The sori of ferns are produced on the underside of the fronds. They may be in discrete groups, and round or elongate in shape, or spread along the veins and margin of the fronds. Some ferns have elongated marginal sori formed by the overlapping of adjacent sori. These are termed fusion-sori or coenosori. Sori may be marginal and continuous (*Pteris*) or marginal discrete (*Adiantum*). Rounded sori are usually situated at the end of free vein, at a point on its length or at the junction of veins (*Polypodium*), where as an elongated sorus (which is not a fusion sorus) spread along a vein (*Asplenium*). Where sori spread along all the veins of a fertile frond, covering the whole lower surface, the arrangement is known as acrostichoid (*Pyrrosia*). Sori may be marginal or medial (Fig. 7).

2.2.9. Indusium

The sori may be naked (ex-indusiate), when the sporangia are readily visible, or they may be protected by a thin, usually membranous outgrowth from the frond known as an indusium (*Adiantum*). An indusium may be variously shaped, but its shape is usually characteristic of the species or perhaps even of the genus. Indusia shape includes rounded or circular and peltate (*Polystichum*), elongate along vein (*Asplenium*), elongate along margin (*Pteris*), kidney shaped (*Nephrolepis*), cup shaped, pocket shaped (*Davallia*), valvate shaped (*Hymenophyllum*), trumpet shaped (*Trichomanes*), false indusium (*Pteridium* and *Adiantum*). In some genera it may be covered by specialized reflexed margins of the frond and these are called false indusia e.g. *Adiantum* and *Pteris*.

2.2.10. Spores

The spores are produced by subsequent mitotic as well as meiotic division in the sporogenous mother cell. The fate and structure of developing spores are determined by the involved tapetal and callose tissues which serve the nutritional supply to them. The division in each sprogenous mother cell led to form individual spores. Spore features are of tremendous importance to taxonomists because they shed light on evolution and the relationship between various groups of ferns. The spores differ invariably in their shape, size, structure and colour. A remarkable sculpturing on the sporoderm categorizes the spores in to monolete or trilete. Monolete spores are roughly the shape of a bean seed and are angled along one edge where they have been pressed together in the tetrad. Trilete spores have three faces and are roughly triangular in cross sections. The outermost region of the spores called periexine followed by intine. Spores may be winged or simple with entire or dentate margin. The ornamentation of the spores led to exhibit reticulation throughout or restraint to polar or antipolar regions. Some spores contain green chlorophyllous sporoplasm and exhibit the endosporic germination tendency (*Osmunda regalis*). Thus, the shape, size, colour, sculpturing and ornamentation of spore in the ferns and their allies delimit them in to different taxonomic rank. It involves an individual discipline palynology or spore morphology, which has been proven to be useful in the taxonomy of ferns (Devi 1997).

Morphological and anatomical attributes help resolve issues related to phylogeny and interrelationships of different groups with allied and far relative pteridophytes, in addtion to species identification. Above methods and strategies would help more accurately to execute the collection, preservation and identification for preparing floristic and taxonomic account to explore the possibilities for their further prospection.

References

Beddome RH (1892) Handbook to the Ferns of British India, Ceylon and Malaya Peninsula. Thacker Spink. & Co. Calcutta

Chandra S (2000) The Ferns of India (Enumeration, Synonyms & Distribution). International Book Distributors, Dehra Dun, India, pp 1-459

Chandra S and Kaur S (1987) A nomenclatural guide to R. H. Beddome's Ferns of South India and Ferns of British India. Today and Tomorrow's Printers and Publishers New Delhi, India, pp 1-139

Chandra S and Kaur S (1994) Nomenclature of Indian Ferns. Indian F Journal 11:7-11

Clarke CB (1880) A review of Ferns of Northern India. Trans Linn Soc London 12:425-611

Copeland EB (1947) Genera Filicum, the genera of ferns. Ronald Press, New York I - xiv, pp 1-247

Devi S (1977) Spores of Indian Ferns. Today & Tomorrow's Printers & Publishers, New Delhi, India, pp 1-228

Dixit RD (1984) A census of the Indian Pteridophytes. Botanical Survey of India, Howrah, Calcutta, pp 1-177

Hope CW (1899) The Ferns of North Western India, including Afghanistan, the trans-India Protected State and Kashmir. J Bombay Nat Hist Soc 12:315-325, 527-538, 621-633

Hope CW (1900) The Ferns of North Western India. J Bombay Nat Hist Soc 13:25-36, 236-251

Hope CW (1902) The Ferns of North Western India. J Bombay Nat Hist Soc 14:118-127, 252-266, 458-480

Hope CW (1903) The Ferns of North Western India. J Bombay Nat Hist Soc 14:720-749

Hope CW (1904) The Ferns of North Western India. J Bombay Nat Hist Soc 15:78-111

Jones D (1987) Encylopaedia of Ferns. Oregon USA: Timber Press

Khullar SP (1994) An Illustrated Fern Flora of the West Himalaya. Vol. I, International Book Distributors, Dehra Dun, India, pp 1-506

Khullar SP (2000) An Illustrated Fern Flora of the West Himalaya. Vol. II, International Book Distributors, Dehra Dun, India, pp 1-544

Manickam VS and Irudayaraj V (1992) Pteridophytic flora of the Western Ghats, South India. BI Publications Pvt. Ltd., New Delhi, pp 1-653

Nayar BK and Kaur S (1974) Companion to RH Beddome's Handbook to the Ferns of British India, Ceylon and Malaya Peninsula. The Chronica Botanica, New Delhi

Page CN 1979. The diversity of ferns: An ecological perspective in: Dyer AF (ed) The experimental biology of ferns pp 9-56, Academic Press London

Pichi-Sermoli REG (1987) A look at the chromosome number in the families of Pteridophyta. Webbia 41 (2):305-314

Singh HB (2003) Economically viable pteridophytes of India. In: Pteridology in the New Millenium, Chandra S and Srivastava M (eds.), Kluwer Academic Publishers, Netherland, pp 421-446

Singh S and Panigrahi G (2005a) Ferns and Fern-Allies of Arunachal Pradesh, Vol. I. Bishen Singh Mahendra Pal Singh, Dehradun, India, pp 1-426

Singh S and Panigrahi G (2005b) Ferns and Fern-Allies of Arunachal Pradesh, Vol. II. Bishen Singh Mahendra Pal Singh, Dehradun, India, pp 427-881

Chapter – 10

Diversity and Systematics of Gymnosperms

Baleshwar, Bhaskar Datt and T.S. Rana

1. Introduction

The gymnosperms constitute a distinct subdivision of spermatophytes, the seed bearing plants. The term 'gymnosperm' is derived from Greek words, *gymnos*, naked and *sperma*, seed, which indicates that these are naked-seeded plants, i.e., those in which the seeds are not enclosed within the fruits but are borne freely exposed on an open carpel. In contrast in the angiosperms the seeds are enclosed within a closed carpel (ovary). The first known use of the term gymnosperm was given by Theophrastus in 300 BC in his book "Enquiry into Plants" to describe "plants with unprotected seeds". Goebel (1905) rightly described gymnosperms as "Phanerogams without ovary". The Scottish botanist, Robert Brown in 1825 first distinguished gymnosperms from angiosperms. At one time they were considered to be single class of seed plants, called Gymnospermae, but taxonomists are now tend to recognise four distinct divisions of extant gymnospermous plants - coniferophyta (Pinophyta), Cycadophyta, Ginkgophyta and Gnetophyta. Geological history reveals that all divisions of gymnosperms have been distinct groups since their origin.

Among the gymnosperms are plants with stems that may barely project above the ground and others develop into the largest trees on the earth. The extant gymnosperms are a small group of plants comprising ca. 1026 species belonging to 83 genera and 12 families (Christenhusz *et al.* 2011). They are distributed throughout the world, with extensive latitudinal and longitudinal ranges. Economically, gymnosperms are of outstanding value, fulfilling the greater proportion of our timber as well as resin requirements. They are also the dominant forest makers of the world and in addition they play a leading role in the preservation of the environment (Sahni 1990). By far the largest group of living gymnosperms is the conifers (pines, cypresses and relatives), followed by cycads and Gnetales (*Gnetum*, *Ephedra* and *Welwitschia*).

1.1. Salient features of gymnosperms

- Gymnosperms are predominantly woody plants represented by trees, shrubs and rarely vines.
- In general stems are aerial, erect, branched (except *Cycas* and *Zamia*) and woody.
- Plants are unisexual and may be monoecious (*Pinus*, *Cedrus*, *Juniperus*) or dioecious (*Cycas, Ginkgo, Ephedra*).
- Gymnosperms, like all vascular plants have a sporophyte-dominant life-cycle. The gametophyte (gamete-bearing phase) is relatively short-lived.
- Leaves may be monomorphic or dimorphic.
- Leaves or needles are adapted to harsh environmental conditions by having hypodermis, thick cuticle, and sunken stomata without air space system in mesophyll to conserve water.
- Gymnosperms develop tap root system from radicle for better anchorage and absorption like angiosperm trees. In some cases roots show symbiosis with certain algal cells as in coralloid roots of *Cycas* or fungus as in mycorrhizal roots of *Pinus.*
- Wood of gymnosperms lack fibres and vessels in xylem (except *Gnetum*) and companion cells in phloem and is classified as softwood (hardwood in case of angiosperms having these cells). In gymnosperms annual rings in wood are clear and easily countable. Resin canals, the characteristic feature of Pinaceae prevent fungal growth and insect feeding on wood.
- Gymnosperms have cones or strobili in place of flowers as in angiosperms.
- Gymnosperms are heterosporous like angiosperms while majority of pteridophytes are homosporous.
- In *Ginkgo* and Cycads the sperms are motile, large (about 180 μm) and multiflagellated whereas sperms are non motile in Conifers and Gnetales as in angiosperms. Sperms are carried to the archegonia by pollen tubes.
- In gymnosperms archegonia (absent in angiosperms) lack neck canal and neck canal cells, whereas pteridophytes possess them.
- During pollination the pollen grains directly settle on the micropyle whereas in angiosperms pollens fall on stigma and have to travel through the stylar canal to reach ovary and ovule.

- The endosperm is haploid in gymnosperms but is triploid in angiosperms.
- Cleavage polyembryony is prevalent in gymnosperms but absent in angiosperms.
- Gymnosperms produce seeds on the surface of sporophylls or similar structures instead of being enclosed within a fruit as in flowering plants.
- The number of cotyledons may be one or two (*Cycas*) or a whorl of many (*Pinus*)

2. Origin and Evolution of Gymnosperms

Gymnosperms unlike angiosperms are an ancient group of plants. While some taxa of gymnosperms are extinct, others with many primitive features are known both in fossil as well as living forms. There are still some taxa with only living gymnosperms that are distributed throughout the temperate, tropical and even in arctic regions. Geological history reveals that gymnosperms made their appearance in Palaeozoic era, some 300 mya (million years ago). Alike present day angiosperms some of the gymnosperms were dominant during past geological ages as known by their fossil records. Certain groups of primitive gymnosperms such as Cycadofilicales, Bennettiatales and Cordaitales are now extinct whereas Cycadales, Ginkgoales and Coniferales are represented both by living as well as fossil forms. Cycadofilicales or the seed ferns form one of the most important and interesting evolutionary line with plants bearing fern like fronds with ovules that mature into primitive type of seeds. These early gymnosperms that appeared during Carboniferous period resembled so much with ferns that the period is often called "Age of Ferns". Out of two other groups of fossil gymnosperms, Cordaitales, often regarded as the ancestral stock for modern Coniferales and Gnetales, were well developed during Palaeozoic era, while the Bennettitales that showed close resemblances with modern Cycads were dominant during Mesozoic era. Diverse early gymnosperms were the major vegetation present during the Mesozoic era, the era also known as the "Age of Dinosaurs". Some groups of primitive gymnosperms became extinct before or as a result of the Cretaceous–Tertiary (K/T) Extinction event at the end of the Cretaceous period 65 mya. The modern gymnosperms are represented by cycads (Cycadophyta); Ginkgo (Ginkgophyta) with sole surviving member i.e. *Ginkgo biloba*; conifers (Coniferophyta); and Gnetophytes consisting of morphologically three diverse taxa; *Gnetum*, *Ephedra* and *Welwitschia* (Gnetophyta). These groups display distinctive reproductive features. Gnetales are probably, a recent group with a very few fossil records. Towards the end of the Mesozoic era, during Cretaceous, gymnosperms gradually gave way to angiosperms, the dominant modern flowering plants. The precise evolutionary connections between different gymnosperm groups and that of primitive angiosperms are depicted in Fig. 1.

Gymnosperm seeds develop either on the surface of scales or leaves, often modified to form cones, or at the end of short stalks as in *Ginkgo*. These cones, however, are not the same as fruits in flowering plants. During pollination, the immature male gametes, or pollen grains, shift among the cone scales and land directly on the ovules, which contain the immature female gametes, rather than on elements of a flower (the stigma and carpel) as in angiosperms. Furthermore, at maturity, the cone expands to reveal the naked seeds. Modern gymnosperms are woody plants that occur as shrubs, climbers or trees. Seeds and wood are adaptations that allow gymnosperms to cope with global climate changes and to live in relatively harsh environments i.e. cold, dry, and even in marshy habitats. Fossil records help us to explain important gymnosperm traits like the structure, reproduction, and ecological roles of modern gymnosperms.

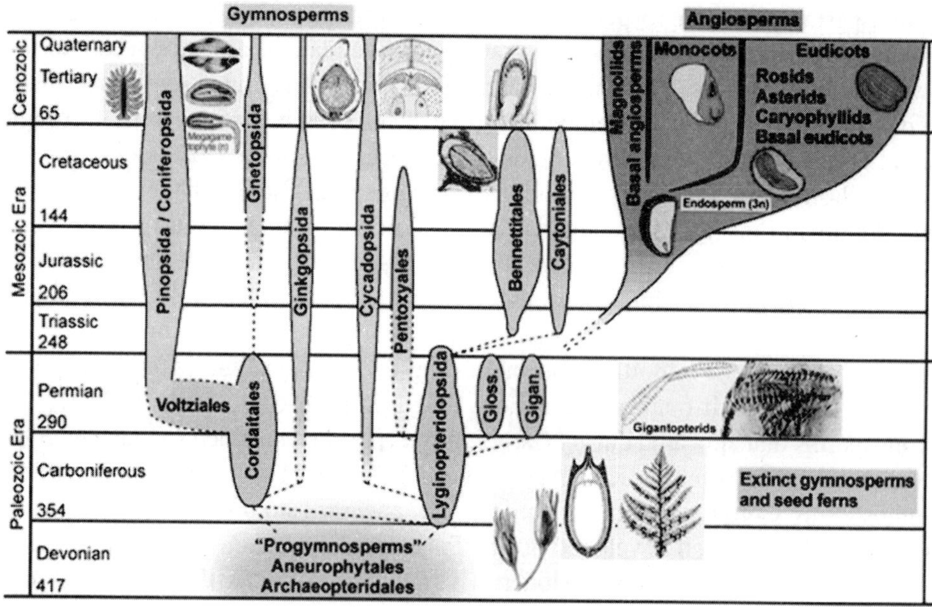

Fig. 1. Evolutionary tree considering major gymnosperm and angiosperm clades. (Source: Leubner 2007) Time scale: geological eras, periods, time in mybp (million years before present)

Wood first appeared in a group of plants known as the progymnosperms. *Archaeopteris*, a progymnosperm, reached a height of 20 m and had fern-like leaves. It was heterosporous, produced wood but not seeds, and lived roughly 380 mya. Progymnosperms were able to produce a vascular cambium and wood because their vascular tissue was arranged in a ring around a central pith of nonvascular tissue (in contrast, the vascular tissue of earlier tracheophytes was arranged differently). This ring of vascular tissue, known as a eustele, contained cells that were able to develop into the vascular cambium as seedlings grew into saplings. The vascular cambium then produced wood, allowing saplings to grow into tall trees. *Sequoiadendron giganteum* is among earth's largest organisms, weighing as much as 6,000 tons and reaching an amazing height of 100 m. The large size of sequoias and other species based on the presence of wood, a tissue composed of numerous pipe-like arrays of empty, water-conducting cells whose walls are strengthened by an exceptionally tough secondary metabolite known as lignin. These properties enable woody tissues to transport water upward for great distances and also to provide the structural support needed for trees to grow tall and produce many branches and leaves. In modern seed plants, a special tissue known as the vascular cambium produces both thick layers of wood inside and thin layers of inner bark outside. The inner bark transports watery solutions of organic compounds. Vascular cambium, wood, and inner bark are critical innovations that helped gymnosperms and other seed plants to compete effectively for light and other resources needed for photosynthesis.

Modern seed plants inherited the eustele, explaining why many gymnosperms and angiosperms are also able to produce vascular cambia and wood. Although progymnosperms were woody plants, fossil evidence indicates that they did not produce seeds. This observation reveals that wood originated before seeds evolved.

3. Brief Description of Extant Gymnosperms

3.1. Cycads

Approximately 300 species of cycads under 11 genera are distributed primarily in tropical and subtropical regions. However, many species of cycads are rare, and their tropical forest homes are increasingly threatened by human activities. The structure of cycads is so interesting and attractive that many species are cultivated for use in outdoor plantings or as house plants. The non woody stems of some cycads emerge from the ground much like tree trunks, some reaching 15 m in height, whereas the stems of other cycads are not conspicuous because they are subterranean. Cycads display spreading, palm like leaves (cycad comes from a Greek word, meaning palm) (Fig. 2). Mature leaves of the African cycad *Encephalartos laurentianus* can reach an astounding 8.8 m in length. In addition to underground roots, which provide anchorage and uptake water and minerals, many cycads produce coralloid roots. Such roots extend aboveground and have branching shapes resembling corals. Coralloid roots harbour light-dependent photosynthetic cyanobacteria within their tissues. The cyanobacteria, which form a light blue-green ring beneath root surfaces, are nitrogen fixers that provide their plant hosts with nitrogen crucial to their growth.

Fig. 2. *Cycas rumphii* bearing female cone

Reproduction in Cycads is distinctive in several ways. Individual cycad plants produce conspicuous cone like structures that bear either ovules and seeds or pollen. When mature, both types of reproductive structures emit odours that attract beetles. These insects carry pollen to ovules, where the pollen produce tube that deliver sperm to egg for fertilization.

3.2. Ginkgo

The beautiful tree *Ginkgo biloba* is the only living representative of the order Ginkgoales (division Ginkgophyta). Over one hundred million year ago, *Ginkgo biloba* was not "single" but a member of large group (Ginkgoales) with many genera and species. All other genera and species of the group disappeared from the earth and *Ginkgo bilba* survived and became the only living descendant (Hsieh 1992). During Mesozoic era, Ginkgoales occupied a considerable area, with over 15 genera of which *Baiera* had a wide distribution on earth as evidenced by fossil records. The other genera such as *Phenecopsis*, *Hartzia* and *Ginkgo* were the important trees of the Northern Hemisphere. They were as common as dinosaurs living on the earth at that period.

Fig. 3. Forking leaves of *Ginkgo biloba*

Fig. 4. *Ginkgo biloba* bearing ovule cones

Ginkgo biloba is referred to as maidenhair tree for its superficial foliar resemblance to the maidenhair fern (*Adiantum* spp.). This is also sometimes referred to as living fossil for its close resemblance to fossilised extinct species. *G. biloba* takes its species name from the lobed shape of its leaves, which have unusual forked veins (Fig. 3). It is a native of China and has been planted in Chinese and Japanese temple gardens since ancient times, and is now valued in many parts of the world as an attractive, fungus- and insect-resistant ornamental tree. It tolerates cold weather and unlike most gymnosperms, withstand the adverse atmospheric conditions of urban areas.

G. biloba trees are widely planted along city streets because of their ornamental value and also tolerate cold, heat, and pollution better than most trees. In addition, these trees are long-lived (individuals can survive for more than a thousand years) and grow up to 30 m in height. Individual tree produce either ovule cones and seeds or pollen cones, based on a sex chromosome system much like that of humans. Ovule-producing trees have two X chromosomes whereas pollen-producing trees have one X and one Y chromosome. Wind disperses pollen to ovules, where pollen grains germinate to produce pollen tubes. These tubes grow through ovule tissues for several months, absorbing nutrients that are used for sperm development. Eventually the pollen tubes burst, delivering flagellate sperm to egg cells. After fertilization, zygote develops into embryo, and the ovule integument develops into a fleshy, bad-smelling outer seed coat and a hard, inner seed coat (Fig. 4). For street-side or garden plantings, people usually select pollen-producing trees to avoid the foul smelling ovules.

3.3. Conifers

The conifers are a lineage of trees named for their seed cones, of which pine cones are familiar examples (Fig. 5). Conifers have been important components of terrestrial vegetation since the end of the Carboniferous period, some 300 mya. Modern conifers are most diverse and dominant group of gymnosperms being represented in many parts of the world by 74 genera and 684 species. Magnificent conifer forests determine to a great extent the climate of the area and its environment (Mehra1988). The leaves of many conifers are long, thin and needle-like, whereas in Cupressaceae, Taxaceae (Fig. 6A,B) and some taxa of Podocarpaceae, the leaves are flat and triangular, taxa of scale-like. *Agathis* (Araucariaceae) and *Nageia* (Podocarpaceae) have broad, flat, strap-shaped leaves. Conifers, particularly *Pinus, Abies, Picea, Cedrus, Junipers*, *Cupressus* and *Cryptomeria japonica* make pure stands in mountains and high-latitude forests, whereas *Taxus*, and *Tsuga* occur in mixed broad leaved forests in temperate zone. Conifers produce simple pollen cones and more complex ovule-bearing cones. The pollen cones of conifers bear many leaflike structures, each bearing a microsporangium in which meiosis occurs and pollen grains develop. The ovule cones, also called seed cones, are composed of many short branch systems that bear ovules. Ovules contain female gametophytes within which egg cells develop (Fig. 7A,B,C).

Fig. 5. *Pinus wallichiana* bearing ovule cones

Fig. 6. Female cone bearing twigs of (A) *Cephalotaxus griffithii*, (B) *Taxus contorta*

Fig. 7. (A) *Cedrus deodara*, (B) *Juniperus indica*, (C) *Cupressus sempervirens*, bearing female cones

When conifer pollen is mature, it is released into the wind, which transports pollen to ovules. When released from pollen tubes, one of the sperm cells fuses with egg cell, to form zygote that will grow into the embryo within the ovule. Altogether, it takes nearly two years for pines to complete the process of male and female gamete development, fertilization, and seed development. The seeds of pines and some other conifers develop wings that aid in wind dispersal for long distances (Fig. 8). Other conifers, such as yew and junipers, produce seeds or cones with bright-coloured, fleshy coatings that are attractive to birds, which help disperse the seeds.

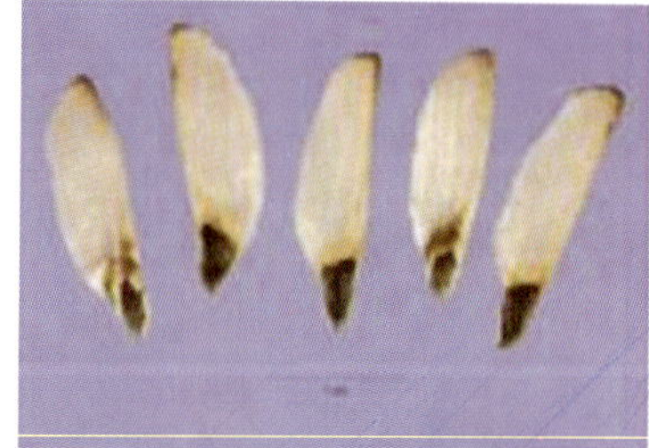

Fig. 8. Pine seeds having large wings

Conifer wood contains many water transport cells known as tracheids that are adapted for efficient conduction even in dry conditions. Like the tracheids of other vascular plants, those of conifers are devoid of cytoplasm and occur in long columns that function like plumbing pipelines. Tracheid side and end walls possess many thin-walled, circular pits through which water moves both vertically and laterally from one tracheid to another. Conifer pits are unusual in having a porous outer region that lets water flow through and a nonporous, flexible central region called the torus (plural, tori) that functions like a valve. If conifer tracheids become dry and fill with air, they are no longer able to conduct water. In this case, the torus presses against the pit opening, sealing it. The torus valve thereby prevents air bubbles from being spreading to the-next tracheid. This conifer adaptation localizes air bubbles, preventing them from stopping water conduction in other tracheids. The presence of tori in their tracheids helps to explain why conifers have been so successful for hundreds of millions of years. Conifers wood (and leaves) may also display conspicuous resin ducts, passage ways for the flow of syrup-like resin that helps to prevent attack by pathogens and herbivores. Resin that exudes from tree surfaces may trap insects and other organisms, then harden in the air and fossilize, preserving the inclusions in amber.

A number of conifers occur in cold climates and thus display adaptations to such environments. Their conical shape and flexible branches help conifer trees shed snow, preventing heavy snow accumulations from breaking branches. Conifer leaf shape and structure are adapted to resist damage from drought that occurs in both summer and winter, when liquid water is scarce. Conifer leaves are often scale-like or needle-shaped; these shapes reduce the area of leaf surface from which water can evaporate. In addition, a thick, waxy cuticle coat of conifer leaf surfaces, help to prevent water loss and attack by pathogens.

Most of the conifers are evergreen; that is, their leaves live for more than one year before being shed, and all leaves do not shed during the same season. Retaining leaves through winters help conifers to start up photosynthesis earlier than deciduous trees, which in spring must replace leaves lost during the previous autumn. Evergreen leaves thus provide an advantage in the short growth season of alpine or high-latitude environments. However, some conifers do lose all their leaves in the autumn. *Larix griffithii* (Himalayan Larch) occurs in eastern Himalaya and *L. himalaica* (Nepal Larch) endemic to central Nepal and south Tibet, are the only deciduous conifers in the Indian subcontinent. *Taxodium distichum* (Bald Cypress) of southern U.S. floodplains, *Larix laricina* (tamarack) of Canada and northeastern U.S., and dawn redwood (*Metasequoia glyptostroboides*), a native of the Sichuan-Hubei region of China, are other examples of deciduous conifers. Fossils indicate that dawn redwoods once grew abundantly across wide areas of the Northern Hemisphere until a few million years ago and then disappeared. However, in the 1940s, a forester found a living dawn redwood growing in a remote Chinese village, and subsequent expeditions located forests of the conifer. Since then, dawn redwood trees have been widely planted as ornamentals, prized for their attractive foliage and cones. As recently as in 1994, David Noble, a botanist found a previously unknown conifer species, *Wollemia nobilis*, in an Australian National Park (Jones *et al.* 1995). Like the dawn redwood, *Wollemia* is an attractive tree that is likely to become more widely distributed as a result of human cultivation.

3.4. Gnetophytes

Related to conifers, the modern gnetophytes consist of three genera, *Gnetum*, *Ephedra*, and *Welwitschia* with distinctive adaptations (Fig. 9A,B,C). *Gnetum* is unusual among modern gymnosperms in having broad leaves similar to those of many tropical plants. Such leaves foster light capture in the dense forest habitats. *Gnetum* occur as vines, shrubs, or trees in tropical Africa and Asia. *Ephedra*, a native to arid regions of the southwestern America, has tiny brown scale-like leaves and green, photosynthetic stems. *Welwitschia*, a strange-looking plant that grows in the coastal Namib Desert of southwestern Africa, one of the driest places on Earth, is represented by the only species, *W. mirabilis*. A long taproot anchors a stubby stem that barely emerges from the ground. Two very long leaves grow from the stem but rapidly become wind-shredded into many strips. The plant is thought to obtain most of its water required from coastal fog, explaining how it can grow and reproduce in such a dry place.

Fig. 9. (A) *Gnetum*, (B) *Ephedra* and (C) *Welwitschia mirabilis* (introduced at NBRI)

The principal point of contention in the taxonomic debate is whether the gnetophytes share a common ancestor with the angiosperms or the conifers (which are usually, but not always, assumed to share a common ancestor with the cycads and *Ginkgo*). The former opinion is called the "anthophyte" hypothesis, where anthophytes are defined to be a clade including both angiosperms and gnetophytes. The latter opinion is the "gnetifer" (or the variant "gnepine", gnetophytes plus Pinaceae) hypothesis, where gnetifers are a clade including both gnetophytes and the conifers. A succinct online review of the subject is provided by Castillo (2001), and cladograms clarifying the anthophyte and gnetifer hypotheses are provided by Burleigh and Mathews (2004). Every year or so a new paper is published supporting one hypothesis or another, and the accumulation of molecular data seems to be swinging in favour of the gnetifer or gnepine hypothesis. But there are still recent morphological studies supporting the anthophyte hypothesis (Friis *et al.* 2007). Christenhusz *et al.* (2011) simply assigned each of the four gymnosperm groups to a subclass under the seed plants, Equisetopsida. They also considered the gnepine hypothesis and put the gnetophytes and Pinaceae in one subclass, and all other conifers in another subclass. It appears likely that Darwin's "abominable mystery", the origin of the angiosperms, will remain a mystery for a few more years. Until it is resolved, systematic position of the gnetophytes will remain unclear.

4. Brief History of Classification

In early classification schemes, the gymnosperms (Gymnospermae) were regarded as a "natural" group. There are conflicting evidences on the question of whether the living gymnosperms form a clade. The fossil records of gymnosperms include many distinctive taxa that do not belong to the four modern groups, including seed-bearing trees that have a somewhat fern-like vegetative morphology (the so-called seed ferns or pteridosperms). When fossil gymnosperms such as Bennettitales, Caytonia and the Glossopterids are considered, it is clear that angiosperms are nested within a larger gymnosperm clade, although which groups of gymnosperms are their closest relatives remains unclear.

Bentham and Hooker (1862-1883) placed the gymnosperms between Dicotyledons and Monocotyledons. Hofmeister's (1851) remarkable investigation on the embryology of the plants, promoted Van Tieghem (1898) to remove them from this intermediate position and placed them, as one of the two divisions of the Spermatophyta: the gymnosperms and the angiosperms. The discovery of the motile sperms in Ginkgo by Hirase (1895, 1896 and 1898) led Engler (1897) to create a new order Ginkgoales. Engler and Prantl (1926) divided the gymnosperms into seven classes or orders: (1) Pteridospermae, (2) Cycadales, (3) Bennettitales, (4) Cordaitales, (5) Ginkgoales, (6) Coniferales, and (7) Gnetales. Sahni (1920a,b,) recognised two phyletic lines in the Gymnospermae: (1) Phyllosperms consisting of (a) Pteridospermae, (b) Cycadales, (c) Bennettitales; and (2) Stachyosperms having (a) Cordaitales, (b) Ginkgoales, (c) Coniferales, and (d) Taxales (which he separated from Coniferales and placed *Taxus*, *Torreya* and *Cephalotaxus* in this order).

Melchior and Werdermann (1954) proposed another treatment and gymnosperms were treated as a division of the plant kingdom equal in rank with bryophyta, pteridophyta, and angiosperms. Presently, the extant gymnosperms are divided into four groups: Conifers, Cycads, *Ginkgo*, and Gnetales. Recently, Christenhusz *et al.* (2011), assigned each of the four gymnosperm groups to a subclass under the seed plants, Equisetopsida (Table 1).

Table 1. Classification of gymnosperms (Christenhusz *et al.* 2011)

Subclass	Order	Family	Genera
Cycadidae	Cycadales	Cycadaceae	***Cycas***
		Zamiaceae	*Bowenia, Ceratozamia, Dioon, Encephalartos, Lepidozamia, Macrozamia, Microcycas, Stangeria, Zamia*
Ginkgoidae	Ginkgoales	Ginkgoaceae	*Ginkgo*
Gnetidae	Welwitschiales	Welwitschiaceae	*Welwitschia*
	Gnetales	Gnetaceae	***Gnetum***
	Ephedrales	Ephedraceae	***Ephedra***
Pinidae	Pinales	Pinaceae	***Abies***, *Cathaya*, ***Cedrus***, *Keteleeria*, ***Larix***, *Nothotsuga*, ***Picea***, ***Pinus***, *Pseudolarix, Pseudotsuga*, ***Tsuga***
	Araucariales	Araucariaceae	*Agathis, Araucaria, Wollemia*
		Podocarpaceae	*Acmopyle, Afrocarpus*, ***Dacrycarpus***, *Dacrydium, Falcatifolium, Halocarpus, Lagarostrobos, Lepidothamnus, Manoao, Microcachrys*, ***Nageia***, *Parasitaxus, Pherosphaera, Phyllocladus*, ***Podocarpus***, *Prumnopitys, Retrophyllum, Saxegothaea, Sundacarpus*
	Cupressales	Sciadopityaceae	*Sciadopitys*
		Cupressaceae	*Actinostrobus, Athrotaxis, Austrocedrus, Callitris, Calocedrus, Chamaecyparis, Cunninghamia*, ***Cupressus***, *Cryptomeria, Diselma, Fitzroya, Fokienia, Glyptostrobus*, ***Juniperus***, *Libocedrus, Microbiota, Metasequoia, Neocallitropsis, Papuacedrus, Pilgerodendron, Platycladus, Tetraclinis, Thuja, Thujopsis, Sequoia, Sequoiadendron, Taiwania, Taxodium*
		Taxaceae	***Amentotaxus***, *Austrotaxus*, ***Cephalotaxus***, *Pseudotaxus*, ***Taxus***, *Torreya, Widdringtonia*

*Genera in bold occur in India in wild.

5. Distribution and Diversity of Extant Gymnosperms

Distribution of gymnosperms in the distant past was much more extensive than at present. Although since the Cretaceous Period_(~146 mya - ~65.5 mya) gymnosperms have been gradually displaced by the more recently evolved angiosperms, they are still successful and abundant in many parts of the world and occupy large areas of the Earth's surface. Conifer forests, for example, cover vast regions of northern temperate lands in North America and Eurasia. In fact, they grow in more northernly latitudes than do angiosperms. Species that occupy areas of the world with severe climatic conditions are adapted to conserve water and exhibit enormous xerophytic habits; the leaves are covered with a heavy waxy cuticle and pores (stomata) are sunken below the leaf surface to decrease the rate of transpiration. *Cupressus* or the true cypress is a genus with ca 25 species distributed in Mediterranean region, Asia and North America; 2 species are native to the Himalaya. On the basis of molecular studies, Adams *et al.* (2009) splitted the genus *Cupressus* into four genera, namely *Cupressus* with ca 9 species in the Old World; *Callitropsis* with 1 species in north western America; *Chamaecyparis* with 2 species in northern America and 3 species in east Asia; and *Hesperocyparis* with 16 species in west northern America to Columbia. However, *Cupressus* sensu lato includes *Callitropsis*, *Hesperocyparis* and *Xanthocyparis* (Mao *et al.* 2010). Genus *Juniperus* consists of ca 60 species distributed in Northern Hemisphere; 6 species confined to the Himalaya. *Cupressus* and *Juniperus*, are most difficult genera to identify in field and in the herbarium, as both of them exhibit enormous diversity between and among the species. The genus *Taxus* is comprised of 24 species and 55 varieties (Sputj 2007) distributed across the northern temperate and subtropical regions as far south as El Salvador in Central America and Sumatera in Southeast Asia. The species are classified into three groups by differences in leaf epidermal and stomatal features. These are: (1) the *Wallichiana* group with 11 species, occurring from central Himalayas to Indonesia and the Philippines, in north America in the Pacific northwest and from Mexico to central America with an isolated occurrence in Florida, (2) the *Baccata* group with nine species in temperate Eurasia, northern Africa and eastern north America, and (3) the *Sumatrana* group with four species overlapping in distribution with the *Wallichiana* subgroup in Asia, but absent from north America. The *Wallichiana* group is further divided into subgroups *Wallichiana* and *Chinensis* and the *Baccata* group is divided into the *Baccata* and *Cuspidata* species alliances. Recently, Shah *et al.* (2008), based on both the morphological and molecular data delimited *Taxus contorta* (*T. fauna*) from *T. wallichiana* or *T. baccata.* They maintained the broader concept for a northwest Himalayan species. The diversity of *Taxus* in India is represented by *T. contorta* (*T. fauna)* in north western Himalaya, *T. wallichiana* in central Himalaya and *T. mairei* (*T. wallichiana* var. *mairei*) in the north east India (Poudel *et al.* 2012).

Cycads are distributed throughout the world but are concentrated in equatorial regions. They are primarily tropical and subtropical in distribution with about 300 species constituting 11 genera and 2 families, Cyacadaceae and Zamiaceae. *Ginkgo biloba*, is the only representative (monotypic) species of Ginkgophyta that is native to north west Zhejiang province in eastern China, in the Tian Mu Shan Reserve (Del Tredici *et al.* 1992) and Dalou Mountains, southwestern China (Tang *et al.* 2012). The Gnetophytes, the most recent gymnosperms include three morphologically diverse genera, *Gnetum*, *Welwitschia* and *Ephedra*. *Gnetum* comprises of more than 40 species that occur as vines, shrubs, or trees in tropical Africa or Asia. About 67 species of *Ephedra* occur worldwide and are adapted to semiarid and desert conditions (Pearson 1929). They are distributed over wide areas of the northern hemisphere, including southwestern north America, Europe, north Africa, and southwest and central Asia, and in the southern hemisphere, in South America from Ecuador to Patagonia (Price 1996). In temperate climates, most of the *Ephedra* species grow on shores or in sandy soils of cold deserts with direct sun exposure. *Welwitschia* has only one living representative (monotypic) species, *W. mirabilis* that grows in the coastal Namib Desert of southwestern Africa, one of the driest places on Earth.

The gymnosperms are one of the dominant and most abundant groups of plants. In spite of their wide range of distribution, they relatively exhibit less species diversity than the angiosperms. It is obvious by the fact that the gymnosperms are represented by ca. 1026 species belonging to 83 genera and 12 families (Christenhusz *et al.* 2011), while several larger genera such as *Astragalus, Dendrobium* and *Acacia* individually comprised of 1000 or even more species. The conifers constitute the most diverse group of gymnosperms with 684 species under 74 genera and 7 families, followed by the cycads. Sahni (1990) reported 60 species under 17 genera and 11 families from India and adjacent countries, of which 54 species occur in the present territories of India. A critical analysis of extant gymnosperms in India based on review of available literature revealed that ca. 73 taxa (species and infraspecific categories) placed under 17 genera and 7 families. North-eastern region of India harbour more than 40 species representing the maximum diversity in India. About 97 species of 27 genera are introduced in India from different parts of the world (Matthew 1981, Sahni 1990, Chaudhuri 1993, Karnataka Forest Department Report 2004, Dar and Dar 2006, Tripathi *et al.* 2009, Tewari *et al.* 2010, Government Botanical Gardens, Ooty 2013) (Fig. 10A). *Ephedra* is most species rich genus (12 species) in India, occurs primarily in the western Himalayan regions from temperate to cold deserts (Sahni 1990; Sharma and Uniyal 2008, 2010), except *E. foliata* which is found in arid and semi arid parts of Rajasthan, Gujarat, Punjab and Haryana. *Cycas,* is the next most dominant genus, with 10 species distributed mostly in south and south-eastern parts of India (Lindstrom and Hill 2002), followed by *Juniperus* (8 species) and *Gnetum* (7 species) (Table 2). The life form diversity of Indian gymnosperms is shown in Fig. 10B.

Table 2. Current status of gymnosperm genera of wild occurrence in India *vis-a-vis* world

S.No.	Genera	No. of species in India	No. of species in World
1.	*Ephedra*	12	40
2.	*Cycas*	10	107
3.	*Juniperus*	8+1 var.	85
4.	*Gnetum*	7+4 var.	40
5.	*Pinus*	6+1 var.	157
6.	*Abies*	4+1	49
7.	*Amentotaxus*	1	6
8.	*Cedrus*	1	4
9.	*Cephalotaxus*	2	11
10.	*Cupressus*	2+1	25
11.	*Dacrycarpus*	1	9
12.	*Larix*	1	11
13.	*Nageia*	1	5
14.	*Picea*	3	33
15.	*Podocarpus*	1+1	104
16.	*Taxus*	3	24
17.	*Tsuga*	1	8

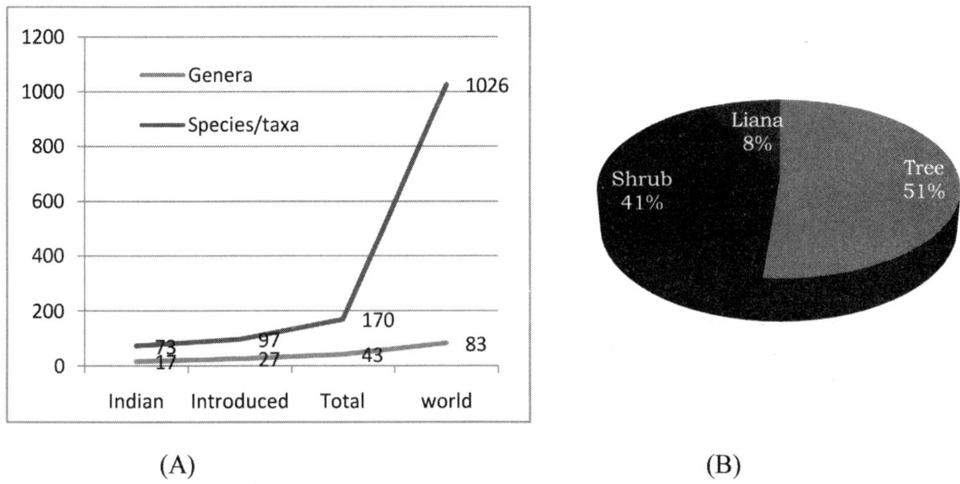

Fig. 10. (A) Current status of gymnosperms in India, (B) Life form diversity of Indian taxa

The gymnosperms exhibit great diversity in habit comprising species with stems that may barely project above the ground on one hand and the largest of trees on the other. *Sequoiadendron giganteum* is the tallest living gymnosperm nearly 100 meters high with a maximum girth of about 15 meters. *Pinus aristata*, (Bristle cone pine) known as "father of forest" is the largest and the longest-lived organism on the earth (5000 years).

Cycads resemble palm trees, with fleshy stems and leathery, feather-like leaves. The tallest cycads reach upto 19 metres in height. *Zamia pygmaea*, which is the smallest gymnosperm and endemic to Cuba and the Isla de Pinos, generally grow in open dry habitats (Donaldson 2003). *Ephedra* (joint fir) is a shrub or undershrub, while some species of *Gnetum* are vines. The unusual *Welwitschia* has a massive, squat stem that rises a short distance above the ground and from the edge of the disk-shaped stem apex, arise two leathery strap like leaves that grow from the base and survive for the life of the plant. The leaves may be large and pinnately compound (cycads); small, simple and needle like in *Pinus*, *Cedrus*, scale like in *Ephedra* and broad angiosperms like in *Gnetum*.

5.1. Endemism in Indian gymnosperms

Endemism of any taxa can be taken as a criterion to measure the centres of origin, diversity and speciation of a group. There are 15 taxa of gymnosperms endemic to India, of which the *Cycas* constitutes the largest group of endemics (Table 3).

Table 3. Endemic gymnosperms in India

S. No.	Species Name	Phytogeographical Region
1.	*Cycas annaikalensis*	Western Ghats (Kerala: Palaghat)
2.	*C. beddomei*	Eastern Ghats (Andhra Pradesh: Cuddapah Hills)
3.	*C. circinnalis*	Western Ghats (Karnataka, Kerala, Maharashtra, Tamil Nadu)
4.	*C. indica*	Western Ghats (Karnataka: Hassan district)
5.	*C. sphaerica*	Eastern Ghats (Andhra Pradesh, Karnataka, Orissa, Tamil Nadu)
6.	*C. swamyi*	Western Ghats (Karnataka: Nagmangala, Mandya District
7.	*C. zeylanica*	Andaman and Nicobar Islands (extended to Sri Lanka)
8.	*Abies pindrow* var. *brevifolia*	Western Himalaya (Himachal Pradesh, Jammu and Kashmir, Uttarakhand)
9.	*Amentotaxus assamica*	Eastern Himalaya (Arunachal Pradesh)
10.	*Cephalotaxus griffithii*	Western Himalaya (Uttarakhand)
11.	*Ephedra kardangensis*	Western Himalaya (Himachal Pradesh: Lahul Spiti; Jammu-Kashmir or extended to Pakistan)
12.	*E. khurikensis*	Western Himalaya (Himachal Pradesh: Lahul Spiti)
13.	*E. sumlingensis*	Western Himalaya (Himachal Pradesh: Lahul Spiti)
14.	*Gnetum contractum*	Western Ghats (Kerala, Tamil Nadu)
15.	*G. ula*	Eastern and Western Ghats (Andhra Pradesh, Goa, Karnataka, Kerala, Maharashtra, Orissa)

5.2. Current status of rare, endangered and threatened (RET) species in India

Although several species of gymnosperms are quite common with substantial share of forest cover in India, primarily in the temperate regions, some species are facing varying degrees of natural and anthropogenic threats. According to IUCN 2013 Red List of Threatened Species, *Cycas annaikalensis* is Critically Endangered; *Amentotaxus assamica, Cycas beddomei, C. circinnalis, Taxus contorta* and *T. wallichina* are Endangered; *Cycas nathorstii, C. pectinata, C. swamyi, C. zeylanica*; *Gnetum contractum, Picea brachytyla, Pinus merkusii* and *T. meiri* are Vulnerable; *Cephalotaxus mannii* and *Cupressus cashmeriana* are Near Threatened. These species, therefore, require urgent measures for their conservation. *Ephedra gerardiana* and *Ephedra foliata* are also on road to rarity in Indian region. Concerted efforts are required to identify and monitor rare and endangered gymnosperms.

6. Methods and Approaches to study Gymnosperms

A herbarium is a collection and a database of dried and pressed plant specimens. Each specimen is a voucher deposited for future reference because a herbarium is a repository of information for geographical distributions, taxonomic, biological and ecological data. To study gymnosperms important methods and approaches are:

6.1. Plant collection and preservation of vouchers

Two main approaches of plant collection are: (i) to collect as much as one possibly can- a method desirable for writing flora of a particular area/ region; (ii) to collect certain groups like particular family, genus or species for revisionary, monographic, phytochemical or molecular studies. Specimens of most of the gymnosperms are prone to lose their leaves quite early, so it is advisable to poison the gymnosperm plants before pressing in order to retain their leaves intact. Poisoning kills the plant and thereby inhibits abscission layer formation. It is generally done by shortly dipping the whole specimen in a saturated solution of mercuric chloride in ethyl alcohol. Disposable gloves must be used while poisoning because mercuric chloride is a deadly poisonous chemical. In those places where the day's collections are bulky, they are spread in old newspapers and bundled up. Each bundle is then placed in a large polythene bag and formalin solution (10%) is poured over the bundles, so that the bundle just gets soaked thoroughly. The bags are then tied airtight. Killing a specimen by dipping it in boiling water for a while is also useful to hold on the leaves. Bulky materials especially cones and some vegetative parts are not suited for mounting directly on herbarium sheets. They can be placed in boxes designed to fit into herbarium case pigeonholes.

6.2. Identification of plant specimens

The collected specimens are identified with the help of regional Floras or revisions/ monographs, if availble. The different taxa can be segregated with the use of dichotomous keys, followed by descriptions, illustrations and photographs. Their identity should be confirmed by matching with already identified authentic specimens preserved in the herbarium. A microscope is essential for the observation of many diagnostic features. The most distinguishing morphological characters are leaves (shape - acicular or needle-like in Pinaceae, fern-like in Cycads, scale-like or awl-shaped in *Cupressus*, *Juniperus*, *Araucaria*, etc., or broad-leaved in *Podocarpus*, *Agathis*, *Ginkgo*, *Gnetum* etc.; indumentums-glabrous or hairy; arrangement-opposite, alternate or spiral, free or in bundles); nature of stem/trunk (erect or rambling/climbing, simple or branched, scaly or smooth, glabrous or hairy); arrangement of reproductive organs or strobili (solitary, fascicled or panicled; axillary or terminal; shape and size; number, shape and size of bracts, etc.). An attempt has been made here to provide artificial keys for identification of gymnosperms in India up to families and genera level.

Key to the families

1a. Leaves pinnately compound, palm like..**Cycadaceae**

1b. Leaves simple...2.

2a. Leaves fan-shaped, with conspicuous parallel veins. Ovules 2, at ends of long stalks...**Ginkgoaceae**

2b. Leaves scale-like, acicular, linear or ovate-elliptic, without conspicuous parallel veins. Ovules 1 to many on ovuliferous scales or on short stalks...........................3.

3a Vessels present in xylem. Perianth present. Male flowers in bracteate spikes..4.

3b. Vessels absent in xylem. Perianth absent. Male flowers in catkins, spikes or in subglobose heads...6.

4a. Leaves only two, very long strap shaped. Stem unbranched or very shortly forked...**Welwitschiaceae**

4b. Leaves many. Stem branched..5.

5a. Leaves scale-like, reduced to sheaths. Stem performs photosynthesis ....**Ephedraceae**

5b. Leaves broad, with distinct reticulate venation. Leaves perform photosynthesis. ....**Gnetaceae**

6a. Ovule one to each ovuliferous scale..**Araucariaceae**

6b. Ovule two or more to each ovuliferous scales...7.

7a. Monoecious. Wood and leaves with resin canals. Seeds winged; cotyledons many..8.

7b. Dioecious or monoecious. Wood and leaves without resin canals (except *Cephalotaxus*). Seeds wingless; cotyledons two..9.

8a. Tall trees. Female cones woody; scales spirally arranged, ovules two. Leaves needle like, alternate or fascicled on short shoots. Wings of seeds large...**Pinaceae**

8b. Shrubs or trees. Female cones more or less succulent when ripe; scales opposite, ovules many. Leaves scale like and decurrent, short shoots absent. Wings of seeds small (absent in *Juniperus*)........................**Cupressaceae** (including **Taxodiaceae**)

9a. Leaves usually 2-ranked at least in branchlets. Seeds usually surrounded by aril; ovules erect..**Taxaceae** (including **Cephalotaxaceae**)

9b. Leaves usually spiral. Seeds borne on a swollen fleshy receptacle; ovules inverted...**Podocarpaceae**

Keys to the genera (wild and commonly cultivated)

Araucariaceae

1a. Dioecious. Leaves awl shaped or lanceolate, less than 5 cm long, spirally and closely arranged. Seeds connate with bracts...***Araucaria***

1b. Monoecious. Leaves ovate-elliptic, 5-12 cm long, opposite or nearly so, distant. Seeds detached from bracts...***Agathis***

Pinaceae

1a. Leaves in clusters ....2.

1b. Leaves not in clusters..4.

2a. Leaves more than 5 cm long, 2-5 (-7) in a cluster (bundle), base enclosed by sheath..***Pinus***

2b. Leaves less than 5 cm long, many in a cluster, base not enclosed by sheath..3.

3a. Evergreen trees with stiff leaves. Branches horizontal. Female cones more than 10 cm long; seeds triangular...***Cedrus***

3b. Deciduous trees with soft leaves. Branches pendulous. Female cones less than 10 cm long; seeds obliquely obovoid...***Larix***

4a. Leaves needle like, quadrangular. Female cones pendulous. .... ***Picea***

4b. Leaves flat, bilateral. Female cones erect..5.

5a. Leaves variously arranged on the branchlets, either pectinate, spiral or crowded. Female cones 6-18 cm long, cylindrical..***Abies***

5b. Leaves 2-ranked on branchlets, Female cones ovoid, up to 2.5 cm long..***Tsuga***

Cupressaceae

1a. Female cones succulent, indehiscent or slightly dehiscent when mature; seeds wingless..***Juniperus***

1b. Female cones woody or leathery, dehiscent when mature; seeds usually winged (except Taxodium)..2.

2a. Female cones with peltate scales, maturing in 1st or 2nd year................................3.

2b. Female cones with flattened scales, maturing in 1st year..5.

3a. Deciduous trees. Leaves 4-14 mm long. Seeds irregularly triquetrous, wingless..***Taxodium***

3b. Evergreen trees. Leaves 3 mm. Seeds subglobose to ovoid, with lateral narrow wings..4.

4a. Branchlets usually not arranged in a plane. Female cones maturing in 2nd year; scales with 3–numerous seeds...***Cupressus***

4b. Branchlets arranged in a plane. Female cones maturing in 1st year; scales with (1-) 2 (–5) seeds...***Chamaecyparis***

5a. Leaves tapering from base to a long-acuminate tip. .... ***Cunninghamia***

5b. Leaves not as above, scale like, acicular or awl-shaped..6.

6a. Lateral leaves 4–7 mm long, with conspicuous white stomatal bands abaxially. Female cone scales each with 3–5 seeds..***Thujopsis***

6b. Lateral leaves usually less than 4 mm long, without conspicuous white stomatal bands abaxially. Female cone scales each with 1 or 2 seeds.................................7.

7a. Female cones with only middle pair of scales fertile; seeds with 2 subapical unequal wings..***Calocedrus***

7b. Female cones with middle 2 or 3 pairs of scales fertile; seeds with 2 lateral, narrow wings or wingless..8.

8a. Female cones with 8 or 10 thin scales; bracts almost completely enveloped by cone scales, free apex a very short mucro; seeds with 2 lateral narrow wings ....***Thuja***

8b. Female cones with 6 or 8 thick scales; bracts partly enveloped by cone scales, free apex a long recurved cusp; seeds usually wingless ....***Platycladus***

Taxaceae

1a. Leaves with resin canal on the lower side between midrib and epidermis. Seeds completely enclosed by aril. ....*Cephalotaxus*

1b. Leaves without resin canal. Seeds partly enclosed by aril.....................................2.

2a. Leaves 7-11.5 cm long. Pollen cones arranged in terminal spikes. Seed-bearing structures long pedunculate; aril saccate, almost enclosing seed leaving extreme apex, purple when ripe...*Amentotaxus*

2b. Leaves up to 4 cm long. Pollen cones solitary in leaf axils, not forming spikes. Seed-bearing structures shortly pedunculate or subsessile; aril cupular, enclosing only basal part of seed, red when ripe..*Taxus*

Podocarpaceae

1a. Leaves not more that 2.5 cm long, dimorphic; juvenile leaves 2-ranked, arranged in 1 plane, linear; adult leaves needle like or scalelike, falcate .... *Dacrycarpus*

1b. Leaves 10 – 17 cm long, monomorphic...2.

2a. Leaves linear with a single raised midvein on one or both surfaces........*Podocarpus*

2b. Leaves elliptic without an obvious midvein but with many, slender, longitudinal veins ....*Nageia*

7. Economic Importance of Gymnosperms

Like angiosperms, gymnosperms are also important for humankind because of various kinds of products, such as wood, paper, beverages, food, cosmetics, and medicines are obtained from them (Table 4). The conifers occupy only 3.3% of the total forest area, but with this meagre representation they yield 35% of the total wood used in India (Mehra and Jain 1976). Most of the timber used in modern buildings, is derived from conifers like *Cedrus*, *Pinus*, *Abies*, *Picea,* distributed in western Himalaya and *Cupressus*, *Larix*, *Juniperus* in Eastern Himalaya, due to their soft wood and straight trunks with uniform texture. *Cryptomeria japonica* introduced from Japan in Darjeeling hills and Sikkim, which is now one of the most widely growing trees in the area, where it is called *Dhuppi* and is most favoured for its light wood, extensively used in house building and other construction work. Valuable resins such as turpentine, tars, etc. are also obtained from conifers. *Sabudana*, an important edible starch is obtained from sago cycad, *Cycas revoluta*, on other hand *Pinus gerardiana,* an evergreen tree of inner north west Himalaya yields almond like seeds (Chilgoja or pinenuts) which are largely eaten by the natives and stored for winter use. Several species of *Juniperus, Abies, Cupressus, Cedrus, Picea, Thuja and Tsuga* are sources of valuable essential oils used as scents in soaps, air fresheners and perfumes. Hemlock, Spruce and Larch yield tannin, while *Taxus* and *Pinus wallichiana* yield useful dyes. Pines, firs and spruces are also notable source of pulpwood for paper industry. Some species of *Taxus*, *Ephedra, Pinus* and *Juniperus* spp. have tremendous medicinal value and are used both in indigenous as well modern system of medicines. Over exploitation of *Taxus* (for Paclitaxel-a mitotic inhibitor used in cancer chemotherapy) and *Ephedra* spp. (for ephedrine and pseudoephedrine used for treatment of asthma, bronchitis, allergies, cold and flu) from their natural habitats have pushed them to threatened state. Pseudoephedrine sales are now restricted because this compound can be used as a starting point for the synthesis of illegal drug Methamphetamine (Remburg and Stead 1999). Ephedrine has also been used to enhance sports performance, a practice that has elicited medical concern. Cycads are used as ornamentals and for landscaping due to their slow growth and shinny glossy leaves have high commercial value. Some of the species i.e. *Thuja*, Spruce, Cypress and Pines are cultivated in gardens and lawns. Conifer forests are typical of the Himalaya which provides a cool and soothing environment for recreation and health. The fossil gymnosperms contributed greatly to the formation of fossil fuels like most of the coal seams of the Peninsular Lower Gondwana of India.

All the extant gymnosperms have great scope, for scientific exploitation as *Gnetum* is considered next to dicotyledonous, to understand the process of evolution of Angiosperms. In India diversity and bioprospection of only some economically important gymnosperms have been carried out and large numbers of taxa are yet to be exploited scientifically.

Table 4. Economic importance of gymnosperms in India

S. No.	Economic value	Name of taxa	Part used
1.	Construction, furniture, packing cases and other wood work	*Cedrus deodara, Pinus* spp., *Abies* spp., *Picea smithiana, Cupressus torulosa, Cryptomeria japonica, Juniperus polycarpos*	Trunk/timber
2.	Resin (turpentine, tar, etc.)	*Pinus roxbughii, P. wallichina*	Trunk exudes
3.	Tannin	*Tsuga dumosa, Picea smithiana, Larix griffithii*	Bark
4.	Purfumery, soaps, incense, air-freshners	*Abies* spp., *Cedrus deodara, Pinus* spp. *Thuja* spp., *Tsuga dumosa, Cupressus* spp.	Leaves/ heartwood
5.	Dye	Taxus spp., *Pinus wallichina*	Bark
6.	Paper pulp	*Pinus* spp., *Abies* spp., *Picea smithiana*	Wood
7.	Food	*Cycas revoluta/ Pinus gerardiana, P. roxburghii*	Stem starch/ seeds
8.	Medicinal	*Ephedra* spp., *Taxus* spp., *Ginkgo biloba, Juniperus* spp., *Cedrus deodara, Pinus roxsburghii, P. wallichina*	Leaves, stem, resin (*Pinus*), berries
9.	Ornamental/ aesthetic	*Cycas* spp., *Cupressus* spp., *Thuja* spp., *Pinus* spp., *Podocarpus spp., Agathis robusta, Araucaria spp., Ginkgo biloba, Juniperus* spp., *Picea smithiana*	Whole Plant

Source: Anonymous 1948-76, Ambasta 1986, Jain 1991.

8. Future Prospects

Both fossil as well as extant gymnosperms received adequate attention of several workers (Hooker 1883; Sahni 1928, 1931, 1948; Biswas 1933; Bharadwaja 1957; Dogra 1964, 1985; Maheshwari and Biswas 1970; Pant 1973; Mehra and Jain 1976; Sahni 1990). Whereas specific determination of those belonging to genera *Pinus, Cedrus, Larix, Picea, Tsuga, Cephalotaxus, Taxus* and *Podocarpus* do not raise any controversial issue, the taxonomy of the species of the genera like *Juniperus, Abies, Cupressus, Gnetum* and *Ephedra* are in utter confusion and needs to be addressed by applying modern tools of molecular biology. Most of the complex taxa have not so far been studied intensively in India except for their taxonomic and nomenclature treatment. There is an urgent need to investigate the genetic diversity and population structures of such taxa. Molecular markers both dominant and co-dominant have been widely used to investigate population structures and genetic diversity of different plants species (Williams *et al.* 1990), and to propose phylogenetic hypothesis in plants (Rath *et al.* 1998). The molecular characterization and quantification of population differentiation of the wild populations of taxonomically complex as well as endangered taxa of Indian gymnosperms would be an interesting prospective arena of research.

Gymnosperms are one of the fascinating groups of plants to study. Evolutionary tendencies reveal that they exhibit close affinities with higher cryptogams, the pteridophytes on one hand and with angiosperms on the other. Apart from this they have a high economic potential and they also play a significant role in the preservation of the environment. Various developmental activities, particularly urbanization and construction of roads coupled with mining activities and over-exploitation are the main causes for dwindling the number and size of the population of some of the gymnosperms.

References

Adams RP, Bartel JA, Price RA (2009) A new genus, *Hesperocyparis*, for the cypresses of the western hemisphere. Phytologia 91:160–185

Ambasta SP (1986) The Useful Plants of India. Publication and Information Directorate, CSIR, New Delhi, India

Anonymous (1948-76) The Wealth of India—Raw Materials Vols. 1-11. Council of Scientific and Industrial Research, Publication and Information Directorate, CSIR, New Delhi, India

Bentham G and Hooker JD (1862-1883) Genera Plantarum, Lovell Reeve, London

Bharadwaja RC (1957) Genus *Gnetum* Linn. in India, Pakistan and Burma. J Ind Bot Soc 36:408-420

Biswas K (1933) Distribution of wild gymnosperms in the Indian Empire. J Ind Bot Soc 12(1):24-47

Burleigh JG, Mathews S (2004) Phylogenetic signal in nucleotide data from seed plants: implications for resolving the seed plant tree of life. Am J Bot 91:1599–1613

Castillo GRH (2001) Seed plant phylogeny and the anthophyte hypothesis. http://www.biology.ualberta.ca/courses.hp/biol606/OldLecs/Lecture2001.03.HC.html, accessed on 26 November 2007

Chaudhuri AB (1993) Forest plants of eastern India. APH Publishing Corporation, pp 583-593

Christenhusz MJM, Reveal JL, Farjon AK, Gardner MF, Mill RR, Chase MW (2011) A new classification and linear sequence of extant gymnosperms. Phytotaxa 19:55–70

Dar AR, Dar GH (2006) The wealth of Kashmir Himalaya-Gymnosperms. Asian Journal of Plant Sciences 5(2):251-259

Davis PH (1961) Hints for hard-pressed collectors. Watsonia 4(6):283-289

Del Tredici P, Ling H, Yang G (1992) The Ginkgos of Tian Mu Shan. Cons Biol 6:202-209

Dogra PD (1985) Conifers of India and their wild gene resources in relation to tree breeding. Indian For Nov 85 (Special issue):935-955

Donaldson JS (2003) *Zamia pymaea*, 2006 IUCN Red List of Threatened Species, accessed on 24 August 2007

Engler A (1897) Die natürlichen Pflanzenfamilien, Leipzig: Wilhelm Englemann.

Engler HGA, and Prantl KAE (1926) Die natürlichen Pflanzenfamilien, 13e 2, Gymnospermae 6, Pilger RK, Leipzig: Wilhelm Englemann

Fergoson KD (1985) A new species of *Amentotaxus* (Taxaceae) from northeastern India. Kew Bull 40(1):115-119

Florin R (1956) Nomenclatural notes on genera of living gymnosperms. Taxon 5(8):188-92

Friis EM, Crane PR, Pedersen KR, Bengtson S, Donoghue PCJ, Grimm GW, Stampanoni M (2007) Phase-contrast X-ray microtomography links Cretaceous seeds with Gnetales and Bennettitales. Nature 450: 549-553

Goebel K (1905) Organography of Plants, Englished, Oxford

Government Botanical Gardens, Ooty (2013) (http://en.wikipedia.org/wiki/Government Botanical Gardens, Ooty, accessed on October 17, 2013

Gupta S, Padalia H, Chauhan N, Porwal MC (2002) Rediscovery of *Podocarpus wallichianus*: A rare gymnosperm from tropical rain forests of Great Nicobar Island. Curr Sci, 83(7):806-807

Hirase S (1898) Etudes sur la fécondation et l´embryogénie du *Ginkgo biloba* (second mémoire). J Coll Sci Imp Univ Tokyo 12:103-149

Hofmeister W (1851) Vergleichende Untersuchungen der Keimung, Entfaltung und Fruchtbildung höherer Kryptogamen und der Samenbildung der Koniferen, Leipzig

Hooker JD (1883) Flora British India, Vol. 5, Reeve and Co. Ltd, London

Hsieh L (1992) Origin and distribution of Ginkgo biloba. The Forestry Chronicle 68(5):612-613

Jain SK (1991) Dictionary of Indian Folk Medicine and Ethnobotany. Deep Publication, New Delhi, India

Jones WG, Hill KD, Allen JM (1995) *Wollemia nobilis*, a new living Australian genus and species in the Araucariaceae. Telopea 6(2-3):173-176

Karnataka Forest Department Report (2004) http://karnatakaforest.gov.in/Karnataka Forest Department Research, Training & Publications.mht, accessed on October 17, 2013

Leubner G (2007) The Seed Biology Place, http://www.seedbiology.de, accessed on October 17, 2013

Lindstrom AJ, Hill KD (2007) The genus *Cycas* (Cycadaceae) in India. Telopea 11(4):463–488

Maheshwari P and Biswas C (1970) *Cedrus.* Bot Monogr No 5, CSIR, New Delhi

Mao K, Hao G, Liu J, Adams RP, Milne RI (2010) Diversification and biogeography of *Juniperus* (Cupressaceae): variable diversification rates and multiple intercontinental dispersals. New Phytologist 188:254-272

Matthew (1981) Exotic Plant species documented in Pulney hills Western Ghats of Tamil Nadu (Enlist of species collected from Flora of Palani Hills). http://thewesternghats.indiabiodiversity.org/biodiv/content/documents/document-ab507042-3a0e-4879-bc20-ebfb92f52d88/57.pdf

Mehra PN (1988) Indian conifers, gnetophytes and phylogeny of gymnosperms, New Delhi

Mehra PN, KK Jain (1976) *Abies* and *Juniperus* complexes in the E Himalayas with observations on *Larix griffithii* Hook. f. and *Tsuga dumosa* Eichler, Sree Saraswaty Press, Calcutta

Melchior H (1964) Engler's Syllabus der Pflanzenfamilien, 12e, Vol. 2, Berlin, Gebrüder Borntraeger

Melchior H and Werdermann E (1954) Engler's Syllabus der Pflanzenfamilien, 12e Vol. 1, Berlin, Gebrüder Borntraeger

Pant DD (1973) *Cycas* and Cycadales, Central Book Depot, Allahabad

Pant DD, Nautiyal DD (1963) Cuticular and epidermal studies of some modern Cycadean leaves, etc. Senck Biol 44(4):257-347

Parker RN (1927) The Himalayan Silver Firs and Spruces. Indian For, 52(12): 683-693

Pearson HH W (1929) Gnetales, Cambridge University Press, Cambridge, UK

Poudel RC, Möller M, Gao LM, Ahrends A, Baral SR, Liu J, Thomas P, Li DZ (2012) Using morphological, molecular and climatic data to delimitate yews along the Hindu Kush-Himalaya and adjacent regions. *PLoSONE* 7:e46873

Price R A (1996) Systematics of the Gnetales: a review of morphological and molecular evidence. Int J Pl Sci 157(Suppl. 6):S40-S49

Rath P, Rajasegar G, Chong JG, Kumar PP (1998) Phylogenetic analysis of *Dipterocarpus* using random amplified polymorphic DNA markers. Ann Bot 82:61-65 (RAPDs)

Remburg B and Stead AH (1999) "Drug characterization/impurity profiling, with special focus on methamphetamine: recent work of the United Nations International Drug Control Programme". Bulletin on Narcotics 51:1-2, (UNODC)

Sahni B (1920a) On the structure and affinities of *Acmopyle pancheri* Pilger. Phil Trans Roy Soc London 210(B):253-310

Sahni B (1920b) On certain archaic features in the seed of *Taxus baccata* with remarks on the antiquity of the Taxineae. Ann Bot 34:117-133

Sahni B (1928) Revision of Indian Fossil Plants Pt I, Coniferales (a Impressions and Incrustations). Mem Geo Surv Indi Palaeont Indica (ns) 11:1-49

Sahni B (1931) Revision of Indian Fossil Plants Pt II Coniferales (b. Petrifications. Mem Geo Surv Indi Palaeont Indica (ns) 11:51-120

Sahni KC (1990) Gymnosperms of India and Adjacent Countries. Bishen Singh Mahendra Pal Singh, Dehradun

Shah A, Li D-Z, Moller M, Gao L-M, Hollingsworth M and Gibby M (2008) Delimitation of *Taxus fuana* Nan Li & R.R. Mill (Taxaceae) based on morphological and molecular data. Taxon 57(1):211-222

Sharma P, Uniyal PL and Hammer O (2010) Two New Species of *Ephedra* (Ephedraceae) from the Western Himalayan Region. Syst Bot 35(4):730-735

Spjut RW (2007) Taxonomy and nomenclature of *Taxus* (Taxaceae). J Bot Res Inst Texas 1(1): 203–289

Tang CQ, Yang Y, Ohsawa M, Yi S R, Momohara A, Su W H, Wan H C, Zhang Z Y, Peng M C and Wu Z L (2012) Evidence for the persistence of wild *Ginkgo biloba* (Ginkgoaceae) populations in the Dalou Mountains, southwestern China. Am J Bot 99(8):1408-1414

Tewari LM, Jalal JS, Kumar S, Pangtey YPS Kumar R (2010) Wild and Exotic Gymnosperms of Uttarakhand, Central Himalaya, India. EJBS 4 (1): 32-36

The IUCN Red List of Threatened Plants (2013) http://www.iucnredlist.org/details/32311/0, accessed on October 17, 2013

The Plant List (2013) The Plant List: A working list of all plant species. http://www.theplantlist.org, accessed on 9 September 2013

Tripathi P, Tewari LM, Tewari, Kumar S4, Pangtey YPS Tewari Geeta (2009) Gymnosperms of Nainital. Report and Opinion, 1(3):82-104

Van Tieghem Ph (1898) Elements de botanique ed 3, Masson et Cie, Paris

Williams JGK, Kubelik AR, Livak, KJ, Rafalski JA and Tingey, SV (1990) DNA polymorphism ampliphied by arbitrary primers are useful as genetic markers. Nucleic Acids Res. 18(22):6531-6535

Chapter – 11

Different Methods of Plant Identification

L.B. Chaudhary

1. Introduction

Among three major functions identification, classification and nomenclature of taxonomy (Fig.1), identification is a core activity and one of the major objectives of taxonomy, which determines whether other two elements are the same or different. Classification and nomenclature are two separate methods in taxonomy, however, both are totally dependent on the correct identification of the objects. The identification of an unidentified plant with a correctly identified specimen also involves classification. So, both identification and classification, involve comparison and judgment of similarities and differences. Therefore, the identification is a primary function in classification involving nomenclature which performs an essential role as a means of communication. According to Blackwelder (1967) "identification enables us to retrieve the appropriate facts from the system (classification) to be associated with some specimen at hand" and is "better described as the recovery side of taxonomy." The identification of plants is very important for proper assessment and characterization of biodiversity on which the whole life of human beings is dependent starting from birth to death. The use of biodiversity for various purposes like food, nutrition, pharmacognosy etc. requires accurate and reliable identification and naming of plants (Nesbitt *et al.* 2010).

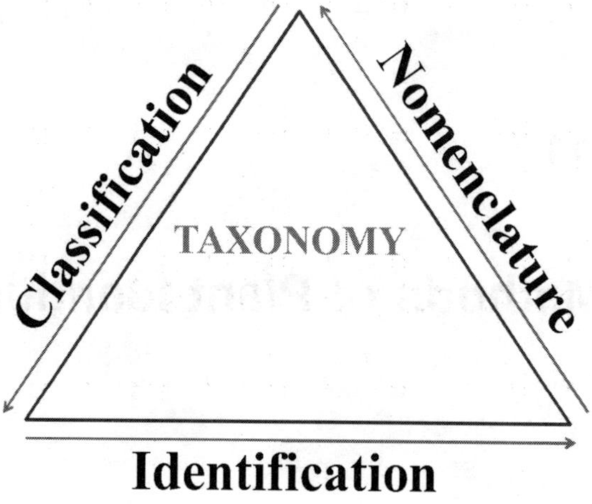

Figure 1. Function of Taxonomy

2. Plant Identification Methods

The plants can be identified by recognition, expert determination, comparison and the use of botanic keys and other similar devices (Geesink 1987). For a thorough and detailed discussion of specimen identification, Sneath and Sokal (1973) may be consulted.

2.1. Recognition

For scientific purpose, the first thing one needs to know about a plant species is its scientific name, as this functions as a key word to knowledge about it. The easiest way to obtain the scientific name of a plant is to show the specimens to an experienced taxonomist, he/she will give the name after one intense glance (Fig. 2). This kind of recognition is also called 'gestalt perception'. This is based on extensive, past experience of the identifier with the plant group in question. According to Morse *et al.* (1968, 1971) the reliability of the identification through this process totally depends on the knowledge of the identifier.

2.2. Expert determination

An expert determination is one of the best methods of identification as far as the reliability or accuracy are concerned. In general the experts prepare treatments like monographs, revisions, synopses, etc. of the group in question. Generally the more recent floras or manuals include the expert's concepts of taxa. Although it is a reliable

Fig. 2. An unidentified *Oxytopis* sp. is being identified by an expert

method of identification, but non availability of experts for all taxa everywhere and requiring the valuable time of experts create delays for identification.

2.3. Comparison

In this method the unknown elements are compared with identified specimens, photographs, illustrations or descriptions. Although, it seems an easy method of identification, but it may be very time consuming or virtually impossible if correctly and suitable materials for comparison are not available. The reliability of identification through this method totally depends on the accuracy and authenticity of the specimens,

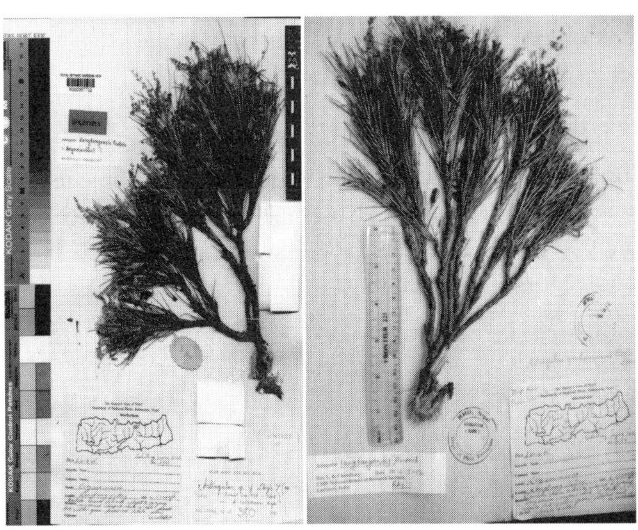

Fig. 3. Comparison of unidentified plant of *Astragalus langtagensis* Podl. with identified/ type specimen

illustrations or descriptions used for the comparison. In Figure 3 an unidentified specimen of *Astragalus* L. has been compared with a type specimen of *Astragalus langtagensis* Podl.

2.4. Botanical keys

This is one of the best and widely used methods of identification which does not require much time, materials, or experience involved in comparison and recognition. Botanical keys are considered a type of taxonomic literature. Keys are devices consisting of a series of contrasting or contradictory statements prepared by identifier. The first modern dichotomous key was designed by Lamarck in his 'Flore Francoise' in 1778. Voss (1952) has provided an interesting account of the history of keys and phylogenetic trees in systematic biology. The evolution of keys has been the result of subsequent taxonomists who study the characteristics of organisms at some taxonomic level (= category) and often develop keys for their identification.

Types of taxonomic keys

There are many types of botanical Keys (e. g. dichotomous key, punch card key etc.). However, here the main emphasis has been given only on dichotomous key for practical purposes.

2.4.1. Dichotomous key

A dichotomous key is constructed of a series of couplets, each consisting of two separate statements (leads). For example:

couplet 1. Seeds round........ sp.

1. Seeds oblong........ 2 (this statement indicates that you go to couplet '2')
couplet 2. Seeds white........ sp.

2. Seeds black.........sp.

The design of dichotomous keys differs in various works. Some use numbers to separate the couplets while others use letters. Also, some taxonomists place each couplet together, while others may separate couplets (see examples below). It is important to indent every other couplet for ease in reading.

Example A: Numerical key with couplets together

1. Seeds round...........sp.

1. Seeds oblong............2

2. Flowers white.........sp.

2. Flowers red ....sp.

Example B: Alphabetical key with some couplets separated

A. Seeds oblong ....B

B. Seeds white ....sp.

B. Seeds black ....sp.

A. Seeds round ....sp.

The dichotomous keys are usually of two types, the indented key and the bracketed key. The bracketed keys are preferred in the modern work. The comparison between two keys has been provided in table 1.

2.4.1.1. Bracketed key

Bracketed keys are those where the two alternative leads of a couplet are written in adjacent lines, giving reference by numbers to their successive subordinate leads. Example of a bracketed key is given below:

Bracketed Key to some *Astragalus* species

1a. Stipules connate 2

1b. Stipules free (occasionally slightly connate at base in *A. khasianus*) 3

2a. Stipules hairy; bracts 10 – 14 mm long, ciliate along margins; peduncles 5 – 17 cm long; calyx lobes *ca.* 4 mm long ....4. **A. isabellae**

2b. Stipules glabrous; bracts 6 mm long, glabrous; peduncles 2 – 6 cm long; calyx lobes *ca.*1 mm long... 2. **A. concretus**

3a. Pods pubescent or pubescent only in young condition....................................4

3b. Pods glabrous 5

4a. calyx lobes *ca.* 2 mm long; ovary glabrous; young pods glabrescent with white, adpressed hairs ....3. **A. emodi**

4b. calyx lobes *ca.* 4 mm long; ovary hairy; pods pubescent with blackish – brown, half spreading hairs ....6. **A. maxwellii**

5a. Calyx lobes minute 6

5b. Calyx lobes 1.5 to 4 mm long 7

6a. Inflorescence 7 – 30 cm long, longer than subtending leaf; pod stipe double than calyx length ....1. **A. chlorostachys**

6b. Inflorescence 6 – 10 cm long; pod stipe equal to calyx length...........8.**A. xiphocarpus**

7a. Stipules 18 – 21 x 5 – 6 mm, lanceolate, glabrous to glabrescent; bracts hairy only along margins; calyx lobes 1.5 mm long, pod stipe longer than calyx length ....5. **A. khasianus**

7b. Stipules *ca.* 7 mm long, setaceous, sericeous; bracts hairy throughout; calyx lobes 2 – 4 mm long; pod stipe more or less equal to calyx length............7. **A. stewartii**

2.4.1.2. Indented Key

Indented keys are not always placed immediately under each other but are grouped together with their consequent couplets in an indented typographic setting. They are called indented because usually the couplets are printed with gradually increasing indenting. Example of an indented key is given below (adopted from Jain and Rao 1977).

1a. Flowers reduced to single pedicelled stamens enclosed in an involucres:

- 2a. Glands on involucres connate into a continuous ring; inner bracts connate forming a tube round the female. 3. **Synadenium**
- 2b. Glands on involucres 1 – 5, discrete; inner bracts not connate:
  - 3a. Leaves alternate or opposite, if opposite, then not inaequilateral at base and without chlorenchyma sheathed veins; main axis not aborting; branching monopodial at least below. 1. **Euphorbia**
  - 3b. Leaves entirely opposite, distinctly inaequilateral at base, with chlorenchyma sheathed veins; main axis aborting just above cotyledons; branching sympodial throughout. 2. **Chamaesyce**

1b. Flowers distinct, not reduced to a single stamen or single ovary contained in an involucres:

- 4a. Cells of the ovary 2-ovuled; plants not laticiferous:
  - 5a. Style present, bifid; fruit capsular, keeled 4. **Phyllanthus**
  - 5b. Style absent, or if present, then united into a stylar column; fruit fleshy not keeled. 5. **Kirganellia**
- 4b. Cells of the ovary 1–ovuled; plants laticiferous, at least in younger parts:
  - 6a. Plants with stinging hairs, climbing; stamen 3. 6. **Tragia**
  - 6b. Plants without stinging hairs, erect; stamens many 7. **Croton**

Table 1. Comparison between Bracketed and Indented keys

Bracketed key	Indented key
• Contrasting characters in a couplet are placed together.	• Contrasting characters in a couplet are placed at different places

• All leads are of approximately the same line length and produce a maximum of efficiency of page.	• Sloping and shortening of lines to the right with a resultant loss of economy of page space.

2.4.1.3. Visual key

The identification of objects with the help of photographs and images is considered easiest methods. Besides bracketed and indented keys, the advances in digital imaging have made possible the creation of completely visual keys for the creation of easy-to-use identification tools. The older keys are text based, the visual key is based almost exclusively on images and minimal amount of text. So, the characters in visual keys are visually defined.

For the construction of visual key the photographs are obtained from internet and herbarium databases or are taken specifically for this purpose. The images are printed and then sorted into hierarchical groups. These hierarchical groups of images are used to create the 'couplets' in the key. A reciprocal process of key creation and testing are used to produce the final keys. For details see http://www.ncbi.nlm.nih.gov/pmc/articles/PMC3072766/.

2.5. Suggestions for construction of keys

1. Identify all groups to be included in a key.
2. Prepare analytical description of each taxon.
3. The description and measurements of similar character to be used in the key should be based on several specimens and not single specimen or a single leaf. Decision should not be taken on a single observation. It is always advised to examine several specimens for single species.
4. The terminology used in the key should be consistent in meaning and should be uniformly used. The use of alternative terms for the same concept to achieve more lively prose should be avoided.
5. Select constant characters rather than variable characters.
6. Select key characters with a contrasting character stated.
7. As far as possible, use morphological characters which can be studied with naked eye or lens.
8. Use positive characters, and cover the whole variability of the group to be keyed out, e. g.

    1\. Leaves opposite.

    1\. Leaves either in whorls, or spirally arranged or distichous.

And not:

1. Leaves opposite.

1. Leaves not opposite.

9. Quantitative statements like 10 – 12 cm, or 5 – 10 m are preferable to words like 'large' or 'small' or 'big' etc.
10. Use the most obvious terminology and not the most correct one, if different and putative source of confusion.
11. Mention the name of the plant part before descriptive phrases, e.g. leaves or flowers blue not blue flowers, leaves alternate not alternate leaves.
12. Start both leads of a couple with the same word if at all possible and successive leads with different words.
13. Use at least two characters per lead when possible.
14. If one alternative has fewer taxa than the other, bring out the couplet having lesser taxa first, and then treat the other.
15. After having finished your work, ask a colleague (preferably somebody without knowledge of that group) to identify some materials with it.
16. Never confuse a key with a classification.

The genera or species in the key are sometime given a serial number, which tallies with the sequence in the text. For example given above in the bracketed key number 1 has been given to *Astragalus chlorostachys* and number 8 to *Astragalus xiphocarpus*, it indicates that in the treatment of the genus *Astragalus* the species has been arranged alphabetically. Similarly, in indented key, number 1 given before *Euphorbia* shows that this is the first genus in the family treated in the text and *Croton* with 7 number being the last.

2.6. Suggestions for using keys

The following points may help in the use of keys:

1. An appropriate keys either from flora, manual, guide, handbook, monograph, or revision should be selected for the materials to be identified. If the locality of an unknown plant is known, a flora, guide or manual treating the plants of that geographic area may be taken for identification. If the family or genus is recognized, one may choose to use a monograph or revision. If locality is unknown, in this case one should select a general work. If materials to be identified were cultivated, any manual treating such plants may be considered as majority of the floras do not include cultivated plants unless naturalized.

2. Before using the key, one must read the introductory comments on the format details, abbreviations, etc. used in the key Always both leads of a couplet should be read before making any decision. Even though the first lead may seem to describe the unknown material, the second lead may be even more appropriate.

3. To properly understand the terminology used in the key, one should always use a standard glossary to check the meaning of the terms. If the both leads of a couplet are not clear or when information provided are insufficient then the decision as to which of the two answers best fits the descriptions should be taken with utmost care.

4. Once the elements are identified with a key, the results should be verified by reading description, comparing the specimen with an illustration or an authentically named herbarium specimen.

2.7. Common problems in key usage

Key users must overcome many practical problems, such as:

1. *Variant forms*. The problems arise in the identification and use of key when the keys do not include all forms of the species.

2. *Incomplete coverage*. Sometimes unidentified species and groups that are difficult to identify or that have been poorly characterized are not include in the key or may be mentioned only in introductory text.

3. *Lighting and magnification*. When keys do not give details of how the specimen was viewed (the magnification, lighting system, angle of view etc.), it creats problems for users. This is very important to compare the characters provided in the keys. The author may, for instance refer to tiny bristles, hairs or chaetae—but how tiny?

4. *Language*. Very few keys are multilingual. Translations of a key may be incorrect or misleading. Many keys contain vague words that do not translate.

5. *Obsolescence*. The oldest keys do not contain more recently described taxa. They may also use outdated taxon/species names, which may not be appropriate today.

References

Blackwelder RE (1967) Taxonomy, A Text and Reference Book. John Wiley & Sons, Inc. New York

Geesink R (1987) Structure of keys for identification. In: Manual of Herbarium Taxonomy (Theory and Practice). Vogel EFD (ed.), UNESCO, Indonesia pp 77-86

Jain SK and Rao RR (1977) A handbook of field and herbarium methods, Today and Tomorrow's Printers and Publishers, New Delhi

Morse LE, Beaman JH, Shetler SG (1968) A computer system for editing diagnostic keys for Flora North America. Taxon 17:479-483

Morse LE, Peters JA, Hamel PB (1971) A general data format for summarizing taxonomic information. BioScience 21:174-18

Nesbitt M, McBurney RPH, Broin M, Beentje HJ (2010) Linking biodiversity, food and nutrition: The importance of plant identification and nomenclature. J Food Comp Analysis 23 (6):486-498

Sneath PHA and Sokal RR (1973) Numerical Taxonomy: The Principles and Practice of Numerical Classification. W H Freeman and Company, San Francisco

Voss EG (1952) The history of keys and phylogenetic trees in systematic biology. Journal of the Scientific Laboratories of Denison University 43:1-25

Chapter – 12

Herbarium: Techniques and Management

Bhaskar Datt, Baleshwar and T.S. Rana

1. Introduction

Herbarium is simply a dried and pressed collection of plants arranged in an accepted sequence of classification (Lawrence 1951, Shetler 1969). It can be a very useful teaching aid or an absorbing hobby (Cronquist 1966). According to Fosberg and Sachet (1965) "A modern herbarium is a great filing system for information about plants, both primarily in the form of actual specimens of plants and secondarily in the form of published information, pictures and recorded notes". It is also a research, training and service institution that serves as a reference centre, documentation facility and data store house (Radford 1986). It is now well recognized that a herbarium is a vast reservoir of facts about plants. The word 'Herbarium' was originally applied, not to a collection of plants but to a book dealing with medicinal herbs. Tournefort in about 1700 used the term as an equivalent to 'Hortus Siccus' which was later on adopted by Linnaeus (Stearn 1957). The present concept and development of herbarium is due to efforts of botanists for more than four centuries. The concept of herbarium has been changing continuously, which in turn, has affected the gradual development of herbarium and herbarium practices.

For practical reasons, the classification of the world's flora is primarily based on herbarium material and the literature associated with it. Despite its limitations, a herbarium has certain advantages over living collections. It is usually only in the herbarium that we can compare all the related species of a genus in the same place, in the same state and at the same time (Davis 1961).

The herbarium material, therefore, formed the foundation of all botanical studies. We should, therefore, realise that herbaria are the custodians of enormous data pertaining to plants. They are usually seen as old-fashioned and expensive "stamp collection" without realising their paramount role in biological sciences. However, the world-wide rise of interest in the field of conservation, biodiversity and bioprospection offer a new chance for herbaria. Furthermore, computers offer a golden opportunity to raise the status of herbarium taxonomy by making routine curatorial procedures more efficient.

Although Jain and Rao (1976) elaborately discussed the methods of field collections, identification and preservation in the herbarium, the salient features of herbarium in up dated form with various procedures followed in Indian context are discussed in this chapter.

2. Historical Background

Since the dawn of civilization man has been dependent on the plants for all his basic needs like food, clothing and shelter. The thought of sharing his knowledge and experience about the utility of plants with his fellow men must have struck to his mind. He must have even felt the need to preserve these plants for future reference. Perhaps this must have been the beginning of the concept of the development of herbarium and herbarium collections. In addition to the plants used by man for his material comforts, a large number of plants of curious features attracted the attention of early botanists. For example, some plants had abnormal morphological structures; some had curious means of nutrition or curious habit and habitat. This led the early botanists to carry out research in all aspects of plant life such as their morphology, growth, multiplication and evaluation of medicinal and other properties. The concept of preserving plant specimens in dried form, as practiced today, is ca 450 years old. Although one cannot say with certainty that who discovered this method of preserving plants. However, the oldest preserved herbarium specimen is kept in Rome, collected by the naturalist Gherardo Cibo, a pupil of Luca Ghini in 1532. The beginning of the herbarium as a dried collection affixed to paper is attributed to Luca Ghini (1490-1556) who perhaps made the first herbarium specimen. He made several plant collection trips in Italy, and subsequently, recommended this method of plant preservation which paved the ways for the establishment of the first herbarium of the world in 1545 at the University of Padua, Italy, along with the establishment of the first botanic garden. During the same period Greeks were known for the use of plants in treating various ailments. Dioscorides' "Materia Medica" provides an account of the medicinal use of about 100 plants, following Theophrastus in this regard. As the Renaissance developed in Italy, the Italians began teaching botany and developed the first ever botanical gardens. They also prepared "Books" of dried mounted plants, which they called "*Horti Sicci*" or "Dry Gardens". In the beginning the specimens were mounted on sheets of paper which were bound in the form of a book. Although this practice was fairly common during the time of

Linnaeus, he departed from the convention of the day in mounting on single sheets and arranging them according to groups as is the practice today. In spite of the efforts of Linnaeus and his immediate successors, the binding of the herbarium sheets into volumes continued as late as 1830 when Asa Gray sold her many bound volumes of grass and sedge herbarium specimens (De Wolf 1968). The present procedure of pressing and drying specimens for storage has been amazingly successful, and has stood the test of time. Plants so preserved provide a concrete basis for past, present and future reference. The present concept of herbarium collections alongside detailed field data is also due to the experience of botanists over four centuries. Early botanists recorded only scanty data or none at all to accompany their collections. Frequently it was difficult to trace the origin of collections. Similarly the present size of herbarium sheets (28 x 42 cm ± 1 cm) is also the result of the combined efforts and experience of many scientists. In the early days the size of sheets varied with the size of plants.

3. Functions or Utility of a Herbarium

The potential role of herbaria is rarely realised among a section of modern biologists. Hence it is necessary to highlight the diverse functions of a modern herbarium and prove how it acts as a connecting link between various disciplines. The following are a few important functions of a herbarium.

- Serves as a repository of voucher specimens and related data on which varieties of botanical researches are carried out.
- Serves as a secure repository for "Type" specimens which are valuable for taxonomic researches.
- Serves as a fundamental resource for identification of all plants.
- Provide material and data for floristic, revisionary as well as monographic studies.
- Provide accurate locality data for planning field trips.
- Facilitate the documentation of flowering and fruiting times of all plants.
- Facilitate and promote the exchange of new material among institutions.
- Aids in teaching and in all biological researches.
- Most estimates on global biodiversity today are based on herbarium collections only.
- Aids in biodiversity monitoring: carrying out scrutiny of herbarium collections to obtain quantitative baseline data on the distribution and abundance of keystone species is essential for all monitoring programmes.
- Vast collections of a particular species in a herbarium aid in assessing the variations (diversity) exhibited by a species in its distributional range and hence help in population biology studies.

- Aids in assessment of conservation status of a taxon (helps in cataloguing of rare and endangered species for conservation programme).
- Serves as a repository of historic collections and at times acts as the only record of past vegetation.
- Serves as a source for search of new genetic material for improvement of our cultivated stock.
- Aids in assessment and cataloguing of all species of economic potential; these may include commercial species, medicinal herbs, food plants and so on.
- Serves as a source of material for investigations on anatomy, morphology, palynology, ecology, chemistry, molecular biology, pharmacognosy and for environmental impact assessments.
- Seeds of herbarium specimens can be used to resurrect species extinct in the wild using modem technologies.
- Herbarium, like a library, is a resource centre; solves all queries about plants from scientists, administrators, lawyers, corporate personnel, industrialists, environmentalists and foresters.
- Helps in development of computer databases on plants and maintains active links to international network of systematic resources and electronic knowledge base.

4. Important Herbaria of the World and India

Herbaria range from small personal collections to large collections of universities and government institutions. According to a recent census there are about 3990 recognized herbaria in the world. Shetler (1969) estimated that the herbarium resources of the world may include as many as 250 million specimens; of this estimate, 78 million are in European herbaria and 36 million in American herbaria, by about the middle of the 20th Century. Herbaria were so numerous throughout the world that there arose a need to publish a consolidated list of all world herbaria for assisting the botanists. This important task was accomplished by the International Association for plant Taxonomy (IAPT) by publishing Ist edition of "*Index Herbariorum*" in 1952. The *Index Herbariorum* is such an extremely useful publication for locating different herbaria of the world, size and type of collections, important historical collections, name of the curators etc., this compilation is greatly helpful for loan and exchange of specimens etc., at international level. Some important world and Indian herbaria are listed in Table 1 and 2.

Table 1. Ten largest herbaria in the world

S.No.	Name	Acronym	Specimens (ca.millions)
1.	Muséum National d'Histoire Naturelle, Paris, France	P, PC	9.5
2.	New York Botanical Garden, New York, USA	NY	7.2
3.	Komarov Botanical Institute, St. Petersburg, Russia	LE	7.2
4.	Royal Botanic Gardens, Kew, London, England	K	7
5.	Conservatoire et Jardin Botaniques de la Ville de Genève, Geneva, Switzerland	G	6
6.	Missouri Botanical Garden, St. Louis, USA	MO	5.9
7.	The National Herbarium Nederlands (NHN), The Netherlands	L, U, WAG	5.5
8.	British Museum of Natural History, London, England	BM	5.2
9.	Harvard University Herbaria, Cambridge, USA Massachusetts,	GH, A, AMES, ECON, FH	5
10.	Naturhistorisches Museum Wien, Vienna, Austria	W	5

Source: http://www.flmnh.ufl.edu/herbarium/herbariaandspecimens.htm

Table 2. Ten largest herbaria in India

S. No.	Name	Acronym	No. of Specimens
1.	Central National Herbarium, Hawrah	CAL	2515000
2.	*Forest Research Institute & Colleges Herbarium, Dehra Dun	DD	340000
3.	Botanical Survey of India, Southern Circle, Coimbatore	MH	274863
4.	Botanical Survey of India, Eastern Circle, Shillong	ASSAM	271000
5.	National Botanical Research Institute, Lucknow	LWG	269513
6.	*Blatter Herbarium, St. Xavier's College, Mumbai	BLAT	200000
7.	Botanical Survey of India, Western Circle, Pune	BSI	170000
8.	Botanical Survey of India, Northern circle, Dehra Dun	BSD	127 000
9.	Presidency College, Chennai	PCM	100000
10.	University of Agricultural Sciences, University of Agricultural Sciences, Bengaluru	UASB	86 954

Sources: Botanical Survey of India, accessed on October 07, 2013 (http://164.100.52.111/index.asp); and * Index Herbariorum, accessed on October 07, 2013 (http://sweetgum.nybg.org/ih/)

5. Herbarium Making

5.1. Sources of materials

The influx of material into a herbarium is usually from the following sources:

- **Collections made by researchers/ students, and staff.** These are the major sources of material which vary according to herbarium goals. Collections may be for floristic, revisionary, monographic, or taximetric studies.
- **Exchange.** Duplicates or special collections are exchanged among various institutions, usually on a one-for-one basis. This was one of the important means

of adding to a collection, but due to strict enforcement of CBD, this practice however, is now discouraged by almost all herbaria.

- **Gifts.** These may range from an entire herbarium to a few specimens sent to a staff specialist within a continent/ country.
- **Loans.** Loans are generally either temporary (short term; e.g., a study of specific taxa for monographic study) or indefinite (permanent; e.g., the loan of an entire herbarium from one institution to another).
- **Identification service.** In most herbaria specialists are willing to determine specimens for other research institutions or for plant based professionals as service identifications. In general the specimens sent for identification are kept by receiving institutions.

5.2. Collection of plant materials

The collection of plant specimens is the first and foremost step for making a herbarium. Since the primary classification of angiosperms is largely based on herbarium material, therefore, the success of classification depends on the quality and quantity of the material used in the study. Good collection methods are the corner-stone of the whole edifice. We, therefore, need to enrich our herbaria, not to fill them with poorly annotated scraps that sheerly waste the space and can only tantalise, mislead or confuse taxonomists in their efforts to achieve a stable basic classification (Davis and Heywood 1963). The methods of collection adopted can vary from group to group, and must be modified according to climate and facilities available. The nature of plant collection varies with the purpose such as, for establishment of new herbarium, for enrichment of existing herbarium, for writing a flora, for collecting plant specimens/material of particular genus/species a for revisionary/ monographic studies, for collecting large quantities of a particular plant/ part of that plant as crude drug, for ethnobotanical studies and for collecting live material for introduction in gardens. Generally following tools are necessary for plant collection work.

- **Vasculum.** It is metallic case generally in size of 50x30x15 cm with a lid, used for keeping plants immediately after collection. For the sake of convenience in place of vasculum, polythene bags can be used for carrying the specimens in the field, the opening of the bags should be tightened with rubber band.
- **Field press.** It is made up of simple hard board or light plywood of 30 x 42 cm, tied with two cotton or leather straps on both ends (Fig. 1A).

  Field book. These are small books with 100 leaves (could be more depending upon the requirements) each with 6 tags or tickets containing same number (Fig. 1B).

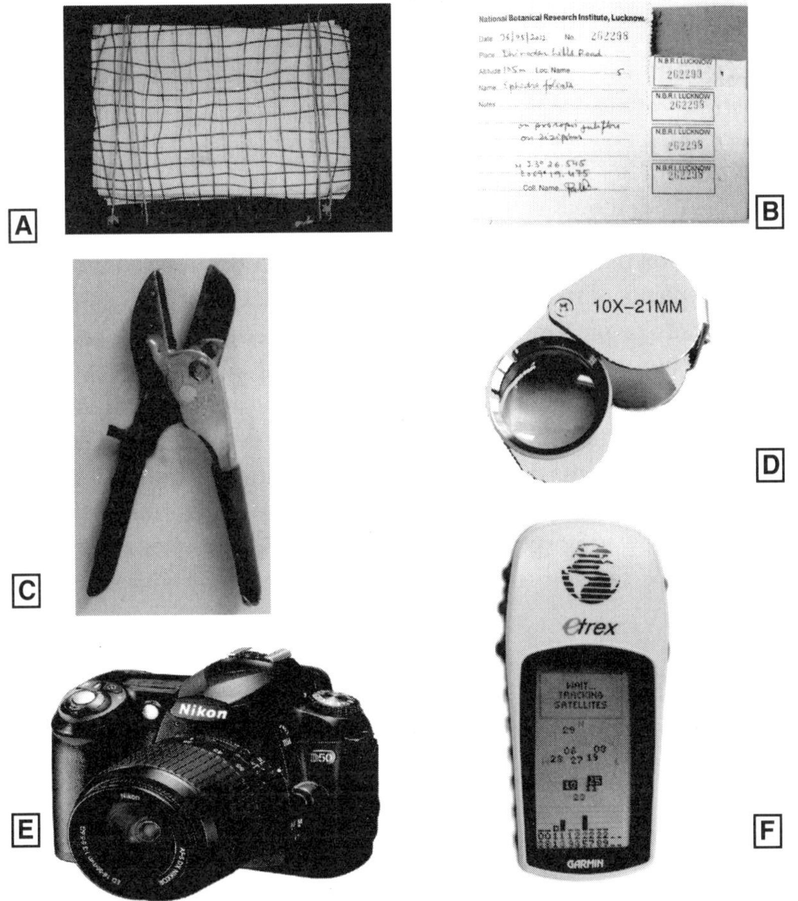

Fig. 1. Tools used for plant collection (A) Field press with blotters, (F) Field Book, (C) Scateur, (D) Magnifying Lens, (E) Digital Camera, (F) GPS

- **Blotting papers/ old newspapers.** These are used for drying and pressing the collected specimens.
- **Specimen tubes.** Glass or plastic tubes with tight caps are used for bringing the specimens in fixative.
- **Secateur or cutter.** Used for cutting small twigs from trees and shrubs (Fig. 1C).
- **Digger.** Used for uprooting the herbs or bulbous plants.
- **Magnifying lens (10X).** This is useful in documenting micro-morphological characters of plant (Fig. 1D).
- **Digital camera.** Having high resolution with zoom for photography of plants and its surroundings (Fig. 1E).

- **Shoulder bags (Haversacks).** Very useful for carrying collection kits.
- **Global Positioning System (GPS).** Very useful for recording geographical coordinates of particular location of plant (Fig. 1F).

5.3. What to collect?

What plants shall be collected depends on the purpose of study. If the objective of study is the preparation of a flora of the region, than the collection should be exhaustive and samples of all plants of the area should be collected. Collections should contain at least flowers or fruits or preferably both. In case of grasses, sedges, and other herbs, the whole plant including the underground parts should be collected. Collection of a number of plants from a population to show the range of variation in characters is better than collecting a single plant. In recent years, for correct description of species, much emphasis is laid on mass collections or population studies. It is particularly valuable that, if possible, both flowering and fruiting twigs are taken from the same plant. In precocious flowering species, generally flowers of only one sex are present at a time. Care should be taken to search and collect flowers of both sexes, sometimes from separate plants.

5.4. How to collect?

Some hints on collections for herbarium and taxonomic study with special reference to India are provided below.

5.4.1. Size of plant specimens

The size of the herbarium-sheet is approximately 28x42 cm, and this limits the size of the plants or twigs for collection. In case of small herbs, the whole plant with roots or underground parts can be accommodated on one mounting sheet. But in case of larger herbs, though it is possible to collect the whole plant, yet it has to be cut into two or three parts. In such cases the field number should be same for all the parts and should be marked A, B, C, etc. For woody plants, such branches, twigs or specimens should be collected as to give a fair representation of the species, and suitable to fill one sheet.

5.4.2. Number of specimens of each species

To facilitate distribution and exchange, at least six specimens of each plant are to be collected. Also, unmounted duplicates are useful for certain detailed studies, such as in palynology, anatomy, etc. Duplicates should be deposited in as many herbaria as possible, where they can be maintained safely.

5.4.3. Field number

The numbers given to collections are a very important record. Even the most valuable specimens like type specimens are referred to by collection numbers. Some collectors give their own individual serial numbers and maintain their sequence even if they change the institutions. On the other hand, some institutions maintain one continuous serial number for collections by their different scientific or technical staff. It is desirable to maintain a continuous serial either of the institute or of an individual for exploration in a particular area till that work is completed.

Each specimen should be numbered before it is put in the collection bag. Even if a species was collected previously during the trip, a fresh number should be given for its each fresh collection. Only one series of numbers should be used and sequence maintained. For each specimen the number should be entered in the field-book and a tag bearing that number should be attached to the specimens. As previously stated, as far as possible at least six duplicates of the same species should be collected and the same number be given to all duplicates. It will be seen that each page of field note-book provides tags for six specimens.

5.4.4. Field notes

A very important part of the plant collection work is the record of field notes. Detailed notes should be entered in the field note-book at the time of collection in the field itself. Generally, the following details should be recorded in the field note-book.

- **Date of collection.** On the date plant is collected.
- **Vernacular names and uses.** The vernacular names should be obtained from the local guide or coolie; care should be taken to avoid cooked-up names. The names should be very carefully heard and translated into Roman. It helps if the name is written also in the local or some Indian script.
- **Locality.** The name of the place and also the distance and direction with reference to a familiar or known place which can be easily located in a map should be given. A note like '3 km N. E. of Shillong' is much better than simply writing name of the place. It is better to give both.
- **Habitat.** This denotes the condition under which the plant is growing, such as marsh, grassland, thickets, deciduous forest, rock crevices, steep slopes or flat bottom of valley, etc. The purpose is to record the environment in which the plant was found growing. Notes on associated species should be provided, as they are very useful. If their names are not known their collection number can be given.
- **Description.** It should be brief and should include chiefly those characters which cannot be observed in pressed specimens, such as habit - erect or pendant; inflorescence-spreading or compact; colour of flowers, fruits and leaves; pubescence; presence of aroma and latex. Significant variation in size of plants should be noted. Notes on any other structure of plant which might be of interest can be added, such as shape, size and colour of bulbs and tubers. Notes should

also be made of the relative abundance and frequency of the plants. This information is valuable for commercial collections, and after some field experience, one can make reasonably fair assessment. Sketches of special or curious characters can be useful in identification. These can be recorded on the back of page of the field note-book.

- **Collectors' name.** Collector's name should be written in full, and not mere initials or surname.

After tying the number and recording notes in the field note-book, the specimens are kept in the vasculum or polythene bags or in other containers. They may even be placed in the field press. On days of heavy rainfall, or when there is large amount of collections, pressing in the field is difficult and the work of pressing is done in the camp. Sometimes, for want of time or other exigencies, even numbering of duplicates has to be done in the camp. But this must be avoided as it can involve mixing of collections from different spots, and in case of grasses and sedges, etc. even mixing of species.

5.5. Pressing and drying of specimens

Pressing is the process of placing specimens between the absorbents under moderate pressure. Herbaceous specimens should be washed to remove mud from roots. All plant parts such as leaves and flowers etc, are spread out neatly. This needs considerable patience. Some leaves are placed facing up and others facing down to show the characters on both surfaces. This is especially important in case of ferns which have sori on the abaxial side. In case of gamopetalous flowers, if possible, one flower should be split open longitudinally and pressed with corolla spread out to show androecium and gynoecium. If the specimens are longer than the size of mounting sheet, they can be folded like V, N, M or even W to hold the point of bend(s) in position, a slit card can be inserted during the process of drying and poisoning. After gluing and stitching, the plant parts should easily stay in position. If there are too many leaves or branches, a few are removed, so that there is as little overlapping as possible and all parts are easily visible. The extra leaves and branches are cut (and not torn away) with a secateur or scissors a little above the base, leaving a small portion, so as to show their position of attachment.

The main object of pressing is to flatten and dry the specimens. This is done by keeping the straps tight and by changing the blotters every day for 6-10 days depending on weather. The plants gradually lose their moisture and finally become completely dry. The used and moist blotters should be dried properly, so that they can be used over and over again.

6. Collection of Special Groups/Kinds of Plants

6.1. Collection of succulents

Succulent plants like Cacti, Euphorbias, members of Crassulaceae, etc. present unusual difficulties in making herbarium specimens, and their thick, succulent tissues take very long time to dry and so they require special attention. Unless dried by artificial heating or frequent changing of dryers, they are prone to rot and catch fungal infection. Hence, either the tissues should be killed by dipping in boiling water, or excess of tissues removed by hollowing out the thick organs. The tissues can also be killed by treating with alcohol or strong formalin. Dipping in boiling water for few seconds is most suitable for killing the tissues. Cacti may be handled by splitting the joints, killing the tissues and then drying them by artificial heat.

While recording the characters of succulents, especially the spiny succulents, details like shape, size and the arrangement of spines and joints should be noted.

6.2. Collection of aquatic plants

Free floating minute aquatic plants like *Lemna, Wolffia, Azolla* etc. are collected by inserting a wire press or sieve plate with white paper or muslin cloth below the specimen in water and taken out; the paper or cloth is lifted slowly with both hands and placed between the dryers. To collect slender aquatic herbs like *Zannichellia*, *Utricularia Naja* and *Ceratalophyllum* species a piece of blotting paper hould be submerged under the plant and the paper with plant lifted carefully slopping it to facilitate water runoff. Put the blotter containing plant in the plant press alongwith some extra dryers are provided below and above. The plants should be changed along with the paper or muslin cloth. For the first one or two days, the changing should be more frequent.

Some aquatic lithophytes like the members of Podostemaceae cannot be detached from the rocky substratum and a portion of the rock needs to be broken apart for collecting them. The rock with the plant has to be either dried or preserved in liquid medium.

6.3. Collection of plants having mucilage, gums and resins

Some plants (like *Hibiscus, Dodonaea,* etc.) contain mucilage or resin; they stick to the dryers and cause difficulty while changing. Such specimens should be placed in a folder of muslin or any other thin cloth and pressed. Only the dryers should be changed and not the muslin cloths, until the plants are fully dry.

6.4. Collection of aroids

Aroids frequently have a virtually impenetrable epidermis, and unless killed properly, they may continue to grow even in the press. The same is true for some bulbous orchids. The killing is done either by heating at high temperature and under proper ventilation, or by use of alcohol and formalin.

The tubers of the aroids are often essential for correct identification, but their preservation is tedious. Sometimes whole tubers are collected and kept in museum; else notes are taken on their shape and size, and their small sections are kept with herbarium specimens.

6.5. Collection of very large plants like bamboos, palms and bananas

Due to their large size, bamboos, palms and bananas and some ferns like *Cyathea* and *Angiopteris* require special methods for collection. There are very elaborate methods and instructions developed by specialists for these groups. The habit of the plant and the approximate size of the calm (or psuedo - stem), leaves and inflorescences must be recorded in the field note-book. Either the whole plant or its main parts should be photographed (with a scale) or their sketches must be made on the spot. Effort should be made to collect leaves (or portion of leaves) noting the actual size. In case of bamboos, few ligules must be collected, noting the serial number of the nodes from base, from which they are taken. In case of bananas it is helpful if spathes with flowers are collected.

6.6. Pressing of bulky specimens

Bulky fruits like those of *Capparis, Artocarpus, Diospyros* and *Balanites* also require special methods for pressing. They can be pressed by slicing the parts sufficiently thin to go into the press. If necessary, the excess tissue is removed after dipping in boiling water. Bulbs normally dry well without splitting. The thick parts make the bundles uneven in pressing, and it is managed by providing pads of paper.

6.7. Collection of seeds

It is advisable to collect seeds along with the collection of specimens. Seeds should be mature. These should be dried and placed in a packet and mounted along with herbarium specimens. Collection of seeds is important as they aid in identification. Also with modern technologies herbarium seeds can be used to resurrect species extinct in the wild.

7. Poisoning and Preservation of Specimens

7.1. Poisoning

The specimens are poisoned either immediately in the camp or after reaching the laboratory. It is advisable to poison the plants immediately after collection; poisoning kills the plants and thereby the formation of abscission layer is prevented. Following methods are generally used for poisoning.

7.1.1. Mercuric chloride

The poisoning is generally done by dipping the whole plant in a saturated solution of mercuric chloride in ethyl alcohol. The plant is again put in dryers and pressed till it get completely dried. Mercuric chloride is corrosive for metals, so enamel trays should be used. Disposable gloves must be used while poisoning. Mercuric chloride is a deadly poisonous chemical and its effect on human beings is cumulative. Therefore, most herbaria have banned the use of Mercuric chloride.

7.1.2. Formalin

This method is highly suited for tropical countries. In those places where the day's collections are so numerous that it is not always possible to dry the collection by changing of blotters; this method is adopted. The collections are spread in ordinary old newspapers and bundled up. Each bundle is then placed in a large polythene bag. 10% formalin is poured over the bundles, so that the bundles just get soaked thoroughly, without however leaving excess of formalin in the bags. The bags are then tied airtight. No further change of folders is necessary till reaching the laboratory. By this method it is possible to bring the collections made even over 3-4 months. On reaching the laboratory, the bundles are opened out; the specimens are exposed to the atmosphere to drive away the excess of formalin fumes. Then the specimens are spread out for pressing and drying as usual.

7.2. Fumigation

The herbarium needs to be fumigated at regular intervals for killing the microbes and pests on mounted specimens. Following processes are used for this purpose:

7.2.1. Chemical treatment

This is done for killing pests in mounted as well as unmounted duplicate specimens. This process involves use of any one of the volatile poisonous liquids like methyl bromide, carbon disulphide or carbon tetrachloride. These are placed in small saucers or petri dishes in each herbarium case and the cases kept closed for about a week. Sometimes paradichlorobenzene (PDB) is used. Small cloth bags with paradichlorobenzene are placed in all or many of the pigeon-holes. This chemical can also be sprinkled on sheets or in bundles, but PDB should not be used along with naphthalene. The naphthalene is also a suitable fumigant in place of PDB.

7.2.2. Heating

Some herbaria use electric heat instead of chemical fumigation. This requires use of specially insulated chambers with heating elements which can keep specimens at 44^0 C for some hours is an effective method. But this method is not practiced in India. Specimens may also be treated in a microwave.

7.2.3. Freezing

Freezing the specimens is the technique least dangerous to human health, and is very simple. The specimens must be frozen to -18°C or colder and kept at that temperature for at least 48 hours. In practice, when specimens are frozen in deep-freezers in bulk and/or in boxes, it is necessary to freeze them for 72 hours (3 days and 3 nights) to ensure to kill all germs and pests. Bundles of specimens should be sealed in plastic bags to avoid moisture condensing on the sheets. This method is also commonly not practiced in our country.

7.3. Mounting

After the specimen is pressed, dried and poisoned, it is to be affixed (along with a label) on a mounting sheet. The mounting sheets are made from heavy longlasting white card sheet in uniform size of 28x42 cm ± 1 cm). While mounting, the specimens should be neatly and uniformly spread and fixed on the sheet, and all parts of the plant should be easily visible for study. For achieving this, before the plant is glued on sheet, it should be placed on the sheet in various positions so as to judge which position will give the best display. Some important precautions in this regard are:

- Only one specimen is placed on one sheet.
- As far as possible, the lower part of the plant, such as root is at the base of the sheet.
- There is some space left on the right hand bottom for label; if label is already printed on mounting sheet, no part of the plant should cover the label.
- The field number tagged with the plant is pasted in a proper straight position, along with the thread.

Mounting is done by various methods, and these can be classified into two main categories.

7.3.1. Gluing the plant on the sheet and stitching

The common technique now in use in our country is pasting specimens sheets on with glue. The common animal glue used for book-binding, joinery work, etc. and available in market as flakes, or pieces, is employed. The glue paste is made by adding flakes of glue to boiling water, gradually and in small quantities, till it becomes a thin syrupy paste. As the paste tends to be thick and hard on cooling, the pot is kept on low heat during mounting work. To give this glue some insect repellant property, small quantity of mercuric chloride, thymol crystals or copper sulphate (Blue vitrol) is added.

In some circumstances, when glue is not available or not considered desirable, ordinary white gum paste (available in bottles for office use) can be employed, but this is not satisfactory. Fevicol also can be used but this may be slightly expensive.

7.3.2. Strapping

In this process the specimen is not glued to the sheet, but only loosely strapped to the sheets by means of thread stitches or by some other device such as gummed cloth or paper tapes or by liquid plastic method.

7.4. Herbarium labels

Mounting of the specimens is followed by pasting of herbarium labels (Fig. 2). The size and design of herbarium labels slightly vary according to need and it should be around 8 × 12 cm. In general the labels should contain the following data.

- Name of the family
- Name of the genus and species
- Locality of collection
- Date of collection
- Description/remarks/notes
- Collector's name and number
- Vernacular name and local uses

Herbarium label is fixed on the bottom right hand corner about 1 cm away from edges of the mounting sheet. It should be fixed with paste or glue.

पादपालय, वै०औ०अ०प० रा□ट्रीय वनस्पति अनुसंधान संस्थान, लखनऊ
Herbarium, CSIR-National Botanical Research Institute, Lucknow (LWG)
Flora of ....

नाम..
Name..
Family...
Loc. Name ....Col. No ....
Loc. & Alt...
Lat/Long ....
Collector ....
Date .... Det ....
Notes ....
....

Fig. 2. Label of Herbarium sheets

8. Identification and Determination of Plants

Usually, identification is considered to be the process through a which specimen whose name is not known is recognised by its characters, to be similar to some known plant, and accordingly given a name. But, it is now felt that since no two individual plants are exactly identical; this process should be called determination istead of identification. Therefore, the annotation slips are called as *'Determinavit'* (or abbreviated to *Det.*) slips. However, the word identification is now so universally employed that it clearly signifies the entire process.

For the purpose of identification, the scientific method is to study the macro and micro-morphological characters of the plant. Macro-morphological characters can be studied with naked eyes or by hand len. To study the micro-morphological characters, like indumentation, structure and arrangement of floral parts, number and position of ovules in ovary, number and type of stamens, type and position of glands/hairs, structure of seeds, etc., a compound microscope is essential (Fig.3). These characters should be checked with the Flora, revision/monograph by going through various keys. This should be followed by the comparison with full description and illustration. Thereafter, it is to be carefully matched with the authenticated specimens already available in the herbarium.

For a satisfactory identification, the full process given above is essential. Experienced workers are often able to guess fairly up to the genus and some even up to the species. In such cases, it is not necessary to run through the keys, but may directly be checked and matched with the already identified sheets of that species.

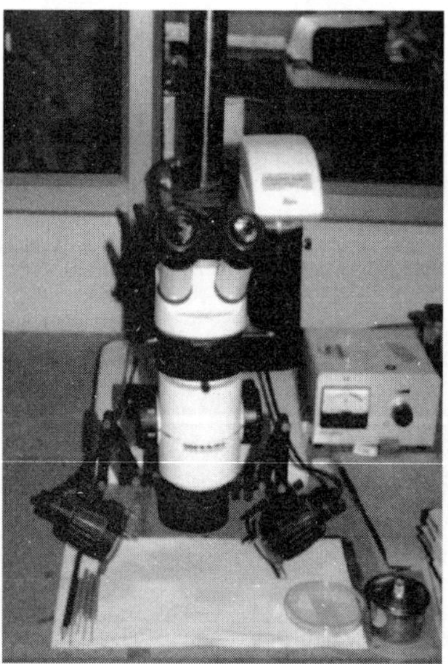

Fig. 3. Compound Microscopes

If the plant does not satisfactorily fit in the keys or match in the herbarium, effort should be made to compare it with species of adjacent floras in larger herbaria. It is the process which can eventually lead to the discovery of a new taxon.

The next and equally important task in the process of identification is the use of correct nomenclature. Every effort should be made to label the sheets with the latest accepted name for the taxon.

9. Filing of Specimens

9.1. Accessioning

Before the specimens are incorporated in the herbarium they are stamped with the distinctive mark of the herbarium or institution and a number is given. Stamping is usually done on the top right hand corner with a rubber stamp. This stamp carries the name of the institution, accession number and date of accession.

The sheets are also listed in an accession register (Fig. 4) along with their accession number and then incorporated in the main herbarium. The mounted, identified and accessioned herbarium sheets are sorted out in accordance with family, genus and species. All the sheets of the same species are placed in lighter covers called the 'species cover' or folders, and all the species (with species covers or folders) belonging to one genus are placed in one or more folders of heavy paper, called the 'genus cover'. The genus and species covers serve many purposes such as:

- Protection of specimens
- Ease in arrangement according to some system
- Convenience in handling

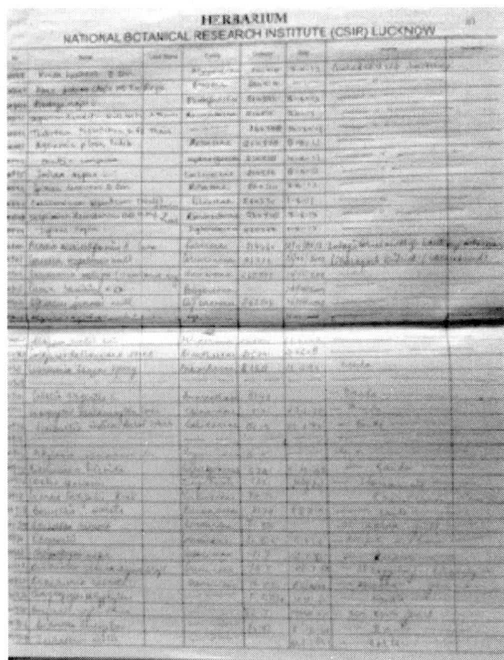
HERBARIUM
NATIONAL BOTANICAL RESEARCH INSTITUTE (CSIR) LUCKNOW

Fig. 4. Accession Register

In larger herbaria RBG like kew, species from different continents are placed in different genus covers, and the name of the continent are noted on the cover.

9.2. Arrangement of specimens in the herbarium

The specimens are usually arranged in the herbarium according to some recognised system of classification. In many Indian herbaria, the order and numbering of families and genera is according to Bentham and Hooker's *Genera Plantarum*. This is primarily due to historical reasons as well as for convenience. In case of ferns, arrangement is generally according to Copeland's *Genera Filicum*.

The pigeon-hole, where bundles of a new family start, is marked by a fixed label or by hinged flap-board separator cardboard. The name of the family is printed or written on this in bold letters.

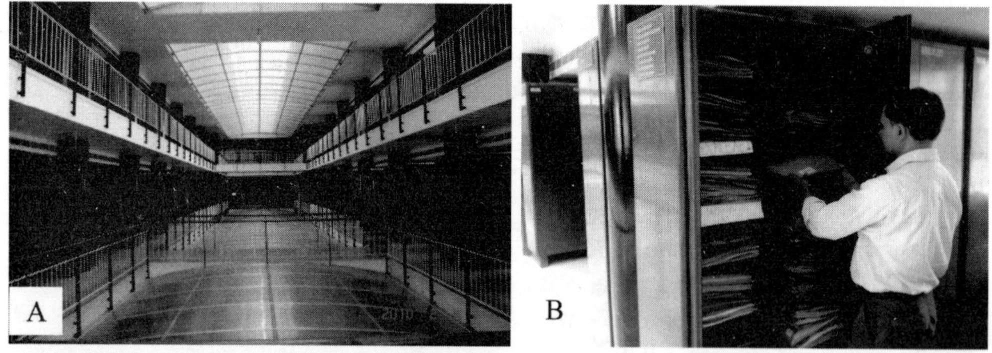

Fig. 5. (A) NBRI Herbarium, (B) Arrangement of specimens

9.3. Special arrangements for certain sheets

9.3.1. Segregation of Types

Since Type sheets are very valuable and are not supposed to be referred during routine work, therefore, they should be separated from the other sheets and kept preferably in fire-proof steel almirahas. In this way these irreplaceable Type sheets are saved from unnecessary handling and damage. For easy distinction, each Type sheet should be wrapped carefully in a folder, marked conspicuously with red label, red margin or band.

9.3.2. Segregation of cultivated plants

Specimens of cultivated plants of a region should be separated from others and kept in the general herbarium at the end of each family. It is better if cultivated species are kept in genus covers of a different colour.

9.3.3. Segregation of sheets for students and beginners

If a herbarium is frequently visited by inexperienced students and other beginners; it is advisable to maintain a set of collections from the local region, separately.

9.3.4. Undetermined specimens

If some specimen's are not identified, these should not be mixed with others, but should be kept in separate bundles labelled 'Dubia' ('indet') for study by specialists of those groups whenever such persons visit the herbarium.

9.3.5. Bulky herbarium

Many large specimens, such as bamboos or large rhizomes and fruits cannot be mounted or displayed on mounting sheets, and require special drawers or cupboard. In the U.S. National Herbarium (Smithsonian Institution) in Washington such collections are lodged in large drawers and special shelves and this section of the herbarium is called 'Bulky Herbarium'.

9.3.6. Handling or sorting of unmounted duplicates

After mounting 1 or 2 specimens of each collections, the remaining duplicates should be preserved carefully, as these may be needed for various purposes. The bundles of these duplicates should also be fumigated at regular intervals, and stored in closed or open racks. The specimens should be packed in old newspapers and bundles arranged according to serial numbers. The bundles should bear the labels or tags showing the contents (collection number .... to..........). This will later on help in tracing the duplicates specimens.

10. The Virtual Herbarium (Online Digital Herbarium)

The computerisation can make taxonomists and their herbaria more effective. Using a personal computer may make routine curatorial procedures more efficient. In addition to housing plant collections, many herbaria have initiated computerised data information system to record and access the information on plant specimens, as well as to access information from other herbaria worldwide (Simpson 2006). A virtual herbarium, also known as online digital herbarium, concerns with the collections of digital images of preserved plant specimens or other related collections. Such herbaria are established to improve accessibility of specimens to taxonomists, students, researchers, and plant based professionals worldwide through internet. Additional information such as botanical name, common/vernacular name (s), if any, locality of collection, date of collection, the collector, field notes, etc., are attached to every specimen. Today, large numbers of online digital herbaria have come up throughout the world, particularly in western countries. Some of the important virtual herbaria in the world are as follows.

- Royal Botanic Garden, Kew (http://apps.kew.org/herbcat/navigator.do)
- Missouri Botanical Garden (http://www.tropicos.org/)
- The Linnaean Herbarium (http://www.linnean.org/specimen-collections/Linnaean+Herbarium)
- Australia's Virtual Herbarium (http://avh.chah.org.au/)
- Botanic Garden and Botanical Museum Berlin-Dahlem, Freie Universität Berlin. (ww2.bgbm.org/herbarium/default.cfm)
- New Zealand Virtual Herbarium (http://www.virtualherbarium.org.nz/home)
- Seagrass-Watch, Australia (http://www.seagrasswatch.org/home.html)
- The Virtual Herbarium at the New York Botanical Garden. (http://sciweb.nybg.org/science2/VirtualHerbarium.asp)
- U.S. Virtual Herbarium Online. (http://uaes.usu.edu/htm/aes-news/us-virtual-herbarium-online/)

The need for the development of virtual herbaria in India was more realised during the last one decade, and digitization of several Indian herbaria initiated. Following herbaria have already stepping ahead in this direction.

- Janaki Ammal Herbarium, IIIM Jammu (Formerly RRL, Jammu) (http://www.iiim.res.in/herbarium.htm)
- Kerala Forest Research Institute Herbarium, Trichur (http://kfriherbarium.org/)
- FRLH-Herbarium and Raw Drug Repository, Bangalore (FRLHT- ENVIS Centre on Medicinal Plants) (http://envis.frlht.org/digital-herbarium-main.php)
- National Botanical Research Institute Herbarium, Lucknow, UP (http://www.nbri.res.in/herbarium/)
- National Institute of Oceanography (NIO), Dona Paula, Goa (http://www.nio.org/index/option/com_url/task/frame/title/Digital%20Herbarium/tid/2/sid/18/thid/160)
- Botanical Survey of India, Kolkata (bsi.gov.in; http://164.100.52.111/cnh/majorprojects.shtm)

11. Instructions for Herbarium Workers

As each specimen is an irreplaceable material, therefore good practice in the herbarium is a necessity in order to preserve the specimens. The following are instructions that any herbarium staff member or visitor should follows while handling herbarium material.

- Never eat or drink inside the herbarium.
- Do not open windows and the air conditioners, if necessary, must stay on throughout the working time.
- Never leave the cabinet doors open. Always make sure that they are firmly closed in order to prevent the influence of dust and insects in the herbarium.
- Never reach up to remove material from high shelves, always use steps or a ladder. Some specimens may have irritating hairs and dust that might get into your eyes.
- When removing material from the shelf, hold specimens by both sides and take them out as one whole block.
- Never remove any material from the herbarium to study elsewhere except under special need with permission of the curator.
- Insert a hanging label in the cabinet to indicate the taxon removed, date of removal, the name of the person who removed it and its present whereabouts.
- Never turn the specimens as though they were pages of a book. Keep the folder flat on a table and lift the sheets one at a time with their faces up.
- Never rest books, heavy objects or elbows on a herbarium folder.
- Never bend a specimen to examine it under a microscope. It is advisable to use a long – armed microscope or a hand-lens.
- Old determinavit slips must never be removed, new ones can be added with the date and botanist's name in full using the annotation labels provided. Never write directly on the herbarium sheets.

12. Current Problems in Management of Herbaria

In the current era of biotechnology and molecular biology the classical subjects like the taxonomy and herbarium have witnessed a great debacle. Unless immediate measures are taken to restore the herbaria and associated taxonomic activity the entire biological community would soon end up in a state of confusion with regard to their own taxa of investigation and may have to knock the doors of scientists outside the country for seeking basic information on identity, nomenclature and even distribution of those taxa which occur at their door steps (Rao 1995). Although herbarium is a facility of a data bank on plants to be used by all biologists only a handful of surviving taxonomists are putting their efforts to nourish and maintain the fragile heritage of herbaria in the country. This has resulted in erroneous conclusion that herbaria are the establishments solely of taxonomists. Rather, herbaria are national facilities and their maintenance should be a national responsibility.

Some herbaria developed over several decades of efforts of taxonomists are today at the verge of collapse due to wrong notion among the policy makers that herbaria are merely a storehouse of collections of dead plants which cannot contribute to the national development nor can generate funds for research forgetting that herbaria are simply a facility of a database on plants from which all biologists draw their basic information directly or indirectly about the plant on which they carry out all advanced researches (Rao 1995). The voucher specimens, on whom varieties of researches are carried out, are rather significant. It is only the herbaria which maintain such voucher specimens for posterity. With the establishment of the National Biodiversity Authority (NBA), a statutory autonomous body under the Ministry of Environment and Forests, Government of India in 2003, the importance of voucher specimens has increased many folds. Recently several herbaria have been designated as National Repositories for the Indian flora by NBA where voucher specimens are to be essentially deposited before sending papers for publication or submission of dissertations/ thesis. In the absence of voucher specimens vast amount of new data/new discoveries on a particular plant species quite likely may not pertain to the plant species studied. Even if the species is well known it may contain a number of ecotypes, cytotypes, varieties or strains and hence it is absolutely necessary to maintain the 'voucher specimens' in any herbarium so as to allow the subsequent researchers to evaluate statements made in the literature about the characteristics of a given taxon (Rao 1995).

The editors of various scientific journals should also ensure that all authors cite the accession numbers of their voucher specimens and the herbaria where they have been deposited. While this enhances the value of scientific findings, at the same time strengthens the justification of the importance of the herbaria. Herbaria require large buildings and staff for maintenance of vast collections, laboratories for associated researches and funds for continuous explorations for enriching the herbaria, particularly in a developing' country like India where the coverage of holdings is incomplete. The knowledge of variations within species is another aspect which is very limited. But the increasing financial squeeze and prioritization of cutting-edge research programmes has greatly affected the overall health of the herbaria in the country.

A rational economic basis for maintenance and furtherance of herbarium research must consider the fact that botanists with all their concerted efforts have known only 1/ 10th of what exists in our tropical forests and still less is known of the economic utility of those species which are recorded. Infraspecific biodiversity and population variations of the rich tropical flora are little understood. But certainly maintaining and enriching a herbarium is expensive. A strict monitoring of the quality of the incoming collections and zones from where they come is essential in order to maintain the quality and not the size of the herbarium. More than one herbaria if any, within a city or town (e.g. Dehra Dun, Lucknow) can be considered for merging in view of the increasing cost of maintenance, manpower and space on buildings etc., rather than 'killing' a herbarium due to wrong policies and apathy towards such classical subjects upon which many future solutions depend (Rao 1995).

References

Cronquist A (1966) "Herbarium", in Encyclopaedia Brittanica. Inc. Chicago. 11:409-410

Davis PH (1961) Hints for hard- pressed collectors. Watsonia 4:283-289

Davis PH and Heywood VH (1963) Principles of Angiosperms Taxonomy. Van Nostrand, Princeton, New Jersey

De Wolf GP (1968) Notes on making a herbarium. Arnoldia 28:69-111

Fosberg RR, Sachet MH (1965) Manual for tropical Herbaria. Reg Veg Vol 39, Utrecht, Netherland

Jain SK and Rao RR (1978) A Handbook of Field and Herbarium Methods. Today and Tomorrow's Publ, New Delhi

Lawrence GHM (1951) Taxonomy of Vascular Plants. The MacMillan Company, New York

Radford AE (1986) Fundamentals of Plant Systematic. Harper and Row, Inc., New York

Rao RR (1995) Are herbaria redundant? Curr Sci 69(12):968-969

Shetler SG (1969) The herbarium: past, present and future. Proc Biol Soc Amer 82:687-758

Simpson MG (2006) Plant Systematics, Elsevier, UK p 525

Stearn WT (1957) An introduction to the "Species Plantarum" and cognate botanical works of Carl Linnaeus. Prefixed to the Ray Society facsimile of Linnaeus's Species Plantarum, 1, London

Chapter – 13

Preparation of Flora, Revision and Monograph

L.B. Chaudhary and T.S. Rana

1. Introduction

The increasing size and activities of humans have mounted severe pressure on the survival of other life forms on the earth. This is a matter of great concern for the humanity. Therefore, a critical study to inventory all plant species has become very essential and urgent before they are lost or become extinct. In the face of extinction of many of these species, we need to learn as much about them as quickly as possible. In the recent years there has been a great concern for biological diversity and conservation of biological resources. Although much emphasis has been given to the ways and means for inventories (Flora) of biological diversity, there is also need for more in depth studies that deal with basic relationship of species. Monographic, revisionary and systematic studies that provide basic descriptive information about organisms and their relationship are required to document the information and knowledge about the plant wealth of a country. This chapter is to highlights the usefulness of the Flora and revisionary studies, and provides the relevant information to the beginners and the budding plant taxonomists about the preparation of an 'Ideal Flora' and 'Creative Revision' to meet the requirement of the present time.

2. Flora

Systematic enumeration of plants of a particular area is termed as 'Flora' (Dewolf Jr. 1964, Stace 1989; Palmer *et al.* 1995). In the last few decades a large number of Floras have been published at national and international levels. In the present biodiversity crisis the importance of Floras has increased many folds. Now-a-days Floras have got immense values even for non-taxonomist/ botanists. Therefore, it has become necessary to rethink about the quality or contents of Floras and the methods of preparation of an 'Ideal Flora' for the 21st century (Frodin 1984; Krishnamurthy *et al.* 1995).

2.1. Aim for local flora studies

A good Flora is not only a listing of plants, but is a storehouse of taxonomic knowledge. An amateur researcher must get acquainted with any previously published Flora before starting the work. According to Radford (1986) following information can be obtained from any Flora.

- How to use a manual?
- How to collect and prepare specimens?
- How to describe the plants?
- How to use a key for identification?
- How to recognize families, genera, species and infraspecific taxa?
- How to name the taxa of plants?
- How to relate species diversity to habitat diversity in a region?
- To become acquainted with fundamental taxonomic concepts and principles
- To become acquainted with the historical development in plant taxonomy
- To become acquainted with at least one system of plant classification

2.2. Uses of flora

Previously it was thought that one taxonomist writes a Flora for the other taxonomist. This concept has now completely changed. Day by day the taxonomy is becoming a more synthetic and applied branch of botany. In the present time the utility of Floras has increased many folds as they are required by various sections of people in our society. The Floras may be useful in many ways to many users:

- College and University Teachers
- Agricultural development
- Architects and landscape designs
- Civil and Mining Engineering
- Customs Officials
- Biodiversity assessment and management
- Conservation programme

- Environmental impact assessment
- Land Management
- Ecosystem Management
- Forest Management
- Basic information for plants for future research work
- Forensic Science
- Development of Botanic Gardens/Parks
- Phylogeny of Plants
- Discovery of drug plants for pharmaceutical and Ayurvedic companies
- Seed companies
- Village, Town and City planning
- Bioprospecting
- Protected area Management
- Weed and pest control
- Vegetation study
- Assessment of rare and endangered species
- Nursery
- Plant migration and dispersal
- Evaluation of Phytogeography pattern
- Climate change study

2.3. Selection of area for floristic study

The selection of area for floristic study depends upon a few factors like the number of man power to be involved in the study, amount of funds available, period of time, geographical/botanical significance of the area etc. Depending upon the size of area the Flora may be country Flora, state Flora, Flora of a major phytogeographic zones and District Flora. While selecting the area for floristic study it is also decided that expected Flora would cover only angiosperm or entire plant groups including cryptogams.

2.4. Survey and collection of plants

Survey and collection are the major tasks in floristic studies. The area should be divided into 3 - 4 ecological zones and each zone should be surveyed thoroughly in all seasons. Even the common or same species growing throughout the areas should be collected from different ecological zones in different seasons for recording diversity. All information which can be seen only in the fields should be recorded during collections. In the present time the importance of herbarium specimens deposited in herbaria has been realized by many non-traditional users for all sort of biological sciences and therefore, it is suggested that information like pollinators, ecology, soils, uses, seedling and sapling data, epiphylls, insect plant association etc. should also be recorded for each taxon (Stern and Eriksson 1996). The procedure for collection of plants and preparation of herbarium specimens should be in accordance with the standard protocols mentiond in (Jain and Rao 1976).

2.5. Contents of modern flora

- Title
- Geography (site information)
- Environmental information
- Keys to the families, genera and species
- Taxonomic treatment (for each species)
    - Nomenclature with citations and synonyms
    - Vernacular names, if any
    - Concise and clear description
    - Phenology
    - Distributional data with maps
    - Ecological data
    - Uses/Ethnobotanical data
    - Interaction/association data
    - Conservation/status
    - Exotic or native origin
    - List of voucher specimens
- Summary Statistics
- Complete bibliography

- Images/Illustrations
- Index
- Electronic copies

2.5.1. Title

The Title of flora should be clear, descriptive and unambiguous. Palmer *et al.* (1995) suggest that in all Flora the title should contain the word Flora which is self descriptive and frequent in use. They consider that titles such as 'an assessment of the plants of................' or 'species composition of ....' as ambiguous. The taxonomic scope of Flora (e.g. Angiosperm Flora, Lichen Flora, Bryophyte Flora, Pteridophyte Flora or Cryptogam Flora etc.) and site of Flora must be included in title.

2.5.2. Geography (site information)

The geography of study area must be defined in the Flora. The information like site name, state (s), country (ies), latitude, longitude, boundaries, etc. should be provided. A map with detailed information of site is highly desirable in Flora. If site is not well known, a clear cut direction for reaching it should be given along with map.

2.5.3. Environmental information

Although it is not possible to provide a detailed account of ecology of flora area, however, a brief summary of environment is desirable to be mentioned in the Flora with some essential data like minimum and maximum elevation (in meters), physiographic regions, names of river system draining the site, major habitats, water bodies, if available, soils, annual precipitation, temperature, etc. The past and present disturbances and impact of human activities on the study area should also be described in the Flora.

2.5.4. Nomenclature in flora

The name of taxa to be included in the Flora should follow previously published monographs, revisions, Flora and International Code of Botanical Nomenclature (Currently the International Code of Nomenclature for Algae, Plants and Fungi, McNeil *et al.* 2012). For correct nomenclature various electronic sites/data bases like TROPICOS, IPINI, plant list etc. may also be now consulted. Each taxon name should be followed by authority name (or names) and reference to the place of publication. All expected names of taxa should be typed in bold face. In addition to first references a few previous references with the work on the area may be cited. Only a few relevant or common synonyms should be given in italics. Usually, Type specimens are not provided in the flora.

2.5.5. Vernacular names

Vernacular names, if any, used within the area of Flora, should be given. Sometimes several Vernacular names may be available for the same taxon if flora area is large and inhibited by several kinds of people who use different local languages. Therefore, the language of vernacular names must be provided.

2.5.6. Taxonomic description in flora

The taxonomic work is impossible without a thorough knowledge and experience in the art of making description (Veldkamb 1987). The good description can be written by practice. The description of a species should be concise, clear and logical. All descriptions should be written in a definite sequence for all taxa. They should be easily comparable one with another, particularly within a genus. The plants should be described from base to top, from outside to inside and from the total to the details and each sentence should be started with subject and not with adjectives. The characters given in the family and genera should not be repeated in the species and infraspecific taxon description. The description should be made from the materials collected by author or native to the area of the flora. It is always desirable to prepare a fresh description rather than copying from the previously published monographs and Floras. The description should be based on the examination of several specimens, collected from different zones to record the range of variations.

2.5.7. Methods of plant identification

The identification of plants can be done by three ways - keys, comparison and experts. The best means for identification is the keys. The plants can be identified based a previously published keys in Flora, manuals, monographs or revisions. Once the plants are identified by a key it must be compared with the identified authentic specimens housed in the herbaria. Sometimes help can be taken from experts in the identification of plant specimens. But the situation is quite different among present youngsters. Now everybody wants to get identified their plant specimens directly asking from experts/ colleagues without describing and preparing keys to the plants. Even they copy the description and keys from the Flora of other regions which may not include the description of their plants. It should be highly discouraged. Elaborate description of plant identification has been provided in the chapter 11 of this book.

2.5.8. Suggestions for construction of keys for flora

Following steps may be useful in constructing a good key:

- Identify all groups to be included in the key
- Prepare a description

- Use macroscopic morphological characters and non-variable character states.
- Avoid characteristics that can be seen only in field or on herbarium specimens.
- Prepare a comparative chart
- Construct strictly dichotomous keys
- Use parallel construction and comparative (contrasting) terminology in each lead of a couplet
- Put a number of characters (polythetic, not monothetic) in each lead
- Follow a key format
- Start both lead of a couplet with the same world and successive lead with different words
- Start lead with name of plant part (noun) then follow with descriptor (adjective)
- Test key's reliability and consistency for identification with the specimens
- Tell your friends/ colleagues to identify your plants with the help of keys which you have prepared

2.5.9. Specimens examined

Except a few critical Flora (Flora based on a monographic or revisionary work), usually, the list of specimens examined is not provided in the general Flora. However, in some Floras (like State Floras or District Floras) only selective specimens collected by author(s) are cited with locality, collector's name and number. If specimens are not cited in the text, then in that case, a list of herbaria consulted for herbarium specimens examined should be provided somewhere in the introductory part of the Flora (Turrill 1964).

2.5.10. Index

Index is very much essential in the Flora. There should be only one index for all binominals (families, genera, species and infra-specific taxa) appeared in the Flora. The vernacular names should also come in the same index. The accepted names should be in normal face while synonyms and vernacular names should be typed in italics in the index. However, sometimes only index to families and genera is provided to save the space.

2.5.11. Abbreviation

Standard abbreviation can be used in the Flora for the books, journals etc. who repeatedly appear in different places. It should be used uniformly throughout the Flora and its number should not be more.

2.5.12. Glossary

It is not advisable to publish a glossary in the Flora. Flora writers should always use very common technical words in their descriptions.

2.5.13. New taxa

New taxa should not be described in the Flora (Turrill 1964). They should be published either in monograph/revision or separately in any taxonomic journal as they require diagnosis, detailed description, relationship with allied species etc.

2.5.14. Illustrations

Turrill (1964) has very correctly mentioned that the Illustrations enhance the value/ quality of a Flora. Illustrations may be black and white figures, photographs etc. They may be of whole habit or floral parts or of both. Scales should be provided for all Illustrations. In large Floras due to space problem only a few Illustrations are published in a separate volume. Along with illustrations the distributional maps should also be given. Either each taxon can be shown on a separate map or 2-3 or more taxa on a single map.

2.5.15. Exotic or native origin

The origin of each species whether it is native or exotic must be mentioned. The geographic origin of exotic species may also be indicated.

2.5.16. Conservation status

An assessment of the abundance/density of each species should be carried out with the help of any abundance scale. An abundance scale is provided here for the use of the Flora writers.

Density	Score	Description
Abundant	5	Dominant or codominant in one or more common habitats
Frequent	4	Easily seen or found in one or more common habitats but not dominant in any common habitat
Occasional	3	Widely scattered but not difficult to find
Infrequent	2	Difficult to find with few individuals or colonies but found in several location.
Rare	1	Very difficult to find and limited to one or very few locations or uncommon habitats
Absent	0	Not found but found in a previously survey from the ame or similar site or was otherwise suspected to occur

(*Source*: Palmer *et al.* 1995)

2.5.17. Summary statistics

A summary of entire work (2-3 pages) must be included in Flora. The statistical analysis of Flora (total number of families, genera, species, percentage, number of exotic and native species, rare species, endemic species etc.) should be given in tabular form or in graphic forms with the help of computer.

2.5.18. Electronic copies

Usually Floras consist of about hundred to thousand of species. Now with the help of computer it has become very easy to prepare a comparative floristic account of all flora included in the work than manually. Therefore, it is suggested to make available entire data in the electronic form for future workers. It should be mentioned somewhere in the text, if Flora is available electronically, with all details (the nature of file, the medium, the operating system, the cost and contact person).

2.5.19. The plan of preparation of a flora

All expected names to be included in the Flora should be listed, preferably on index cards. Single card should be used for each taxon and every data related with the taxon should come on that card. After describing all taxa keys to the families, genera and species are made. The families should be arranged in the Flora according to any system of classification which is more acceptable. It is not necessary to provide a new system of plant classification in the Flora. The genera and species can be arranged alphabetically. All accepted names in italics. The sub-heading under each taxon like distribution, phenology, ecology etc. should be typed in bold italics. Index should be placed at end of the text and illustrations either between the species or at the end of the Flora. Each illustration should be refereed to the related species. Any standard pattern could be followed for the preparation of bibliographic list.

3. Revision/Monograph

Although, there is no clear cut definition of revision, however, a formal definition of revision has been provided by Radford (1986). According to him "A revision is a treatment of selected taxa throughout at least a major portion of their range, including a study of nomenclature and classification along with descriptions based on several types of evidence". Maxted (1992) states that "A revision is a novel analysis of the variation patterns within a particular taxon, considered in conjunction with information from the literature, which results in the generation of primary (novel classification) and secondary (descriptions, keys, synonymised lists, taxon illustrations, critical notes, etc) products.

Davis and Heywood (1963) divided revision into complete and partial revisions. A complete revision is one where the taxa studied are investigated throughout their range, whereas a partial revision is associated with research for a Flora of a particular political or geographic area and is thus more restricted.

Some fundamental differences between a Flora and Monograph revision

Flora	Monograph/Rivision
1. Taxonomically less account of information and data.	1. Comprehensive account of the taxonomic data.
2. Of a defined geographical area (on account of a particular region or often arbitrarily defined region).	2. Usually the geographical scope is worldwide or a major zone.
3. Does not cover any variation of taxa even within its own jurisdiction.	3. Provides total variation of a Taxon.
4. Simply a means of identification of taxa.	4. Although provide a means of identification, monographs account for a systematic synthesis of all available data of taxa.
5. Biosystematics information and infra-specific variation are not dealt in detail.	5. Biosystematics information and infra-specific variation are dealt in detail after a distillation of all taxonomic information concerning the species.
6. Mostly very conservative and do not provide result of new taxonomic research. They are results aimed at providing a standard point of reference.	6. More innovative and present new taxonomic results

3.1. Structure of monographs

The taxon based herbarium research can essentially be discussed under two headings i.e. monographs and revisions. The difference between these activities is essentially one of scale and taxonomic focus. A monograph is a collation of all available taxonomic information relating to a particular plant group, whereas a revision is more limited and is largely confined to a collation of morphological and geographic data for a plant group, possibly from a particular area. The precise difference between a monograph and a revision is, however, of limited semantic importance, as in practice the one grades into the other. In addition, a third category synopsis is also used to refer to herbarium researches; however, it appears to be used as a synonym for the work undertaken during a revision (Table 1).

Table 1. Difference between Revision, Monograph and Synopsis (Stuessy 1975, 1993)

Revision	Monograph	Synopsis
A revisionary study is none complete, and it can be regarded as the basic comprehensive statement which presents descriptive material and information about the group including a predictive classification	A monographic study is even more elaborate with usually more types of data and many more hypotheses of relationship and evolutionary patterns and process	The synopsis is a minimal statement about relationship; it may not solve all nomenclatural aspects, nor will it answer many evolutionary questions

3.2. Significance of monographic studies

- Base line for further investigation
- Nomenclature stability
- Classification
- Stimulus for floristic studies
- Ecological studies and conservation
- Evolutionary studies

3.3. Process involved in revision and monograph

The primary objective of a revision is to take an established taxonomic classification and based on novel data analysis, a revised classification can be synthesized. Thus during the course of a revision a fundamental alternation occurs between the initial and the revised classification, i.e. new taxa may be formed, taxonomic limit redefined, taxa become submerged in other taxa etc. For the first time Davis and Heywoods (1963) systematically provided the following nine steps involved in the process of revision.

- Survey literature to delimit the taxon and clarify the associated taxonomic problems.
- Observe representative specimens and select characters which are discontinuous within the taxon.
- Group specimens on correlated characters.
- Identify diagnostic characters which distinguish specimen group.
- Record the pattern of character variation both within and between specimen groups.
- Relate specimen group to any geographical or ecological peculiarities and give taxonomic rank to each specimen group.

- Group taxa on the basis of similarities, so that natural series, sections, etc. build up into a natural classification of the whole taxon.
- Comminicate the revision results in the appropriate stylized form of nomenclature, synonymslist, keys, description, scatter diagrams, maps, etc.
- Infer and discuss phylogenetic trends among taxa.

Maxted (1992) suggested following model to produce the more explicit revision.

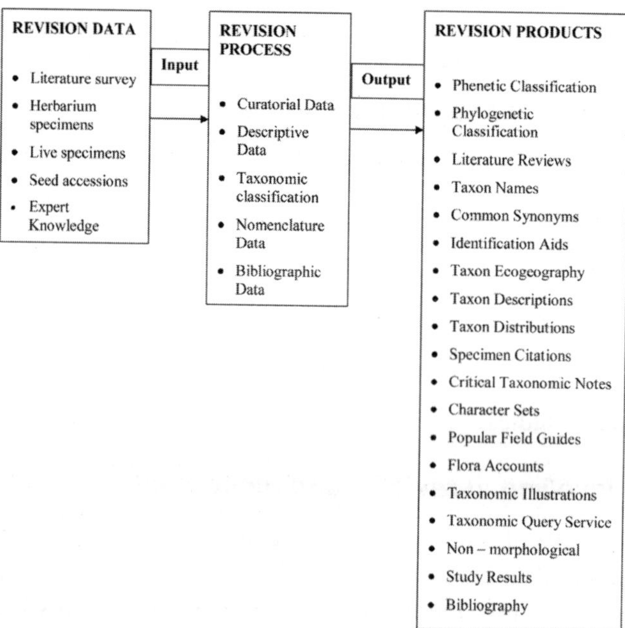

Some of the important points that should be considered during revision are as follows:

• Selection of group

A small genus of 50-60 species is considered ideal for a beginner for revisionary work. The size of the taxon also depends upon the size of geographical area and time duration available.

• Historical account

The taxonomic history of the group is usually mentioned so that the reader has an understanding of how the concept of the genus has changed over the years.

• Index to the name of taxa

All binomials pertaining to a species (may be valid or invalid) should be recorded with full citations from the Index Kewensis and its supplements or from other standard publications and different databases.

• Literature collection

All relevant literature related with the taxon must be collected from different sources. Some literature which may not be available in India can be procured from Kew or any other important botanical libraries.

• Protologue

Protologues are those materials connected with the original/first description of the species. It may be a description, illustrations and the specific herbarium specimen on which the new species was founded by the founding author. Location and collection of protologues is basic for taxonomic revisions/monographs.

• Type specimens

Type specimens and the names of herbaria where they have been preserved should be located so that these are borrowed for critical study. It is also necessary to visit all important herbaria for this purpose. Collect Type material attached to all binomials under a taxon. Study critically to ascertain that all Types belong to same species or if different, they need to be segregated along with the attached binomials to elevate it as a different taxa following the type methods.

• Field work

The extensive survey and collection of plant specimens from different zones are also essential as this provides a firsthand knowledge of the group. The collection and preservation of specimens should be in accordance with some standard procedure (Jain and Rao 1977).

• Study of materials

Select all materials that is needed for work. Make groups of specimens according to their overall resemblances and differences without looking at the name on the label. Make an analytical description of plants in different piles and then compare the description with the protologue. Also match the specimens with the type. Depending upon the similarities or differences with the protologue segregation or amalgamation of specimens can be done, and accordingly change their nomenclature. If a part of a species after the study of Type material proves to fit well under another validly published taxon as new combinations can be effected following ICN. Such studies require very critical examinations of types as well as general collections, thorough knowledge of ICN and a capacity for taxonomic analysis.

Following procedures can be adopted for examination of specimens.

- Select all materials that are needed for work.
- Make groups of specimens according to their overall resemblances. Look at each

specimen and make piles of the plant which look the same. While doing this do not look at the name on the label.

- Take a large pile and select a complete specimen.
- Make an analytical description of plant.
- Then continue with the other specimens of the pile.
- Make another sheet of paper notes on the details in which each subsequent plant differs from the first one. Keep these separate for each collection number.
- If specimen appears to be very different at closer look, remove these from the pile to study when you have finished the other pile.
- Stop working with the first pile after 10 specimens. Other specimens can be finished later on.
- Continue now with another pile. Make again a preliminary description and compare with the first one. Note down the differences.
- Finish all piles.
- At the end, check all preliminary descriptions mutually and make them consistent.
- From description remove all generic characters.
- Assign a correct name of the taxon with the help of protologue and Type specimen.
- After that construct a key to the taxa.

3.4. Problems in revisionary study

- Non availability of reasonably sufficient funds
- Location of type specimens mostly in Foreign herbaria
- Loaning of specimens from different herbaria
- Non availability of trained manpower
- Time consuming process
- Poor collection of plant materials, particularly of bigger plant groups
- Collection of plants from throughout the study area
- Non availability of old literature in the botanical libraries
- Problems in publications and sale

References

Davis PH and Heywood VH (1963) Principles of angiosperm Taxonomy. Oliver and Boyd Ltd, Edinburgh pp 556

DeWolf GP Jr (1964) On the size of Floras. Taxon 11(5):149-153

Frodin DG (1984) Guide to standard Floras of the world. Cambridge University Press, Cambridge

Jain SK and Rao RR (1976) A handbook of field and herbarium methods, New Delhi

Krishnamurthy KV, Kumar TS and Nandagopalan V (1995) Floristics for the 21st Centuary. In: Pandey AK (ed.), Taxonomy and Biodiversity, CBS Publishers and Distributors, New Delhi pp. 73-80

Matthew KM (1978) The urgent need of monographs in Indian taxonomic botany. Indian J For 1:319-328

Matthew KM (1979) A methodology for botanical revisions. Indian J For 2(3):204-217

Maxted N (1992) Towards defining a taxonomic revision methodology. Taxon 41:653-660

Mcneill J *et al.* (2012) International Code of Nomenclature for algae, fungi, and plants (Melbourne Code), Regnum Vegetabile 154. A.R.G. Gantner Verlag KG. ISBN 978-3-87429-425-6 (adopted by the Eighteenth International Botanical Congress Melbourne, Australia, July 2011)

Palmer MW, Wade GL, Neal P (1996) Professional Biologist: standard for the writing of Flora. Bioscience 45(5):339-345

Radford AE (1986) Fundamentals of Plant Systematics. Harper & Row, Publishers, Inc., New York

Rau MA (1977) Future strategies in Botanical exploration and Flora writing activity in India. Bull Bot Surv India 19(1-4):176-178

Stace CA (1989) The Plant Taxonomy and Biosystematics. Edward Arnold, London

Stern MJ, Erikson T (1996) Symbioses in herbaria: recommendation for more positive interactions between plant systematics and ecologists. Taxon 45:49-58

Stuessy TF (1975) The importance of revisionary studies in plant systematics. Sida 6(2):104-113

Stuessy TF (1993) The role of creative morphology in the biodiversity crisis. Taxon 42:313-321

Turrill WB (1964) Flora. In: Turrill WB (ed.), Vistas in Botany, Vol. 4, Pergamon Press, New York pp. 225-238

Veldkamp JF (1987) Manual for the description of flowering plants. In: Manual of herbarium taxonomy (Theory and Practice) Vogel EF de (ed.), UNESCO, Indonesia pp. 20-25

Chapter – 14

A Brief Introduction to The International Code of Plant Nomenclature

L.B. Chaudhary

1. Introduction

The scientific naming of plants is called botanical nomenclature. Although, nomenclature is included in taxonomy, but it distinctly differs from taxonomy. Plant taxonomy is concerned with grouping and classifying plants, however, the botanical nomenclature provides names for the results of this process. Names are a means of reference to all living and non-living things. The common names of pants and animals vary from place to place and language to language. But, scientific names of plants and animals are universal. For uniformity and consistency, the scientific names are given to plants according to certain internationally accepted rules provided in *International Code of Nomenclature*. It was Linnaeus who for the first time proposed some rules for generic names of plants in his *Fundamenta Botanica*, 1736 and *Critica Botanica*, 1737. However, the first proper set of rules of plant nomenclature was drafted by A. P. de Candolle and passed by the International Botanical Congress at Paris in 1867. The present Botanical nomenclature is governed by the *International Code of Nomenclature for algae, fungi, and plants* (*ICN*), which replaces the *International Code of Botanical Nomenclature* (*ICBN*) in 2012.

Botanical nomenclature has a long history, going back to the period when Latin was the scientific language throughout Europe, and perhaps further back to Theophrastus. The starting point for modern botanical nomenclature is Linnaeus' *Species Plantarum* (1753). In the nineteenth century it became increasingly clear that there was a need for rules to govern scientific nomenclature and initiatives were taken to produce a body of laws. These rules were published in successively more sophisticated editions after every six years. The most recent is the *Melbourne Code*, adopted in 2012.

All formal botanical names are governed by the *ICN*, and within the limits set by that code there is another set of rules, the *International Code of Nomenclature for Cultivated Plants (ICNCP)*. The latter Code applies to plant cultivars that have been deliberately altered or selected by humans and require separate recognition

The *Code* may be modified only by action of a plenary session of an International Botanical Congress on a resolution moved by the Nomenclature Section of that Congress. Permanent nomenclature committees are established under the auspices of the International Association for Plant Taxonomy. Members of these committees are elected by an International Botanical Congress. The committees have power to co-opt and to establish subcommittees.

Here a brief synopsis of the recent Code (i. e. *Melbourne Code* 2012) is provided in simplified way including some examples (chiefly from Spermatophytes) either taken from various sources (Jeffrey 1968, Jain and Rao 1977; Henry and Chandrabose 1979; McNeill *et al.* 2012) or on which the author has personally worked.

2. Melbourne Code (2012)

The XVIII International Botanical Congress held in Melbourne, Australia in July 2011 made a number of very significant changes in the rules governing botanical nomenclature, although always covering algae and fungi as well as green plants. The present Code made some substantial changes including its title. Since the VII International Botanical Congress in Stockholm in 1950, successive editions of the *Code* have been published as the *International Code of Botanical Nomenclature*, commonly abbreviated as *ICBN*. In Melbourne, reflecting the view, particularly amongst mycologists, that the word "Botanical" was misleading and could imply that the *Code* covered only green plants and excluded fungi and diverse algal lineages, it was agreed that the name be changed to *International Code of Nomenclature for algae, fungi, and plants (ICN).*

The rules that govern the scientific naming of algae, fungi, and green plants are revised at Nomenclature Section meetings at successive International Botanical Congresses. It supersedes the *Vienna Code* (2006), published six years ago and like its immediate predecessors, it is written entirely in (British) English. However, the overall presentation and arrangement of the remaining text of the *Melbourne Code* remains broadly similar to that in the *Vienna Code*.

In addition to the change in the title of the *Code* and the separation of the Appendices, there were five other major changes to the rules of nomenclature adopted in Melbourne. (i) the acceptance of certain forms of electronic publication; (ii) the option of using English as an alternative to Latin for the descriptions or diagnoses of new taxa of non-fossil organisms (iii) the requirement for registration as a prerequisite for valid publication of new names of fungi (iv) the abolition of the provision for separate names for fungi with a pleomorphic life history (v) and the abandonment of the morphotaxon concept in the nomenclature of fossils.

Apart from the above revision, it was also decided at Melbourne that the Appendices (other than Appendix I on the nomenclature of hybrids) will not be published further along with the main text, and indeed may be published only electronically. Therefore, the current Code comprises only the main text of the *Code* that is Preamble, Division I- Principles, Division II- Rules and Recommendations, Division III- Provisions for the Governance of the *Code*, Appendix I- Names of Hybrids, the Glossary, Index of scientific names and Subject index.

3. Outline of Melbourne code (2012)

Preamble

Division I. Principles (1-6)

Division II. Rules and Recommendations (Articles 1-62)

Chapter I. Taxa and their rank (Articles 1-5)

Chapter II. Status, typification and priority of names (Articles 6- 15)

Section 1. Status definitions (Article 6)

Section 2. Typification (Articles 7- 10)

Section 3. Priority (Articles 11-12)

Section 4. Limitation of the principle of priority (Articles 13-15)

Chapter III. Nomenclature of taxa according to their rank (Articles 16-28)

Section 1. Name of taxa above the rank of family (Articles 16-17)

Section 2. Names of families and subfamilies, tribes and subtribes (Articles 18-19)

Section 3. Names of genera and subdivisions of genera (Articles 20-22)
Section 4. Names of species (Article 23)

Section 5. Names of taxa below the rank of species (infraspecific taxa) (Articles 24-27)

Section 6. Names of organisms in cultivation (Article 28)

Chapter IV. Effective publication (Articles 29-31)

Section 1. Conditions of effective publication (Article 29-30)

Section 2. Dates of effective publication (Article 31)

Chapter V. Valid publication of names (Articles 32-45)

Section 1. General Provisions (Articles 32-37)

Section 2. Names of new taxa (Articles 38-40)

Section 3. New combinations, names at new rank, replacement names (Article 41)

Section 4. Names in particular groups (Articles 42-45)

Chapter VI. Citations (Articles 46-50)

Section 1. Author citations (Articles 46-50)

Chapter VII. Rejection of names (Articles 51-57)

Chapter VIII. Names of anamorphic fungi or fungi with a pleomorphic life cycle (Article 59)

Chapter IX. Orthography and gender of names (Articles 60-62)

Section 1. Orthography (Articles 60-61))

Section 2. Gender (Article 62)

Division III. Provision for the Governance of the Code

Appendix 1. Names of hybrids (Articles H.1- H.12)

Glossary

4. Principles

Principle I

The nomenclature of algae, fungi, and plants is independent of zoological and bacteriological nomenclature.

Principle II

The application of names of taxonomic groups is determined by means of nomenclatural types.

Principle III

The nomenclature of a taxonomic group is based upon priority of publication.

Principle IV

Each taxonomic group with a particular circumscription, position, and rank can bear only one correct name, the earliest that is in accordance with the rules, except in specified cases.

Principle V

Scientific names of taxonomic groups are treated as Latin regardless of their derivation.

Principle VI

The rules of nomenclature are retroactive unless expressly limited.

5. Taxa and Their Rank (Articles 1-5)

Taxonomic groups of any rank will be referred to as 'taxa' (singular: 'taxon'). The concept of term 'taxon' first appeared in Stockholm Code (1952). The term 'taxon' is not a category in the classification system but it is a very useful general term for a category on taxonomic group of any rank.

The principal ranks of taxa in descending sequence are: kingdom (regnum), division or phylum (divisio or phylum), class (classis), order (ordo), family (familia), genus (genus), and species (species). Thus, each species is assignable to a genus, each genus to a family, etc.

The secondary ranks of taxa in descending sequence are tribe (tribus) between family and genus, section (sectio) and series (series) between genus and species, and variety (varietas) and form (forma) below species.

If a greater number of ranks of taxa is desired, the terms for these are made by adding the prefix "sub-" to the terms denoting the principal or secondary ranks.

The principal ranks of hybrid taxa (nothotaxa) are nothogenus and nothospecies. These ranks are the same as genus and species. The prefix "notho" indicates the hybrid character.

For purposes of standardization, the following abbreviations have been recommended in the code: cl. (class), ord. (order), fam. (family), tr. (tribe), gen. (genus), subg. (subgenus) sect. (section), ser. (series), sp. (species), subsp. (subspecies), var. (variety), f. (forma), nothosp. (nothospecies).

6. Status Definition (Article 6)

6.1. Effective publication

Effective publication is publication in accordance with Art. 29-31.

6.2. Valid publication

Valid publication of names is publication in accordance with Art. 32-45 A name of a taxon has no status under the Code unless it is validly published. A validly published name only can be treated as legitimate or illegitimate.

6.3. Isonyms

When the same name, based on the same type, has been published independently at different times perhaps by different authors, then only the earliest of these has nomenclatural status. The name is always to be cited from its original place of valid publication, and later isonyms may be disregarded.

Ex. Baker (Summary New Ferns: 9. 1892) and Christensen (Index Filic.: 44. 1905) independently published the name *Alsophila kalbreyeri* as a replacement for *A. podophylla* Baker (1881) non Hook. (1857). As published by Christensen, *A. kalbreyeri* is a later isonym of *A. kalbreyeri* Baker without nomenclatural status.

6.4. Legitimate name

A validly published name or epithet that is in accordance with all rules.

6.5. Illegitimate name

A validly published name or epithet that is contrary to one or more articles of code. An illegitimate name must be rejected. Some of the common illegitimate names are: Superflous names, Later homonyms, Tautonyms

A name that according to the *Code* was illegitimate when published cannot become legitimate later unless it is conserved.

6.6. Correct name

A legitimate name at the rank of family or below that must be adopted under the rules for a taxon with a particular circumscription, position and rank. A legitimate name may be correct or incorrect according to different concepts of the taxa.

Ex. The generic name *Vexillifera* Ducke (1922), based on the single species *V. micranthera,* is legitimate. The same is true of the generic name *Dussia* Krug & Urb. ex Taub. (1892), based on the single species *D. martinicensis.* Both generic names are correct when the genera are thought to be separate. Harms (in Repert. Spec. Nov. Regni Veg. 19: 291. 1924), however, united *Vexillifera* and *Dussia* in a single genus, the latter name is the correct one for the genus with that particular circumscription. The legitimate name *Vexillifera* may therefore be correct or incorrect according to different taxonomic concepts.

6.7. Synonym

Synonyms are the different names used for the same taxonomic group or taxon. There are two kinds of synonyms - Nomenclatural synonyms and Taxonomic synonyms.

6.7.1. Nomenclatural synonyms

Nomenclatural synonyms are the names based on the same type. They are also called homotypic synonyms.

Ex. For example, *Miliusa tomentosa* (Roxb.) Sinclair, *Saccopetalum tomentosum* (Roxb.) Hook. f & Thoms. and *Uvaria tomentosa* Roxb. are names based on the same type, hence they are nomenclatural synonyms.

6.7.2. Taxonomic synonyms

Taxonomic synonyms are the names based on different types considered belong to the same taxon. They are also called heterotypic synonyms.

Ex. Hibiscus vitifolius L. (1753) and *Hibiscus obtusifolius* Willd. (1801) are names based on two different types. The study of both the types has proved them identical. This is entirely based on taxonomic judgement. Hence, *Hibiscus obtusifolius* Willd. is a taxonomic synonym of *Hibiscus vitifolius* L.

6.8. Homonym

A name spelled exactly like a validly published name for a taxon of the same rank based on a different type.

Ex. Jatropha heterophylla Steudal, Nom. Bot. ed. 2: 1: 1799 and *Jatropha heterophylla* Heyne ex Hook. f., Fl. Brit. India 5: 382. 1887 are based on two different types. The later homonym i.e. *Jatropha heterophylla* Heyne ex Hook. f. is illegitimate and is replaced by the new name *Jatropha heynei* Bal. in Bull. Bot. Surv. India 3: 41. 1961. The name *Jatropha heterophylla sensu* Heyne ex Hook. f., Fl. Brit. India 5: 382. 1887 (auct. non Steudal 1840) will become synonym of *Jatropha heynei* Bal.

6.9. Tautonym

A binary name in which the specific epithet exactly repeats the generic name. Tautonyms are treated as illegitimate in nomenclature.

Ex. Cordifolia cordifolia Hook. *Strobus strobus* (L.) Small

6.10. Autonym

Autonyms are such names as can be established automatically, whether or not they actually appear in the publication in which they are created. For example, the recognition of a subspecies or variety that does not contain the type of the species automatically establishes a subspecies or variety that does contain the type. Hence the final epithet of an autonym at the infraspecific level is that of specific name unaltered, but without citation of an author's name.

Ex. Aconitum heterophyllum Wall. ex Royle var. *heterophyllum*

The publication of the name *Crotalaria willdenowiana* DC. subsp. *glabrifoliolata* Ellis (1965), automatically established the name of another subspecies *Crotalaria willdenowiana* DC. subsp. *willdenowiana*, the type of which is that of the name *Crotalaria willdenowiana* DC.

6.11. Nomen novum (nom. nov.)

The name of a new taxon (e.g. genus novum, gen. nov., species nova, sp. nov.) is a name validly published in its own right, i.e. not based on a previously validly published name, it is not a new combination, a name at new rank, or a replacement name.

Ex. Cannaceae Juss. (1789), *Canna* L. (1753), *Canna indica* L. (1753), *Heterotrichum pulchellum* Fisch. (1812), *Poa sibirica* Roshev. (1912), *Solanum umtuma* Voronts. & S. Knapp (2012), *Oxytropis sanjappae* Chaudhary (2013).

6.12. Combinatio nova **(comb. nov.)**

A new combination or name at new rank (status novus, stat. nov.) is a new name based on a legitimate, previously published name, which is its basionym. The basionym provides the final epithet, name, or stem of the new combination or name at new rank. The new name is typified by the type of basionyms.

When a taxon of lower rank than genus is transferred to another genus or species, with or without alteration of rank, based on taxonomic judgement, the original name or epithet if legitimate (Basionym) is retained and the author of the basionym is cited in parentheses followed by the name of the author who effected the alteration (the author of the new combination).

Ex. The basionym of *Centaurea benedicta* (L.) L. (1763) is *Cnicus benedictus* L. (1753), the name that provides the epithet.

Ex. The basionym of *Crupina* (Pers.) DC. (1810) is *Centaurea* subg. *Crupina* Pers. (Syn. Pl. 2: 488. 1807), the name of which the epithet provides the generic name.

Ex.. The basionym of *Anthemis* subg. *Ammanthus* (Boiss. & Heldr.) R. Fern. (1975) is *Ammanthus* Boiss. & Heldr. (1849), the name that provides the epithet.

A new combination can at the same time be a name at new rank (*comb. & stat. nov.*).

Ex. Aloe vera (L.) Burm. f. (1768), based on *A. perfoliata* var. *vera* L. (Sp. Pl.: 320. 1753), is both a new combination and a name at new rank.

6.13. Replacement name

A replacement name (avowed substitute, nomen novum, nom. nov.) is a new name based on a legitimate or illegitimate previously published name, which is its replaced synonym. The replaced synonym, when legitimate, does not provide the final epithet, name, or stem of the replacement name.

Ex. Caulerpa pinnata C. Agardh (1817), based on the illegitimate *Fucus pinnatus* L. f. (1782), a later homonym of *F. pinnatus* Huds. (1762).

Ex. Centaurea chartolepis Greuter (2003), based on *Chartolepis intermedia* Boiss. (1856), the epithet *intermedia* being unavailable in *Centaurea* because of *Centaurea intermedia* Mutel (1846).

6.14. Diagnosis

Diagnosis is a brief statement of characters which differentiates the taxon from its nearest relatives. The diagnosis of a species or an infraspecific taxon may be very short, merely a statement of one or two characters not possessed by the other species in the same genus. According to ICN (2012) the diagnosis of a new species may be written either in Latin or English for new taxa.

7. Typification (Articles 7-10)

The Code gives great emphasis on typification of various taxa in order to achieve stabilization of names. The Code provides that all taxonomic groups will be based on nomenclatural types. 'A nomenclatural type is that element to which the name of a taxon is permanently attached, whether as the correct name or as a synonym'. This means that all names are permanently attached with some taxon or specimens designated as type. For species (and infraspecific taxa) the type is a specimen or sometimes only an illustration. This is that specimen on which species was based and originally described. Therefore, the first name given to this species by founding author is permanently attached, whether this name remains correct or is later found to be synonym to some other name. The nomenclatural type is not necessarily the most typical or representative element of a taxon. Although, the concept of type started in America in 1892, it was internationally accepted in code as an article in 1930 in Cambridge Code and in Paris Code (1956) it was made mandatory to mention type in the publications of new species published from January 1958. Gradually, several changes have been brought out in subsequent Codes and finally following 'Types' have been recognised in current ICN (2012).

It is strongly recommended that the material on which the name of a taxon is based, especially the holotype, be deposited in a public herbarium or other public collection with a policy of giving bona fide researchers access to deposited material, and that it be scrupulously conserved.

7.1. Holotype

A holotype is the one specimen or illustration used by the author or designated by him as the nomenclatural type. Any designation made by the original author, if definitely expressed at the time of the original publication of the name of the taxon, is final. If the author used only one element, it must be accepted as the holotype even the term "type" or its equivalent is not used in the protologue.

7.2. Isotype

An Isotype is any duplicate of holotype; it is always a specimen. These are fragments from the same plant from which the holotype was made or plants forming part of the same gathering as the holotype and bearing the same field number.

7.3. Paratype

A paratype is a specimen cited in the protologue other than the holotype, isotype(s) or syntypes by the founding author at the time of describing a new taxon. The paratype specimens bear numbers different from the holotype, and can even be from very different localities and by different collectors, but they must have been studied by the author of taxon.

7.4. Syntype

A syntype is any one of two or more specimens cited by the author when no holotype was designated, or anyone or two or more specimens simultaneously designated as types. If the author studies collections from different localities and by different collectors and decides to establish a new species on them, and labels all of them as types, all those specimens become syntypes.

7.5. Lectotype

A lectotype is a specimen or illustration designated from the original material as the nomenclatural type, if no holotype was indicated at the time of publication, or if the holotype is missing, or if a type is found to belong to more than one taxon. If a holotype is lost or destroyed, the lectotype is chosen from the isotype. If no holotype was designated and if syntypes exist, one of them must be chosen as the lectotype. If neither an isotype, nor a syntype is extant, a paratype, if such exists, may be chosen as the lectotype.

7.6. Neotype

A neotype is a specimen or illustration selected to serve as nomenclatural type, if no original material is extant, or as long as it is missing. A neotype may be designated only when all the originally cited material is believed lost or destroyed.

7.7. Epitype

An epitype is a specimen or illustration selected to serve as an interpretative type when the holotype, lectotype, or previously designated neotype, or all original material associated with a validly published name, is demonstrably ambiguous and cannot be critically identified for purposes of the precise application of the name to a taxon. Designation of an epitype is not effected unless the holotype, lectotype, or neotype is explicitly cited.

8. Priority and Limitation of the Principle of Priority (articles 11-15)

Each family or taxon of lower rank with a particular circumscription, position, and rank can bear only one correct name, special exceptions being made for eight families and one subfamily for which alternative names are permitted.

A name has no priority outside the rank in which it is published.

Ex. *Campanula* sect. *Campanopsis* R. Br. (Prodr.: 561. 1810) when treated as a genus is called *Wahlenbergia* Roth (1821), not *Campanopsis* (R. Br.) Kuntze (1891).

For any taxon from family to genus inclusive, the correct name is the earliest legitimate one with the same rank, except in cases of limitation of priority by conservation.

Ex. When *Aesculus* L. (1753), *Pavia* Mill. (1754), *Macrothyrsus* Spach (1834), and *Calothyrsus* Spach (1834) are referred to a single genus, its correct name is *Aesculus* L.

The principle of priority does not apply above the rank of family.

A name of a taxon has no status under the *Code* unless it is validly published. Valid publication of names of plants of different groups is treated as beginning at some prescribed dates corresponding to the publication of some works. For example, Spermatophyta and Pteridophyta, 1 May 1753 (Linnaeus, Species Plantarum ed. 1) is considered starting point.

The Code does not recognise priority of position, provided both names were published on the same date. The two volumes of Linnaeus's Species Plantarum ed. 1 (1753), which appeared in May and August respectively, are treated as having been published simultaneously on the former date (i. e. 1 May 1753).

Ex. The generic names *Thea* L. (Sp. Pl.: 515. May 1753) and *Camellia* L. (Sp. Pl.: 698. Aug 1753) are treated as having been published simultaneously on 1 May 1753. The combined genus bears the name *Camellia,* since Sweet (Hort. Suburb. Lond.: 157 1818) who was the first to unite the two genera, chose that name and cited *Thea* as a synonym.

9. Nomenclature of Taxa According to Their Rank (Articles 16-27)

9.1. Name of a taxon above the rank of family

The name of a taxon above the rank of family is treated as a noun in the plural and is written with an initial capital letter. Such names may be either *(a)* automatically typified names, formed from the name of an included genus in the same way as family names by adding the appropriate rank-denoting termination.

Ex. Automatically typified names above the rank of family: *Lycopodiophyta,* based on *Lycopodium; Magnoliophyta,* based on *Magnolia.*

Automatically typified names end as follows: the name of a division or phylum ends in *–phyta,* (in algae or fungi it ends in *–phycota* or *–mycota,* respectively), the name of a subdivision or subphylum ends in *–phytina,* (in algae or fungi it ends in *–phycotina* or *–mycotina,* respectively), the name of a class in the algae ends in *–phyceae,* and of a subclass in *–phycidae,* the name of a class in the fungi ends in *–mycetes,* and of a subclass in *–mycetidae,* the name of a class in the plants ends in *–opsida,* and of a subclass in *–idea,* names of orders or suborders are to end in *–ales* and *–ineae*, respectively.

9.2. Name of a family and subfamily

The name of a family is a plural adjective used as a noun, it is formed from the stem of a genus with the termination *aceae.* The generic name from which the name of a family is formed provides the type of the family name, but is not a basionym of that name. *Ex. Rosaceae* (from *Rosa*), *Salicaceae* (from *Salixs*), *Plumbaginaceae* (from *Plumbago*), etc.

The following names, of long usage, are treated as validly published: *Compositae* (nom. alt.: *Asteraceae,* type: *Aster* L.); *Cruciferae* (nom. alt.: *Brassicaceae,* type: *Brassica* L.); *Gramineae* (nom. alt.: *Poaceae,* type: *Poa* L.); *Guttiferae* (nom. alt.: *Clusiaceae,* type: *Clusia* L.); *Labiatae* (nom. alt.: *Lamiaceae,* type: *Lamium* L.); *Leguminosae* (nom. alt.: *Fabaceae,* type: *Faba* Mill. [= *Vicia* L.]); *Palmae* (nom. alt.: *Arecaceae,* type: *Areca* L.); *Papilionaceae* (nom. alt.: *Fabaceae,* type: *Faba* Mill.); *Umbelliferae* (nom. alt.: *Apiaceae,* type: *Apium* L.). When the *Papilionaceae* are regarded as a family distinct from the remainder of the *Leguminosae,* the name *Papilionaceae* is conserved against *Leguminosae*.

The name of a subfamily is a plural adjective used as a noun, it is formed in the same manner as the name of a family but by adding the termination *–oideae* instead of *–aceae*.

9.3. Name of tribe and subtribe

A tribe is designated in a similar manner, with the termination *–eae,* and a subtribe similarly with the termination *–inae*.

9.4. Name of a genus

The name of a genus is a noun in the nominative singular, or a word treated as such, and is written with an initial capital letter. It may be taken from any source whatever, and may even be composed in an absolutely arbitrary manner. *Ex. Convolvulus, Gloriosa, Hedysarum, Impatiens, Rhododendron, Rosa, Trifolium,* etc.

The name of a genus may not coincide with a Latin technical term in use in morphology at the time of publication unless it was published before 1 January 1912 and was accompanied by a species name published in accordance with the binary system of Linnaeus.

Ex. *"Radicula"* (Hill, 1756) coincides with the Latin technical term "radicula" (radicle) and was not accompanied by a species name in accordance with the binary system of Linnaeus. The name *Radicula* is correctly attributed to Moench (1794), who first combined it with specific epithets.

The name of a genus may not consist of two words, unless these words are joined by a hyphen.

Ex. *"Uva ursi"*, as originally published by Miller (1754), consisted of two separate words unconnected by a hyphen, and is not therefore validly published; the name is correctly attributed to Duhamel (1755) as *Uva-ursi* (hyphenated when published).

Authors forming generic names should comply with the following:

(a) Use Latin terminations insofar as possible.

(b) Avoid names not readily adaptable to the Latin language.

(c) Not make names that are very long or difficult to pronounce in Latin.

(d) Not make names by combining words from different languages.

(e) Indicate, if possible, by the formation or ending of the name the affinities or analogies of the genus.

(f) Avoid adjectives used as nouns.

(g) Not use a name similar to or derived from the epithet in the name of one of the species of the genus.

(h) Not dedicate genera to persons quite unconnected with botany, mycology, phycology, or natural science in general.

(i) Give a feminine form to all personal generic names, whether they commemorate a man or a woman.

(j) Not form generic names by combining parts of two existing generic names, because such names are likely to be confused with nothogeneric names.

The name of a subdivision of a genus is a combination of a generic name and a subdivisional epithet. A connecting term (subgenus, sectio, series, etc.) is used to denote the rank. The epithet is either of the same form as a generic name, or a noun in the genitive plural, or a plural adjective agreeing in gender with the generic name, but not a noun in the genitive singular. It is written with an initial capital letter. Ex. *Astragalus subg. Phaca.*

When it is desired to indicate the name of a subdivision of the genus to which a particular species belongs in connection with the generic name and specific epithet, the subdivisional epithet should be placed in parentheses between the two; when desirable, the subdivisional rank may also be indicated. *Ex. Astragalus (Cycloglottis) contortuplicatus, Loranthus* (sect. *Ischnanthus*) *gabonensis.*

9.5. Names of species

The name of a species is a binary combination consisting of the name of the genus followed by a single specific epithet. If an epithet consists of two or more words, these are to be united or hyphenated. An epithet not so joined when originally published is not to be rejected but, when used, is to be united or hyphenated. The epithet in the name of a species may be taken from any source whatever, and may even be composed arbitrarily. Ex. *Astragalus himachalensis*, *Adiantum capillus-veneris.*

In forming specific epithets, authors should comply also with the following:

(a) Use Latin terminations insofar as possible.

(b) Avoid epithets that are very long or difficult to pronounce in Latin.

(c) Not make epithets by combining words from different languages.

(d) Avoid those formed of two or more hyphenated words.

(e) Avoid those that have the same meaning as the generic name (pleonasm).

(f) Avoid those that express a character common to all or nearly all the species of a genus.

(g) Avoid in the same genus those that are very much alike, especially those which differ only in their last letters or in the arrangement of two letters.

(h) Avoid those that have been used before in any closely allied genus.

(i) Not adopt epithets from unpublished names found in correspondence, travellers' notes, herbarium labels, or similar sources, attributing them to their authors, unless these authors have approved publication.

(j) Avoid using the names of little-known or very restricted localities unless the species is quite local.

9.6. Names of taxa below the rank of species (infra-specific taxa)

The name of an infraspecific taxon is a combination of the name of a species and an infraspecific epithet. A connecting term is used to denote the rank. Ex. *Parochetus communis* subsp. *africanus*

10. Effective Publication (Articles 29-31)

The effective publication should contain following conditions.

(a) Publications should be in printed matter. Publication can also be made on or after 1 January 2012 electronically in Portable Document Format (PDF) as online publication with an International Standard Serial Number (ISSN) or an International Standard Book Number (ISBN). The content of a particular electronic publication must not be altered after it is effectively published. Corrections or revisions must be issued separately to be effectively published.

(b) Printed matter should be widely distributed to the general public or at least to scientific institutions with generally accessible libraries.

(c) Publication on or after January 1953 of a new name should be in scientific paper. Publication on or after 1 January 1953 in a thesis submitted to a university or other institute of education for the purpose of obtaining a degree does not constitute effective publication.

(d) The date of effective publication is the date on which the printed matter or electronic version is available.

11. Valid Publication of Names (Articles 32-45)

The following conditions are to be followed for valid publication.

(a) The name of a taxon must be effectively published.

(b) The name of a taxon must be accompanied by a description or diagnosis.

(c) A name of new taxon of plants published between 1 January 1935 and 31 December 2011, inclusive, must be accompanied by Latin description or diagnosis. Authors publishing names of new taxa on or after 1 January 2012 should give or cite a full description or diagnosis in Latin or English or a reference to a previously and effectively published Latin or English description or diagnosis. A new combination published on or after 1 January 1953 for a previously and validly published name, basionym should be clearly indicated.

(d) A clear identification of the rank of taxon concerned must be given with a new name or combination on or after 1 January 1953. It is also recommended that when publishing nomenclatural novelties, authors should indicate this by a phrase including the word "novus" or its abbreviation, e.g. genus novum (gen. nov., new genus), species nova (sp. nov., new species), combinatio nova (comb. nov., new combination), nomen novum (nom. nov., replacement name), or status novus (stat. nov., name at new rank).

(e) Publication of the name of a new taxon of the rank of family or below on or after 1 January 1958 is valid only when the nomenclatural type is indicated.

(f) The illustration or figure required should be prepared from actual specimens, preferably including the holotype.

12. Citation of Author's Name and General Recommendations on Citation (Articles 46-50)

The name of taxon should be followed by the name of the author(s) who first validly published the name. Authors' names put after the names of plant may be abbreviated, unless they are very short. Approved publications should be used to get the standard abbreviated form of author's name.

Ex. *Astragalus* L. (L. for Linnaeus).

12.1. Use of et (and) or et al. (and others)

When a name is published jointly by two authors, the names of both are connected by *et* or by an ampersand (&).

Ex. *Trigonella upendrae* Chowdhery *et* R. R. Rao

or

T. upendrae Chowdhery & R. R. Rao

When name is published jointly by more than two authors, the citation of the taxon should be restricted to that of the first author followed by *et al.*

Ex. *Indotristicha tirunelveliana* Sharma, Karthikeyan and Shetty should be written as *I. tirunelveliana* Sharma *et al.*

12.2. Use of ex

When an author who first validly publishes a name ascribes it to another person (who has earlier proposed but not validly published the name) the correct author citation is the name of actual validly publishing author, but the name of other person followed by the connecting word 'ex' may be inserted before the validly publishing author, if desired.

Ex. *Aconitum heterophyllum* Wall. ex Royle

or

A. heterophyllum Royle

12.3. Use of in

When a name with a description supplied by one author is published in another author's work, the word 'in' should be used to connect the name of two authors.

Ex. *Tricholepis tibetica* Hook. f. & Thoms. in Clarke, Com. Ind. 241. 1876.

12.4. Use of emend

If an original description of an taxon is incomplete and subsequently corrected, or the circumscription of a taxon is altered retaining the same type the citation of the original author is retained but followed by a word emend and the name of the author responsible for the change.

Ex. *Medicago polymorpha* L. emend Shinn.

12.5. Use of pro parte *(*p. p.*)*

It means partly or in part. When an alteration of the circumscription of a taxon has been considerable, resulting in its division into two or more taxa, in the citation of newly formed taxa the misapplied original name is indicated by adding *pro parte (p.p.)*

Ex. – *Ficus abelii* Miq., Ann. Mus. Bot. Lugd.-Bat. 3: 281. 1867.

F. pyriformis sensu Hook. f., Fl. Brit. India 5: 533. 1888, p. p.

12.6. Use of sensu amplo (s. ampl.) *&* sensu stricto (s. str)

Sensu amplo or *sensu lato* means in the broader sense and the *sensu stricto* means in the narrow sense.

Ex. *Eugenia* L. *s. lat.* (when it includes *Syzygium, Cleistocalyx, Eugenia* – proper, etc.)

Eugenia L. *s. str.* (excluding *Syzygium, Cleistocalyx,* etc.)

12.7. Use of nomen nudem

Nomen nudem means naked names, i.e. a name without description etc. Wallich in his Catalogue (1828-49) published many new names without any description or reference to an earlier publication. Hence these names are not valid.

Ex. *Anagyris nepalensis* Wall., Cat. No. 5340. 1831, *nom. nud.*

12.8. Use of auct. *(*auctorum, *author)* non

A misapplied name is indicated by the word *auct non* followed by the name of the original author and the bibliographic reference of the misidentification.

Ex. *Phyllanthus fraternus* Webster

=*P. niruri auct. non* L., Hook. f. Fl. Brit. India 5: 298.1887 (or *P. niruri* sensu Hook. f.)

This plant has been treated as *P. niruri* L. in most of our Indian Floras. Webster's work proved that the Indian plant is *P. fraternus.* The word *auct. non* in the citation means that *P. niruri a*s treated by all authors, except Linnaeus is *Phyllanthus fraternus* Webster.

12.9. Use of nom. cons.

If a taxon is conserved the words *nom. cons.* (*nomen conservandum*) should be added to the citation.

Ex. Leguminosae *nom. cons.*, Tectona L. f. *nom. cons.*

12.10. Use of parentheses

When a genus or a taxon of lower rank is altered in rank but retains its name or the final epithet in its name, the author of that earlier name, if it is legitimate (i.e. if it is the basionym), is cited in parentheses, followed by the name of the author who effected the alteration (the author of the name). The same provision holds when a taxon of lower rank than genus is transferred to another genus or species, with or without alteration of rank.

Ex. Medicago polymorpha var. *orbicularis* L. (1753) when raised to the rank of species becomes *M. orbicularis* (L.) Bartal. (1776).

Ex. When Parochetus africanus Polh. is reduced as subsp., it becomes *Parochetus communis* Buch. – Ham. ex D. Don ssp. *africanus* (Polh.) Chaudhary & Sanjappa.

13. Rejection of Names (Article 51-57)

A legitimate name must not be rejected merely because it, or its epithet, is inappropriate or disagreeable, or because another is preferable or better known or because it has lost its original meaning.

Ex. The name *Scilla peruviana* L. (1753) is not to be rejected merely because the species does not grow in Peru.

Ex. *Richardia* L. (1753) is not to be rejected in favour of *Richardsonia,* as was done by Kunth (1818), merely because the name was originally dedicated to Richardson.

A name, unless conserved, is illegitimate and is to be rejected if it was nomenclaturally superfluous when published.

Ex. The generic name *Cainito* Adans. (1763) is illegitimate because it was a superfluous name for *Chrysophyllum* L. (1753), which Adanson cited as a synonym. Hence the name *Cainito* Adans. published at a later is to be rejected.

A name of a family, genus, or species, unless conserved, is illegitimate if it is a later homonym, that is, if it is spelled exactly like a name based on a different type that was previously and validly published for a taxon of the same rank.

Ex. *Astragalus rhizanthus* Boiss. (1843) is a later homonym of the validly published name *A. rhizanthus* Royle (1835) and is therefore illegitimate and hence it was renamed *A. cariensis* Boiss. (1849).

14. Orthography and Gender of Names

The original spelling of a name or epithet is to be retained, except for the correction of typographical or orthographical errors

Ex. The generic names *Mesembryanthemum* L. (1753) and *Amaranthus* L. (1753) were deliberately so spelled by Linnaeus and the spelling is not to be altered to *"Mesembrianthemum"* and *"Amarantus"*, respectively.

Ex. *Globba "brachycarpa"* Baker (1890) is typographical errors for *Globba trachycarpa* Baker.

15. Names of Hybrids (Articles H.1- H.12) (Appendix I)

Hybrids between two species are indicated by inserting a hybrid sign (x) between the species names of the two parents. The (x) is not italicised and the names are written in alphabetical order.

Ex. *Medicago* x *varia (M. falcata* x *M. sativa)*

References

Henry AN and Chandrabose N (1979) An Aid to the International Code of Botanical Nomenclature, Today and Tomorrow's Printers and Publishers, New Delhi

Jain SK and Rao RR (1977) A handbook of field and herbarium methods, Today and Tomorrow's Printers and Publishers, New Delhi

Jeffrey C (1968) An Introduction of Plant Taxonomy, J. and A. Churchill Ltd., London

McNeill J and Turland NJ (2011) Major Changes to the Code of Nomenclature-Melbourn, July, 2011. Taxon 60 (5):1495-1497

McNeill J *et al.* (2012) International Code of Nomenclature for algae, fungi, and plants (Melbourne Code), Regnum Vegetabile 154. A.R.G. Gantner Verlag KG. ISBN 978-3-87429-425-6 (adopted by the Eighteenth International Botanical Congress Melbourne, Australia, July 2011)

Chapter – 15

Taxonomy and Nomenclature of Cultivated Plants

K.N. Nair

1. Introduction

Plant taxonomy and the cultivated plants have a close relationship of great antiquity. This ancient relationship can be traced back to the invention of incipient agriculture (neolithic revolution), about 10-13000 years BCE (Stearn 1965; Balter 2007). The early foundation of plant taxonomy was thus laid largely on humans' knowledge on cultivated plants. This body of rudimentary knowledge on identifying the properties of different types of plants for food, medicine, clothing, shelter, and other uses led to the development of plant taxonomy- the first aspect of plant science to emerge as botany (Constance 1957). The fundamental principles of modern plant taxonomy with regard to the natural vegetation and flora were originally derived from cultivated plants (Li 1974).

The active process of cultivation of a plant, primarily through the intentional activity of humans has been considered the main criterion for categorising cultivated plants (Mansfeld 1959). According to Mansfeld's approach, the number of cultivated plant species of the world was estimated to be about 7000 species, excluding partly the forestry plants, ornamentals, and plants used in amenity horticulture (Henelt and IPK 2001). Khoshbakht and Hammer (2008) expanded this list to include a total of 35000 species of cultivated plants, including 28000 ornamental plant species connected to gardening and landscaping. About 2500 plant species have undergone domestication worldwide, with over 160 families contributing one or more crop species (Zeven and de Wet 1982; Dirzo and Raven 2003).The growing numbers of cultivated plant species and the enormous diversity found in each of these species offer plenty of challenges as well as opportunities to the taxonomists to take up cultivated plant taxonomy as an active scientific pursuit. Despite this, there has been a lack of interest among taxonomists towards cultivated plant taxonomy.

2. History of Cultivated Plant Taxonomy

The development of cultivated plant taxonomy from the Pre-Linnaean period to the present time has been marked with several milestones, indicating the emergence of new concepts, postulates and publications on classification and nomenclature of cultivated plants. The following papers present a retrospective review and discussion on the key developments in cultivated plant taxonomy: Stearn (1965); Hetterscheid *et al.* (1996); Spooner *et al.* (2003); Hammer and Morimoto (2011). The most important milestone developments as discussed in the above papers are summarised here in Table 1.

Table 1. Historical milestones in cultivated plant taxonomy

Pre- Linnaean Period	
Cato the Elder (234-149 BCE; Stearn 1986)	The first record of a named cultigen occurs in *De Agri Cultura* in a list that included 120 kinds (cultivars) of figs, grapes, apples and olives.
Valerius Cordus (1515-1544 CE; Stearn 1986)	Published many named cultivars including 30 apples and 49 pears, presumably local German selections.
John Parkinson (1629)	In *Paradisi in Sole paradiscus terrestris*, Parkinson listed 57 apple cultivars, 62 pears, 61 plums, 35 cherries and 22 peaches.
Philip Miller (1724, 1768)	Miller produced a two-volume compendium of garden plants called *The Gardeners and Florists Dictionary or a complete System of Horticulture*. The first edition in 1724, subsequently revised and enlarged until the last and 8th edition in 1768.
Linnaean Period	
Linnaeus (1753)	Linnaeus considered cultivated plants as creation of humans and treated as horticultural variants created from wild plants.
Post-Linnaean Period 19th Century	
Alefeld (1866)	Promoted infraspecific classification of cultivated plants in "Varietaeten Gruppen" (a precursor of 'convariety').
AP de Candolle (1867)	Published *Lois de la Nomenclature Botanique* - a precursor of subsequent editions of the *Regles de la Nomenclature Botanique*. The 1905 edition (Vienna Code) includes the following recommendation and articles pertaining to cultivated plants: Recommendation I: 'Modification of cultivated plants should be associated as far as possible, with the species from which they are derived'. Article 11: 'In many species, varieties ('varietas') and forms (forma) are distinguished and in some cultivated species, modifications still more numerous'. Article 30: 'Forms and half – breeds among cultivated plants should receive fancy names, in common language, as different as possible from the Latin name of the species and varieties. When they can be traced back to their

(Contd.)

	species, a subspecies or a botanical variety, this is indicated by a succession of names'.
20th Century	
Bailey (1918)	Argued for separate recognition of entities of cultivated plants. Proposed the term 'cultigen' for those species whose origin is unknown and are only known in cultivation.
Bailey (1923)	Defined the principal fundamental subdivision of the cultigens and named it 'cultivar'.
Vavilov (1926,1935)	Introduced the differential phyto-geographical method for infraspecific classification of cultivated plants, which included the formal taxonomic approach for naming infraspecific taxa.
Pangalo (1948)	Proposed the term 'nidus'as one of the infraspecific unit of a cultigen.
Grebenscikov (1949)	Proposed the term 'convariety'as an infraspecific unit of a cultigen.
Lanjouw *et al.* (1952)	Proposal for an independent International Code of Nomenclature for Cultivated Plants (ICNCP) published as an Appendix III to the Stockholm Code of ICBN (1952). Under Section H- following special categories of cultivated plants were defined: 'line, clone, hybrid group (grex), line-hybrid and convariety'.'Convariety' defined as 'a closely allied cultivars, somewhat analogous to the subspecies'.
Mansfeld (1953, 1954)	Proposed a hierarchical system in which all elements at one particular level are exhaustively classified. Opted for a closed classification and argued for more than one system of classification of cultivated plants, but all must start from the 'cultivar' as the fundamental category. Proposed three general classification principles for cultivars based on: (i) Position of their wild progenitors in the taxonomic hierarchy (ii) Uses (iii) Ways of maintenance in cultivation. Following six categories were recognized and all to be complied with the rules of ICBN: *'specoid', 'subspecoid', 'convar', 'provar', 'nidus', 'cultivar'.*
Stearn (1953)	In the first independent edition of the International Code of Nomenclature for Cultivated Plants (ICNCP), the category 'convariety' defined and the term used for designating cultivars grouped on the basis of similarity of characters important to cultivation. Stipulated that 'convarieties' with Latin names ought to be published with Latin description. Introduced hybrid nomenclature.
Helm (1957)	Criticised Mansfeld's multilevel classification system as presenting inflation of the number of Latin epithets.
Jirasek (1958, 1961)	Proposed a general term for all systematic categories of cultivated plants- the 'taxoid' ('taxonoid') - a collective description for individual taxonomic categories for the systematic classification of cultivated plants. Favoured for an artificial classification based on few characters for cultivated plants than multicharacter (natural) classification. Proposed five subordinate categories of 'taxoids', including 'subcultivar'.
ICNCP (1958)	'Convariety' classified as a supplemental category to be complied with ICBN.
Jeffrey (1968)	Recognized that formal botanical classification is unsuited to

(*Contd.*)

	classify cultivated plants. Proposed a classification system between two fixed points: 'cultivar' (lower limit) and 'species' (upper limit). Opted for a multidiscipline analysis of cultivated plants for resolving the relationships of cultivated plants to their wild relatives. Proposed categories 'provar' and 'convar' between 'cultivar' and 'species' – but without their names in Latin forms. Suggested to use 'subspecoid' to designate the cultivated aspect of a species.
ICNCP (1969)	No mention of 'convariety', but the term supplanted by 'Group'.
Harlan and De Wet (1971)	Based on the concept of biological species and crossability, defined the total gene pool of cultivated plants in to 'primary', 'secondary' and 'tertiary' gene pools depending on the degree of crossability between individual cultivar plants. Proposed to establish 'subspecies' to accommodate the cultivated race of a species.
Baum (1981)	Denounced the hierarchical classification and concept of taxa in cultivated plant taxonomy, but advocated classifications based on multicharacter analysis for cultivated plants as used for taxa.
Pickersgill (1986)	Supported the usefulness of 'subspecoid' category in cultivated plant taxonomy. Addressed the consequences of ephemeral character distribution in cultivated plants due to random selection activities of humans and also by the process of secondary domestication, and undermined the rationale of focusing attention to hierarchy in the taxonomy of cultivated plants.
Hanelt (1986)	Proposed the following schemes for classifying infraspecific variations in crop plants: (i) *Formal taxonomic classifications* under the rules of ICBN using (a) Diagnostic-morphological (b) Phenetic –numerical and (c) Ecogeographic methods (ii) *Informal classifications* using non-standard categories or categories as proposed in ICNCP with the following approaches: (a) Diagnostic-morphological (b) Phenetic –numerical and (c) Genetic (iii) *Mixed classifications.*
Hetterscheid and Brandenburg (1995)	As opposed to the general term 'taxon', they proposed the term 'culton' as a systematic group of cultivated plants based on one or more user-driven criteria.
21st Century	
Hanelt and IPK (2001)	Published an encyclopaedia on cultivated plants (excluding ornamentals and forest trees) and discussed the taxonomic framework of crop taxonomy and evolution; The Encyclopaedia provides taxonomy, updated scientific names, common names, economic use, and distribution of economically important plants.
ICNCP 2004- Brickell *et al.* (2004)	Seventh edition of ICNCP.
ICNCP 2009- Brickell *et al.* (2009)	The current and eighth edition, replacing the previous ICNCP (2004) with as many as 59 novel provisions included under Divisions II – VI. The term 'taxon' has been reinstated to replace the phrase 'distinguishable group of plants' throughout the text of the Code. In addition to 'cultivar' and 'Group', a third category 'grex' has been recognized, although the latter is allowed to be applied only to orchids. This change permits the possible recognition of character-based Groups within a parent-based 'grex'.This Code also introduced changes as to naming, formation and application of the Rules and Recommendations for the categories of 'cultivar', 'Group' and 'grex' under Articles 27 and 29.

3. Cultivated Plant Taxonomy: Conceptual Frameworks

The main objective of a taxonomist, who attempts to study a given taxon at any rank, is to arrive at a predictive classification supported with a stable nomenclature and practically useful identification keys or manuals. In order to achieve this goal, the taxonomist follows a certain species concept (s), adopts a suitable methodological framework of choice and availability, and uses a currently accepted Code of nomenclature. This often results in different classification systems with different circumscription for the same taxon by different taxonomists. Taxonomic literature is replete with examples of such differential taxonomic treatments, especially of the cultivated plants (Spooner *et al.* 2003; Hammer and Morimoto 2011). Some such case studies and reviews on taxonomy of complex cultivated plant groups include that of Potato (Dodds 1962; Bukasov 1978; Hawkes 1990; Contreras and Spooner 1999; Huaman and Spooner 2002; Spooner *et al.* 2007; Spooner 2009; Ovchinnikova *et al.* 2011), Tomato (Rick 1979; Rick *et al.* 1990; Child 1990; Olmstead and Palmer 1992, 1997; Spooner *et al.* 1993; Bohs and Olmstead 1997, 1999; Peralta and Spooner 2001; Peralta *et al.* 2008;); Wheat (Thellung 1918; Stebbins 1956; MacKey 1966, 1968, 1981; Dorofeev and Korovina 1979; Hanelt 2001); Lettuce (Lindqvist 1960; De Vries and Raamsdonk 1994; Frietemade Vries, 1996; De Vries 1997; Koopman *et al.* 1998, 2001; Dolezalova *et al.* 2004; Kroistkova *et al.* 2008), *Brassica* (U 1935; Harberd 1972; Snogerup 1980; Palmer *et al.* 1983; Song and Osborn 1992; Hanelt and IPK 2001), *Prunus* (Rehder 1960; Hanelt and IPK 2001), *Citrus*: Swingle and Reece 1967; Tanaka 1977; Mabberley 2004, 2008; Nicolosi 2007).

The taxonomy of cultivated plants differs from wild plants in the following ways: (1) distinction made on the habitat, cultivated vs. wild, (2) distinction on the mode of origin, the cultivated plants from the intentional actions of mankind vs. natural selection in the wild without the aid of human assistance, (3) distinction in the classification categories, the cultivated plant taxonomy concerned with the use of special classification categories that do not conform to the nested hierarchy of ranks implicit in the Botanical Code (ICN-International Code of Nomenclature for algae, fungi and plants), but adopts the categories of cultivar, Group and grex as stipulated in the *Cultivated Plant Code* (ICNCP-International Code of Nomenclature for Cultivated Plants), (4) distinction in the user groups of the ICN and ICNCP, the former focuses on the need of all plant taxonomists and the latter for the needs of people requiring names of plants used in commercial world of agriculture, forestry and horticulture, and (5) the differential purpose of cultivated and wild plant taxonomy, which is essentially 'plant-centered' in the ICN and 'human-centered' in the ICNCP (McNeill 2008; Brickell *et al.* 2009; McNeill *et al.* 2012).

The taxonomy of cultivated plants is complicated by the occurrence of outcrossing, selfing, apomixis, clonal propagation, polyploidization, interspecific or intraspecific hybridization and introgression. The complex phenotypic and genetic variation patterns in cultivated plants caused by the long term influence of human-assisted selection, domestication and improvement are additional impediments to the taxonomy of cultivated plants (Baker1970). While natural selection has a profound effect in modifying or eliminating certain variants in natural wild plant systems, the artificial selection under cultivation is an extremely important factor contributing to the complex variation in plants under cultivation. Human aided selection and domestication of cultigens significantly alter the morphology, physiology, and breeding behaviour such as the switch from cross-pollination to selfing, out- crossing to apomixes or vegetative propagation (Baker 1970; Meyer *et al.* 2012). The plants in cultivation do not have a natural population structure. Scattered distribution of crop plants over the globe makes adequate sampling of cultigens difficult for taxonomic or other purposes. Poor documentation of herbarium materials of cultivated plants and their plant parts also makes the taxonomic study of cultivated plants a difficult task (Heiser 1969). Lack of precise knowledge and information on the actual centres of origin, wild ancestry, wild relatives and progenitors of the present day cultivars is also a major stumbling block in conducting detailed taxonomic studies on cultivated plants. Hybridization and introgression are other major constraints in cultivated plant taxonomy. Occurrence of polyploidy (autoploids and alloploids), as reported in many crop – weed complexes, is another important aspect to be closely examined in the context of cultivated plants. Taxonomy of cultivated plants, therefore, requires special attention and treatment (Harlan and de Wet 1971).

4. Species Concepts and Methodologies in Cultivated Plant Taxonomy

There are as many as 22 different species concepts used in taxonomy (Mayden 1997). The six major classes of species concepts used in plant taxonomy are the (1) Morphological species concept, (2) Biological species concept, (3) Ecological species concept, (4) Cladistic species concept, (5) Eclectic species concept, and (6) Nominalistic species concept (see Stuessy 1990; Spooner *et al.* 2003; Wilkins 2003, 2011 for detailed discussion and examples of these species concepts). A synopsis of the widely discussed and debated species concepts in contemporary taxonomy is presented in Table 2.

The methodologies applied in cultivated plant taxonomy do not differ from those used in systematic and taxonomic studies of wild plants. However, the complexity in origin, diversification through domestication, discontinuous evolutionary mechanisms, and other peculiarities in crop plants do not justify that the same philosophy of species concepts and methodological frameworks be applied for the taxonomic and systematic study of cultivated and wild plants with uniform standards. The often–used morphological or taxonomic species concept is the widely applied criterion for collection and preliminary identification of germplasms of cultivated plants. But morphology is not enough by itself to be used in open, objective and accurate taxonomic decisions in cultivated plants. The biological species concept, which takes into account the degree of gene exchange and reproductive isolation within and amongst populations, is also

Table 2. Major Species Concepts in Taxonomy

Concept	Definition	Proponents
Morphological Species Concept	Species defined on morphological or anatomical characters.	Linnaean taxonomy school/ traditional taxonomic revisions, Monographs, Floras
Taxonomic Species Concept	Species are the smallest groups that are consistently and persistently distinct, and distinguishable by ordinary means.	Cronquist 1978
Biological Species Concept	Originally proposed by Ernst Mayr in 1940, and subsequently redefined in 1963 and 1970:	
	Species as group of actually or potentially interbreeding natural populations, which are reproductively isolated from such other groups.	Mayr (1942)
	Species is a reproductive community, an ecological unit and a genetic unit.	Mayr (1963)
	Species are group of interbreeding natural populations that are reproductively isolated from such other species (Interbreeding Species Concept).	Mayr (1970)
Evolutionary Species Concept	An evolutionary species (Evospecies) is a lineage (ancestral-descendant sequences evolving separately from others with its own unitary evolutionary role and tendencies.	Simpson (1961)
Genetic Species Concept	Species is the most inclusive group of organisms having the potential for genetic and/or demographic exchangeability.	Templeton (1985)
Ecological Species Concept	Species as a lineage or closely related set of lineages, which occupies an adaptive zone minimally different from that of any other lineage in its range and which evolves separately from other lineages outside its range.	Van Valen (1976)

(*Contd.*)

Concept	Definition	Proponents
Cladistic/Phylogenetic Species Concept	Species delimitation based on phylogenetic history/ cladistic criteria (irrespective of morphology, interbreeding behaviour, or ecological considerations, except as they may be used for reconstructing phylogenetic history). Species is defined based on the concept of monophyly, and apomorphies (unique characters that are not shared by other groups). If one or more taxa share an apomorphy, it is called synapomorphy and if only one taxon carries an apomorphy , it is termed as autapomorphy. An apomorphy shared by only one set of taxa is called plesiopmorphy.	Hennig (1966)
Eclectic Species Concept	Species are defined, formed and maintained by a variety of biological factors, including morpho logical, interbreeding, ecological and phylogenetic factors.	Doyden & Slobobchikoff (1974) Templeton (1989): Cohesive Species concept Ereshefsky (2000): Pluralistic Species View
Nominalistic Species Concept	A philosophy that questions the very existence of species, and believes that individuals or interbreeding populations are the only population system with an objective reality. Species are Abstract Fictions.	Burma (1954)
Other Species Concepts		
Agamospecies	Includes asexual taxa resulting from parthenogenesis in animals and apomixes in plants.	Cain (1954)
Nothospecies	Species formed from hybridization of two sexual species.	Wagner (1983)

(Contd.)

Concept	Definition	Proponents
Compilospecies	Species that plunders genetic resources of other species through introgressive hybridization. This category includes genetically aggressive, highly polymorphic species mostly of hybrid origin, with more than one ploidy level, very weedy, obscuring other species boundaries.	Harlan and De Wet (1963)
Phenospecies/OTUs	A concept and approach involving grouping of species based on phenetic similarity between the individuals and on phenotypic gaps. This implies that species are not unique and represent Operational Taxonomic Units (OTUs). A concept in numerical taxonomy and applied in multi-character evaluation of OTUs.	Sneath and Sokal (1973)

found deceptive in many cases of cultivated plant species (Baker 1970). Jeffery (1968), therefore, proposed that the formal botanical categories do not work well at infraspecific levels in cultivated plants.

Harlan and de Wet (1971) pointed out that the lack of understanding of the special modes of origin and different selection pressures operating in cultivated pants often tempts the traditional taxonomists to over classify the crop plants. Harlan and de Wet (1971) provided a genetic perspective and genetic focus for cultivated plant classification. They proposed three informal categories of (1) primary gene pool-GP1, (2) secondary gene pool-GP2, and (3) tertiary gene pool- GP3. These categories were identified on the basis of the degree of crossability, fertility of hybrids, chromosome pairing, and gene transfer within and among cultivars, spontaneous races, and close or remote wild relatives of the crop plant (Fig. 1).

Although Harlan and De Wet's scheme seems applicable to plant genetic resources, including crop-weed complexes, it cannot be applied to many of the present- day cultivars that have been evolved through crossing several species. Such cultivars cannot be classified as a part of any single species (Hetterschied *et al.* 1996). Inadequate knowledge on the mode and directions of crossing among crop and weed populations is also a handicap in applying Harlan and de Wet's gene pool concept as a fool proof method in many crop plant taxa.

5. Multidisciplinary Approaches in Cultivated Plant Taxonomy

The approach to crop plant taxonomy has changed drastically with the application of modern techniques and methods in molecular systematics, cladistics and population genetics, which eventually paved the ways for adoption of phylogenetic or cladistics, biosystematic or ecological species concepts in crop plant classification (Stuessy 1990; Spooner *et al.* 2003).

Molecular taxonomic/systematic approaches will be critically important in understanding the taxonomy of crop-complexes, featured by the problems of apomixis, inter and intraspecific hybridization, genetic introgression, polyploidy, etc. Molecular data, especially DNA/RNA sequence data have been useful in elucidating origin and relationships among crops and their weedy relatives in several plant genera of cultivated plants (Weising *et al.* 2005). Molecular genetic data provide sufficient insights into unravelling the pattern and process of geographic distribution, origin, adaptive radiation, and population bottlenecks within and among species, based on geographically significant genetic signals (e.g. Demesure *et al.* 1996; Dumoline *et al.* 1997; Grivet and Petit 2002).

Phylogeography is yet another modern approach that aims at studying the principles and historical process governing the geographic distribution of individuals of species/ populations, using geographic distributional and gene genealogy data (Avise 2000).

Phylogeographic studies can be based on information from nuclear, mtDNA and cpDNA. The possibility of conducting nuclear phylogeographic studies in plant species was first demonstrated by Olsen and Schaal (1999), who sequenced 962 bp of the single copy nuclear gene *G3pdh* in the cultivated cassava (*Manihot esculenta)* and its wild relatives, *M.esculenta* ssp. *flabellifolia* and *M.pruinosa*. Phylogeography is also significant in understanding the pattern and distribution of genetic diversity in wild and cultivated/domesticated populations of a plant/ animal species and thereby to trace its geographic/genetic origin from gene geneology data.

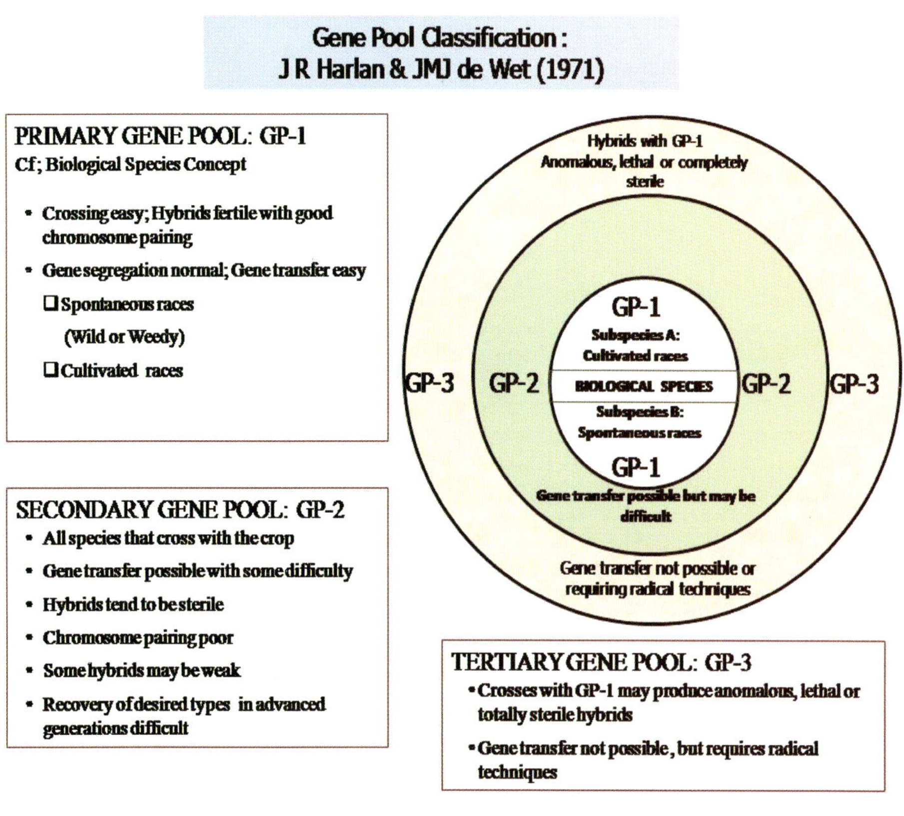

Fig. 1. Harlan and de Wet's gene pool classification of crop plant and wild relatives

Box 1. Evolution of wheat

Wild diploid wheat (*Triticum urartu*, 2n = 2x = 14, genome AuAu) hybridized with the B genome ancestor that is the closest relative of goat grass (*Aegilops speltoides*, 2n = 2x = 14, genome SS) 300,000–500,000 years before present (BP) to produce wild emmer wheat (*T. dicoccoides,* 2n = 4x = 28, *genome AuAuBB*). Cultivation of wild emmer by hunter-gatherers about 10,000 BP, and a subconscious selection gradually created a cultivated emmer (*T. dicoccum*, 2n = 4x = 28, genome AuAuBB) that spontaneously hybridized with another goat grass (*Aegilopsis tauschii*, 2n = 2x = 14, genome DD) (around 9,000 BP) and produced an early spelt (*T. spelta*, 2n = 6x = 42, genome AuAuBBDD). About 8,500 BP, natural mutation changed the ears of both emmer and spelt to a more easily threshed type that later evolved into the free-threshing ears of durum wheat (*T. durum*) and bread wheat (*T. aestivum*). Durum wheat (*T. durum*) derived from *T. dicoccum* is a free-threshing naked wheat and is widely cultivated today for pasta production.

Source: Peng *et al.* (2011)

There are many examples of multidisciplinary studies on the origin, domestication pathways and crop plant taxonomy (Weising *et al.* 2005; Hammer and Morimoto 2011; Meyer *et al.* 2012). As an example, the origin of hexaploid wheat as analysed by Peng *et al.* (2011) is summarized in Box1, Fig. 2.

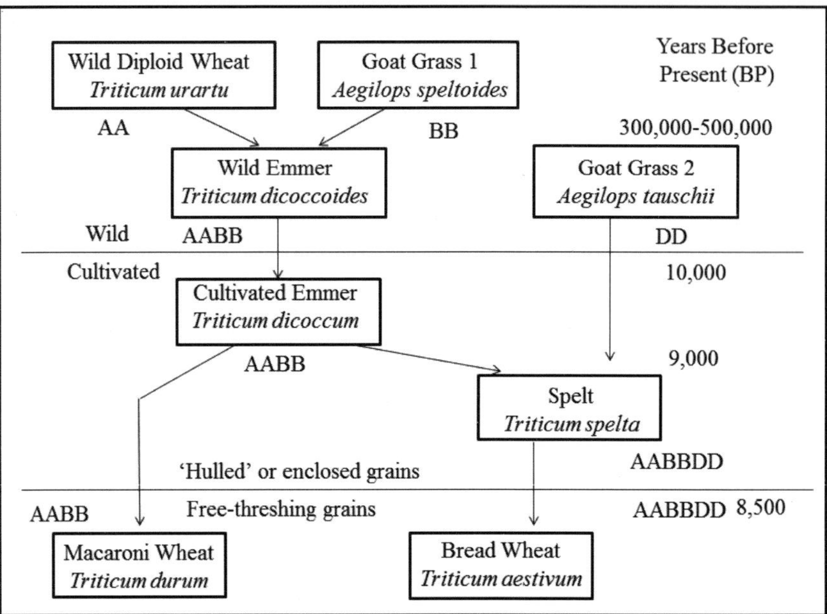

Fig. 2. Evolution of wheat from the prehistoric Stone Age grasses to modern macaroni and bread wheat; *Source*: http://www.newwhallmill.org.uk/wht-evol.htm; Peng *et al.* (2011).

6. Citrus: An Example of a Crop with a Complex Origin and Taxonomy

The Citrus fruit classification presents a chequered history, ever since the domestication and dispersion of the edible citrus fruits across the world (Mabberely 2004). The genus *Citrus* presents a typical example for the intricacies in unravelling the origin, evolution and classification of crop plants having complicated life history traits and reproduction. Citrus fruits in cultivation and commerce encompass a large complex array of cultivars, natural apomicts, basal species, hybrids, bud sports, and a few wild species. A brief on the complex taxonomy of *Citrus* is presented in Box 2, Table 3 and 4, Fig. 3 and 4.

Box 2. Taxonomy of Citrus Fruits

The genus *Citrus* L. belongs to the family Rutaceae. It includes some of the major fruit crops of the world, such as the citrons, lemons, limes, mandarins, sour oranges, sweet oranges, pummelos, grapefruits, kumquats, etc. *Citrus* is believed to have its primary centre of origin in south and south-east Asia, particularly in the region extending from northeast India, eastward through the Malayan Archipelago to China and Japan, and southward to Australia.

The taxonomy of *Citrus* and allied genera (*Eremocitrus* Swingle, *Fortunella* Swingle, *Microcitrus* Swingle and *Poncirus* Raf.) has been complicated by several factors, such as hybridization (intrageneric and intergeneric), apomixis (adventitious nucellar polyembryony), polyploidy, and bud mutations. Consequently, the genus has been circumscribed variously in different classification systems with the number of recognized species varying from 16 (Swingle and Reece 1967), 20–25 (Mabberley 2008) to 162 (Tanaka 1977).

Two principal systems of *Citrus* taxonomy in current use are that of Swingle (1943–revised by Reece in 1967) and of Tanaka (1977). Swingle's system recognized 16 species under two subgenera: 10 under subgenus *Citrus* and six under subgenus *Papeda*, while Tanaka (1977) included a total of 162 species under two subgenera viz. *Archicitrus* and *Metacitrus*. Both the systems differed mainly in their basic concepts. Swingle (1943) did not accord species status to bud sports and hybrids, whereas Tanaka (1977) accepted most of the hybrids, cultivars, and bud sports as true botanical species. Subsequent classifications on *Citrus* by various other workers were modification of either Swingle's or Tanaka's systems (e.g. Hodgson 1965; Singh 1967; Singh and Nath 1969).

Barrett and Rhodes (1976) and Scora (1988) considered that *Citrus medica* L. (citron), *C. maxima* (Burm.) Merr. (omelo) and *C. reticulata* Blanco (mandarin) constituted the three basal species of edible citrus and frequent hybridization and apomictic reproduction among these three species produced several other unique biotypes, including *C. aurantiifolia* (Christm.) Swingle (lime), *C. aurantium* L. (Sour orange), *C. limon* (L.) Osb. (Lemon), *C. paradisi* Macdf. (Grapefruit) and *C. sinensis*

L. (Sweet orange) (Fig. 3). The three basal species in edible citrus was supported by several molecular studies. Bayer *et al.* (2009) in a molecular study of nine cpDNA genes or spacers concluded that *Citrus sensu lato* also included the related genera viz. *Clymenia*, *Eremocitrus*, *Feroniella*, *Microcitrus*, *Oxanthera* and *Poncirus*

The origin of sweet orange, sour orange and the grapefruit from reciprocal crosses and back crosses between pomelo (*Citrus maxima*) and mandarin (*Citrus reticulata*) is an interesting example to demonstrate the intricacies in the origin of Citrus fruits and their taxonomic disposition (Fig. 3). The recent report on the whole genome analysis of sweet orange and an associated SSR analysis revealed the hybrid origin of the sweet orange (Xu *et al.* 2012) (Fig. 4). Today there are many named and unnamed varieties of sweet oranges and grape fruits in horticulture and trade. Morphological descriptors alone will not be sufficient to classify the innumerable Citrus varieties either in the sense of the formal ICN -based classification or the informal ICNCP –based classification.

Table 3. Tanaka's (1977) classification of the genus *Citrus*

Genus	Subgenus	Section	Subsection
Citrus	*Archicitrus*	Papeda	
		Limonellus	
		Citrophorum	Citrioides
			Limoniodes
			Documenoides
		Cephalocitrus	Decumana
			Intermedia
		Aurantium	Medioglobossa
			Aurantioides
			Sinensioides
			Osmocitrioides
	Metacitrus	Osmocitrus	Euosmocitrus
			Pseudoacrumen
		Acrumen	Euacrumen
			Microacrumen
		Pseudofortunella	

Table 4. Citrus classification by Swingle (1943), Revised by Swingle and Reece (1967)

GENUS CITRUS		
Subgenus Papeda		**Subgenus Citrus**
Section *Papeda*	Section *Papedocitrus*	*C.medica*
C. micrantha	*C. ichangensis*	*C.limon*
C. celebica	*C. latipes*	*C.aurantiifolia*
C. macroptera		*C.aurantium*
Citrus hystrix		*C.sinensis*
		C.reticulata
		C.grandis (= *C.maxima*)
		C.paradisi
		C.indica
		C.tachibana

Fig. 3. The origins of oranges, grape fruit, lemons and bergamot (sour orange)
Source: Mabberley (2004).

Fig. 4a. Origin of sweet orange from reciprocal cross between pomelo and mandarin
Source: Xu *et al.* (2012)

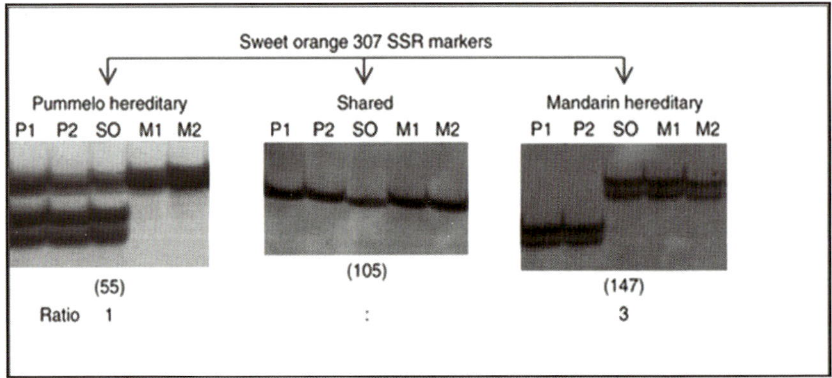

Fig. 4b. SSR marker genotyping of two pummelo cultivars (P1 and P2), sweet orange (SO) and two mandarin cultivars (M1 and M2). Of the 307 SSR markers, 105 were shared among pummelo, orange and mandarin (middle), 55 were pummelo hereditary (left) and 147 were mandarin hereditary (right). *Source*: Xu *et al.* (2012)

7. International Code of Nomenclature for Cultivated Plants (ICNCP)

The complex nature of variations and diverse mode of origins of cultigens warrant a separate set of rules of nomenclature for cultivated plants. The nomenclature of cultivated plants was governed by the provisions of the then International Code of Botanical Nomenclature (now the ICN for algae, fungi and plants; McNeill *et al.* 2012) until a new Code, the International Code of Nomenclature for Cultivated Plants (ICNCP) came into force in 1953 (Stearn 1953). The first *Cultivated Plant Code* (Wageningen), which was published in 1953, has been followed by eight subsequent editions – in 1958 (Utrecht), 1961 (update of 1958), 1969 (Edinburgh), 1980 (Seattle), 1995 (Edinburgh), 2004 (Toronto) and 2009 (Wageningen).

Amendments to the *Cultivated Plant Code* are prompted by international symposia for cultivated plant taxonomy which allow for rulings made by the *International Commission on the Nomenclature of Cultivated Plants.* Since 1985 (Wageningen Symposium), six such international symposia have been organised, the last one held at Beijing during 15-19 July 2013. Each new *Cultivated Plant Code* includes a summary of the changes made to the previous versions.

The ICN provides for naming of all plants, whether in wild or cultivated, in a nested hierarchy of ranks (known as taxon or taxa), whereas the ICNCP does not require any hierarchical categories, but provides rules for cultivated plants or cultigens (cultivar, Group, grex; the grex applied only to orchids).

7.1. ICNCP 2009

ICNCP 2009 (Brickell *et al.* 2009) is the current and eighth edition adopted by International Union of Biological Sciences (IUBS) – International Commission for the Nomenclature of Cultivated Plants. The principal aim of the ICNCP is to provide uniformity, accuracy, and stability in naming agricultural, forestry and horticultural plants. The Rules and Recommendations of this Code are applicable to all organisms traditionally treated as plants (including algae and fungi), and whose origin or selection is primarily due to intentional human activity.

ICNCP-2009 replaces the previous ICNCP-2004 with as many as 59 novel provisions included under Divisions II–VI. The major changes in the current Code includes: (a) the reinstatement of the term 'taxon' to replace the phrase 'distinguishable group of plants' throughout the text of the Code; (b) Recognition of a third category 'grex' in addition to 'cultivar' and 'Group', although the category 'grex' is applied only to orchids.

This change permits the possible recognition of character-based Groups within a parent-based grex; (c) introduction of changes as to naming, formation and application of the Rules and Recommendations for the categories of cultivar, Group and grex under Articles 27 and 29.

7.2. Outline of ICNCP 2009

The essential Rules and Recommendations governing the nomenclature of cultivated plants are included in the Code in six divisions in the following order: Division I-Principles; Division II-Rules and Recommendations [6 Chapters, 35 Articles]; Division III-Names of hybrid genera; Division IV-Registration of names; Division V-Nomenclatural standards; Division VI-Modifications of the Code. The Code begins with a Preamble that outlines the statements on the philosophy and purpose of the various provisions included in the Cultivated Plant Code. There are eleven appendices (Appendices-I-XI) annexed to the Code, which include (i) Directory of International Cultivar Registration Authorities (ICARs: www.ishs.org/icra.index.htm), (ii) Directory of Statutory Plant Registration Authorities (www.upov.int), (iii) Places Maintaining Nomenclatural Standards, (iv) Libraries Holding Significant Collections of Nursery Catalogues, (v) Special Denomination Classes, (vi) List of Conserved and Rejected Epithets, (vii) The Latin Names of Plants, (viii) The Nomenclatural Filter, (ix) Quick Guide for New Cultivar names, (x) Trade Designations, (xi) Flow Chart of Nomenclatural Bodies and processes.

7.2.1. Division I: Principles

This Division includes statements on 11 principles, which form the very foundation of this Code for naming plants in cultivation. The following are the essence of the principles 1-11 as stated in the Code:

Principle 1: ICNCP aims to promote uniformity, accuracy and stability in the naming of agricultural, forestry and horticultural plants.

Principle 2: ICN governs the names in Latin form of plants, except for the generic names of intergeneric graft-chimaeras, which are entirely governed by ICNCP; Taxa whose origin or selection is primarily due to intentional human activity may be given names as per provisions of ICNCP; With the exception of any Latin component with their names (governed by ICN), the nomenclature of names in the categories of cultivar, Group and grex is governed by ICNCP alone.

Principle 3: Naming of taxa under ICNCP to be governed by the rule of priority of publication.

Principle 4: Names of plants governed under ICNCP must be universally and freely available for use by any person to denote a taxon.

Principle 5: Regulation of terminology to be applied to be used for the categories of taxa governed by ICNCP and the names to be applied to those individual taxa; Terminologies of names established under National /International Listing or Plant Breeders Rights may take precedence of those established under ICNCP as the Code does not regulate use of names established under the former categories.

Principle 6: Applying Trade designations as marketable devices are not to be recognized under ICNCP.

Principle 7: Common names of plants are not regulated under ICNCP.

Principle 8: Recognition of importance of international registration of cultivar, Group and grex names and publication and promotion of the lists of such names for promoting uniformity, accuracy and stability in the naming of cultivated plants.

Principle 9: Recommendation on selection, preservation and publication of nomenclatural standards for stabilizing the precise application of cultivar and Group names.

Principle 10: Endorsement of application of Rules and Recommendations of the ICNCP by all those responsible for the formation and use of names for cultivated plants.

Principle 11: Provisions of ICNCP are retroactive unless stated otherwise.

7.2.2. Division II: Rules and Recommendations

All the major rules and recommendations governing the naming of cultivated plants are dealt within this Division under nine chapters and 35 articles. Some of the most important articles in these chapters are presented below with suitable examples.

7.2.2.1. Relationship with the ICBN (now ICN) (Article 1)

This article in its three clauses spells out the relationship of ICNCP with ICN in the following ways:

(1) Plants in cultivation may be named in accordance with the ICN at least to the level of genus or to the level of species or below.

(2) Plants in cultivation meeting the criteria of being recognised as cultivars, Groups or grexes may be given epithets in accordance with ICNCP and assigned to a named taxon under the ICN.

(3) Hybrids between taxa, including, if it so wished, those arising from cultivation, to be named as per Appendix I of ICN, or in addition, cultivated plants through hybridization may be named as cultivars, Group, grexes as per ICNCP.

Ex. 1. *Solanum* × *procurrens* is the name under ICN for the hybrid between the European *S. nigrum* and the South American *S. physalifolium* that occurred in cultivation.

Ex. 2. *Hypericum* ×*inodorum* is the name under ICN for the hybrid between *H. androsaemum* and *H. hircinum* when both species meet in wild and cultivation. Cultivar and Group names can be assigned to this hybrid name, if desired.

Ex. 3. The progeny of the repeated cross between *Victoria amazonica* (male) × *V. cruziana* (female) is a cultivar and named as *V.* 'Longwood Hybrid', and another cultivar resulting from the reciprocal cross *Victoria cruziana* (male) × *V. amazonica* (female) is named *V.* 'Adventure' in accordance with the ICNCP.

7.2.2.2. Definitions (Chapter 2, Articles 2-13)

Cultivar (Article 2)

The basic category of cultivated plants whose nomenclature is governed by ICNCP is the *Cultivar*. A cultivar is an assemblage of plants that (a) has been selected for a particular character or combination of characters, (b) is distinct, uniform and stable in these characters, and (c) when propagated by appropriate means, retains those characters (Article 2.3).

Cultivars differ in their mode of origin and reproduction, and whatever the means of propagation, only those plants which maintain the characters of a particular cultivar may be included within that cultivar (Article 2.4). Articles 2.5 - 2.19 describe the provisions for assigning cultivar status for plants with different modes of propagation.

Group (Article3)

The formal category which may comprise cultivars, individual plants, or combinations thereof on the basis of defined character-based similarity is the Group (Article 3.1). Criteria for forming and maintaining a Group vary according to the required purposes of the users. All members of a Group must share the character (s) by which the Group is defined (Article 3.2).

Ex.1. In *Primula,* the cultivars 'Mac Watt's Blue', 'Old Irish Scented', and 'Osborne Green' are the best suited for outdoor cultivation and have been assembled under *Primula* Border Auricula Group.

Ex.2. *Iris* Dutch Group includes the complex of early flowering cultivars arising from *I.tingitana*, *I.xiphium* var. *lusitanica*, and *I.xiphium* var. *praecox*.

Ex.3. The cultivars of *Festuca rubra* have been allocated three Groups: Hexaploid Non- creeping Group, Hexaploid Creeping Group, and Octoploid Creeping Group.

Grex (Article 4)

The formal category for assembling plants based solely on specified parentage is the grex. This is only to be used in orchid nomenclature (Article 4.1).

It may also be noted that (i) in current usage parents of a grex restricted to the rank of species or another grex, and (ii) a grex name applies to a cross and its reciprocal.

Ex. 1. The grex name for the cross *Paphiopedilum* Atlantis grex × *P*. Lucifer grex is *P.* Sorel grex. The latter is also the name for the reciprocal cross.

Graft- Chimaera (Article 5)

A Graft –Chimaera is a plant resulting from grafting the vegetative tissues of two or more plants belonging to different taxa, and is thus not a sexual hybrid. Graft Chimaeras below the rank of genus may be recognized as cultivars (Art. 5.1). (Please see for rules at the rank of genus Art 24; below genus to be recognized as cultivars).

Denomination Class (Article 6)

A denomination class is the unit within which the use of a cultivar, Group or grex epithet may not be duplicated except when the re-use of an epithet is permitted in accordance with Article 30 (Art 6.1). Under the ICNCP, a denomination class is a single genus or hybrid genus unless a special denomination class has been provided by ISHS Commission for Nomenclature for Cultivar Registration (Art 6.2).

Ex .1. *Hibiscus rosasinensis* has been designated as a denomination class. According to the provisions of Art 6.1, the cultivar epithet may not be duplicated in that species, but it can be used once in the remainder of the genus which forms the second denomination class.

Ex.2. The ISHS Commission for Nomenclature and Cultivar Registration designated *Hylocerreae* as a denomination class to designate the genera in the tribe *Hylocerreae* (Cactaceae), which are all freely hybridizing and have an uncertain taxonomic status.

7.2.2.3. Names and epithets (Article 8)

Cultivar and Group Name

In accordance with the provisions of ICNCP, the name of a cultivar or Group consists of a combination of the name of the genus or lower taxon to which it is assigned with a cultivar or Group epithet (Article 8.1). The following examples show that the name of a same cultivar or Group can be written in any of the following ways:

Ex.1. *Fragaria* 'Cambridge Favorite', *Fragaria* × *ananassa* 'Cambridge Favorite', strawberry 'Cambridge Favorite','Cambridge Favorite' strawberry (English), Erdbeere 'Cambridge Favorite'(German), fraiser 'Cambridge Favorite'(French), 'Cambridge Favorite' morangueiro (Portugese).

Ex.2. *Alcea rosea* Chater's Double Group, *Alcea* Chater's Double Group, hollyhock Chater's Double Group (English), rose tremiere Groupe Chater's Double (French),

Stockrose Chaters Doppelte Gruppe (German), stokroos Chaters Dubbele Groep (Dutch).

Grex name

Name of a grex consists of the name of a genus to which it is assigned together with a grex epithet (Article 8.2).

Ex. *Spiranthes* Awful grex , lady's tresses Awful gx (English), schroeforchis Awful grex (Dutch), Drehwurz Awful grex (German).

7.2.2.4. Established name (Article 10)

An established name is one that is in accordance with Article 27 of ICNCP.

7.2.2.5. Accepted names (Article 11)

An accepted name is the earliest one that must be adopted for a cultivar, Group, grex, or the generic name of an intergeneric chimaera (Article 11.1).

If an accepted cultivar name becomes rejected or replaced with a new name by a statutory plant registration authority, or is otherwise replaced by a name designated by such an authority, the earlier accepted name becomes a synonym of the newer name (Article 11.2).

7.2.2.6. Conserved names (Article 12)

A conserved name is one that, although otherwise contrary to the Rules of ICNCP, must be adopted for a cultivar, Group or grex by a ruling of the IUBS International Commission for Nomenclature of Cultivated Plants.

7.2.2.7. Conventions for Presentation of names (Articles 14-18)

Cultivar Status

The status of a cultivar is indicated by enclosing the cultivar epithet within single quotation marks. The usage of abbreviations such as cv. and var. are not to be used within a cultivar name. The quotation mark can be expressed by using (‘) at the beginning and (’) of the cultivar epithet, or by using apostrophe (') or (`) on each side of the epithet.

Ex. *Iris* ‘Cantab’, *Iris* 'Cantab', *Iris* ´Cantab´

Cultivars that are thought to be graft-chimaeras are not to be indicated by the addition sign (+) before the cultivar epithet (Article 14.2)

Syringa ‘Correlata’ is the name for *Syringa* × *chinensis* + *S. vulgaris*, and must not be written as *Syringa* + ‘Correlata’. Similarly cultivars that are thought to be of hybrid origin are not to be indicated by (×) before the cultivar epithet.

Ex.1. *Digitalis* 'Mertonensis' – not Digitalis × 'Mertonensis'.

Ex.2. *Distichis* ' Mrs Rivers' originated from *D. buccinatoria* × *D. laxiflora* must not be written as *Distichis* × ' Mrs Rivers'.

Group status

Formal Group status is indicated by the use of the word Group or its equivalent in another language as the first or final word in the Group epithet (Article 15.1).

Ex. *Begonia* Elatior Group, *Brassica oleracea* Sabellica Gruppe (German), *Hydrangea macrophylla* Groupe Hortensis (French), Tulipa Grupo Darwin (Spanish).

Grex status

This is indicated by the use of the word "grex" or the standard contraction "gx". The epithets of grex are not to be placed in parenthesis. When a grex epithet and a Group epithet are to be cited in the same name, the grex epithet should be followed by "grex" or "gx" to distinguish it from the Group epithet (Articles 16.1-16.3).

Ex.1. *Cymbidium* Alexandri gx 'Westonbirt' – not to be written as *Cymbidium* (Alexandri gx) 'Westonbirt'.

Ex.2. *Paphiopedilum* Sorel grex and *Cymbidium* Alexandri gx can be written as *Paphiopedilum* Sorel and *Cymbidium* Alexandri, respectively.

Ex.3. When Francis Suzuki Group is established within the grex × *Rhyncosohrocattleya* Marie Lemon Stick, the name should be written as × *Rhyncosohrocattleya* Marie Lemon Stick grex Francis Suzuki Group or × *Rhyncosohrocattleya* Marie Lemon Stick gx Francis Suzuki Gp to distinguish the grex and Group epithets.

7.2.2.8. Starting points in nomenclature (Article 18)

Articles 18.1-18.3 of this Code provides the following publications and dates as the starting point for naming cultivar, Group or grex in any denomination class:

(a) A list or publication designated for a denomination class by the ISHS Commission for Nomenclature and Cultivar Registration preferably on application from the relevant International Cultivar Registration Authority, or in the absence of such an authority, in consultation with appropriate organizations.

(b) In the absence of such an approved list or publication, Linnaeus's *Species Plantarum* (both volume 1 and 2) published on 1 May 1753 is considered as the starting point.

(c) The starting point for grex names is 2 January 1858, the date on which publication of the first artificial orchid cross reported in Europe was recorded.

(d) The starting point for generic names for intergeneric chimaeras is Linnaeus's *Species Plantarum* (both volume 1 and 2) published on 1 May 1753.

7.2.2.9. Name of wild plants brought in to cultivation (Article 20)

Plants from wild brought in to cultivation and subsequently not classified as cultivars or Groups retain the names that are applied for the same plants growing in nature.

7.2.2.10. Condition for publication (Article 25)

Publication is effected under this Code by distribution of printed or similarly duplicated materials, including indelible autograph (though sale, exchange or gift) to the general public or at least to botanical, agricultural, forestry and horticultural institutions with libraries accessible to botanists, agriculturists, foresters and horticulturists generally.

The following as per this Code does not constitute effective publication:

(a) Communication of new names at a public meeting , (b) placing of labels in collections or gardens open to the public, (c) issue of microform made from manuscripts, typescripts or other unpublished materials, (d) publication via electronic media (e) publication in confidential trade lists that are not made generally available.

7.2.2.11. Conditions for establishment (Article 27)

In order to establish names under this Code , the names must (a) be published on or after the starting point date for the relevant denomination class, (b) appear in a dated publication, (c) have a form that complies provisions of Art 21-25 for a cultivar; Art 22.4-22. 6 for a Group; Art 23 for a grex; Art 24.3 for a generic name of an intergeneric graft chimaera, (d) for a cultivar or Group published after 1 January 1959, be accompanied by a description or by reference to a previously published description (Article 27.1).

7.2.2.12. Description

As per this Code, a description is a word or one or more words that (a) indicate one or more recognizable character of a cultivar or Group, (b) distinguishes the new cultivar or Group from one whose name has been previously or simultaneously published (Art 27.2).

7.2.2.13. Re-use of epithets (Article 30)

The epithet of a cultivar, Group, or grex must not be re-used within the same denomination class for any other cultivar, Group, or grex unless re-use of the cultivar,

Group, or grex epithet is accepted by an appropriate International Cultivar Registration Authority (Article 30.1).

An International Cultivar Registration Authority may only accept reuse of a cultivar, Group, or grex epithet if that authority is satisfied that the original cultivar, Group, or grex (a) is no longer in cultivation, and (b) has ceased to exist as breeding material, and (c) may not be found in a gene bank, and (d) is not known component in the pedigree of other cultivars, Groups, or grexes, and (e) the name has rarely been used in publications, and (f) reuse is unlikely to cause confusion (Article 30.2)

7.2.2.14. Rejection of names (Article 31)

Names governed by this Code are to be rejected and not to be used if they are contrary to the Rules of this Code, except in cases under articles 11.4- 11.7, 19.1 and 29.2 -29.3 (Article 31.1).

The name of a cultivar accepted and published by a statutory plant registration authority, even if using alternative terms must not be rejected (Article 31.2).

Name of a cultivar, Group, or grex is to be rejected if its publication is against the expressed wish of its raiser or breeder (Article 31.4), and also when a name has been rejected and replacement name provided under previous Codes, such name should be rejected in accordance with this Code (Article 31.5).

7.2.2.15. Citation of author's names

It is not necessary to cite the name of the author who has established name governed by this Code. However, if the author citation is desired for the name of a cultivar, Group, or grex, the name of the author may be placed following the epithet of the cultivar, Group, or grex, and in case of generic name of intergeneric chimaeras, the author name is to be placed following the generic name.

7.2.2.16. Nomenclatural standards

A nomenclatural standard as governed by this Code is preferably a herbarium specimen to which the name of the cultivar or Group is permanently attached. It is more or less equivalent to the Nomenclature Type designated for a name as per the ICN. An image, other than one maintained digitally, can be designated as the Nomenclatural Standard when a specimen is not made available or when essential characteristics best recognized from a suitable illustration. The specimen of new cultivar or Group, either living or dried, should be sent to the appropriate International Cultivar Registration Authority or to a public herbarium that maintains collections of nomenclatural standards along with any colored photographs, illustrations of the specimen.

Grex names have no nomenclatural standard as they are solely based on statement of their parentage.

7.2.2.17. Names of hybrids

The hybrids between taxa, including those arising from cultivation, are to be named as per Appendix I of ICN, or in addition, cultivated plants through hybridization may be named as cultivars, Group, grexes as per ICNCP. Division III of ICNCP provides a summary of the provisions for naming intergeneric hybrids as outlined in ICN Appendix 1.

The Appendix 1 of ICN includes 12 articles on the rules governing the naming of hybrids. The most important articles in Appendix are reproduced here with examples.

Article H.1

H.1.1 Hybridity is indicated by use of the multiplication sign × or by addition of the prefix "notho-" to the term denoting the rank of the taxon.

Article H.2: hybrid formula

H.2.1 A hybrid between named taxa may be indicated by placing the multiplication sign between the names of the taxa; the whole expression is then called a hybrid formula.

Ex.1. *Agrostis* L. × *Polypogon* Desf.; *Agrostis stolonifera* L. × *Polypogon monspeliensis* (L.) Desf., *Mentha aquatica* L. × *M. arvensis* L. × *M. spicata* L.; *Polypodium vulgare* subsp. *prionodes* (Asch.) Rothm. × *P. vulgare* L. subsp. *vulgare;*

It is usually preferable to place the names or epithets in a formula in alphabetical order. The direction of a cross may be indicated by including the sexual symbols ♀: female; ♂: male) in the formula, or by placing the female parent first. If a non-alphabetical sequence is used, its basis should be clearly indicated (H.2A.1).

Article H.3: Hybrids arising from two or more taxa

H.3.1. Hybrids between representatives of two or more taxa may receive a name. For nomenclatural purposes, the hybrid nature of a taxon is indicated by placing the multiplication sign × before the name of an intergeneric hybrid or before the epithet in the name of an interspecific hybrid, or by prefixing the term "notho-" (optionally abbreviated "n-") to the term denoting the rank of the taxon. All such taxa are designated nothotaxa.

Ex.1. *Agropogon* P. Fourn. (1934); × *Agropogon littoralis* (Sm.) C. E. Hubb. (1946); *Melampsora* × *columbiana* G. Newc. (2000); *Mentha* × *smithiana* R. A. Graham (1949); *Polypodium vulgare* nothosubsp. [or nsubsp.] *mantoniae* (Rothm.) Schidlay (in Futák, Fl. Slov. 2: 225. 1966).

Article H.6: Names of intergeneric Hybrids

H.6.1 A nothogeneric name (i.e. the name at generic rank for a hybrid between representatives of two or more genera) is a condensed formula or is equivalent to a condensed formula (but see Art. 11.9).

H.6.2 The nothogeneric name of a bigeneric hybrid is a condensed formula in which the names adopted for the parental genera are combined into a single word, using the first part or the whole of one, the last part or the whole of the other (but not the whole of both) and, optionally, a connecting vowel.

Ex.1. ×*Agropogon* P. Fourn. (1934) (*Agrostis* L. × *Polypogon* Desf.); ×*Gymnanacamptis* Asch. & Graebn. (1907) (*Anacamptis* Rich. × *Gymnadenia* R. Br.); × *Cupressocyparis* Dallim. (1938) (*Chamaecyparis* Spach × *Cupressus* L.); ×*Seleniphyllum* G. D. Rowley (1962) (*Epiphyllum* Haw. × *Selenicereus* (A. Berger) Britton & Rose).

Ex.2. ×*Amarcrinum* Coutts (1925) is correct for *Amaryllis* L. × *Crinum* L., not " ×*Crindonna*". The latter formula was proposed by Ragionieri (1921) for the same nothogenus, but was formed from the generic name adopted for one parent *(Crinum)* and a synonym (*Belladonna* Sweet) of the generic name adopted for the other *(Amaryllis)*. Being contrary to Art. H.6, it is not validly published under Art. 32.1(c).

Ex.3. The name ×*Leucadenia* Schltr. (1919) is correct for *Leucorchis* E. Mey. × *Gymnadenia* R. Br., but if the generic name *Pseudorchis* Ség. is adopted instead of *Leucorchis,* ×*Pseudadenia* P. F. Hunt (1971) is correct.

Ex.4. Boivin (1967) published ×*Maltea* for what he considered to be the intergeneric hybrid *Phippsia* (Trin.) R. Br. × *Puccinellia* Parl. As this is not a condensed formula, the name cannot be used for that intergeneric hybrid, for which the correct name is ×*Pucciphippsia* Tzvelev (1971). Boivin did, however, provide a Latin description and designate a type; consequently, *Maltea* B. Boivin is a validly published generic name and is correct if its type is treated as belonging to a separate genus, not to a nothogenus.

H.6.3 The nothogeneric name of an intergeneric hybrid derived from four or more genera is formed from the name of a person to which is added the termination *ara;* no such name may exceed eight syllables. Such a name is regarded as a condensed formula.

Ex.5. ×*Beallara* Moir (1970) (*Brassia* R. Br. × *Cochlioda* Lindl. × *Miltonia* Lindl. × *Odontoglossum* Kunth).

When a nothogeneric name is formed from the name of a person by adding the termination *ara,* that person should preferably be a collector, grower, or student of the group (H.6A.1).

Article H.7: Name of hybrids between subdivisions of a genus

H.7.1 The name of a nothotaxon that is a hybrid between subdivisions of a genus is a combination of an epithet, which is a condensed formula formed in the same way as a nothogeneric name (Art. H.6.2-4), with the name of the genus.

Ex.1. *Ptilostemon* nothosect. *Platon* Greuter (in Boissiera 22: 159. 1973), comprising hybrids between *P.* sect. *Platyrhaphium* Greuter and *P.* sect. *Ptilostemon; P.* nothosect. *Plinia* Greuter (in Boissiera 22: 158. 1973), comprising hybrids between *P.* sect. *Cassinia* Greuter *and P. sect. Platyrhaphium.*

Article H.9: Valid publication of nothogenus or nothotaxon

H.9.1 In order to be validly published, the name of a nothogenus or of a nothotaxon with the rank of subdivision of a genus (Art. H.6 and H.7) must be effectively published (see Art. 29-31) with a statement of the names of the parent genera or subdivisions of genera, but no description or diagnosis is necessary, whether in Latin, English, or any other language.

Ex.1. Validly published names: ×*Philageria* Mast. (1872), published with a statement of parentage, *Lapageria* Ruiz & Pav. × *Philesia* Comm. ex Juss.; *Eryngium* nothosect. *Alpestria* Burdet & Miege (pro sect.) (in Candollea 23: 116. 1968), published with a statement of its parentage, E. sect. *Alpina* H. Wolff × E. sect. *Campestria* H. Wolff; ×*Agrohordeum* E. G. Camus ex A. Camus (1927), published with a statement of parentage, *Agropyron* Gaertn. × *Hordeum* L., and its later synonym ×*Hordeopyron* Simonet (1935, *"Hordeopyrum"*; see Art. 32.2), published with an identical statement of parentage.

Note 1. Since the names of nothogenera and nothotaxa with the rank of a subdivision of a genus are condensed formulae or treated as such, they do not have types.

Article H.10: Names of nothotaxa at species or infraspecific ranks

H.10.1 Names of nothotaxa at the rank of species or below must conform with the provisions (a) in the body of the Code applicable to names in the same ranks (see Art. 32.4) and (b) in Art. H.3. Infringements of Art. H.3.1 are treated as errors to be corrected (see also Art. 11.9).

Ex.1. The nothospecific name *Melampsora* ×*columbiana* G. Newc. (in Mycol. Res. 104: 271. 2000) was validly published, with a Latin description and designation of a holotype, for the hybrid between *M. medusae* Thüm. and *M. occidentalis* H. S. Jacks.

Note 1. Taxa previously published as species or infraspecific taxa that are later considered to be nothotaxa may be indicated as such, without change of rank, in conformity with Art. 3 and 4 and by the application of Art. 50 (which also operates in the reverse direction).

Article H.11: Name of nothospecies with parents from different genera

H.11.1 The name of a nothospecies of which the postulated or known parent species belong to different genera is a combination of a nothogeneric name with a nothospecific epithet.

Ex.1. × *Heucherella tiarelloides* (Lemoine & É. Lemoine) H. R. Wehrh. is considered to have originated from the cross between a garden hybrid of *Heuchera* L. and *Tiarella cordifolia* L. (see Stearn in Bot. Mag. 165: ad t. 31. 1948). Its basionym, *Heuchera* ×*tiarelloides* Lemoine & É. Lemoine (1912), is therefore incorrect.

H.11.2 The final epithet in the name of an infraspecific nothotaxon of which the postulated or known parental taxa are assigned to different species may be placed subordinate to the name of a nothospecies (but see Rec. H.10B).

Ex.2. *Mentha* × *piperita* L. nothosubsp. *piperita* (*M. aquatica* L. × *M. spicata* L. subsp. *spicata*); *M.* ×*piperita* nothosubsp. *pyramidalis* (Ten.) Harley (in Kew Bull. 37: 604. 1983) (*M. aquatica* L. × *M. spicata* subsp. *tomentosa* (Briq.) Harley).

Article H.12: Names of subordinate taxa within nothospecies

H.12.1 Subordinate taxa within nothospecies may be recognized without an obligation to specify parent taxa at the subordinate rank. In this case non-hybrid infraspecific categories of the appropriate rank are used.

Ex.1. *Mentha* × *piperita* f. *hirsuta* Sole; *Populus* × *canadensis* var. *serotina* (R. Hartig) Rehder and *P.* ×*canadensis* var. *marilandica* (Poir.) Rehder (see also Art. H.4 Note 1).

Note 1. When there is no statement of parentage, Art. H.4 and H.5, governing the circumscription and appropriate rank of hybrid taxa, do not apply.

Note 2. Art. H.11.2 and H.12.1 cannot both be applied simultaneously at the same infraspecific rank.

H.12.2 Names published at the rank of nothomorph are treated as having been published as names of varieties (see Art. 50).

8. Nomenclatural Discrepancies in Cultivated and Wild Plants

Cultivated plant nomenclature is still linked to the provisions of ICN. The main differences of ICNCP from those of ICN are that in the ICNCP (i) 'cultivar' is the basal unit and cannot be subdivided (ii) the cultivar names and their circumscriptions are fixed to standards (not by types) (iii) the basic binomial consists of a (notho) genus name plus a cultivar epithet (iv) the use of denomination class as an extra nomenclature device, and (v) reuse of cultivar names in certain cases (see Spooner *et al.* 2003; ICNCP-Brickett *et al.* 2009). The retention of Hybrid Appendix in ICN, which is applicable to ICNCP, is a matter of ongoing debate among specialists of the Botanical Code and Cultivated Plant Code (Hetterschard and Brandenburg 1995; Spooner *et al.* 2003; McNeill 2004 2012).

Table 5. Differences between ICN and ICNCP

International Code of Nomenclature for Algae, Fungi and Plants (ICN)	International Code of Nomenclature for Cultivated Plants (ICNCP)
Nomenclature rules for taxa (group proposed on the basis of evolutionary classification criteria)	Nomenclature rules for cultivated plants (man-made entities)
Exclusively devised for objects classified in a closed classification system	Exclusively devised for objects classified in an open classification system
A potentially infinite number of categories	A limited number of categories, presently the cultivar and cultivar-group
Categories are not defined	The cultivar is defined
No basal rank	The cultivar is the basal unit and cannot be subdivided
Names are fixed to types	A cultivar's name and circumscription are fixed to standards
Basic binomial consists of a genus name plus a species epithet	Basic binomial consists of a (notho-) genus name plus a cultivar epithet
No nomenclature devisces apart from the ranked categories	The denomination class as an extra nomenclature device
Reuse of names forbidden (homonymy)	Reuse of names allowed in certain cases

(*Source*: Spooner *et al.* 2003)

9. The Way Ahead

Modern taxonomic studies on crop plants demand a combination of different methods and strategies aimed at a practical special purpose classification and nomenclature of cultigens, inference on their mode of origin and pattern of variations within and among the crop gene pools, actual centres of origin and /or diversity, and a multiuser-oriented, global taxonomic information retrieval systems and databases of crop species and their gene banks and *ex situ* and *in situ* genetic reservoirs.

10. Important Publications and Websites on Cultivated Plants

Food and Agricultural Organization of the United Nations (FAO) for basic information on crop plants statistics. www.fao.org.

Gardenweb for Horticultural information and horticultural and botanical terms. www.gardenweb.com

Germplasm Resource Information Network (GRIN) for information on US germplasm holdings, nomenclatural lists of economically important plants, noxious weeds, rare and endangered plants and references. www.ars-grin.gov/npgs/tax

Henelt P and Institute of Plant Genetics and Crop Plant Research (eds.) (2001) Mansfeld's encyclopaedia of agricultural and horticultural crops, 6 vols, Springer, Berlin

Hortax – The Horticultural Taxonomy Group- a committee of European plant taxonomists and horticulturists involved in classification and nomenclature of cultivated plants. www.hortax.org.uk

Index herbororiarum for index of all public herbaria in the world. www.nybg.org/bsci/ih/ih.Index

International Association of Plant Taxonomy (IAPT) for information on botanical nomenclature, official publications (Taxon and Regnum Vegietabile), ICN and ICNCP. www.botanik.univie.ac.at/iapt

International Cultivar Registration Authorities (ICRAs) for all lists of ICRAs, and nomenclatural codes. www.ishs.org/sci/icra.htm

International Society for Horticultural Sciences (ISHS) for all horticulture related information. www.ishs.org

Mansfeld's World Database of Agricultural and Horticultural Crops. An online database developed at IPK since 1998, and reflects the contents of "Mansfeld's Encyclopedia of Agricultural and Horticultural Crops" (Hanelt and IPK 2001) and contains information on 6,100 crop plant species, excluding forestry and ornamental plants. www.mansfeld-ipk-gatersleben.de/

Multilingual Multiscript Plant Name Database for an exhaustive list of common names of plants in many languages. www.gmr.landfood.unimelb.edu.au/Plant names/

Royal Horticultural Society (RHS) for information on many aspects of horticulture, including plant finder providing updated information on plants. www.rhs.org.uk

TROPICOS for current information on plant names. www.tropicos.org

Union for the Protection of New Varieties of Plants. www.upov.int

Wiersema JH and Leon B (1999) World Economic Plants: A standard reference. CRC Press, Boca Raton, FL.

References

Alefeld F (1866) Landwirtschaftliche Flora. Brandenburg, W.A., Oost, E.H. & Van den Vooren,Wiegandt & Herpel, Berlin

Andrews S, Leslie A and Alexander C (eds.) (1999) Taxonomy of Cultivated Plants. RBG, Kew, London.

Avise JC (2000) Phylogeography:The history and formation of species. Harvard Univ Press, Cambridge, MA

Bailey LH (1918) The indigen and cultigen. Science 47:306-309

Bailey LH (1923) Various cultigens, and transfers in nomenclature. Gentes Herb 1:113-115

Baker HG (1970) Taxonomy and the biological species concept in cultivated plants. In: Frankel OH, Bennet E (eds) Genetic Resources in Plants- Their exploration and conservation, IBP handbook No.11, Blackwell Scientific Publications, Oxford

Balter M (2007) Multiple Birth: People in many different parts of the world independently began to cultivate and eventually domesticate plants. Science 316:1830-1835

Barrett HC, Rhodes AM (1976) A numerical taxonomic study of affinity relationship in tultivated Citrus and its close relatives. Syst Bot 1:105-136

Baum BR (1981) Taxonomy of the infraspecific variability of cultivated plants. Kulturpflanze 29:209-239

Bayer RJ, Mabberley DJ, Morton CM, Cathy H, Sharma IK, Pfeil BE, Rich S, Hitchcock R, Sykes S (2009) A molecular phylogeny of the orange subfamily (Rutaceae: Aurantioideae) using nine cp DNA sequences. Am J Bot 96:668–685

Bohs L, Olmstead RG (1997) Phylogenetic relationships in *Solanum* (Solanaceae) based on *ndh*F sequences. Syst Bot 22:5–17

Bohs L, Olmstead RG (1999) *Solanum* phylogeny inferred from chloroplast DNA sequence phylogeny. In: Nee M, Symon DE, Lester RN, Jessop JP (eds.), Solanaceae IV, Advances in Biology and Utilization. Royal Botanic Gardens, Kew. pp 97–110

Brickell CD, Alexander C, David JC, Hetterscheid WLA, Leslie AC, Malecot V, JinX, Cubey JJ (eds.) (2009): International Code of Nomenclature for Cultivated Plants, incorporating the rules and recommendations for naming plants in cultivation adopted by the International Union of Biological Sciences, International Commission for the Nomenclature of Cultivated Plants. 8th ed., Scripta Horticulturae (ISHS) 10:1-184

Brickell CD, Baum BR, HetterscheidWLA, Leslie AC, McNeill J, Trehane, P, Vrugtman, F, Wiersema, JH (eds) (2004) International Code of Nomenclature for Cultivated Plants, Seventh Edition. Acta Horti 647:1–123, i–xxi

Bukasov SM (1978) Systematics of the potato. Trudy po Prikladnoj Botanike Genetike i Selekcii 62:3–35

Burma BH (1954) Reality, existence, and classification: A discussion of the species problem. Madroño 12:193–209

Cain J (1954) Animal species and their evolution, G Bell, London

Child A (1990) A synopsis of *Solanum* subgenus *Potatoe* (G. Don) D'Arcy [*Tuberarium* (Dun.) Bitter (s.l.)]. Feddes Repert 101:209–235

Constance L (1957) Plant Taxonomy in an age of experiment. Am J Bot 44:88-92

Contreras A, Spooner DM (1999) Revision of *Solanum* section *Etuberosum* (subgenus *Potatoe*). In Nee M, Symon DE, Lester RN, Jessop JP (eds.) Solanaceae IV, Advances in Biology and Utilization, Royal Botanic Gardens, Kew, UK, pp 227–245

Cronquist A (1978) Once again, what is a species? In: Romberger JA (ed.), Biosystematics in agriculture, Allenheld, Osman and Company, Montclair, NJ, pp 3–20

De Candolle AP (1867) Lois de la Nomenclature Botanique. Paris

Demesure B, Comps B, Petit RJ (1996) Chloroplast DNA phylogeography of the common beech (*Fagus sylvatica* L. in Europe. Evolution 50: 2515-2520

De Vries IM (1997) Origin and domestication of *Lactuca sativa* L. Genet Resour Crop Evol 44:165–174

DeVries IM, van Raamsdonk LWD (1994) Numerical morphological analysis of lettuce cultivars and species (*Lactuca* sect. *Lactuca*, Asteraceae). Plant Syst Evol 193:125–141

Dirzo R, Raven PH (2003) Global state of biodiversity and loss. Annual Rev Environ and Resour 28:137-167

Dodds KS (1962) Classification of cultivated potatoes. In: Correll DS (ed.) The potato and its wild relatives. Contributions of the Texas Research Foundation, Botanical Studies 4. Renner, Texas, Texas Research Foundation, pp 517–539

Dolealová I, Lebeda A, Tiefenbachová I, Køístková E (2004) Taxonomic reconsideration of some *Lactuca* spp. Germplasm maintained in world genebank collections. In: Davidson CG, Trehane P(Eds) Proc. XXVI IHC – IVth International Symposium on Taxonomy of Cultivated Plants. Acta Hort 634:193-199

Dorofeev VF, Korovina ON (1979) Wheat. Flora of cultivated plants 1. Kolos, Leningrad

Doyden JT, Slobobchikoff CN (1974) An operational approach to species classification. Syst Zool 23:239–247

Dumolin LS, Demesure B, Fineschi S, Le CV, Petit RJ (1997) Phylogeographic structure of white oaks throughout the European continent. Genetics 146: 1475-1487

Ereshefsky M (2001) The poverty of the Linnaean hierarchy. Cambridge Univ Press, Cambridge, UK.

Frietema de Vries FT (1996) Cultivated plants and the wild flora—effect analysis by dispersal codes. Thesis Leiden Univ

Grebenscikov I (1949) Zur morphologische systematischen Einteilung von Zea mais L. unter besonderer Beriicksichtigung der siidbalkanischen Formen. Ziichter 19:302-311

Grivet D, Petit RJ (2002) Phylogeography of the common ivy (*Hedera* sp.) in Europe : genetic differentiation through space and time. Mol Ecol 11: 1351-1362

Hammer K and Morimoto Y (2011) Classifications of infraspecific variation in crop plants. In: GuarinoL, Rao RV and Reid R (eds.) Collecting Plant Genetic Diversity: Technical Guidelines. CAB International, Wallingford, UK, pp 1-15

Hanelt P (1986) Formal and informal classifications of infraspecific variability of cultivated plants-advantages and limitations. In: Styles BT (ed.) Infraspecific Classification of Wild and Cultivated Plants, Clarendon Press, Oxford, pp. 139-156

Hanelt P and Institute of Plant Genetics and Crop Plant Research (eds.) (2001) Mansfeld's encyclopaedia of agricultural and horticultural crops, 6 vols, Springer, Berlin

Harberd DJ (1972) A contribution to the cyto-taxonomy of *Brassica* (Cruciferae) and its allies. Bot J Linn Soc 65:1–23

Harlan JR, de Wet JMJ (1963) The compilospecies concept. Evolution 17:497–501

Harlan JR, ,de Wet JMJ (1971) Toward a rational classification of cultivated plants. Taxon 20:509-517

Hawkes,JG (1990) The potato: evolution, biodiversity and genetic resources, Oxford: Belhaven Press

Heiser Jr CB (1969) Systematics and origin of cultivated plants. Taxon 18:36-45

Helm J (1957) Versuch einer morphologishsystematischen Gliederung der Art Beta vulgaris L. Ziichter 27:203-222

Hennig W (1966) Phylogenetic systematics. Univ Illinois Press, Urbana, IL

Hetterscheid WLA, Brandenburg WA (1995) Culton versus taxon: conceptual issues in cultivated plant systematics. Taxon 44:161-175

Hetterscheid WLA, Van Den Berg RG, Branderburg WA (1996) An annotated history of the principles of cultivated plant classification. Acta Bot Neerl 45:123-134

Hodgson RW (1965) Taxonomy and nomenclature in citrus fruits. In: Krishnamurthi S (ed) Advances in Agricultural Sciences and their applications. Madras Agric J, Madras, pp 317-331

Huamán Z, Spooner DM (2002) Reclassification of landrace populations of cultivated potatoes (*Solanum* sect. *Petota*). Am J Bot 89:947–965

Jeffrey C (1968) Systematic categories of cultivated plants. Taxon 17:109-114

Jirasek V (1958) Taxonomische Kategorien der Kulturpflanzen. Index sem Hart Bot Univ Carol Praha: 9-16 Jirasek V (1961) Evolution of the proposals of taxonomical categories for the classification of cultivated plants. Taxon 10:34-45

John P (1629) Paradisi in sole paradiscus terrestris (retrieved in 1904). Metheun & Co., London, http://www.archive.org/details/paradisiinsolepa00parkrich

Khoshbakht K, Hammer K (2008) How many plant species are cultivated. Genet Resour Crop Evol 55:925-928

Koopman WJM, Guetta E, Clemens C M, Van De Wiel, Vosman B, Van Den Berg RG (1998) Phylogenetic relationships among *Lactuca* (Asteraceae) species and related genera based on ITS-1 DNA Sequences. Am J Bot 85: 1517–1530

Koopman WJM, Zevenbergen MJ, Van Den Berg RG (2001) Species relationships in *Lactuca* s.l. (Lactuceae, Asteraceae) inferred from AFLP fingerprints. Am J Bot 88:1881–1887

Køístková E, Dole•alová I, Lebeda A, Vinter V, Novotná A (2008) Description of morphological characters of lettuce (*Lactuca sativa* L.) genetic resources. Hort Sci 35: 113–129

Lanjouw J, Baehni C, Merrill ED, Rickett HW, Robyns W, Sprague TA Stafleu FA (1952) International Code of Botanical Nomenclature: Adopted by the Seventh International Botanical Congress; Stockholm, July 1950, Regnum Vegetabile 3, Utrecht. International Bureau for Plant Taxonomy of the International Association for Plant Taxonomy, pp 228

Li HL (1974) Plant taxonomy and the origin of cultivated plants. Taxon 23:715-724

Lindqvist K (1960) On the origin of cultivated lettuce. Heriditas 46:320-350

Linnaeus C (1753) Species Plantarum, ed 1, 2 vols. Stockholm

Mabberley DJ (2004) Citrus (Rutaceae) A Review of recent advances in etymology, systematics and medical applications. Blumea 49:481-498

Mabberley DJ (2008) Mabberley's plant-book: A portable dictionary of plants, 3rd ed, Cambridge University Press, Avon, UK

MacKey J (1966) Species relationships in *Triticum*. Hereditas, Suppl. 2:237–276

MacKey J (1968) Relationships in the Triticinae, In: Finland KW, Shepherd KW (eds.), Proc. Third Int. Wheat Genetics Symp., Canberra, Australia. AustralAcad Sci, pp 39–45

MacKey J (1981) Comments on the basic principles of crop taxonomy. Kulturpflanze 29:199–207

Mansfeld R (1953) Zur allgemeinen Systematik der Kulturpflanzen I. Kulturpflanze 1:138-155

Mansfeld R (1954) Zur allgemeinen Systematik der Kulturpflanzen II. Kulturpjlanze 2:130-142

Mansfeld R (1959) Vorlaufiges Verzeichnis Iandwirtschaftlichoder gartnerisch kultivierter Pflazenarten (mit AusschluB von Zierpflazen). Kulterflanze, Beih 2, 659 pp

Mayden RL (1997) A hierarchy of species concepts: The document f the saga of the species problem. In: Claridge MF, *et al.* (eds.) Species: the Units of Biodiversity, Chapman and Hall, New York

Mayr E (1942) Systematics and the origin of species. Columbia UnivPress, New York

Mayr E (1963) Animal species and evolution. Cambridge MA, The Belknap Press, Harvard University Press

Mayr E (1970) Population, species and evolution: an abridgement of animal species and evolution. Cambridge MA, The Belknap Press, Harvard University Press

McNeill J (2004) Nomenclature of Cultivated Plants: a historical botanical standpoint. Acta Hort 634:29-36

McNeill J (2008) The taxonomy of cultivated plants. Acta Hort 799:21-28

McNeill J, Barrie FR, Buck WR, Demoulin V, Greuter W, Hawksworth DL, Herendeen PS, Knapp S, MarholdK, Prado J, Prud'homme van Reine WF, Smith GF, Wiersema JH , Turland, NJ (eds. & comps.) (2012) International Code of Nomenclature for algae, fungi, and plants (Melbourne Code), adopted by the Eighteenth International Botanical Congress Melbourne, Australia, July 2011. Regnum Vegetabile xxx, pp208

Meyer RS, DuVal AE, Jensen HR (2012) Patterns and processes in crop domestication: an historical review and quantitative analysis of 203 global food crops. New Phytologist 196:29-48

Miller P (1724) The gardeners and florists dictionary, or a complete system of horticulture, Vols 1 & 2, London

Miller P (1768) The Gardener's Dictionary. Reprint 1969 (abridged) with an introduction by WT Stearn. Verlag von J. Cramer, New York

Morton AG (1981) History of Botanical sciences: an account of the department of botany from ancient times to the present day. Academic press, London

Nicolosi E (2007) Origin and Taxonomy. In: Khan I (ed) Citrus: Genetics, Breeding and Biotechnology, CAB International, UK & USA, pp 19-43

Olmstead RG, Palmer JD (1992) A chloroplast DNA phylogeny of the Solanaceae: Subfamilial relationships and character evolution. Ann Missouri Bot Gard 79: 346–360

Olmstead RG, Palmer JD (1997) Implications for the phylogeny, classification, and biogeography of *Solanum* from cpDNA restriction site variation. Syst Bot 22:19–29

Olsen KM, Schaal BA (1999) Evidence on the origin of cassava: Phylogeography of *Manihot esculenta* . Proc Natl Acd Sci USA 96: 5586-5591

Ovchinnikova A, Krylova E, Gavrilenko T, Smekalova T, Zhuk M, Knapp S, Spooner DM (2011) Taxonomy of cultivated potatoes (*Solanum* section *Petota*: Solanaceae). Bot J Linn Soc 165:107–155

Palmer J D, Sheilds CR, Cohen DB, Orten TJ (1983) Chloroplast DNA evolution and the origin of *Brassica* species. Theor Appl Genet 65:181–189

Pangalo KI (1948) Novyje principy vnutrividovoj sistematiki kulturnych rastenij [New principles for the infraspecific systematics of cultivated plants]. Bot Zurn 33:151-155 [in Russian]

Peng JH, Sun D, Nevo E (2011) Domestication evolution, genetics and genomics in wheat. Mol Breeding 28:281-301

Peralta IE, Spooner DM (2001) GBBSI gene phylogeny of wild tomatoes (*Solanum* section *Lycopersicon* subsection *Lycopersicon*; Solanaceae). Am J Bot 88:1888–1902

Peralta IE, Spooner DM, Knapp S (2008) The taxonomy of tomatoes: A revision of wild tomatoes (*Solanum* section *Lycopersicon*) and their outgroup relatives in sections *Juglandifolium and Lycopersicoides*. Syst Bot Monographs 84:186, 3 plates

Pickersgill B (1986) Evolution of hierarchical variation patterns under domestication and their taxonomic treatment. In: Styles BT (ed.) Infraspecific Classification of Wild and Cultivated Plants, Clarendon Press, Oxford, pp 191-209

Rehder A (1960) Manual of cultivated trees and shrubs, 2nd ed, Macmillan Publishing Co, New York

Rick CM (1979) Biosystematic studies in *Lycopersicon* and closely related species in *Solanum*. In: Hawkes JG, Lester RN, Skelding AD (eds.), The biology and taxonomy of the Solanaceae. Linn Soc London Symp Ser 7, Academic Press, New York, pp 667–678

Rick CM, Laterrot H, Philouze J (1990) A revised key for the *Lycopersicon* species. Tomato Genet. Coop Rep 40:31

Scora RW (1988) Biochemistry, taxonomy and evolution of modern cultivated Citrus. In: Goren RK, Mendel K (eds) Proc. 6th Int. Citrus Congress, Margraf Scientific Books, Weikesheim, pp 277-289

Simpson GG (1961) Principles of animal taxonomy, Columbia University Press, New York

Singh R (1967) A key of the Citrus fruits. Ind J Hortic 4:71-83

Singh R, Nath N (1969) Practical Approach to the classification of *Citrus*. In: Chapman HD (ed) Proc Interntl Citrus Symp 1:435-440

Sneath PHA, Sokal RR (1962) Numerical taxonomy: The principles and practice of numerical classification. W. H. Freeman and Company, New York

Snogerup S (1980) The wild forms of the *Brassica oleracea* group (2n=18) and their possible relations to the cultivated ones, In: Tsunoda S, Hinata K, Gómez-Campo C (eds.) *Brassica* crops and wild allies, biology and breeding. Japan Scientific Societies Press, Tokyo, pp 121–132

Song KM, Osborn TC (1992) Polyphyletic origins of *Brassica napus*: New evidence based on organelle and nuclear RFLP analyses. Genome 35:992–1001

Spooner DM (2009) DNA Barcoding will frequently fail in complicated groups: an example in wild potatoes. Am J Bot 96:1177–1189

Spooner DM, Anderson GJ, Jansen RK (1993) Chloroplast DNA evidence for interrelationships of tomatoes, potatoes, and pepinos (Solanaceae). Am J Bot 80:676–688

Spooner DM, Hetterscheid WLA, van den Berg RG and Brandenburg WA (2003) Plant nomenclature and taxonomy: an horticultural and agronomic perspective, In: Janik J (ed.) Horticultural Reviews, John Wiley & Sons Inc, pp 1-59

Spooner DM, Núñez J, Trujillo G, Herrera RM, Guzmán F, Ghislain M (2007) Extensive simple sequence repeat genotyping of potato landraces supports a major re-evaluation of their gene pool structure and classification. Proc Natl Acad Sci UA104:19 398–19 403

Stearn WT (1965) The origin and later development of cultivated plants. J Royal Hort Soc 90:279-291, 322-341

Stearn WT (1986) Historical survey of the naming of cultivated plants. Acta Hort 182:18-28

Stearn WT (ed.) (1953) International Code of Nomenclature of Cultivated Plants, IAPT, Taxon, Utrecht.

Stebbins, GL(1956) Taxonomy and the evolution of genera, with special reference to the family Gramineae. Evolution 10:235–245

Stuessy TF (1990) Plant Taxonomy: The Systematic Evaluation of Comparative Data. Columbia Univ Press, New York

Swingle WT (1943) The botany of Citrus and its wild relatives. In: Webber HJ, Batchelor DL (eds) The Citrus Industry, Vol 1, University of California, Berkeley, pp 128-474

Swingle WT, Reece PC (1967) The botany of Citrus and its wild relatives in the orange subfamily. In: Reuther W, Webber HJ, Batchelor DL (eds) The Citrus Industry. Vol 1, University of California, Berkeley, pp 190-340

Tanaka T (1977) Fundamental discussion of Citrus classification. Stud Citrol 14:1-6

Templeton AR (1989) The meaning of species and speciation: a genetic perspective. In: Otte D, Endler JA (eds.) Speciation and its consequences. Sinauer Associates, Inc, Sunderland, MA, pp 3–27

Thellung A (1918) Neuere Wege und Ziele der Botanischen Systematik, erlautert am Beispiele unserer Getreidearten. Naturw. Wochenschr., Neue Folge 17:470

U N (1935) Genomic analysis in *Brassica* with special reference to the experimental formation of *B. napus* and peculiar mode of fertilization. Japan J Bot 7:389–452

Van Valen L (1976) Ecological species, multispecies, and oaks. Taxon 25:233–239

Vavilov NI (1926) Studies on the origin of cultivated plants. Tr po prikl. bot geni sel 16:3-248

Vavilov NI (1935) The phytogeographical basis for plant breeding. (in Russian).In: Theoretical basis for plant breeding, Vol. 1, Sel'chozgiz, Moscow, Leningrad,) pp.17-75.

Wagner WH (1983) Reticulistics: The recognition of hybrids and their role in cladistics, Columbia University Press, New York, pp 63-79

Weising K, Nybom H, Wolff K, Kahl G (2005) DNA Fingerprinting in Plants: Principles, Methods and Applications, 2nd ed. Taylor & Francis group, Boca Raton, FL

Wilkins JS (2003) The origins of species concepts: History, characters, modes, and synapomorphies. PhD Thesis, University of Melbourne, November 2003, pp 242

Wilkins JS (2011) Philosophically speaking, how many species concepts are there? Zootaxa, 2765:58–60

Xu Q, Chen L, Ruan X, Chen D, Zhu A, Chen C, *et al.* (2012) Draft whole genome sequencing supports Hybrid Origin of Sweet orange (*Citrus sinensis* (L.) Osbeck [Rutaceae] Nature Genetics 45:59-68

Zeven AC, de Wet JMJ (1982) Dictionary for cultivated plants and their regions of diversity excluding most ornamentals, forest trees and lower plants. Center for Agricultural Publishing and Documentary, Wageningen, the Netherlands

Chapter – 16

Palynology and Techniques of Pollen Preparation

Arti Garg and Saurabh Sachan

1. Introduction

Among the various well associated morphological units in plants, the reproductive units demand maximum protection which is achieved in pollen grains (also in spores) by encasing the germplasm with a well organized wall with a unique structure and ornamentation. Apart from its functional significance, the wall bears important characteristics which are of immense diagnostic and phylogenetic values. The morphological diversities in pollen and spore organization are greatly exemplified in the vascular plants. The term 'Palynology' was coined by Hyde and Williams in 1944 for the science of pollen studies, which include the study of pollen as well as spores. Although some investigators tend to limit the scope of palynology to morphological investigations but others include pollen analyses of peats, coals, honey, and rocks as well as the re-sulting conclusions drawn from this information, and in a wider sense it is also connected with the pollen chemistry.

The wall of pollen and spores is a very peculiar substance which is resistant to the destructive action of corrosive acids and alkalis and is readily preserved in nature under suitable conditions. The shape, size, and ornamentation or sculpturing of pollen has been the subject of study since the invention of the microscope. Pollen morphology is therefore greatly utilized in interpretation of ancient floras, for determining authentic species phylogeny, distribution and climatic con-ditions under which they grew, based upon the pollen content of the matrices. Pollen which lodges in moss tufts and lichen mats, or falls in water and gradually sink, or those that fall on sphagnum and clays particularly under anaerobic conditions, are well preserved. This pollen is not destroyed in the transformation of peats to lignite or clay in shale and thus has been preserved for hundreds of thousands of years.

During the present century pollen morphological studies are substantiated and elaborated magnanimously with the aid of Electron microscopy. The surface configuration and finer details are being accurately analyzed for their proper application in plant taxonomy as well as in microfossil diagnosis.

2. Techniques of Pollen Preparation

Various methods have been suggested for processing pollen for morphological studies. Of these, the acetolysis method of Erdtman (1952) is most suitable for studying the exine structure as it helps to dissolve the protoplasm and clear the exine. But, thin walled grains such as those of Commelinaceae and Lauraceae cannot resist the acetolysis treatment and therefore they are treated with 70% alcohol and stained with safranin.

Pollen grains for reference samples can be taken from dried (herbarium) material or from fresh, living specimens. For general use the origin of the material is of no importance, but voucher specimens must always be available in a public herbarium and cross-referenced against the sheet and the preparation. Preparation techniques for recent pollen for morphological studies are the same except for the quantity of available sample. Following methods are generally used in pollen preparation:

2.1. Erdtman's Acetolysis method

The most widely adopted technique for pollen slides preparation of most angiospermous families is the Acetolysis method of Erdtman (1952, 1960). The method is considered most useful as it helps to artificially fossilize the pollen and spores facilitating easy comparison with the fossil palynomorphs. A schematic diagram showing procedure for slide preparation of modern pollen grains is given for the benefit of beginners (Fig. 1) and detailed below:

- The anthers are dissected and placed in a centrifuge tube.
- A 9: 1 mixture of acetic anhydride and concentrated sulfuric acid is prepared and added to the pollen contained in centrifuge tube.
- These are heated in a water bath at 98 °C for 1 minute, and stirred gently.
- After being cooled for about 10 minutes, they are centrifuged for 1 minute at 1000 rpm, and then decanted.
- The sediment is rinsed through a bronze wire screen of 200 mesh and centrifuged, decanted and rinsed twice with tap water.
- About 10 drops of 50% glycerine is added to the sediment and left for 15 minutes.
- Then it is again centrifuged, decanted, and inverted on a filter paper and left overnight.

- Then using a fine platinum needle, a small square of glycerine jelly is gently rotated in the sediment and placed on a slide.
- The slide is covered with 18 mm circular cover slip, and placed on a hot plate set at 80°C.
- This is sealed with molten paraffin.
- Finally, it is cooled and the excess paraffin is removed with xylene.

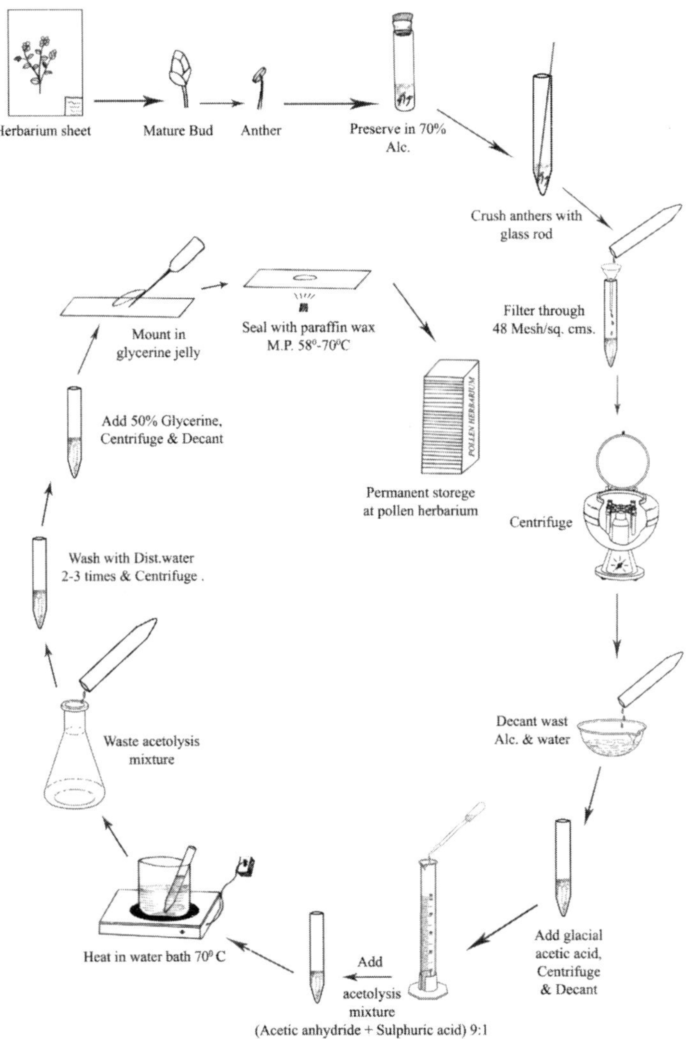

Fig. 1. Schematic diagram showing procedure for slide preparation of pollen grain (Acetolysis Method)

2.2. Ikuse (1956) method

Permanent slides of Acanthaceae *(Thunbergia),* Asclepiadaceae, Cannaceae, Lauraceae, Marantaceae, Orchidaceae, and Zingiberaceaeare are generally prepared by this method as pollen walls of these grains are very weak and are not resistant to the treatment with acid; thus they tend to rupture into fragments, if acetolyzed. The method involes the following steps:

- Anthers are placed on a slide and a drop of 95% alcohol is added.
- The anthers are dissected and the anther wall removed neatly.
- These are washed with xylene and then with 95% alcohol.
- After being stained with a gentian violet water solution, absolute alcohol is added to it until dry.
- Then the sediment is mounted with glycerine jelly, and the cover slip sealed with paraffin wax.

2.3. Nair's (1960) method

This enables a researcher to comprehend the comparative differences between the acetolysed and unacetolysed grains at the same time. In this modified method the pollen grains with protoplasmic contents are stained directly with safranin (unacetolysed) and those treated with a mixture of acetic anyhydride and conc. sulphuric acid to remove protoplasmic contents (acetolysed) are mixed in the same slide.

2.4. Tatzreiter (1985) method

This is an elaborate method which takes about 3-4 hours and is impractical for individual samples. One should prepare batches of about a dozen samples at the same time. The steps are:

(i) Prepare solutions:

A: 84 ml HCL 5%

16 g $MgCl_2$ Ñ $6H_2O$, 98% purity

8 g $NaSO_4$ Ñ $7H_2O$, 9.5% purity

12 g $AlCl_3$, non-acquous, 97% purity

1 g Al phosphate, basic, Merk.

B: 60 g KCNS, 98.5% purity or

50 g NaCNS, 98.5% purity.

Dissolve in aq. Dest to 100 ml.

(ii) Place 0.5 g dry plant material in centrifuge tube with 5 ml solution A and 1 ml solution B. Heat in double boiler, stirring with a glass rod for *c.* 20 min. Hood this as poisonous fumes develop.

(iii) Sieve on ceramic filter 0.5 mm mesh. Wash down with 5% $MgSO_4$ solution from squirt bottle. Centrifuge.

(iv) Wash with 96% alcohol – centrifuge.

(v) Add 3-5 drops conc. lactic acid*c.* 90. And 1-2 ml conc. H_2SO_4 and heat in double boiler for *c.* 15 min. until dark brown-blackish. Stir.

(vi) Cool, add alcohol and shake. Centrifuge. Repeat 3-4 times.

(vii) Squirt 3% $K_2Cr_2O_7$ solution into, or stir up, bottom sediment. Heat in double boiler for few min.

(viii) Heat with aq. Dest. in double boiler 3-4 times.

(ix) Embed or prepare for SEM.

3. Mounting

It is advisable to prepare at least 2 slides for each sample and in case of surplus of polliniferous material, even up to 4 slides can be prepared as sometimes the centrifuge tubes tend to break during centrifugation and the material gets lost. One properly labeled voucher specimen is to be deposited in the Palynological herbarium of a recognized government agency from where it can be easily accessed by future workers for reference slide consultation during pollen identification.

- While mounting a pipette, glass rod, capillary tube or a needle carrying glycerine jelly cube at its tip is used for picking the acetolysed pollen material from centrifuge tubes.
- This pollen is placed on slide.
- A large thin cover slip (2" x 7/8, No. 1) is used to cover the pollen and spread it evenly to the whole cover slip area.
- The sides of cover slip are sealed with molten paraffin and the wax is allowed to cool.
- The slide thus prepared is cleaned with cotton dipped in xylene and labeled appropriately.

3.1. Pollen isolation from Honeys

The pollen content of honey is of considerable interest as is evident from the extensive literature in the field. Pollen source serves as indices of the floral resource of the honeys and are indicators of the geographical origin of honey and their season of production. Honey adulteration or mixtures can therefore be detected by their pollen constituents. Bees also collect pollen, pack it into small pellets and these can be removed by pollen traps in the hives. The color of the pellets is used to identify the plants upon which the bees have been working. If the plants are abundant, the pollen in the pellet is usually that of a single species. However, if plants are scarce, as in early spring or late fall, individual bees may collect from several species of plants as can be seen by the variety of pollen in the same pellet.

For isolation of pollen from honeys procedure is followed.

- About 10g of honey is dissolved in 20 ml of cold, distilled water.
- This is heated in water bath at about 45°C and centrifuge for 3 to 5 minutes at 2500 - 3000 rpm.
- The supernatant liquid is gently decanted.
- The sediment is stirred with a heated platinum loop and one drop of this is smeared on a slide.
- The smear is dried in oven at 35°C or on a hot plate.
- This is the mounted in a drop of liquid glycerin jelly, covered with a large sized cover glass and sealed.

3.2. Pollen preparation from Oils, Asphalts and bitumens

Sittler (19550) gave the following methods for isolation of pollen and spores in oils, bitumens and asphalts. He noted that it was necessary to treat large quantities of material and, for that reason, recommended a centrifuge capable of handling a considerable volume.

3.2.1. Crude oil

- 5 to 10 liters of crude oil is centrifuged and oil decanted.
- The pollen containing residue is washed and centrifuged several times through benzene and then through a mixture of benzene-alcohol-acetone, and finally through 95% alcohol to wash off and remove all oil content.
- The residue is washed with distilled water so that the following chemicals readily act upon the pollen containing sediment.
- The residue is treatedwith dilute Hydrochloric acid and then with dilute Hydroflouric acid.
- The remaining organic matter is either boiled in 10% KOH, soaked in Schulze, or dehydrated with acetic acid.
- The residue thus cleared is acetolysed.

3.2.2. Asphalts

Asphalts also included trunks of cypress trees, pine cones, many animals, and birds in the Rancho La Brea asphalt pits Abraham (1938). Those of petroleum origin were soluble in ethane, pentane, petroleum ether, gasoline, kerosene, whereas those derived from coals were soluble in ben-zol, toulol, xylol, phenol, aniline, and pyridine. Pyridine dissolved about 63% of some asphalts but that carbon disulphide almost completely dissolved them. Ozakerite and Tabbyite were 81% and 61% soluble in petroleum naphtha respectively while carbon disulphide dissolved 99% and 92-94%, respectively. Pollen from these, are extracted and acetolyzed by Settler's (1955) pro-cedure:

- Surface of the specimen is rapidly washed in ben-zene to remove any superficial contamination. About 1 kg of bitumen or asphalt is dissolved in a benzenetechnique filter, or in a mixture of 70% benzene, 15% methanol, and 15% acetone.
- This is centrifuged for a long time, enough to deposit the sediment and produce a clear, supernatant liquid.
- The residue is washed and centrifuged several times through benzene, then a mixture of benzene-alco-hol-acetone, and finally 95% alcohol.
- The residue is washed in water so that the chemi-cal treatment acts upon the sediment.
- The residue is treated with dilute Hydrochloric acid and then with dilute Hydroflouric acid.
- The remaining organic matter is either boiled in 10% KOH, soaked in Schulze for several hours and acetolysed.

3.3. Pollen isolation from Ice

Sufficient pollen is found trapped and preserved in ice, which is useful tools to study the process of glaciations. Pollen are isolated from glacial ice (Vareschi 1935), however sometimes winged conifer pollen get crushed or broken. The process involes:

- Approximately one cubic foot of ice is allowed to melt (Heusser 1954) and the sediment left to settle for 15 to 50 hours and then most of the water is decanted.
- The residue is transferred to beakers and settling followed with decantation is repeated until the sediment could be stored in a 8-dram vial.
- This sediment is treated with dilute hydrochloric acid, centrifuged, washed in distilled water, and re-centrifuged.

- The resi-due is then boiled in 10% potassium hydroxide, strained through sieves, washed until alkali free, and then stained.
- The final sediment is transferred to a slide, the liquid allowed to evaporate up to near dryness.
- A few drops of melted glycerin jelly is placed on the pollen,covered with a cover glass and sealed.

3.4. Pollen isolation from peats

- About 1-2 cc. of the peaty sample is put into a boiling tube with approx. 20 cc. of 10% sodium or potassium hydroxide (NaOH or KOH).
- This is followed by very gentle breaking up with a glass rod and left in boiling water bath for 10 minutes to 12 hours until the sample is thoroughly broken up and no lumps remain.
- The mixture is filtered through fine steel gauze into a centrifuge tube.
- The sediment is washed and left on the gauze towards the middle under a running tap and invert into a petri dish.
- Then centrifuged for 4 minutes at about 2000 rpm.
- The supernatant is decanted and sediment washed 2 or 3 times with distilled water in the centrifuge tube (i.e. until the supernatant remains clear).

3.5. Pollen isolation from muds and clays

The material is mixed with about 10 cc. of distilled water and allowed to stand and settle for two minutes. The supernatant is decanted into another centrifuge tube after which the original material is centrifuged and supernatant decanted. About 1/ 4 cc. of mat-erial is transferred into a 10 cc. platinum crucible with the aid of a little 7% HCl if necessary. This is placed on a tripod in fume chamber. (*Precaution: Close window as far as possible*). Dilute Hydrofluoric acid is added carefully until the crucible is about two-thirds full. This is then boiled gently for 2 or 3 minutes and left to cool about 20 cc. 7% HCl is poured into a small beaker. Using tongs, the crucible contents are gently poured into this HCl (in the beaker). The mixture is transferred into a centrifuge tube. Then centrifuged and washed twice before proceeding for isolation as in peats.

For calcareous samples about 2 gm material is placed in a beaker with 7% HCl. If effervescence occurs, concentrated HCl is added drop by drop until effervescence ceases. This is then filtered as for peats. If there is no effervescence, it is directly centrifuged, washed once and processed as for peats.

4. Techniques of Pollen Preparation for Electron Microscopy

4.1. SEM techniques

Palynology and microscopy are insepar-able. Equipped with nothing but a light microscope (LM) the palynologists have endeavored to build up this science for its valuable contributions to Botany and its vistas are enormous. For-tunately, the pollen and spore size are of an average range between 20 to 50 μm and sometimes up to 110 μm, a range that can ideally be observed under LM. The pores and colpi also fall well within the resolving power of a LM, *c.* 200 nm. While the elements that contri-bute to the exquisite sculpturing of the ectoexine are only 2 to 4 times larger than this resolving power, this very limitation has brought about the best in palynologists who could record an incredible array of patterns. The LM is indispensable during these studies and shall continue to play a leading role in palynological studies.

In modern times the scanning electron microscope (SEM) have witnessed wide acceptance by palynologists because they (1) are useful in revealing surface topo-graphy, (2) operate at resolutions in the range of 10 nm (100 A^0), (3) provide image under field depth of 300- 500 times greater than that of LM, (4) can be used to obtain stereo-pairs that can solve many a vexing problem of struc-ture at the submicroscopic level, (5) can be coupled with acces-sories such as X-ray micro-analyzers and computers to store and process information and (6) requiring lesser time in sample preparation and mounting them on stubs. Since the first commercial production of SEM in 1965, it has found extensive use in all fields of biology, medicine and material sciences. Palynologists have used it mostly for descriptive purposes and as a powerful aid in assessing phylogenetic inter-relations (Walker 1974a, b; Taylor and Levin 1975; Chuang *et al.* 1978). SEM has also been used in the applied fields of palynology as oil prospect-ing, medicine, plant pathology and crimi-nology. There are more than a dozen com-panies in the world that offer a wide variety of SEMs. The Central Electronics Limited is expected to market Indian-made SEM. Various institutes in India are also acquiring SEM machines. The basic principles underlying image production through SEM are highlighted here, which shall provide a guideline to the pollen workers for better understanding the specifications, which must be taken into account while installing the machine specifically for pollen and spore studies.

4.2. Principles and limitations of the Scanning Electron Microscope

Detailed account of the construction or operating principles of SEM can be found in Everhart and Hayes (1972), Hearle *et al.* (1972), Oatley (1972), Black (1974), Muir (1974), Revel (1975) and Dayanandan (1978). However, a brief account of the salient features is presented here for general information. A scanning electron microscope employs electrons to scan the surface of a specimen. Electrons are gene-rated by heating a tungsten filament to about 2900 K. Lanthanum hexaboride can be substituted for tungsten while an altogether new principle (that of pulling electrons off by a strong positive field) is used in Field Emission guns. The instru-ment is normally operated at an accelerating voltage of 15 to 25 KV. The electron beam is de-magnified with the help of two or three electromagnetic lenses to produce a narrow beam of about $100A^0$ diameter. This beam of primary electrons is made to scan the specimen surface in a rectangular raster-fashion, much like in a television tube.

When a beam of primary electrons strikes a specimen surface a number of in-teresting events occur. If the specimen is thin enough, electrons can be transmitted (and used in transmission electron micro-scopy). In thicker specimens, some of the electrons are simply backscattered; some electrons get conducted through the speci-men. But the primary electrons can also excite the atoms in the specimen to emit Auger and secondary electrons. Besides these, the specimen can also respond by emitting photons as X-rays and light (fluorescence). The secondary electrons which have low energies (up to 50 eV) are best suited for image production and are responsible for most of the SEM pictures. As the primary electron beam scans over the surface of the specimen, there is a con-tinuous production of secondary electrons from the specimen surface. The amount of secondary electron generated at every point depends upon the surface topography and chemical composition of that micro area. Secondary electrons are collected by a detector which in turn feeds them to a scintillator to be conducted through a light pipe. This signal is then displayed on a cathode ray tube (CRT). The scan generator that controls the movements of the primary electron beam also controls the scanning in the display CRT. Thus, there is a one to one correspondence between the area scanned and the area displayed on the CRT.

4.3. Resolution, Magnification and Depth of Field

Resolution is primarily a function of the diameter of the electron beam. Production of a narrow beam in turn depends upon several factors including the accelerating voltage, type of the electron gun, tem-perature and quality of the electromagnetic lenses. Specially built experimental SEMs and the recently introduced scanning trans-mission electron microscopes (STEM) have resolutions better than 10sA and less. How-ever, most SEMs operate in the range of $100\text{-}200A^0$ resolution. *Magnification*is simply the ratio of the display screen area to the area scanned on the specimen surface. The beam can be made to scan large or extremely small areas to provide magnifications in the range of 15x to 100,000x. Perhaps, the most unique feature of SEM is its *depth offield.* Because the working distance in an SEM can be as long as 48 mm, and the diameter of the beam remarkably uniform over a considerable distance, well focused images of three dimensional quality can be obtained of specimens that possess com-plex topography. At 20,000x a SEM can retain a depth of field of about 10 μm (with a resolution of about $100A^0$). In a light microscope even at 1200 X the depth of field is only about 0.08 /μm. The improved depth of field in SEM makes it possible to obtain useful stereo-pairs of very minute objects.

4.4. Specimen preparation

4.4.1. Collection of Samples

Collection of spores and pollen for SEM is identical to conventional methods employed in LM. Bacterial and fungal spores may need special preparation proce-dures (Nickerson *et al.* 1974). These and other small algal spores from culture may have to be centrifuged to concentrate them. Fern spores or Pollen samples can be obtained from field or regional herbaria. To yield the pollen, anther lobes of angiosperm can be directly isolated and crushed gently with the help of glass rod on vials or in centrifuge tubes.

In all cases, as much of the debris as possible must be removed before mounting the specimen. It is also possible to directly mount a dry anther or sporangium onto the stage and expose the pollen or spores by dissecting them *in situ*. This is the only advisable technique with compressed fossil sporangia where the availability of material is very limited. Fossil pollen and spores from carbonaceous shales can be obtained after treatment with HNO_3 and KOH. The concentration, duration of treatment and the temperature requirements depend upon the nature of preservation. Hydro-fluoric acid should be employed to remove siliceous material.

4.4.2. Cleaning

Cleaning becomes essential when pollen or spores are obtained from immature anthers or sporangia or when bacterial, yeast or algal spores are obtained from culture media. Sonication in isotonic phosphate buffer is ideal. It is also possible to clean them by alternately cen-trifuging and re-suspending in acetone or even distilled water with a drop of detergent. Acetolyzing the pollen grains following usual procedures (Erdtman, 1960) results in clean preparation. But the possibility that the exines of some pollen may be adversely affected by this treatment should be borne in mind (Southworth, 1974) especially since a SEM can equally magnify imperfec-tions.

4.4.3. Fixing and Drying

In order to facilitate the movement of electrons the column of a SEM is main-tained under high vacuum in the order of 10^{-5} torr. Introduction of wet specimen would release water vapors and contaminate the column. Therefore, all specimens must be critically dried. In exceptional cases, however, it is possible to introduce some material that release very little vapor and examine and photograph it before the electron beam begins to damage them. Such material includes pollen on stigma and fungal spores that are still attached to the mycelia that are growing on leaf surface.

The rigidity offered by the sporopollenin (the resistant wall material) makes it possible to air dry most pollen and spore without much distortion even after prolong drying period. In air drying the sample has to pass through a liquid water inter-face. This creates surface tension forces of great magnitude that can distort the speci-men. Pollen with larger area of pore and colpus can be badly affected. Hence to avoid this, sample need to be gradually dehydrat-ed by passing through increas-ing concentrations of ethanol, acetone or similar solvents. Fixing the material in formaldehyde acetic acid (FAA), FPA or 4% gluteraldehyde for varying periods also helps to make the walls rigid and withstand air drying procedures.

4.4.4. Critical Point Drying

When the spore wall is very thin, as in case of bacterial spores (Nickerson *et al.* 1974), spores of some liverworts (Thomas *et al.* 1974), or in case of immature pollen grains, air drying causes severe distortion. In such cases critical point drying is essential (Cohen 1974). In critical point drying, the speci-men is fixed, dehydrated in alcohol and transferred to a medium such as liquid carbon dioxide kept under pressure. When the alcohol has been replaced by the liquid carbon dioxide the special bomb contain-ing the specimen is heated to produce the critical temperature and pressure of CO_2 at which stage the specimen is instantly dried without getting exposed to the distort-ing forces of surface tension.

4.4.5. Mounting and Heavy Metal Coating

Spores and pollen must be mounted on a special stage (usually an aluminium stud, about 1 cm in diameter). If the material is not firmly attached to the stage, it can be pulled off by the vacuum and lead to specimen loss and contamination of the vacuum system. Adhesives are used to attach the spores to the stage. Again, if there is too much adhesive then the speci-men can be immersed in it. Adhesives have ranged from the grease in one's finger tips to saliva and synthetics like Duco cement and Fevicol. Double sticky scotch tape can also be used giving due consi-deration for the fact that it can release substances into the column. Dried spores and pollen are poor con-ductors. The accumulation of localized electrical charges on the specimen surface generally lead to bright spots ('charging') and poor image production which poses a serious and most frequent problem during scanning electron microsopy. To overcome this, spores and pollen are coated with a thin layer of gold or gold/palladium alloy. The coating must be continuous throughout the entire surface of the specimen and the specimen stage. However, since the spherical material makes minimal contact with the specimen stage, they are difficult to coat. Long term storage of coated material may require additional coating before examina-tion.

4.4.6. Stereopairs

For a truly three dimensional image stereopairs should be obtained. After taking a picture of pollen, the same pollen is tilted (by tilting the stage) to about 7° and re-photographed. This technique pro-vides two images, much like what the two human eyes see of the same object. The two pictures can be viewed with a stereo viewer. With some practice, they can also be viewed with the unaided eyes to understand the three dimensional effect. Instru-ments are also available to carry out mea-surements of depth, distance, etc.

4.4.7. X-ray microanalysis

As mentioned earlier, excitation by primary electrons also results in the emis-sion of X-rays. X-ray yield is propor-tional to the weight concentration of the element in the sample; different elements can be detected because of the differences in the wave length of X-ray emitted or the differences in energy content (Black 1974). Pollen or spore can be sectioned to analyze the distribution of different elements along the entire diameter of the specimen. Ele-ments that are firmly incorporated into the architecture of the cell are easily localized. However, soluble ions are readily lost or dislocated. This can be overcome by freeze-drying the specimen, embedding in plastic and sectioning before carbon coating and examination. With X-ray ana-lysis, quantitative measurements are possible up to a detection limit of about $10\sim^{18}$g.

The SEM offers almost unlimited possi-bilities for experimentation. Individual workers will come up with innovations that will add to the existing methodology or modify it. Even a routine re-examination of all the spores and pollen that have al-ready been described with LM will bring enormous amount or additional and new information. Computerizing the SEM to recognize patterns will lead to a better understanding of the diversity of pollen and spores. Selective techniques of wall degradation (Southworth 1974) combined with SEM can contribute more information about the exine sculpture, thickness and columella details.

5. Scope of Pollen Studies

There is a vast scope for palynological studies in India. Compared with the magnitude of the Indian flora, pollen morphology, thus far known of the Indian plants is negligible. The need to make an extensive study of Indian species. The application of this knowledge in plant taxonomy can further be used as an instrument of multiple scientific research in systematic botany, paleobotany, paleoecology, aeropalynology, agriculture, archaeology, climatology, entomology, geology, glaciology, criminology, allergy, stratigraphic correlation of oil-bearing rocks and coal fields, drugs, and improvement of honey.

5.1. Pollen morphology and taxonomy

Morphologically pollen grains are most conservative plant structures and their parent genus or even species can be recognised on basis of morphology as each plant is unique by their pollen. Hence pollen morphology is of great application in taxa identification and delimitation, and interpretation of relationships among different taxonomic levels. Although palynological basis cannot be laid for separation or merging of the broad divisions of vascular plants, they serve as supplementary tools in solving complexities within different groups at various levels. Further, it may be mentioned that the position of aperture in spores is generally proximal in Pteridophytes, distal in Gymnosperms and monocots and zonal in dicots (Nair 1964a). As a rule most monocots are provided with a single aperture (colpus or pore), while the dicots have at least 3-colpate or 3-colporate apertures. This instantly delimits the pollen identification keys.

5.2. Palynology in angiosperms

The application of pollen morphology in plant taxonomy is best evidenced in angiosperms. Among the vascular plants, the largest variety of pollen morphotypes occurs among the angiospermous plants. Pollen grains are typical in the families; Poaceae, being with 1- pore, an annulus and an operculum; Brassicaceae, being 3-colpate (except for the inaperturate type in *Matthiola*) and reticulate; Ericaceae, being held in tetrads; *Trapa,* being provided with a meridional crest; *Moringa* with sac-like aspides bearing pores, and several others among the angiosperms.

The study of pollen grains of Indian plants has provided important data on the taxonomy and evolution of various taxa of flowering plants. The 3-colpate and pantoporate pollen (Nair 1961) of *Caltha palustris* suggested the species with 2 varieties *normalis* and *alba* respectively. Pollen morphology have provided effective base in interpreting the inter-relationships of various plants such as *Callicarpa longifolia* and *Congea tomentosa* (Rehman 1962), *Zea* and *Sorghum* and *Cocos nucifera* (Nair and Sharma1963). In the family Compositae, the lophate pollen grains typify the tribe Vernoneae and Cichoreae. The generic differences in pollen are evidenced in the eurypalynous families such as Euphorbiaceae, Phytolaccaceae, Betulaceae, and Liliaceae (Erdtman 1952). Differences in pollen at species level are expressed in many instances. Among the species of *Anemone*, pollen grains are 3- colpate in *A. obtusiloba*, pantoporate in *A. alchemillaefolia, A. biflora* etc., pantocolpate in *A. alpina, A. hortensis* (many colpate; Erdtman 1952). In the Indian species of *Bauhinia* pollen grains are either inaperturate as in *B. acuminata*, 4- zonocolpate in *B. tomentosa*, 3- zonocolpate in *B. malabarica* or 3-zonocolporate in other species, of which the exine surface is striate in *B.krugii* and *B. variegata,* verrucate in *B. retusa* and reticulate in *B. racemosa, B. purpurea*etc. (Nair and Sharma 1962). Similarly, the excrescences are of various sizes in species of *Valeriana* from the western Himalayas.

The applications of pollen morphology in stenopalynous families are limited. For distinguishing the New England species of *Betula,* pollen size frequency values have been used, and for that of *Alnus* the ratio between the diameter of aspis and measaspidar area has been found to be of significance. Intraspecific differences in the nature of endo-cracks have been noted in members of Ericaceae. At the varietal level particularly, the use of pollen morphology in plant taxonomy is overlooked. There is of course, no single-character difference in pollen to distinguish one variety from the other. But, a study of several cultivated varieties of the species of *Canna, Bougainvililaea, Hibiscus* and some cereals (Nair 1960, 1961) has helped to lay the principles and methods for using pollen morphology in varietal taxonomy. In the cultivated varieties, there are pollen variations and also morphologically-sterile (without protoplasm) pollen grains. The percentages of the different variation types and sterility, give ample indications of the taxonomy and inter-relationship of the plant varieties. Also, it is of significance to note that the nature of the variation is often specific in the different cultivated plants. For example, the variations relate to the exine strata and excrescences alone in *Hibiscus,* to reticulation in *Euphorbia pulcherrima* and *Bougainvillea,* and to pollen size in the cereals.

Pollen morphology has been effectively applied in solving several problems of taxonomy. Pollen grains of Bombacaceae with 3 colpi and with the characteristic exine pattern are clear indications of its difference from the Malvaceae. Similarly, the differences in pollen have been used to suggest the separation of Butomaceae from Alismaceae. Further, among the genera of Butomaceae, Butomopsis with pantoporate grains merits transfer to the Alismataceae, or to a separate family. The inclusion of *Trapa* in a separate family finds palynological support, grains being provided with a meridional crest, different from other members of Onagraceae, in the system of Bentham and Hooker.

Pollen keys are made to categories plants of a genus or species (Nair and Sharma 1952). These artificial keys are particularly useful in the identification of pollen and spore microfossils, and such other isolated grains as those found in the atmosphere, or in honeys. In providing an artificial key for the Western Himalayan plants (Nair 1964b) the morphological characters of pollen have been considered to have various degrees of importance, namely Primary, Secondary and Tertiary.The primary characters are those of apertures, secondary ones are those of exine pattern, and the tertiary characters include those of size, shape and exine thickness.

5.3. Palynology in cryptogams

Palynological studies of cryptogams have earlier been done as part of investigations on the general morphology, or taxonomy of the various taxa (Maheshwari and Kapil 1963). Subramanian (1962) demonstrated the use of spore morphology in the classification of Hyphomycetes.

5.4. Palynology in gymnosperms

A large number of Himalayan gymnosperms have been considered by Nair (1964b) in his extensive study of the pollen grains of western Himalayan plants. In the saccate sporomorphs of Pinaceae, variations in the number of sacs are known for various species of*Pinus* (Puri 1945; Mittre 1957). The pollen grains of *Ephedra foliata* have been shown to possess tiny, sac-like structures which provides evidence of it's possible evolution from the bisaccate forms (Nair 1964a). Palynological support has helped in sub-division of *Gnetum* into Gnemonomorphi, Micrognemones and Araeognemones (Erdtman 1954).

5.5. Palynology in paleobotany

The differences in the pollen spectrum in peat bogs have contributed to the knowledge of ecological successions and changes in floras. *Palynologists* have postulated past climates based upon the assumption that a given assem-blage of plants in the past had the same general climatic conditions as they do today. These studies have revealed the existence of plants thousands of miles from where they grow today, thus contributing information on plant migrations of interest to the phytogeographer. The occurrence of index fossils in similar combina-tions and abundance has led to the correlation of coal seams as being the same although there is a vertical dis-placement of many feet as the result of faulting. Also, a thick, mineable seam has been identified as the same as a very thin seam, many miles away.

5.6. Palynology in oil exploration

Palynology also works as an important tool in the ex-ploration for oil. Many of the major oil companies now have one or more Palynologists to supplement the work of the paleontologists. The use of the pollen spectrum for the location of drilling sites cannot be ignored. Correlation of rock strata by the pollen content is impor-tant in strata where marine organisms are absent. The botanical identification of pollen in the pre-quaternary sediments provides information about past environments, the plant assemblages, and the climatic conditions under which they grew. Differences in pollen content distin-guish marine from continental deposits. The identifica-tion of ancient floras may, in time, help solve the riddle of oil formation.

5.7. Palynology in archaeology

Palynology is generally used by archaeologists to look at vegetation on a regional level rather than providing site-specific information. Vegetation burning and grazing can also be identified in the pollen record, furthering the understanding of prehistoric land management practices. Hence palynology serves as a useful tool in reconstructing the vegetation cover of landscapes in the past. Only those which are anemophilous are recoverable through archaeological methods (taking sediment cores from marshes or lacustrine areas where pollen is preserved in the waterlogged, anaerobic environment), leading to a preponderance of forest and grassy plants in any sample. It is also possible to recover pollen from coprolites, complementing the information on general vegetation cover with details about animal grazing or fodder practices, as well as human diet – even including whether plant matter was cooked. A deposit of pollen will usually be a combination of local pollen from contemporary vegetation, regional pollen brought via wind, water, or soil erosion, and residual pollen accumulated over time.

Pollen variations have also been found significant in studies of plant geography (Nair and Sharma 1962). Morphological studies of pollen of those medicinal plants in which the flowers or pollen alone are used as drugs, have shown the importance of pollen in pharmacognostic diagnosis of these drugs (Nair 1961).

References

Abraham H (1938) Asphalts and Allied Substances: Their occurrence, Modes of Production, Uses in the Arts, and Methods of Testing (4th edition). New York

Black JT (1974) The scanning electron micros-cope: operating principles. In: *Hayat, MA Principles and Techniques of Scanning Electron Microscopy.* Van Nostrand Reinhold Co., New York. 1-43

Chuang T, Hsieh WC, Wilken, DH (1978) Contribution of pollen morphylogy to syste- matics of *Collomia*(Polemoniaceae). Am J Bot 65:450-485

Cohen AL (1974) Critical point drying. In : Hayat, MA, Principles and Techniques of Scanning Electron Microsoopy. Van Nostrand Reinhold C., New York. 44-112

Dayanandan P (1979) Scanning electron mi-croscopy of Pollen and spores. J Palynol 15(2): 111-119

Ehrenberg CE (1953) Studies on elm pollen. Bot Not (3):308- 316

Erdtman G (1952) Pollen morphology and plant taxonomy, Angiosperms. Almqvist & Wiksell, Stockholm

Erdtman G (1957) Literature on Palynology XIX. Geol Foren Stockh F Brhandl 79(4):601-736

Erdtman G (1960) The acetolysis method. A revised description. Sv Bot Tidsk 54:561-564

Erdtman, (1933) The improve-ment of pollen analysis technique. Sv Bot Tidskr Bd 27, H. 3, pp 347-357

Everhart TE, Hayes TL (1972) The scanning electron microscope. Sci Am 226:54-69

Hearle JWS, Sparrow J T and Cross P M (1972) The use of the scanning electron microscope. Pergamon Press, Oxford

Hyde HA, Williams DA (1945) Palynology. Nature 155:265

Ikuse M (1956) Pollen grains of Japan. Tokyo

Maheshwari P and Kapil RN (1963) Fifty years of science in India. Progress of Botany. Indian Sci. Cong. Association, Calcutta

Mittre V (1957) Abnormal pollen grain in some Indian Gymnosperms with remarks on the significance of the abnormalities. J Ind Bot Soc 36:548-563

Muir MD (1974) Fundamentals of the scanning electron microscope for biologists.Scanning Electron Microscopy Symposia 1974. Om Johari (ed.) IIT Research Institute, Chicago.

Nair PKK (1961) Pollen grains of Indian specimens of *Caltha palusbis* L. Grana Palynol 2:98-100

Nair PKK (1964a) Pollen Morphology in Advances in Palynology. PKK Nair (ed.), Nat Bot Gard, Lucknow

Nair PKK (1964b) Pollen grains of western Himalayan plants, Asia Publishing House, Bombay

Nair PKK, Sharma M (1962) Pollen morphology with reference to the geographical distribution of *Argemone mexicann*. Lloydia 25(2):123-129

Nair PKK, Sharma M (1963) Pollen grains of *Cocos nucfera* L., Grana Palynol. 4:373-379

Nickerson AW, Bulla L AJr. and Kurtzman CP (1974) Spores. In: Hayat, M. A *Principals and Techniques of Scanning Electron Microscopy.* Van Nostrand Reinhold Co., New York. 159-180

Oatley CW (1972) *The Scanning Electron Micros-cope.*Cambridge University Press, London

Puri GS (1945) Puri Some abnormal Pollen grains of *Pinus excels* Wall., Curr Sci 14:255-256

Rehman K (1962) Pollen morphology of *Callicar palongifolia* Lamk. and *Congea tomentosa* Roxb. Curr Sci 31:302-303

Revel JP (1975) Elements of scanning electron microscopy for biologists. SEM Symposia. IIT Research Institute, Johari O (ed.)

Sittler C (1955) Method *et* techniques Physico-chimiques de preparation des sediments en vue de leur analyse pollinque. Rev. Insti.Fr. Petr. Ann. Comb. Liquid.10 (2):103-114

Southworth D (1974) Solubility of pollen exines. Am J Bot 61:36-44

Subramanian CV (1962) The Classification of the Hyphomycetes. Bull Bot Surv India 4:249-259.

Tatzreiter S (1985) Präparation von Pollen un Sporenfür das Raster electron en mikroskoop und Lichtmikroskoop unter Verwendung von Rhodaniden. Grana 24:33-43

Taylor TN and Levin DA (1975) Pollen morphology of Polemoniaceae in relation to systematics and pollination systems: scanning electron microscopy. Grana 15:91-112

Thomas RJ, Wolery MG, Taylor J (1974) Critical point drying of liverwort spores for scanning electron microscopy. Stain Tech 49:261-264

Vareschi V (1935) Pollen analysen aus gletscheris. Ber. Geobot. Inst Rubel 1934:81-99

Walker JW (1974a) Evolution of exine structure in the pollen of primitive angiosperms. Am J Bot 61:891-902

Walker JW (1974b) Aperture evolution in the pollen of primitive angiosperms. Am J Bot 61:1112-1136

Chapter – 17

Systematic Significance of Seed Characteristics

Kanak Sahai

1. Introduction

Seeds have complex and remarkable diversity in shape, size, colour and morphological and anatomical sculpturing details of seed coat, which are valuable for taxonomic significance and can be used as an additional parameter combined with other parameters during taxa identification and classification. Distinctive seed coat morphology usually allowing identification of species even solely on the basis of seed characters. All members of a particular taxon possess variation within acceptable limits of their description and can be used for identification purposes. High structural diversity provides most valuable criteria for classification between species and family level. Takhtajan (1959) pointed out that even for the phylogenetic correlation between families and genera, the structure of the seed coat might be important. Most taxonomists agree that data concerning the structure and microstructure are of great significance mainly for the classification of angiosperm taxa. Consequently, in addition to vegetative and reproductive characters, seed characters are being used in most of the studies for the analysis of taxonomic relationship in wide variety of plant families. It was observed that seed morphology and anatomical features are rather conservatives, which makes them taxonomically important (Esau 1977; Barthlott 1984).

Gaertner (1788-1805) was the first botanist to use seeds and fruit morphology for comparison and identification of different taxa. Later on, Harz (1885) presented a comprehensive account of seed structures and their individual characters of different taxa. In the 20th century interest in legume seed morphology was renewed. Capitaine (1912) studied the seed morphology of entire family and concluded that seed morphological characters are helpful in legume classification and identification at the tribal, generic and specific levels. After 40 years of Capitaine, Jensen (1998) documented 27 publications on seed morphology. Gunn (1981) summerized the seed characteristics for 510 legume genera. Seed characters of seventeen families were compared by Isley (1947). Duke (1961) indicated that seed shape, sculpturing and colour provide a critical indication of the systematic position of species of *Drymaria* (Caryophyllaceae). Obermeyer (1962) pointed out that the only reliable distinction between the two genera *Chlorophytum* and *Anthericum* of Liliaceae appears the number and shape of seeds. Hairy outgrowth on the testa, their length and colour provide useful characters in the distinction of genera/species in Malvaceae, Convolvulaceae, Asclepiadaceae and Acanthaceae. Gunn (1970, 1971) divided one hundred species of *Vicia* into major and minor groups on the basis of seed size, shape and length of the hilum, lens position and seed coat pattern and sculpturing. He successfully employed these characters up to specific level. Gunn and Gaffney (1974) studied the seeds of 42 economically important taxa of Solanaceae, including hilum and surface pattern as key character. Later on Gunn and Barnes (1977) had also used seed characters in the delimitation of *Erythrina* species. Seavey *et al.* (1977) studied the seeds of 210 species of Onagraceae, where seven seed groups were recognized and their evolutionary implications were also discussed. Matthews and Levinus (1986) demonstrated fairly constant seed characters in those species of Portulacaceae which have restricted geographical distribution, whereas the seeds of species with wider geographical distribution show varied morphological structures.

Khalik and Van der Maesen (2002) used macro- and micro-morphological characteristics of seeds and prepared a detailed key of 23 genera including 39 species of 61 tribes from the family Brassicaceae to highlight the importance of seed characters as a criterion for separating genera and species. At present seed coat patterns are used for various purposes to solve classification problems, to establish evolutionary relationship, to elucidate adaptive significance of seed coat and to serve as genetic markers for the identification of genotypes in segregating hybrid progenies (Zeng *et al*. 2004; Bayrakdar *et al.* 2010; Poyraz and Ataslar 2010; Gamarra *et al.* 2012). Seed surface patterns and seed coat anatomy have been shown to provide valuable characters for use in the delimitation of taxa in selected groups of Brassicaceae (Murley 1951; Vaughan and Whitehouse 1971; Barthlott 1981; Koul *et al.* 2000; Bona 2013).

2. Methods

For both macro-and micromorphological characterization at least 4-5 healthy and mature seeds randomly taken from 3-5 plants of each taxon were analyzed in detail. The colour, shape, size and dimensions of the seed were examined under stereomicroscope (Sahai and Singh 2001). For SEM analysis, healthy and mature seeds were cleaned with 90% alcohol, mounted on brass stubs, coated with gold palladium in a sputter coater and finally scanned in the Scanning Electron Microscope at an accelerating voltage of 10KV. For microanatomy cleaned seeds were sectioned longitudinally through the midplane, mounted on stub and scanned by following the same above procedure. Finally photographs were taken at different magnifications to highlight the macro- and micromorphological and microanatomical details of the seeds (Juan *et al.* 1999; Sahai 2001; Sahai and Singh 2001).

3. Applications

Following features of a seed are valuable for taxonomic considerations:

3.1. Morphological structure of the seed

Seed bears various morphological characteristics, which can be used for taxonomic purposes. Position and size of aril, epihilum, funicle, strophiole, hilum, lens and integument are often taxonomically important characters (Fig.1A-F).

3.1.1. Aril

It is a fleshy to hard structure that develops from the funiculus or ovule after fertilization. Arils and aril-like appendages may grow from specific parts of seed surface (e.g. raphe, chalaza) or funicle (e.g. caruncle of many members of Euphorbiaceae).

3.1.2. Epihilum

It is a multicellular conspicuous layer covering the hilar groove.

3.1.3. Funicle

Usually remain attached to the legume seed exposing the hilum when abscise from the mature seed. In mimosoid and caesalpinioid taxa the funicle is permanently attached to the mature seed.

3.1.4. Strophiole

Outgrowth of the hilum region which restricts water movement into and out of some seeds.

3.1.5. Hilum

It is a scar left after the detachment of seed from funicle.

3.1.6. Lens

It is situated between hilum and chalaza and modified in the form of pit, groove, scar or mound.

3.1.7. Integument

An integument is a protective cell layer around the ovule that becomes the seed coat.

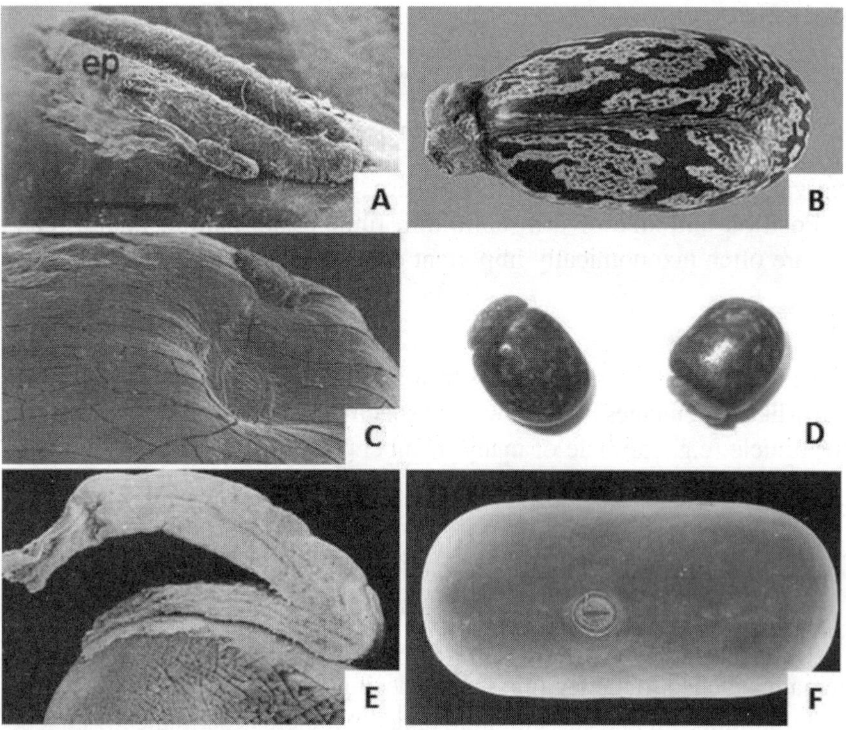

Fig. 1. Micrographs of seed structures. (A) Epihilum, (B) Aril, (C) Lens, (D) Strophiole, (E) Funicle, (F) Hilum

Taxonomic significance of seed morphology

Heywood (1971) drew attention to the importance and impact of scanning electron microscopy technique in the study of systematic resolution. However, application of scanning electron microscopy (SEM) in the field of seed has highlighted new structural criteria for each seed structure in detail and provided many valuable additional macro-morphological diagnostic characters with high taxonomic significance.

Followings are some of the important morphological structures of the seed that can be used as an additional parameter during taxa identification:

Case studies

• Arillar outgrowth and epihilum

Presence of arillar outgrowth and epihilum are taxonomically important in the seeds of Fabaceae (Faboideae). For example though arillar outgrowth is present in the genus *Clitoria* (subtribe Clitorinae) but its shape is different in each species, e.g. in *C. ternatea* it is more or less cock's comb like, whereas in *C. pulchella* it is cap-like (Figs.2A-B). Similarly genera of subtribe Cajaninae have their own characteristic arillar outgrowth (termed as strophiole), while arillar outgrowth is absent in subtribe Phaseolinae. However, presence of epihilum in its variable structure is the characteristic feature of most of the genera of subtribe Phaseolinae (Lackey 1981; Sahai and Singh 2001) (Fig. 2C-D).

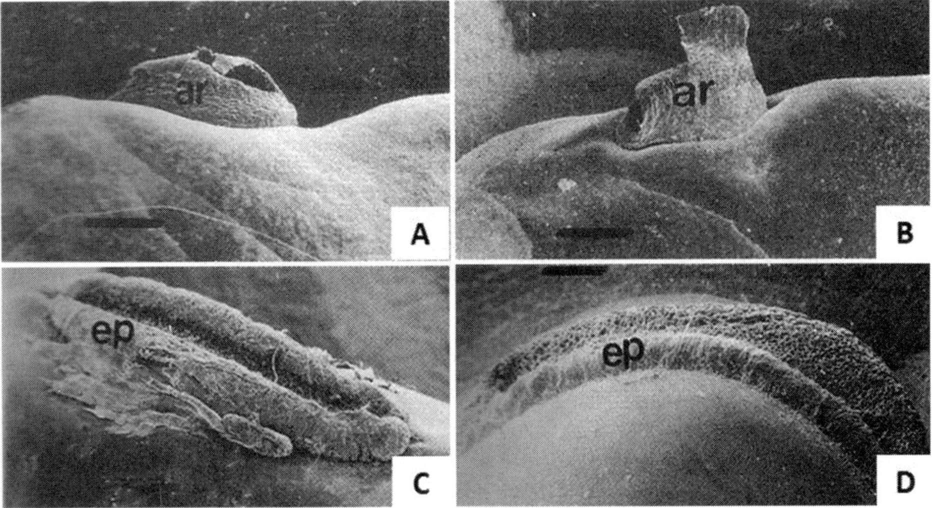

Fig. 2. Microphotographs of Arillar outgrowths and epihilum. (A) *Clitoria pulchella*, (B) *C. ternatea*, (C) *Phaseolus angularis*, (D) *Dolichos lablab* (ar - arillar outgrowth, ep - epihilum)

• Strophiole

The Strophiole is external structure around the hilum. Gunn (1981) prefers the generic name of 'aril' to strophiole, as it is independent of its origin. Van der Maesen (1986) used strophiole along with other characteristics of plant during the grouping of 32 *Cajanus* species into six sections (Fig.1D). Further, he placed *Atylosia* and *Cajanus* under two different genera mainly on the basis of presence of strophiole as it is persistent in the genus *Cajanus*. The major remaining character used in separating *Cajanus* from *Atylosia* is the presence and absence of a seed strophiole. It is frequently used in the taxonomy of Cajaninae. Though in pigeonpea, ripe seeds were usually devoid of strophiole, but the developing seeds do have one that usually shrivels completely. Usher (1966) treated caruncle and strophiole as synonym, defined as an outgrowth near the micropyle and hilum of the seed.

• Lens

It is a part of the seed coat modified in the form of pit, grove, scar or mound, situated between hilum and chalaza. It is also a diagnostic feature of seeds of Fabaceae. A substantial variability is revealed in the lens of each species when it was studied in 14 species of *Cassia*. Each species is characterized by its own lens. Shape, size and position of the lens in each species are genetically constant and serve as distinction from allied species. Presence of conspicuous lens is a constant feature of seeds of a caesalpinioid genus *Cassia* and it has more variation in its gross morphology than the faboid lenses (Fig.3A-D). Earlier it was reported that lens is present only in sub-family Faboideae (Sahai 1999).

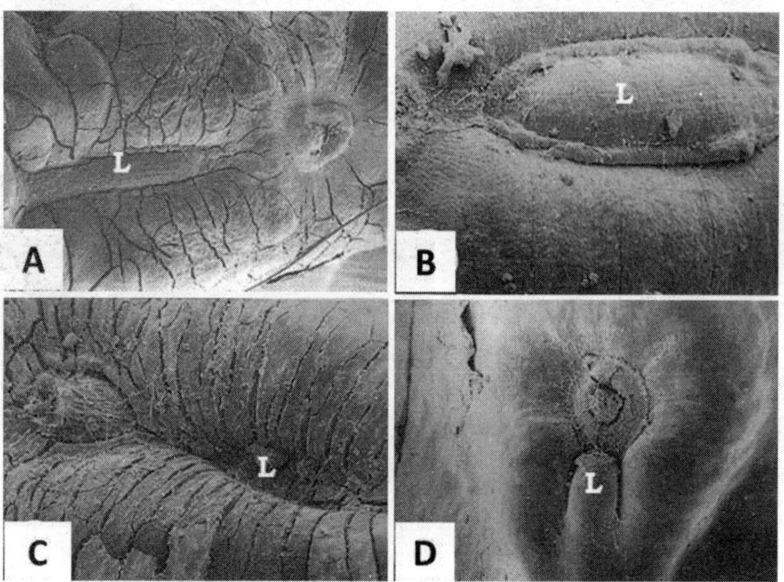

Fig. 3. Microphotographs of the lens of *Cassia*. (A) *Cassia angustifolia*, (B) *C. siamea*, (C) *C. spectabilis*, (D) *C. marylandica* (L – lens)

• Hilum

Shape and position of hilum is also useful to distinguish the taxa of a particular group. Gunn (1969) studied the seeds of the four genera and 20 species of the family Convolvulaceae. All the taxa were keyed and two basic types were recognized on the basis of hilum. Al-Ghamdi (2011) has grouped some species of the genus *Indigofera* (Fabaceae) on the basis of different shape and position (sub-central and central) of their hilum (Fig.4A-C). Similarly Gandhi *et al.* (2011) used characteristic of hilum combined with micro-morphology of seed for identifying the species of *Crotalaria*, *Alysicarpus* and *Indigofera*.

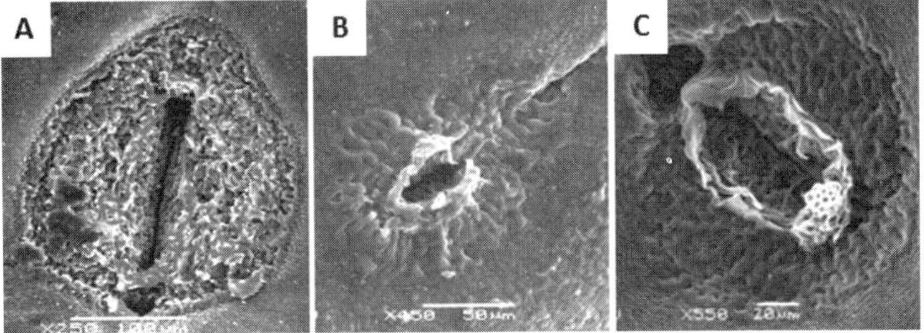

Fig.4. Shape and position of hilum in the seeds of *Indigofera*. (A) *I. Ariculata*, (B) *I. Trita*, (C) *I. Spicata* (After Al-Ghamdi, 2011)

• Funicle

An additional structure present at one end of the seed by which the seed remain attached to the margin of the pod has also taxonomical significance. It is very conspicuous in the seeds of *Acacia* and stays on even after detachment of seed from the pod. Each *Acacia* species is characterized by its own funicle (Fig.5A-F). The substantial variability in the funicular structure serves as a distinctive feature compared to other species of *Acacia* (Sahai, 2003).

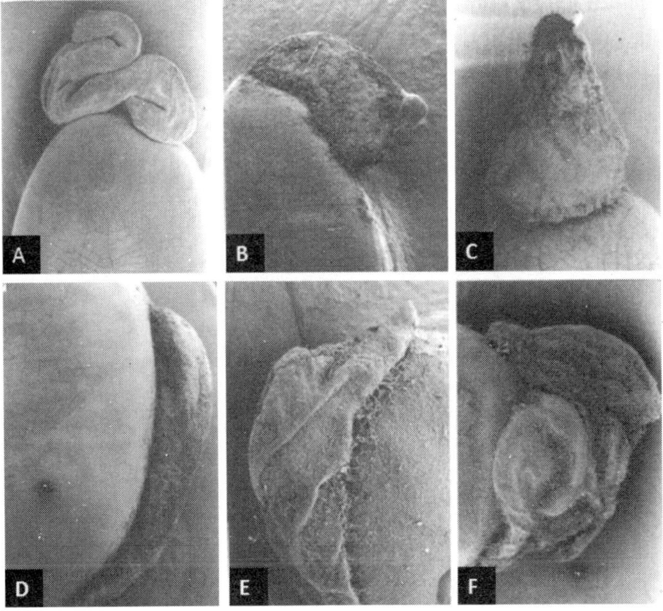

Fig. 5. Funicular structure in the seeds of *Acacia*. (A) *Acacia strongylophylla*, (coiled) (B) *Acacia cyanophylla* (beaked) (C) *Acacia stenoptera* (coned), (D) *Acacia sclereosperma* (arched), (E) *Acacia acuminata* (capped), (F) *Acacia dodonaeifolia* (crowned)

• Integument

In flowering plants number of integuments is generally constant within a family and also orders (Schnarf 1931), thus have a valuable taxonomic importance. Similarly the presence of number of layers in outer and inner integuments is equally valuable for taxonomic consideration.

3.2. Surface structure of the seed

Surface structure is other important features for seed identification. Seed surface varies from very smooth and glossy to rough and fibrous with number of botanical terms e.g., smooth, glabrous, wrinkled, ribbed, reticulate, pulpy, hairy etc. In addition, the colour of the seed, which is usually neglected in taxonomic consideration, is also important as in many cases the colours of the seed are of high diagnostic and significant interest (Berggren 1962; Barthlott 1984).

Scanning Electron microscopy (SEM), with its higher range of magnification, provides an important tool for more precisely characterizing the seed surface. The use of SEM allows for the observation of structures which would be difficult by other means. Some authors have highlighted the importance of this technique as it is very important for the study of seed coats specially for those families in which identification is complicated such as Poltulacaceae (Danin *et al.* 1979) and Scrophulariaceae (Canne 1979; Sutton 1980; Juan *et al.* 1994, 2000), where features of seed morphology have widely used to distinguish the different taxa or to find affinities between them. Bouman and De Lange (1982) and Dietrich (1985) studied the seeds of *Begonia* and *Oenothera* respectively and revealed that testa characters can be used in the delimitation of species as well as hybrids.

According to Barthlott (1984), SEM examination of seed coat features may be grouped into four categories (i) cellular arrangements, (ii) shape of the cells (primary sculpture), (iii) relief of outer cell walls (secondary sculpture), (iv) epicuticular secretions (tertiary sculpture e.g. waxes etc.).

3.2.1. Cellular arrangement

It is the arrangement of seed coat cells and the distribution of trichomes and multicellular appendages can be of considerable and systematic value (Fig.6A). For example, the seed coat of *Eschscholzia* (Papavaraceae) which shows a super-cellular net-like pattern, and appears to be a characteristic of several genera of Papavaraceae.

3.2.2. Shape of the cell

It is the primary and most promising sculpture of the seed surface, particularly the curvature of the outer periclinal wall. Under primary sculpture following groups of micro-characters are considered:-

3.2.2.1. Outline of cells

The shape of the cells can be tetragonal to hexagonal according to their outline (Fig.6B).

3.2.2.2. Anticlinal cell wall boundaries

The superficially visible cells boundaries may be straight, irregularly curved to more or less regularly undulated. Usually they are of high taxonomic significance (Fig.6C). In Orchidaceae, regular undulations are restricted to the Australian and south African Diurideae and Diseae and do not occur within the rest of the family. The anticlinal boundary may be channelled or raised. They exhibit several micro-characteristics e.g. in most of the ancestral genera of Cactaceae they are characteristic feature while completely absent in the derived ones.

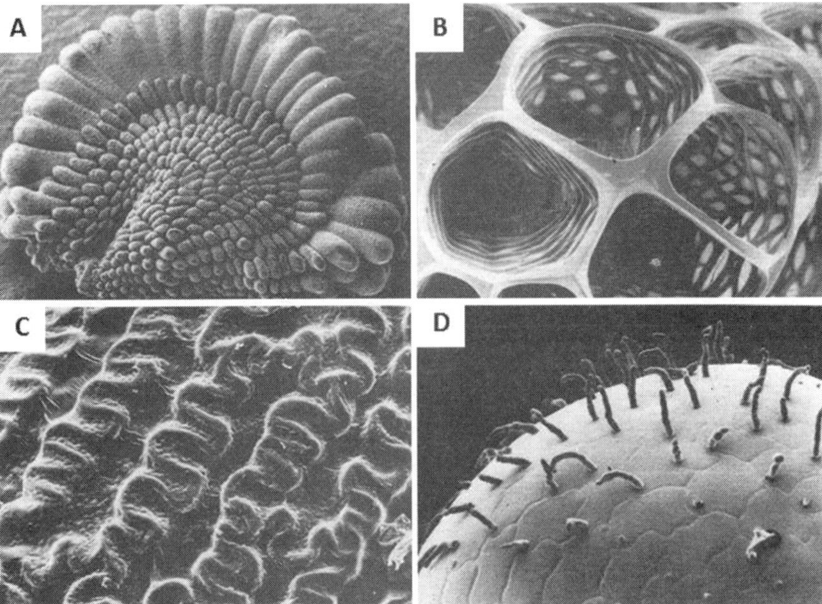

Fig. 6. Different microstructures of the seed surface. (A) cellular arrangement, (B) Outline of the cells, (C) anticlinal cell wall boundries, (D) Curvature of outer periclinal cell wall boundaries (After Barthlot, 1984)

3.2.2.3. Curvature of outer periclinal cell wall boundaries

It can serve as a good diagnostic character. The curvature is responsible for the macroscopically visible roughness of the seed surface. The cells may be flat, concave or convex. These features are of diagnostic value e.g. curvatures may be of high taxonomic interest in the seed coat of *Blossfeldia* (Fig. 6D).Only one very small excentric portion of the outer wall is curved out into a trichome like sculpture. This type of cell shape is restricted to the South American Notocactineae within the Cactaceae. Khalik and Van der Maesen (2002) distinguished some species of the Brassicaceae on the basis of curvature of outer periclinal walls.

3.2.2.4. Secondary cell wall sculpture

If the testa consists of intact cells it means the surface of cuticle may be smooth or may exhibit a micro-ornamentation. Structurally these ornamentations (striations, reticulations, micro-papillations) can result from very different portions of the cell wall (Fig. 6B, 7B).

3.2.2.4.1. *Cuticular sculptures*

Usually occur as irregular to very irregular striations. These cuticular "folds" showing a highly micromorphological diversity are an angiosperm characteristic (Fig.7A). The Cuticular ornamentations are absent from seed coats of all Orchidaceae. Generally cuticular ornamentations may serve as good diagnostic characters but their systematic significance, with some exceptions, is rather limited.

3.2.2.4.2. Secondary wall thickenings

They occur in helical to reticular patterns on the inner side of the outer periclinal walls or on the anticlinal walls. Usually they occur in the form of reticulations or striations (Fig.7B-C). Secondary wall thickenings are always of high taxonomic significance.

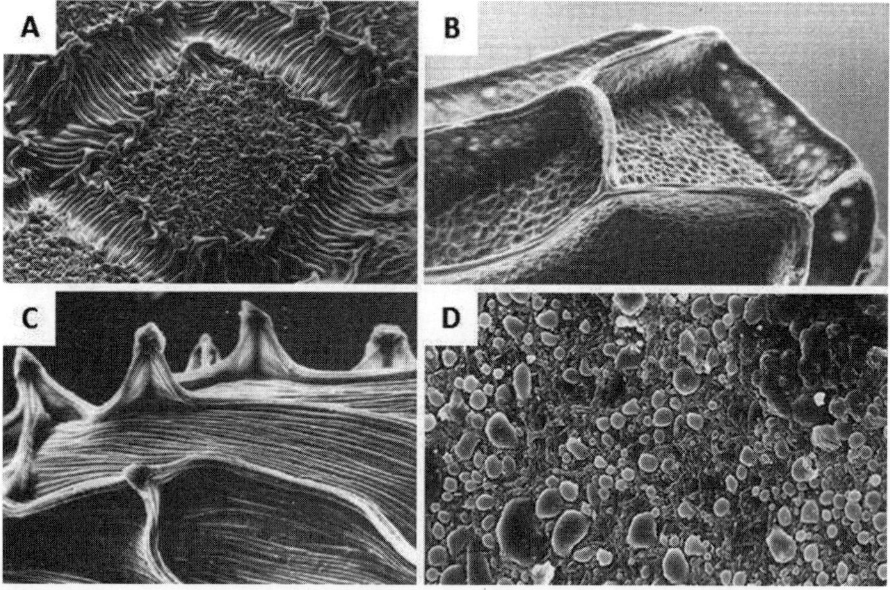

Fig. 7. Different microstructures of the seed surface. (A) cuticular sculptures, (B,C) Secondary wall thickening, (D) Epicuticular secretions (After Barthlot 1984)

3.2.2.5. Epicuticular secretions (tertiary sculpture)

Chemically very different substances may be found on or exude from seed coats (Fig.7D). Mucilaginous adhesive substances are very common. Their nature and origin can be very different and highly specialized. However, may be of taxonomic and systematic significance. Crystalloid waxy secretions occur predominantly on minute water repellent seeds e.g. Droseraceae, Sarraceniaceae and other families.

Technical terms generally used to describe seed spermoderm patterns are: smooth, regulate, tuberculate, reticulate, foveolate, alveolate, areolate, pitted, striations, reticulate-foveolate.

Rugulate: Covered with irregularly arranged elongated structures (Fig. 8H).

Tuberculate: Having small smooth rounded projections or knots (Fig. 8A).

Reticulate: Having a raised network of narrow and sharply angled lines (Fig. 8E).

Alveolate: Excavated in the manner of a section of honey comb (Fig. 8D).

Areolate: Divided into a number of irregular squares or angular spaces (Fig.8C).

Foveolate: Marked with little pits (Fig. 8B).

Pitted: Having numerous small shallow depressions or excavations (Fig. 8F).

Striationss: Covered with longitudinal lines with excavations (Fig. 8G).

Foveate: Pitted or having depressions (Fig. 8I).

Reticulate-Foveate: A type in between reticulate and foveate (Fig. 8J).

3.3. Taxonomic significance of micro-morphology of seed surface

Case studies

• Merger of two genera into one genus

Although the basic pattern of spermoderm (seed surface) of the species of genus *Verbascum* is reticulate-foveolate slight variation placed them under three catagories (i) deep-pitted surface (ii) shallow-pitted surface (iii) slightly depressed surface. As the maximum numbers of species are shallow-pitted thus it is the characteristic feature of *Verbascum* species. There is some controversy about the merger of the two genera *Celsia* and *Verbascum* of the same tribe Verbasceae into one. Some workers concluded their study in support of merger of two on the basis of presence and absence of number of stamen (Fergusion 1971) and type of embryogeny (Kapoor *et al.* 1975; Chandra and Singh 1988). However, seed micro-morphological study involving their seed coat structure and developmental similarities also supports *Celsia* and *Verbascum* should be placed under same genus *Verbascum* (Sahai *et al.* 1994).

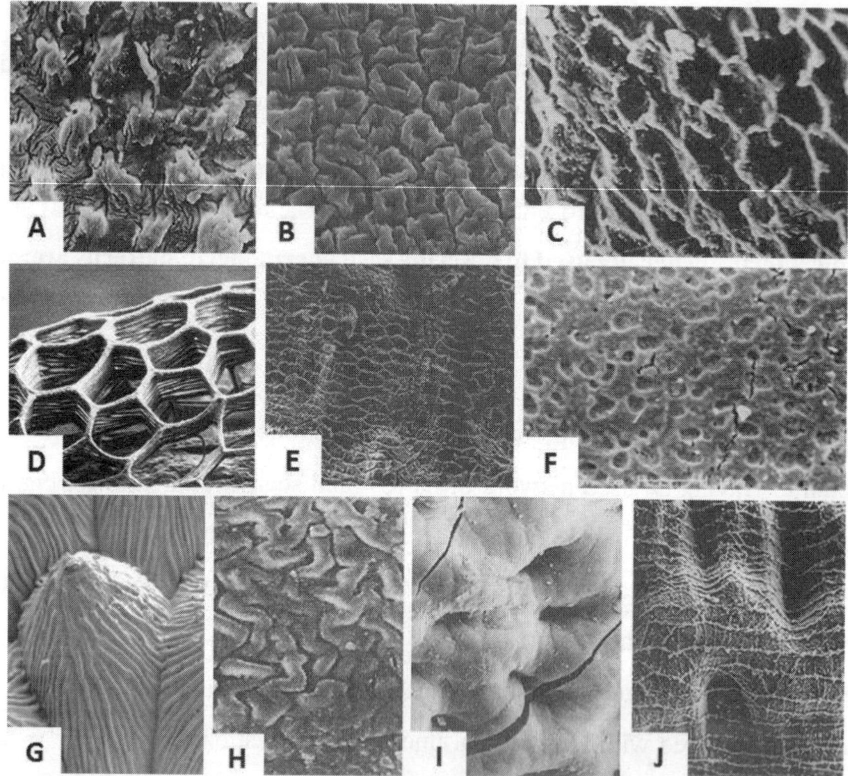

Fig. 8. Different sporoderm patterns. (A) Tuberculate, (B) Foveolate, (C) Areolate, (D) Alveolate, (E) Reticulate, (F) Pitted, (G) Striations, (H) Rugulate, (I) Foveate, (J) Reticulate-foveate

• Deletion / Addition of species

Micromorphological characteristics of the seeds may be useful in taxonomic deletions as well as addition of the species. It was proved after the micromorphology of the seeds of 21 *Cassia* species supported Bentham's (1871) classification because some *Cassia* species have almost same spermoderm pattern (regulate) which are grouped under subgenus *Fistula* by Bentham. Further addition of some other species may be done due to significant similarities in their seed topographical structure (Sahai *et al.* 1997). Similarly, on the basis of considerable differences in seed morphology and seed anatomy along with vegetative and reproductive characters, Bobrov *et al.* (1999) suggested that *Phyllocladus* should be placed in the separate family Phyllocladaceae or Taxaceae or Cephalotaxaceae as suggested by earlier studies.

• Seed micromorphological adaptation to a particular habitat

Among the exotic pines, there are some seed micro-morphological differences in the species, indigenous to the two different habitat/countries. It may be of interest to mention that three different American pine species have great resemblance to each other because of the similarity in their basic pattern (reticulate-foveolate) and secondary ornamentation of seed surface (Fig. 9 C-D). Seed surface of Mexican and Japanese pine, with tuberculate pattern, however, are distinctly different from those of the three American species, having tuberculate as the basic pattern (Fig. 9 A-B). Waxy deposition all over the surface varies in form and quantity seems to be common in the Mexican and Japanese pines. It has been observed in many plant species that their epidermal characters have a great link to their native environment and ecology (Barthlott 1981; Brisson and Peterson 1977). Similarly the seed surface characters besides their use in taxonomic identification are also one of the adaptive features of the plant species. Micromorphology of the exotic pines also showed marked adaptation to their cone habit and seed dispersal behaviour. In American pine species, seeds have thin and reticulate-foveolate seed surface, which may be interpreted as adaptations to wind dispersal over longer distances. On the other hand, Japanese and Mexican pines are closely related to each other due to their seed microstructures and cone habit. Both have fire or serotinous cones i.e. cones which on maturity remain closed and their seed viability may be preserved within the cone for a long period and would disperse following forest-fire. These species have thick waxy depositions on their seed surface (Figs. 9A-B), which help them to maintain their viability for a long period (Sahai 1994).

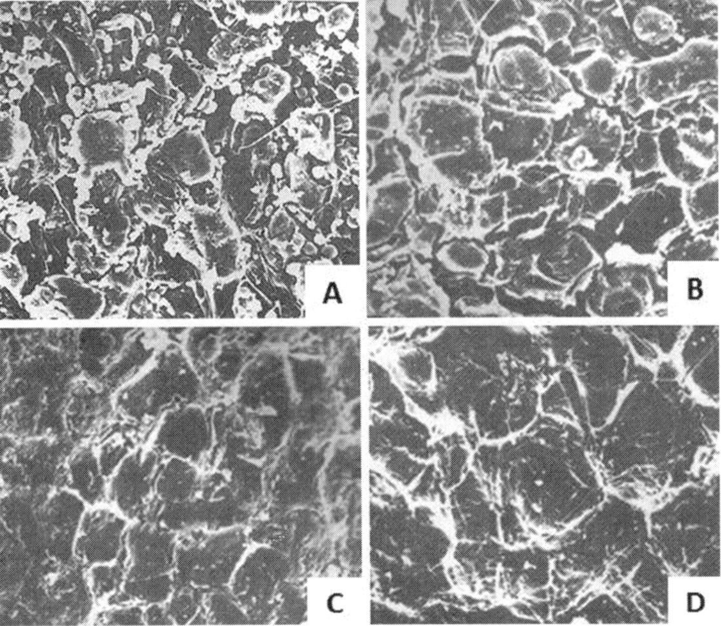

Fig. 9. Micromorphological characters of the seeds of p*inus* species. (A) Japanese pine, (B) Mexican pine, (C, D) American pine

• Seed colour with systematic relationship

The colour of seed is of high diagnostic and systematic interest among taxa. It is also used to distinguish between species or even subspecies. Khalik and Van der Maesen (2002) used colour of the seed to distinguish between subspecies of *Eremobium aegyptiacum* (Brassicaceae). It is yellow-brownish in subsp. *E. aegyptiacum* and *E. longisiliquum* and red-brownish in subsp. *lineare*. Moreover, significant variation in the seed coat pattern of different coloured seeds of *Canavalia virosa* has proved that seed colour can also be used as a significant additional parameter during seed identification (Sahai 2005b) as different coloured seeds of a species even of the same plant possess their own ornamentation (Figs.10A-C).

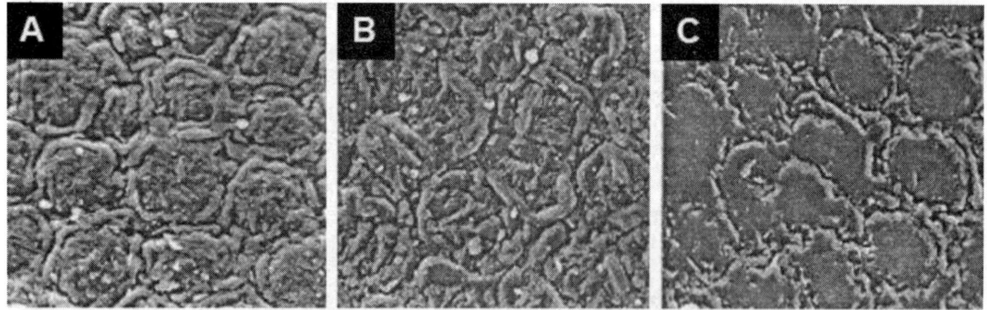

Fig. 10. Micro-morphological characteristics of different coloured seeds of *Canavalia virosa*. (A) Black (B) Mottled (C) Brown

3.4. Anatomical structure of the seed coat

Seed coat anatomy has also of great value in determining taxonomic significance. Carlquist (1961) and Corner (1976) have noted most information on testal histological structure and its taxonomic significance. Vaughan (1968) has suggested that the structure of the testa has provided more taxonomically useful information than other structures in the mature seed. Seed coat anatomy seems to be useful at familial and generic level. Cernohorsky (1947) published most comprehensive account of the seed coat anatomy of cruciferae including 39 genera and 57 species. These published accounts indicate clearly that the number of cell layer in the mature seed coat varies among the genus and species. However, SEM has highlighted new structural criteria and provided many valuable additional diagnostic characters with high taxonomic significance. Eariler workers have shown that variations in structure of the different layers of the testa have taxonomic significance. The epidermis shows marked variation. The sub-epidermal layer of testa is rarely present. However, the palisade layer is always present and its variations have often been utilized in taxonomic studies.

Taxonomic significance of seed coat anatomy

Case studies

Candolle (1825) advocated that anatomical characteristics of seed sometimes exhibit the characteristic feature of a family e.g. Fabaceae are characterized by their upper epidermis containing palisade like cells and a sclerified inner epidermis. At maturity seed exhibits the basic pattern of seed coat differentiation. The common description of each differentiation is as follows:

• Macrosclereids

The outer integument alone forms the seed coat. The outer most epidermal layer is composed of columner palisade-like macrosclereids. The palisade layer is the prominent cell layer of the mature seed coat and its variations have often been utilized in taxonomic studies (Vaughan and Whitehouse 1971). Though they characteristically occur in a single row, a second row i.e. row of counter macrosclereids (palisade cells) is also seen in some legume taxa like *Acacia baileyana* (Fig.10 B, F), *Cassia mimosoides* (Sahai 2001, 2005a).

• Osteosclereids

Below the layer of macrosclereids, there is a layer of osteosclereids. The cells of osteosclereid are biconcave and thick walled and are of the shape of hourglass, therefore also called as hourglass cells. Length and broadness of these cells may vary species to species. A second layer of osteosclereid cells is occasionally present in some taxa after parenchymatous zone. This layer is generally called layer of inner hourglass or osteosclereid cells (Fig.10 B-C).

• Mesophyll zone

Beneath the osteosclereids, mesophyll zone with parenchymatous cell layer is present. The number of parenchymatous cell layer can also vary among the taxa of same genus. In some taxa the second layer of osteosclereids is present after this zone.

Generally histological variability in the seed coat characteristics is considered at species level by many workers. Though seed coat of a particular genus has more or less similar histological characteristics, its each species carries some distinction. In *Cassia* and *Acacia* (Fabaceae) major differences are exhibited in the shape and size of macrosclereids (palisade cells), osteosclereids (hourglass cells) and presence and absence of double layer of macrosclereides and inner osteosclerids. Differences have also been observed in shape and size of parenchymatous cells and degree of their lignifications (Figs. 11A-F). Though parenchymatous zone in each species was multilayered, the number of lignified cell-layer varies species to species (Sahai 2001,

2005a). In addition, the wild relative of crop plants is adapted to greater variations in climatic and edaphic conditions than are the crop plants. For example in *Vigna savi* (Fabaceae) the histological features of hilum is uniform but the structure of the strophiole shows differences between wild and cultivated species of *Vigna*. Presence of Sclerenchymatous tissue in the strophiole region of the wild species indicates their water impermeable dormancy (Kumar and Rangaswamy 1984). However, the usefulness of these diagnostic characters of the testa depends on the levels of variation present among the different species of the same genus which may help in designing a key for identifying the taxa at species level along with other characters.

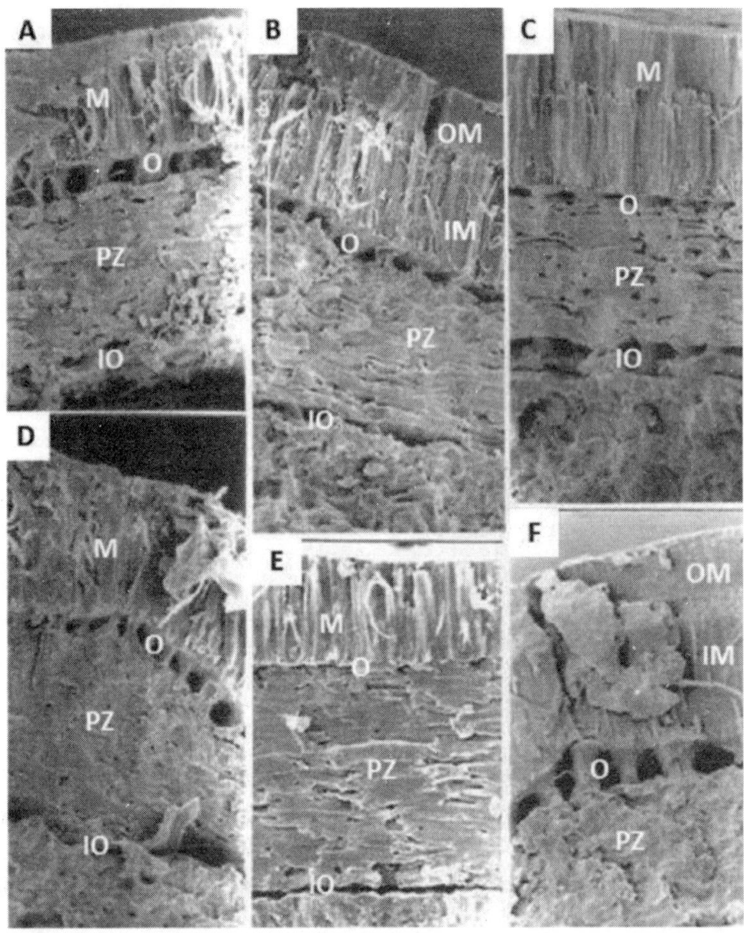

Fig. 11. Micro-anatomical characteristics of seed coat of *Acacia* species under SEM. (A) *A. acuminata*, (B) *A. baileyana*, (C) *A. saligna*, (D) *A. dodonaeifolia*, (E) *A. estrophiolata*, (F) *A. aroma*. (O- osteosclereids, IO – inner layer of osteosclereids, IM – inner layer of macrosclereids, OM – outer layer of macrosclereids, M – macrosclereids, PZ – parenchymatous zone)

References

Al-Ghamdi FA (2011) Seed morphology of some species of *Indigofera* (Fabaceae) from Saudi Arabia (Identification of species and Systematic significance). Amer J Pl Sc 2: 484-495

Barthlott W (1981) Epidermal and seed surface characters of plants: systematic applicability and some evolutionary aspects. Nordic J Bot 1: 345-355

Barthlott W (1984) Microstructural features of seed surfaces. In: Current concepts in plant taxonomy, Heywood VH and Moore DM (eds.), Acadamic Press, London, pp. 95-105

Bayrakdar F, Aytac Z, Suludere Z, Candan S (2010). Seed morphology of *Ebenus* L. species endemic to Turkey. Turk J Bot 34: 283-289

Bentham G (1871) XVIII Revision of the genus *Cassia.* Trans Linn Soc 27: 503-591.

Berggren G (1962) Reviews on the taxonomy of some species of the genus *Brassica*, based on their seeds. Sversk Bot Tidskr 56: 65-135

Bobrov AVF Ch., Melikian AP, Yembaturova EY (1999) Seed morphology, anatomy and ultrastructure of *Phyllocladus* L.C. & A. Rich. ex Mirb. (Phyllocladaceae (Pilg.) Bessey) in connection with the genetic system and phylogeny. Ann Bot 83: 601-618

Bona M (2013) Seed coat microsculpturing of Turkish *Lepidium* (Brassicaceae) and its systematic application. Turk J Bot 37: 662-668

Bouman F, De Lange A (1982). Micromorphology of the seed coats in *Begonia* section Sqamibegonia Warb. Acta Bot Neerl 31: 38-54

Brisson D, Peterson RL (1977) The scanning electron microscopy and x-ray microanalysis in the study of seeds; A bibliographic covering the period of 1967-1976; IITRI/SEM/STEM/2: 117-150

Candolle AP de (1825) Memoires sur la famille des Legumineuses. A. Berlin, Paris

Canne JM (1979) A light and scanning electron microscope study of seed morphology in *Agalinis* (Scrophulariaceae) and its taxonomic significance. Syst Bot 4: 281-296

Capitaine L (1912) Les Graines des Legumineuses. Emile Larose, Paris

Carlquist S (1961) Comparative plant anatomy, Holt, Rinehart & Winston, New York

Cernohorsky Z (1947) Grains des Cruciferes de Boheme. Etude anatomique et morphologique. Opera Bot Cech 5: 5-95

Chandra S, Singh RP (1988) Embryogeny in *Verbascum phlomoides* Linn. (Scrophulariaceae). Curr Sci 57(19): 1075-1077

Corner EJH (1976) The seeds of Dicotyledons, Vol. I & II, Cambridge University Press, Cambridge

Danin A, Baker I, Baker HG (1979) Cytogeography and taxonomy of the *Portulaca oleracea* L. polyploid complex. Israel J Bot 27: 177-211

Dietrich W (1984) *Oenothera.* In: Fl. Il. Catarin., Reitz R (ed), ONAG: 9-31

Duke JA (1961) Preliminary revision of the genus *Drymaria.* Ann Missouri Bot Gard 48: 173-268

Esau K (1977) Anatomy of Seed Plants. Wiley, New York

Ferguson IK (1971) Notes on the genus *Verbascum* (Scrophulariaceae). Rodriguesia 30: 335-344

Gaertner J (1788-1805) De fructibus et seminibus plantarum. Four volumes. Stuttgart: Academiae Carolinae

Gamarra R, Ortúñez E, Cela PG, Guadaño V (2012) *Anacamptis* versus *Orchis* (Orchidaceae): Seed micromorphology and its taxonomic significance. Plant Syst Evol 298(3): 597-607

Gandhi D, Albert S, Pandya N (2011) Morphological and micromorphological characterization of some legume seeds from Gujarat, India. Environ. Exper Biol 9: 105-113

Gunn CR (1969) Seeds of the United States noxious and common weeds in the Convolvulaceae excluding the genus *Cuscuta.* Proc Asso Official Seed Analyt 59: 101-118

Gunn CR (1970) A key and diagrams for the seeds of one hundred species of *Vicia* (Leguminosae). Proc Int Seed Test Asso 35(3): 773-790

Gunn CR (1971) Seeds of native and naturalized vetches of North America. USDA Agricultural Handbook no. 92. USDA Beltsville, Maryland

Gunn CR (1981) Seeds of Leguminosae. In: Advances in Legume Systematics, Polhill RM and Raven PH (eds.), Royal Botanical Garden, Kew, pp. 913-925

Gunn CR, Barnes DE (1977) Seed morphology of *Erythrina* (Fabaceae). Lloydia 40: 454-470

Gunn CR, Gaffney FB (1974) Seed characteristics of 42 economically important species of Solanaceae in the United States. USDA Tech. Bull. No. 1471. Washington: U.S. Government Printing Office

Harz CO (1885) Landwirthschafliche Samenkunde. Paul Parey, Berlin

Heywood VH (1971) Scanning electron microscopy: Systematic and evolutionary applications. London

Isely D (1947) Investigations in seed classification by family characteristics. Iowa Agr Exp Sta Res Bull 351

Juan R, Pastor J, Fernandez I (1994) Seed morphology in *Veronica* L. (Scrophulariaceae) from South-west Spain. Bot J Linn Soc 115(2): 133-143

Juan R, Pastor J, Fernandez I (1999) Morphological and anatomical studies of *Linaria* species from south-west Spain: Seeds. Ann Bot 84: 11-19

Juan R, Pastor J, Fernandez I (2000) SEM and light microscope observations on fruit and seeds in Scrophulariaceae from southwest Spain and their systematic significance. Ann Bot 86: 323-338

Kapoor T, Parulekar NK, Vijayaraghavan MR (1975) Contribution to the embryology of *Celsia coromandelina* Vahl. with a discussion on its affinities with *Verbascum thapsus* L. Bot Notiser 128: 438-449

Khalik KA, Van der Maesen LJG (2002) Seed morphology of some tribes of Brassicaceae (Implications for taxonomy and species identification for the flora of Egypt). Blumea 47: 363-383

Koul KK, Ranjana N, Raina SN (2000) Seed coat microsculpturing in *Brassica* and allied genera (subtribes Brassicinae, Raphaninae, Moricandiinae). Ann Bot 86: 385-397

Kumar D, Rangaswamy NS (1984) SEM studies on seed surface of wild and cultivated species of *Vigna savi*. Proc Indian Acad Sci (Plant Sci.) 93(1): 35-42

Lackey JA (1981) Systematic significance of the epihilum in Phaseoleae (Fabaceae, Faboideae). Bot Gaz 142(1): 160-164

Matthews JF, Levine PA (1986) The systematic significance of seed morphology in *Portulaca* (Portulacaceae) under scanning electron microscopy. Syst Bot 11: 302-308

Murley MR (1951) Seeds of the Cruciferae of northeastern North America. Am Midl Nat 46: 1-81.

Obermeyer AA (1962) A revision of the South African species of *Anthericum*, *Chlorophytum* and *Trachyandra*. Bothalia 7: 669-767

Poyraz IE, Ataslarn E (2010) Pollen and seed morphology of *Velezia* L. (Caryophyllaceae) genus in Turkey. Turk J Bot 34: 179-190

Sahai K (1994) Macro- and micro-morphology of seed surface of some exotic pine species adapted in Indian Himalayan climate. Phytomorphol 44(1&2): 31-35

Sahai K (1999) Structural diversity in the lens of the seeds of some *Cassia* L. (Caesalpinioideae) species and its taxonomic significance. Phytomorphol 49(2): 203-208

Sahai K (2001) Anatomical variability in the seed coat of some *Cassia* L. (Caesalpinioideae) species with taxonomic significance. Taiwania 46(2): 158-166

Sahai K (2003) Funicular diversity in the seeds of *Acacia* Miller with taxonomic significance. Phytotaxonomy 3: 103-107

Sahai K (2005a) Taxonomic relevance of anatomical characteristics of seed coat among some species of *Acacia* Miller. Phytomorphol 55(3&4): 275-281

Sahai K (2005b) Micromorphological variability in the seeds of *Canavalia virosa* (Roxb.) W. & A. in respect to their colour. Phytotaxonomy 5: 58-62

Sahai K, Singh SM (2001) Macro- and micro-morphological features of seeds in six genera of tribe Phaseoleae (subfamily Papilionoideae). Phytomorphol 51(2): 191-197

Sahai K, Kaur H, Pal A (1997) Macro- and micro-morphological characteristics of some *Cassia* species in relation to their taxonomic significance. Phytomorphol 47(3): 273-279

Sahai K, Singh RP, Pal A (1994) Anatomy and surface characteristics of seeds of *Verbascum* species (Scrophulariaceae). Geophytology 24(1): 93-99

Schnarf K (1931) Vergleichende Embryologie der Angiospermen, Berlin

Seavey SR, Raven PH, Magill RE (1977) Evolution of seed size, shape and surface architecture in the tribe Epilobiaceae (Onagraceae). Ann Missouri Bot Gard 64: 18-47.

Sutton DA (1980) A new section of *Linaria* (Scrophulariaceae: Antirrhineae). Bot J Linn Soc 81: 169-184

Takhtajan A (1959) Die Evolution der Angiospermen. Jena

Usher G (1966) A dictionary of Botany. Constable, London, 66: 392

Van Der Maesen LJG (1986) *Cajanus* DC and *Atylosia* W. & A. (Leguminosae). A revision of all taxa closely related to the pigeonpea, with note on other genera within the subtribe Cajaninae. Agri Univ Wageningen Papers 85-4: 1-225

Vaughan JG (1968) Seed anatomy and taxonomy. Proc Linn Soc London 179(2): 251-255

Vaughan JG, Whitehouse JM (1971) Seed structure and the taxonomy of the Cruciferae. Bot J Linn Soc 64: 383-409

Zeng CHI, Wang JB, Liu AH, Wu XM (2004) Seed coat microsculpturing changes during seed development in diploid and amphidiploids *Brassica* species. Ann Bot 93: 555-566

Chapter – 18

Methods and Approaches in Plant Molecular Systematics

T.S. Rana, D. Narzary, S. Verma, K.S. Mahar, Baleshwar, S.A. Ranade and K.N. Nair

1. Introduction

The study of systematics and genetic diversity between or within different species, populations, and individuals is a central task for many disciplines of biological sciences. Systematic studies can indicate which genomes in the plant kingdom to search, sample, and study for the answers to questions relating to the evolution of chemical and physical structures and their synthesis or ontogeny as well as questions about the kinship of the individuals. Such studies would help find new and useful genes and provide information about the phylogenetic relationships among crop species and their wild relatives. The application of molecular markers has been the most active area of research in both animal and plant systematics. The analysis of the genetic variability within and among populations of the species is crucial for understanding their future maintenance and developing improvement as well as conservation programs.

Classical phenotypic characters, such as morphological traits, are still extremely useful but they can sometimes be influenced by environmental conditions. Therefore, during the past decade, classical strategies of evaluating genetic variability such as comparative anatomy, morphology, embryology and physiology have been increasingly complemented by molecular techniques. These include the analysis of chemical constituents (e.g., plant secondary metabolites) and, most importantly, the characterization of macromolecules. The development of molecular markers, which are based on polymorphisms found in proteins or DNA, has greatly facilitated research in a variety of disciplines such as taxonomy, phylogeny, ecology, genetics, and plant breeding (Weising *et al.* 1995). The greater utility of molecular markers derives from following five inherent properties that distinguish them from morphological markers.

Table 1. Differences between molecular and morphological markers

Trait or feature	Molecular markers	Morphological markers
Gross type	Genotypes can be determined at the whole plant, tissue and cellular levels.	Phenotypes of most morphological markers can be distinguished only at the whole plant level.
Diversity of types	A relatively large number of alleles at molecular loci.	Distinguishable alleles at morphological loci occur less frequently and/or must be induced through the application of exogenous mutagens.
Effects of the allele diversity	Usually no deleterious effects are associated with alternate alleles of molecular markers.	For morphological markers, often undesirable phenotypic effects.
Nature of alleles	Alleles of most molecular markers are codominant, allowing all possible genotypes to be distinguished in any segregating generation.	At morphological marker loci usually interact in a dominant-recessive manner, limiting their use in many crosses.
Other genetic effects	Very few epistatic or pleiotropic effects are observed with molecular markers, so that a virtually limitless number of segregating markers can be monitored in a single population.	With morphological marker loci, strong epistatic effects limit number of segregating markers that can be unequivocally scored in the segregating generation.

Source: Tanksley 1983

Under certain experimental conditions, the molecular markers may not be able to distinguish allelic difference of a gene in heterozygous conditions and so behave as "dominant" markers while in other cases these are invariably codominant (allelic difference can be detected). Protein markers (Allozymes, Isozymes) and DNA-based single locus markers (RFLP, SSR, SNPs) are codominant, whereas multilocus DNA fingerprinting markers (e.g., RAPD, DAMD, ISSR, AFLP) are dominant. The numbers and nature of samples to be studied, the population structure where sampling is carried out and the ease and facility with which the markers can be resolved influence whether or not the investigators select single locus (therefore, co-dominant) or multi-locus (therefore, dominant) markers (Table 1). There are two principal advantages of using multilocus markers. The first is that no sequence information about the genome of the organism is needed in order to apply the methods. Thus, marker development costs are minimal and species can be chosen for analysis based on ecological or management criteria rather than the amount of sequence information already known. The second advantage is that genetic differences between individuals can be distinguished with relative economy. Dozens to hundreds of markers are analyzed in a typical study. The overriding disadvantage of such a study is however, the relatively poor quality of genetic information from each individual marker scored since these are not always reliably assigned to independent genetic loci.

For quite a long period of time, biochemical markers (e.g., allozymes, isozymes) have been the molecular markers of choice. In recent years, however attention has increasingly focused on the DNA molecule as a source of informative polymorphisms. Because each individual's DNA is unique, this sequence information can be exploited for any study of genetic diversity and relatedness between organisms. A variety of techniques to visualize DNA sequence polymorphisms have been developed in the past few years, and molecular markers have been derived from these techniques.

Jeffreys *et al.* (1985) introduced the term 'DNA fingerprinting' to describe a method for detecting DNA polymorphism by hybridization of specific multilocus "probes" to electrophoretically separated restriction fragments. In recent years, several modifications of the basic technique have appeared, and related strategies have been developed. Application of molecular markers have blossomed in applied as well as in basic plant sciences.

2. Molecular Methods in Systematics, Genetic Diversity and Phylogenetic Studies

The invention of the polymerase chain reaction (PCR) by Mullis and co-workers (Saiki *et al.* 1988; Mullis *et al.* 1994) revolutionized the methodological repertoire of molecular biology. This technique allows us to amplify any DNA sequence of interest to high copy numbers *in vitro,* thereby circumventing the need for molecular cloning. To amplify a particular DNA sequence, two single-stranded oligonucleotide primers are necessary, which are complementary to motifs on the template DNA. The primer sequences are chosen to allow base-specific binding to the two template stands in reverse orientation. Addition of a thermostable DNA polymerase in a suitable buffer system and cyclic programming of primer annealing, primer extension, and denaturation steps result in the exponential amplification of the sequence between the primer-binding sites, including the primer sequence.

There is an array of molecular markers available in these days with the advent of new technologies. Brief overviews of some relevant and suitable molecular markers that are generally used in the systematics, genetic diversity and phylogenetic studies have been discussed in the continuing paragraphs.

2.1. Protein markers

Protein profiling is the oldest known molecular markers based on protein polymorphisms. The most frequently used technique is the electrophoretic separation of proteins, followed by specific staining of a distinct protein subclass. Using enzymatic and non-enzymatic proteins, numerous investigations have focused on enzyme efficiency, estimating and understanding genetic variability in natural populations, gene flow, hybridization, recognition of species boundaries, and phylogenetic relationships, among other problems (Murphy *et al.* 1996).

Although protein electrophoresis was started 85 years back (Kendall 1928), the modern protocol dates back to 1970 with the combining of PAGE and denaturation of proteins with SDS (Laemmli 1970). The detergent SDS is used to denature proteins into negatively charged polypeptide subunits. After electrophoretic separation on polyacrylamide gels, proteins are visualized with a universal protein stain such as Coomassie Brilliant Blue. Allelic differences are detected as variations in polypeptide size, and alleles are often codominant, although dominant/null allelic systems also occur (Burow and Blake 1998).

Two general forms of protein data can be gathered simultaneously using electrophoretic methods. One is derived from isozymes, which are all functionally similar forms of enzymes, including all polymers of subunits produced by different gene loci or by different alleles at the same locus (Markert and Moller 1959). The other data set consists of allozymes, a subset of isozymes, which are variants of polypeptides representing different allelic alternatives of the same gene locus. Both forms of data are important in molecular systematics, and both involve proteins that can be separated on the basis of net charge and size (Murphy *et al.* 1996). The advent of isozymes as genetic markers in the early 1970's heralded a great advance for genetic studies of plant populations, since only morphological and in some cases cytological markers were available up to that time. Around 20-30 loci per species were available that enabled estimation of genetic diversity and mating system parameters to be made for populations of many species. Even today isozymes are one of the reliable markers to answer many research questions in analysis of genetic variations.

The main advantages of protein markers are their codominant inheritance and the technical simplicity and cost effectiveness. However, there are also a number of limitations. A new allele will only be detected if it affects the electrophoretic mobility of the molecule in question. Only about 30% of the nucleotide substitutions result in polymorphic fragment patterns, and allozymes analysis therefore underestimate the genetic variability (Weising *et al.* 2005). Another problem is that many plant species are polyploids, which can make the interpretation of allozymes patterns quite difficult. Besides these, the plant tissue intended for allozymes studies has to be processed quickly after harvest because proteins are usually quite unstable.

Similarly, allozyme studies have also been widely used to study genetic diversity in plants because they are reproducible (Jenczewski *et al.* 1999), relatively fast and inexpensive (Liu and Furnier 1993), and neutral markers (Cruzan 1998). The main disadvantages of this technique are that it does not produce high levels of polymorphisms to identify a large number of genotypes present in a population (Cruzan 1998) because the requirement for enzyme activity limits the range of loci that can be studied (Beebee and Rowe 2004). DNA based techniques often replace or are used in conjunction with allozymes to generate more polymorphisms.

2.2. Restriction Fragment Length Polymorphism (RFLP)

In the beginning of 1980s genetic marker analysis has shifted from use of protein markers to DNA markers, particularly as a consequence of the enhanced number of potential markers available. The first use of RFLP was in construction of a human genetic map (Botstein *et al.* 1980), and this was suggested as a general method of genetic analysis (Soller and Beckmann 1983). In this procedure, genomic DNA is digested using restriction endonucleases that typically recognize specific six base-pair sequences. Fragments are separated by size electrophoretically on agarose gels, and are denatured and immobilized on nylon or nitrocellulose membranes (blots). Visualization of specific DNA fragments is accomplished by use of DNA probes, synthesized enzymatically as radiolabeled complements to individual cDNA or genomic DNA clones. Blots are exposed to X-ray film, typically from several days to 2 weeks, to expose markers. Now-a-days, non-isotopic methods of detection are also available (Isaac 1994). The basis of RFLP-detected polymorphism is the difference in the restriction enzyme cut sites. Large insertions and deletions also can be detected, but cannot be distinguished from changes in restriction sites. Point mutations in the fragment will not be detected unless in the restriction sites themselves. A variant on the standard method of electrophoresis, denaturing gradient gel electrophoresis (DGGE), makes possible detection of point mutations between restriction sites by separating samples on polyacrylamide gels containing a gradient of urea (Myers *et al.* 1987; Gray *et al.* 1991). RFLP markers are codominant and usually locus specific. Although potentially useful for critical experiments, RFLP analysis is time consuming and labour intensive. Other demerits are requirement of high quality and large amount of DNA, and DNA probes for hybridization.

Cleaved amplified polymorphic sequence (CAPS), also known as PCR-RFLP, is a modified form of conventional RFLP marker (Williams *et al.* 1991; Konieczny and Ausubel 1993). With the advent of PCR technology, amplification of a specific region of a DNA with known sequence has become easier. In the first step of a standard CAPS experiment, defined nuclear or organellar DNA regions are amplified in PCR, using primers complementary to known sequences. In the second step, PCR products are digested with one or more restriction enzymes, and restriction site polymorphisms are displayed by agarose gel electrophoresis and ethidium bromide staining. Alternatively, non-denaturing PAA gels and SSCP gels are also used (Jeandroz *et al.* 2002). As opposed to conventional RFLP analysis, the CAPS approach can be carried out even with a very small amount of template DNA and does not require radioactive treatment or blotting steps. But, to identify suitable combinations of amplicons and restriction enzymes, a wide range of PCR primer pairs and restriction enzymes need to be screened during the initial phase of a CAPS project.

2.3. Random Amplification of Polymorphic DNA (RAPD)

The most common DNA fingerprinting technique is RAPD (Williams *et al.* 1990), where low-stringency annealing of short (10 base) primers of arbitrary sequence is utilized to direct amplification of products of unknown sequence. Unlike other PCR-based systems, only one primer is used. The short sequence of the primers makes possible a multitude of potential primer binding sites throughout the genome, and efficient amplification of DNA fragments may occur when two primer binding sites occur in close proximity.

In RAPD analysis, numerous short (approximately 300 to 2000bp) DNA sequences are amplified from small (approximately 10 ng) genomic DNA samples using one primer. Amplified DNA is typically separated electrophoretically on agarose gels, stained with ethidium bromide, and DNA visualized under ultraviolet light, although radiolabelling of bands, separation on polyacrylamide gels, and detection by autoradiography may detect more, weaker amplification products. Variations on this technique include DAF (DNA amplification fingerprinting, Caetano-Anolles *et al.* 1991) and AP-PCR (arbitrary primed PCR, Welsh and McClelland 1990).

The RAPD technique is technically simple, efficient, and accommodates high throughput (Deragon and Landry 1992), which led to its acceptability among researchers. The PCR amplification of genomic regions, using short oligonucleotide primers of arbitrary sequence, provides an almost unlimited source of informative genetic polymorphisms (Williams *et al.* 1990; Welsh and McClelland 1990). Also, the sensitivity of PCR in DNA amplification minimizes the need for large quantities of tissue and target genomic DNA (Saiki *et al.* 1988). The molecular genetic profiles of individuals are free of confounding environmental effects, which present a certain advantage of RAPD compared to morphological measurements and isozyme analysis. Applications of RAPD have included cultivar distinction (Gidoni *et al.* 1994), marker assisted selection (Kelly and Miklas 1998), linkage mapping (Rajapakse *et al.* 1995; Lacou *et al.* 1998), and conservation biology (Rossetto *et al.* 1999). Experimental work stages are shown in Fig. 1.

The simplicity of the RAPD method is accompanied by inherent limitations. RAPDs are considered to be dominant markers, which is a disadvantage in genetic studies because of the inability to detect heterozygosity as can be done with analysis of co-dominant marker systems. Sources of error have been encountered with the use of RAPDs. Because the generation of RAPDs is accomplished using oligonucleotide primer of anonymous sequence, the amplified fragments are also of unknown nucleotide sequence. These RAPD bands are separated on a gel on the basis of molecular weight (*i.e.* fragment size) and electrical charge. Therefore, a visible band may potentially contain amplified fragments of similar molecular weight that differ in nucleotide sequence. Furthermore, band intensities reflect amplified fragment dosage and less intense bands may be irreproducible (Heun and Helentjaris 1993). Differences in RAPD band intensity have been observed from assay of leaf versus root tissue in soybean (Chen *et al.* 1997). Two associated sources of error commonly attributed to RAPDs are band reproducibility and scoring error (Skroch and Nienhuis 1995). Staub *et al.* (1996) assessed additional sources of error in using RAPDs including age of tissue assayed, pathogen infection, contamination, and variation between stocks of PCR reagents.

EXPERIMENTAL WORK STAGES

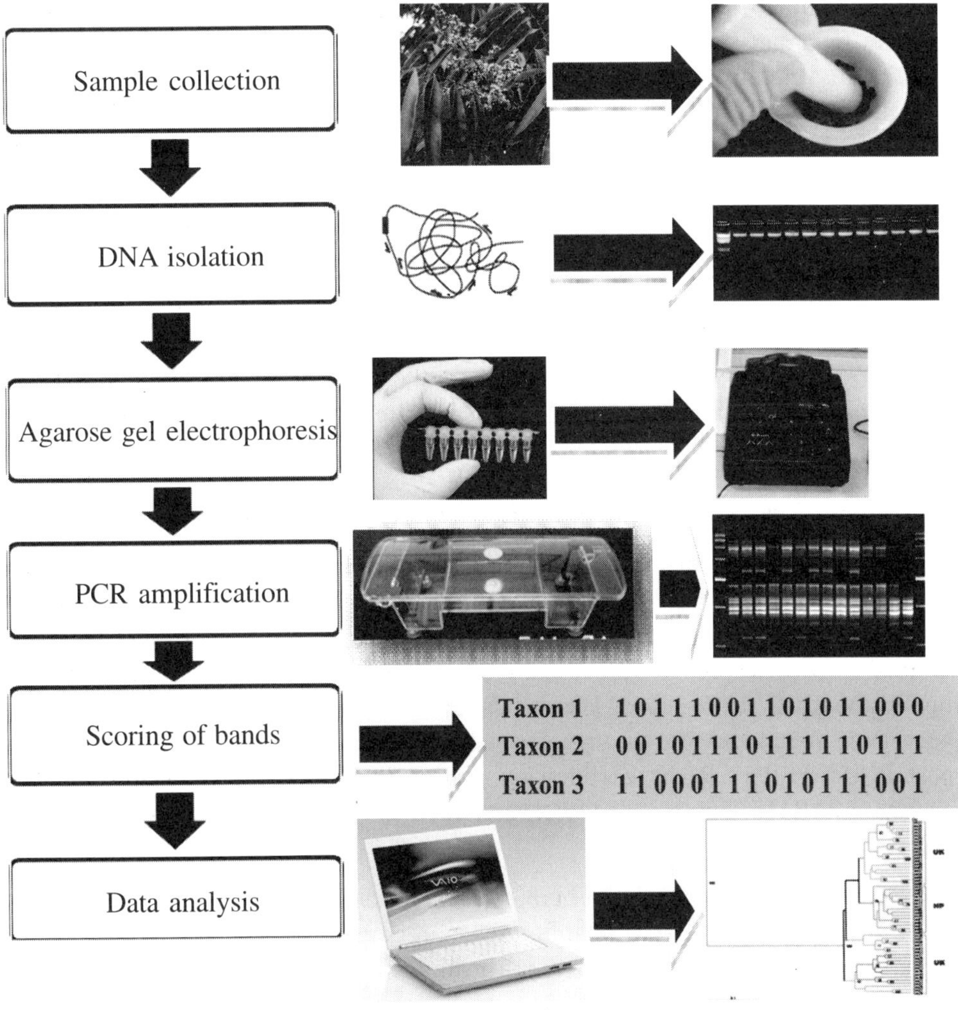

Fig. 1. Different work stages in case of RAPD, DAMD and ISSR methods

2.4. Microsatellite DNA Markers (SSRs)

Microsatellite DNA, also called simple sequence repeats (SSRs) are regions of repetitive DNA that consist of tandem repeats of a core sequence of two to five base pairs. Different alleles at a microsatellite locus differ in the number of tandem repeats of the core sequence. These sequences appear to be ubiquitous in the genomes of eukaryotes, and thousands of potential microsatellite markers could be developed for most species.

Microsatellite DNA markers have tremendous strengths for diversity assessments. They are subject to very high mutation rates relative to single copy nuclear DNA (scnDNA), sometimes producing dozens of alleles. Like allozymes and scnDNA markers, inheritance of microsatellite markers is codominant, so heterozygotes can usually be reliably differentiated from homozygotes. Heterozygosity at microsatellite loci is much more sensitive than other nuclear DNA markers due to the large number of rare segregating alleles at these loci. Loss of these rare alleles actually provides a more sensitive measure of population bottlenecks than does heterozygosity (Leberg 1992). The primary disadvantage of microsatellite DNA markers is development cost.

Microsatellite markers are developed from non-protein coding DNA regions and, therefore are not conserved across taxa, so microsatellite DNA markers developed for one species are often only useful for very similar species. The number of organisms for which microsatellite markers have been developed is increasing rapidly, so it is possible that microsatellite development will be less costly in the near future. In addition, the very large number of alleles present at some microsatellite loci require large sample sizes used to estimate allele frequencies accurately.

2.5. Inter-Simple Sequence Repeats (ISSR)

Inter-simple sequence repeats (ISSR) is a RAPDs-like marker system that does not require any prior knowledge of genome sequence (Godwin *et al.* 1997). The successful application of microsatellite-specific oligonucleotides as PCR primers was first described by Meyer *et al.* (1993), who amplified DNA from different strains of human fungal pathogen *C. neoformans*. The technique was subsequently applied to numerous other organisms, and several acronyms were proposed, including inter-simple sequence repeat PCR (ISSR-PCR) by Zietkiewicz *et al.* (1994), single primer amplification reactions (SPAR) by Gupta *et al.* (1994) and microsatellite-primed PCR (MP-PCR) by Weising *et al.* (1995).

ISSR can access variations in the numerous microsatellite regions by using primers that are anchored at the 5' or 3' end of a repeat region and extended into the flanking region. This technique then allows amplification of the genomic segments between inversely oriented repeats (ISSRs). Generally, a series of single primers are used to generate a series of fragments that are size-separated on either an agarose gel or a polyacrylamide gel (Nagaraju *et al.* 2002). There are three steps involved in ISSR marker analysis: designing oligonucleotide primers, amplifying ISSR segments by PCR, and separating the PCR products. Oligonucleotide primers of ISSR-PCR are usually designed based on the repeat sequence motifs of microsatellites such as (GA)n or (CA)n, following the rule of the adjunction of either the 5' or 3' flanking anchor, such as $(GA)_8$RGY or $(CA)_8$RG. These anchors ensure a higher resolution and better reproducibility of the bands in ISSR profiles.

Being a multi-locus marker, ISSR is useful for fingerprinting, genetic diversity analysis, and genetic mapping. This technique is rapid and can differentiate between closely related individuals. The advantages of this technique include multiple polymorphic loci, high throughput, and low cost. Like RAPDs, ISSRs access variation in the numerous microsatellite regions dispersed throughout the genomes. The initial studies employing ISSR markers (Zietkiewicz *et al.* 1994; Gupta *et al.* 1994) focussed on cultivated species, and demonstrated the hypervariable nature of ISSR makers (Wolfe and Liston 1998; Wolfe *et al.* 1998a, 1998b). Studies of ISSR locus heritability have demonstrated an exceedingly close approximation to classic Mendelian ratios (Tsumura *et al.* 1997). High variability and high "mapping density" as compared with RFLP and RAPD data make these new dominant, microsatellite-based molecular markers ideal for producing genetic maps of individual species (Nagaoka and Ogihara 1997). These features, combined with greater robustness in repeatability experiments and less prone to changing band patterns with changes in constituent or DNA template concentrations, make them superior to other readily available marker systems in investigations of genetic variation among very closely related individuals and in crop cultivar classification (Fang and Roose 1997; Nagaoka and Ogihara 1997).

Disadvantages are that ISSR markers are dominant and less reproducible. Adding an additional PCR re-amplification step into the standard ISSR-PCR protocol can prominently increase reproducibility and productivity (Wiesner and Wiesnerova 2003).

2.6. Directed Amplification of Minisatellite DNA (DAMD)

The term direct amplification of minisatellite DNA (DAMD)-PCR was coined by Gustafson and colleagues and they were the first to apply minisatellite-primed PCR to plants (Somers *et al.* 1996; Bebeli *et al.* 1997; Zhou *et al.* 1997). Minisatellites, also called VNTR (variable number of tandem repeats) are tandemly-repeated DNA sequences. They are ranging in size from a few hundred to several kilobase-pairs while their repeated units vary from 10 to 60bp. A common feature of tandemly arranged repetitive sequences is their high degree of polymorphism, mostly due to variation in the number of tandem repeats but also to the internal structure of individual repeats (Jeffreys *et al.* 1990). The term minisatellites was coined by Jeffreys *et al.* (1985) and initially applied to tandem repeats of 10 to 50bp units, carrying a common GC-rich core sequence of 10 to 15bp, but repeats with longer unit size and higher AT content were also identified. Since these pioneering studies, minisatellite loci have been cloned and sequenced from numerous organisms, including human, cattle, mouse, birds and plants (Weising *et al.* 2005). Most of the minisatellites share a common motif known as the core sequence (Wong *et al.* 1986; Nakamura *et al.* 1988). Studies have shown that minisatellite loci are inherited in a Mendelian fashion and are dispersed throughout the genome in many species, including mouse (Julier *et al.* 1990), bovine (Georges *et al.* 1991), tomato (Broun and Tanksley 1993) and rice (Winberg *et al.* 1993).

The high abundance of minisatellites in eukaryotic genomes allows the use of minisatellite-complementary oligonucleotides as PCR primers to generate numerous polymorphic amplification products. Early studies in this direction were performed with fungi. Meyer *et al.* (1993) showed that the M13 minisatellite repeat unit distinguished and identified different isolates of the human fungal pathogen *Cryptococcus neoformans*, when used as a PCR primer. Similarly, Heath *et al.* (1993) employed various minisatellite core sequences as primers (including M13), to study fish, bird, and human genomes. After electrophoretic separation of the PCR products, differences between species were found, but intraspecific variation was not observed. By using the core sequence of minisatellites as a single primer, this PCR application is capable of producing RAPD-like results for the identification of specific regions in a genome. Since, minisatellite core sequences are longer than RAPD primers, DAMD-PCR can also be effectively carried out at relatively high stringencies, thus yielding highly reproducible results (Heath *et al.* 1993).

2.7. Amplified Fragment Length Polymorphism (AFLP)

AFLP was first described by Vos *et al.* (1995). The technique is based on the selective PCR amplification of restriction fragments from a total digest of genomic DNA. The technique involves three steps, namely, restriction of the DNA and ligation of oligonucleotide adapters, selective amplification of sets of restriction fragments, and gel analysis of the amplified fragments. PCR amplification of restriction fragments is achieved using the adapter and restriction site sequence as target sites for primer annealing. The selective amplification is achieved by the use of primers that extend into the restriction fragments, amplifying those fragments in which the primer extensions match the nucleotides flanking the restriction sites. Using this method, sets of restriction fragments can be visualized by PCR without knowledge of nucleotide sequence (Fig. 2). AFLP is a potential tool for study of biological diversity, varietal identification and genetic mapping (Zhu *et al.* 1998). However, AFLP analysis, like microsatellites, requires a more advanced laboratory infrastructure to detect isotopically or fluorescently labelled PCR products.

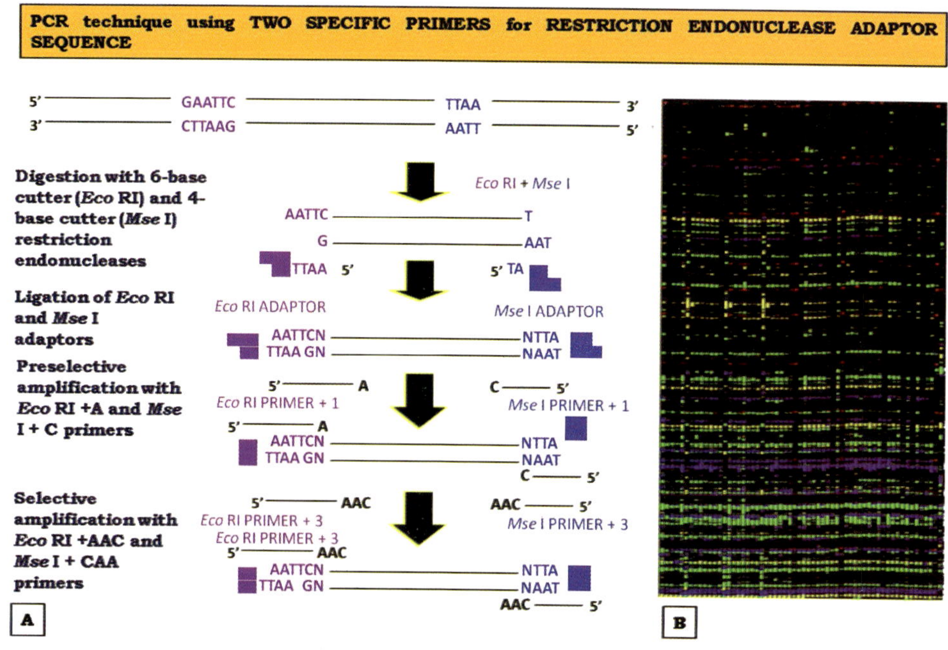

Fig. 2. (A) Shows the different steps in AFLP reaction; (B) A typical AFLP gel profile image generated for fluorescent labeled primers where each primer is labeled with a different fluorochrome

2.8. Gene sequences in phylogeny

In order to infer phylogenetic and taxonomic relationships among plants, both morphological and molecular data are often considered. In recent years advances in technology and knowledge of gene sequences have significantly contributed in angiosperm phylogeny (Hilu *et al.* 2003). Biologists have utilized nuclear (Fig. 3), chloroplast (Fig. 4), and mitochondrial genes to elucidate relationships at all levels of taxonomic rank. Molecular approaches for analyzing phylogeny have become increasingly useful, especially where morphological characters have been insufficient for distinguishing taxa at different levels (Soltis and Soltis 1997). Some of the most commonly utilized genes in molecular systematics and phylogeny of plants have been *rbc*L, a chloroplast gene encoding the large subunit of ribulose-1,5-biphosphate carboxylase (Soltis *et al.* 1990); *atp*B, a plastid gene encoding the beta subunit of ATP synthase (Savolainen *et al.* 2000); *mat*K, a chloroplast gene thought to be involved in splicing introns coding for a maturase (Hilu *et al.* 2003); and portions of the nuclear rDNA cistron unit such as 18S, 5.8S, and 26S, occurring in the nucleolar organizing region of the nucleus (Nickrent and Soltis 1995). These genes have been used in phylogeny because of their ability to be easily amplified by the polymerase chain reaction (PCR), few insertion-deletion events, and their level of evolution and conservation (White *et al.* 1990; Nickrent and Soltis, 1995; Hilu *et al.* 2003).

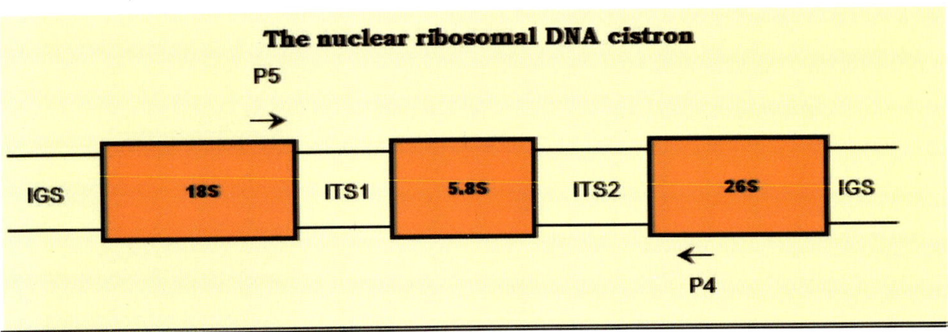

Fig. 3. Schematic structure of the nuclear rDNA of eukaryotes showing the regions involved in the PCR strategy. 18S, 5.8S and 26S are the coding regions for the respective rRNAs. ITS1 and 2 are the internal transcribed spacers while IGS indicates the intergenic spacers. The arrows marked P4 and P5 are primers that are used to amplify the different regions of the rDNA genes and spacers

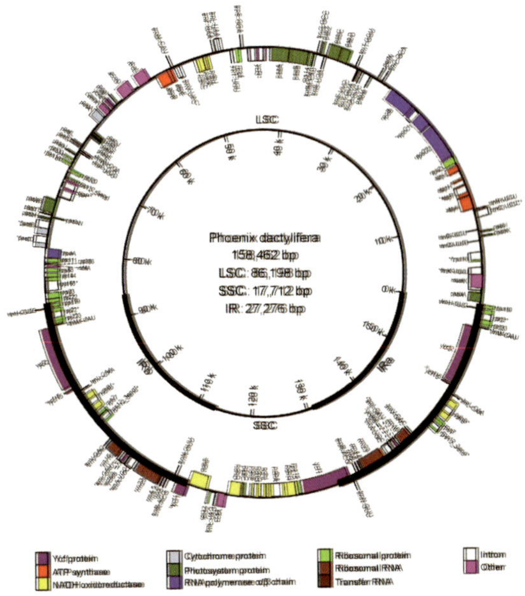

Fig. 4. The map of date palm cp genome sequence
(*Source*: http://www.plosone.org/article/info:doi/10.1371/journal.pone.0012762.)

2.9. Internal Transcribed Spacer (ITS) Region of rDNA

One of the regions that have been extensively used in phylogenetic studies of different organisms is the nuclear ribosomal DNA cistron (Hillis and Dixon 1991). The nuclear genes encoding the cytoplasmic ribosomal RNAs (rDNA) are in most eukaryotes organized into transcriptional rDNA units with a small (18S/SSU), a 5.8S, and a large

(28S/LSU) subunit rDNA region, separated by internal transcribed spacer regions ITS1 and ITS2 (Fig. 3). The DNA sequences for LSU and SSU rDNA are under strong stabilizing selection due to their critical role in ribosome synthesis. Non-coding regions, like the ITS rDNA regions are not under similar functional constraints.

In reported studies, variation between ITS sequences is mostly attributable to point mutations. As a consequence, due to faster accumulation of mutations, these regions usually show higher variability. A relatively minor proportion of sites are affected by insertions or deletions (indels) among sequences that are similar enough to have retained sufficient signal for phylogenetic analysis. Within these limits, sequence alignment is generally unambiguous except in small regions of apparently lower structural constraint. Phylogenetic analysis of combined data sets from both ITS1 and ITS2 spacers yield trees with greater resolution and internal support than analysis based on either spacer alone. The beneficial effect of simultaneous analysis is not surprising based on the low number of useful characters in each spacer. This effect also suggests high complementarities of spacer data, in accordance with similarity in size, sequence variability, and G + C content of ITS1 and ITS2 in most investigated groups of closely related angiosperms (Baldwin 1992; Baldwin *et al.* 1995).

Several general features of the ITS region promote its use for phylogenetic analysis of angiosperms. First, along with the other components of the nrDNA multigene family, the ITS region is highly repeated in the plant nuclear genome. The entire nrDNA repeat unit is present in many thousands of copies arranged in tandem repeats at a chromosomal locus or at multiple loci. This high copy number promotes detection, amplification, cloning, and sequencing of nrDNA. Second, and most importantly from the standpoint of phylogeny reconstruction, is that this gene family undergoes rapid concerted evolution, via unequal crossing-over and gene conversion. This property promotes intragenomic uniformity of repeat-units, in some cases even between nrDNA loci on non-homologous chromosomes, and in general, promotes accurate reconstruction of species relationships from these sequences (Hillis *et al.* 1991; Hamby and Zimmer 1992; Sanderson and Doyle 1992; Wendel *et al.* 1995). As a result, direct sequencing of pooled nrDNA PCR products can be used to extract phylogenetic information in many species. In addition, concerted evolution and sexual recombination may promote nrDNA uniformity within interbreeding populations and thereby minimize the importance of intra-population sampling in phylogenetic studies (Soltis *et al.* 1993). Third, the small size of the ITS region (<700bp in angiosperms) and the presence of highly conserved sequences flanking each of the two spacers make this region easy to amplify, even from herbarium material, using universal eukaryotic primers (White *et al.* 1990). For these many attributes, ITS regions have therefore been used to study wide range of inter family level to intraspecies level genetic variations (Bakker *et al.* 1992; Connell 2000; Lundholm *et al.* 2006) in several plant taxa. A schematic representation of the sequencing of different loci of nrDNA and CpDNA is shown in Fig. 5.

2.10. Single Nucleotide Polymorphisms (SNPs)

SNPs are DNA sequence variations that occur when a single nucleotide (A, T, C, or G) in the genetic code corresponding to a gene, part of a gene, or even a stretch of DNA, that includes more than a single gene or intergenic sequence, has been altered. For a variation to be considered, SNP must occur in at least 1% of the population. SNPs are the most common form of DNA sequence variation and have been used in the study of animal and human genetics, including mapping of the human genome and identification of haplotypes associated with human diseases in medical research (Wang *et al.* 1998). In comparison with other genetic markers, SNPs are more prevalent and conserved within the genome and are used because of their occurrence, more than one per 1,000 base pairs (Osman *et al.* 2003).

With the advent of SNP research, numerous methods of SNP discovery have been proposed, utilized, and are still being developed. Many methods have been adopted for high-thoroughput sequencing. Some of these platforms for detecting SNP include pyrosequencing (Fakhrai-Rad *et al.* 2002), polymorphism ratio sequencing (Blazej *et al.* 2003), degenerate oligonucleotide primer PCR (Jordan *et al.* 2002), ecotilling (Comai *et al.* 2004), and SNPHunter (Wang *et al.* 2005). Due to the influx of SNP data, the application of SNP information, and the varied techniques for obtaining SNPs, bioinformatic techniques have been developed to facilitate the discovery and analysis of SNPs. Internet-accessible tools for data access and display have been developed to help researchers to retrieve data about SNPs based on genes of interest, genetic or physical map locations, or expression pattern (Clifford *et al.* 2004).

Within the last 10 years, SNPs have been extensively studied in medical research and the mapping of SNP markers within the human genome has led to the discovery of many loci associated with predisposition to various diseases (Chen and Sullivan 2003). SNPs have also been used to study population parameters and to estimate divergence within structured human populations (Nicholson *et al.* 2002). Biologists have also investigated the ability of SNPs to estimate population parameters and found that the method of SNP determination may affect the accuracy of predicting genetic occurrences within populations (Kuhner *et al.* 2000). Application of SNPs has varied from studying populations of humans and cattle to bacteria and plants.

Rafalski (2002) reviewed the application of SNPs in crop genetics, discussing linkage disequilibrium, use of expressed sequence tags (ESTs), and outlining discovery procedures, assays, SNPs as markers, and SNP mapping. Studies have been conducted in both major crop plants and in specialized crops, with SNP research in plants accelerating. SNP frequency and haplotype variation were determined in an extensive study involving 25 genotypes of soybean (Zhu *et al.* 2003) and 12 genotypes of wheat (Somers *et al.* 2003). In a similar study, Grivet *et al.* (2003) used ESTs to discover SNPs in sugarcane and suggested that the polymorphisms could serve as potential markers for sugarcane breeding. SNPs have also served as a potential tool to study the genetic diversity within populations of *Eurycoma longifolia* reflecting geographic origin of their individual plants and different natural populations (Osman *et al.* 2003). SNPs are yet to be utilized in plants to infer phylogenetic relationships and therefore, development of SNP-based markers within a taxon would be beneficial in conducting phylogenetic analysis.

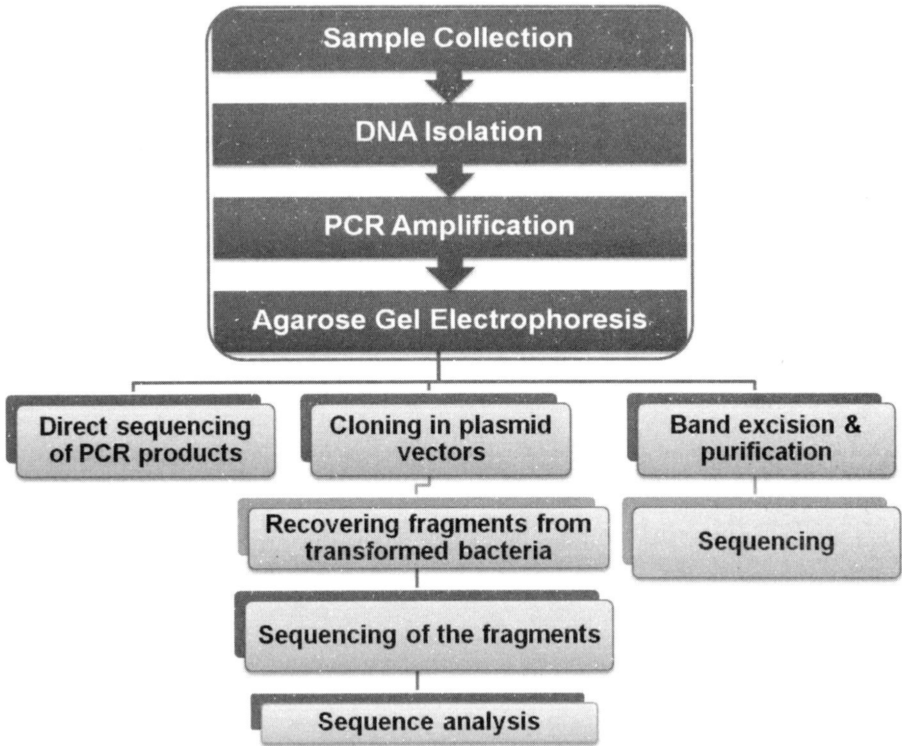

Fig. 5. Schematic representation of the various steps for determining sequences of different loci for ITS and Cp DNA analysis based phylogenetics

2.11. Diversity Array Technology (DArT)

DArT is a new technique of DNA polymorphism analysis. This is hybridisation-based methods using nucleic acids immobilised on solid-state surfaces. Diversity array technology allows simultaneous typing of several hundred polymorphic loci spread over a genome without any previous sequence information about these loci (Jaccoud *et al.* 2001; Wenzl *et al.* 2004). The technique has also been shown to be reproducible and cost effective.

DArT markers are polymorphic segments of genomic DNA that are present in a particular genomic representation and are identified through differential hybridization on a diversity 'genotyping array' that is developed specifically for this purpose. This technology involves initial development of a 'discovery array', which is then used to identify polymorphic DArT markers that are assembled into a 'genotyping array'. The 'discovery array' is developed from a metagenome (pool of genomes representing the germplasm of interest) that is subjected to reduce the level of repetitive DNA (Kilian *et al.* 2005). Diversity arrays generally detect polymorphisms due to single base-pair changes (SNPs) at the restriction sites of endonucleases, and insertion-deletions/ rearrangements within restriction fragments (Jaccoud *et al.* 2001).

There are innumerable PCR generated markers available, but each has its own advantages and limitations. To be a good marker, the following properties would generally be desirable for a molecular marker:

- Moderate to highly polymorphic
- Co-dominant inheritance
- Unambiguous assignment of alleles
- Frequent occurrence in the genome
- Even distribution throughout the genome
- Selectively neutral behaviour (*i.e.*, no pleiotropic effects)
- Easy access (*i.e.*, by purchasing or fast procedures)
- Easy and fast assay (*e.g.*, by automated procedures)
- High reproducibility
- Easy exchange of data between laboratories
- Low cost for both marker development and assay

No single molecular marker fulfils all of these criteria. However, one can choose between a variety of marker systems, each of which combines some – or even most – of the above-mentioned characteristics (Table 2).

Table 2. PCR- based molecular markers used in the molecular systematics of plants

Marker	Co-dominant	Polymorphism	Locus Specificity	Cost	References
Restriction fragment length polymorphism (RFLP)	Yes	Medium	Yes	Medium	Botstein *et al.* 1980
Simple sequence repeats, short tandem repeats or Simple sequence length polymorphisms (SSR, STR, SSLP)	Yes	Very high	Yes	Low	Hamada and Kakunaga 1982; Tautz *et al.* 1986; Litt and Luty 1989
Directed amplification of minisatellite DNA or Variable number of tandem repeats (DAMD ,VNTR)	No	High	No	Low	Jeffreys *et al.* 1985, 1990
Random amplified polymorphic DNA (RAPD)	No	Medium	No	Low	Williams *et al.* 1990
Cleaved amplified polymorphic sequences also called as PCR-RFLP (CAPS)	Yes	Medium	Yes	Medium	Konieczny and Ausubel 1993
Inter simple sequence repeat ISSR)	No	High	No	Low	Meyer *et al.* 1993; Zietkiewicz *et al.* 1994
Sequence characterized amplified regions (SCAR)	Yes/No	Low/Medium	Yes	Medium	Paran and Michelmore 1993; McDermott *et al.* 1994
Single primer amplification reactions (SPAR)	Yes	High	No	Low	Gupta *et al.* 1994
Amplified fragment length polymorphism (AFLP)	No	High	No	Medium	Vos *et al.* 1995
Single nucleotide polymorphism (SNP)	Yes	Very high	Yes	High	Brookes 1999; Cho *et al.* 1999
Diversity array technology (DArT)	No	High	Yes	Very high	Jaccoud *et al.* 2001
Sequencing of different genes from nrDNA and CpDNA	Yes	High	Yes/No	High	Baldwin *et al.*1992, 1995

Source: Weising *et al.* 1995

3. Experimental Protocols Used in the Plant Molecular Systematics

3.1. Collection and preservation of plant material in the field

An important part of any taxonomically oriented sampling procedure is the appropriate labelling of the specimens, and the collection and preservation of herbarium vouchers to document the collected samples. Plant materials should be processed in the herbarium following standard herbarium procedure (Jain and Rao 1977). Voucher specimens should be prepared for all collected materials, and it should be properly dried, labelled, and identified by an expert taxonomist. Identifications should be carried out with the help of Flora, Monographs and other authentic herbarium specimens deposited in the recognized herbaria. Apart from collection of specimens, data should also be gathered on habit, habitat, abundance, frequency, geographical information like latitude, longitude and altitude of the localities visited during the course of the study. Voucher specimens should be deposited in the recognised regional, national and international herbaria for future reference (Fig. 6).

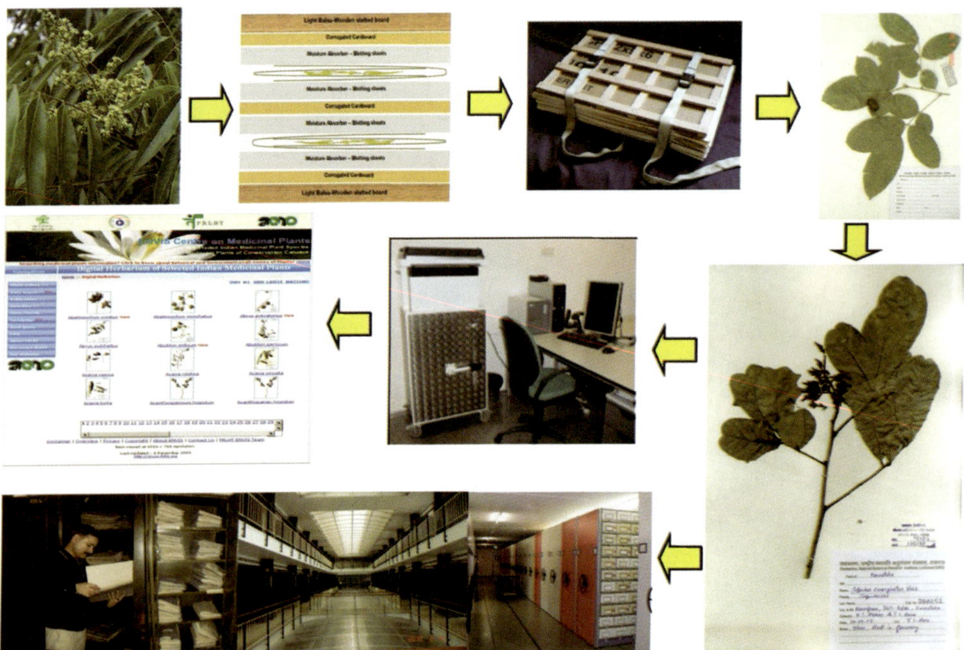

Fig. 6. Different stages during the sampling and processing of plant samples for herbarium archiving

Quality and quantity of plant DNA are greatly influenced by the condition of the starting material. It is always good to have fresh, young tissue available in the botanical garden or in the vicinity of the working place. However, difficulties may arise when the plant species of interest grow at remote locations, and the collected material has to be stored for days, weeks, or even months before they are returned to the laboratory. Large number of methods has been tested to optimize field collection and preservation of plant material in the absence of laboratory facilities (Adams *et al.* 1999; Chase and Hills 1991; Doyle and Dickson 1987; Flournoy *et al.* 1996; Liston *et al.* 1990; Nickrent, 1994; Pyle and Admas 1989; Rogstad 1992; Sytsma *et al.* 1993; Tai and Tansley 1990). Three useful strategies emerged from such experimentation.

(i) Chemical preservation of plant tissue in solutions containing high concentrations of both cetyltrimethyl-ammonium bromide (CTAB) and NaCl (Rogstad 1992; Nickrent 1994)

(ii) Chemical preservation of plant tissue in ethanol (Flournoy *et al.* 1996; Murray and Pitas 1996)

(iii) Rapid drying of plant tissues in silica gel or another desiccating agent (Chase and Hills 1991; Liston *et al.* 1990)

Generally, plants are collected from remote locations for plant systematics studies, therefore, silica gel method (Chase and Hills 1991) has been found to be one of the most suitable methods, and we have been using this method routinely in our laboratory. In this method plant specimen and tissue samples should be collected bearing the same voucher specimen number for DNA studies. About four to six grams of leaves should be placed in a small paper envelope (or wrapped in the tissue paper). Insert the paper envelope containing leaves into a zip-lock plastic bag (25 x 18 cm), and after leaves were placed in bags, 50-60 grams of blue self-indicating (6/20 mesh) silica gel should be added in the bags. The bags should be gently shaken to distribute the silica gel (Fig. 7). After the leaf samples had been in the silica gel for over 12 hours, it should be checked for dryness by bending one leaf. If silica gel turns from a dark violet-blue to pale blue and finally pale white, then fresh silica can be added to the bags. The tissues can remain over silica gel till the time of DNA extraction (Chase and Hills 1991).

3.2. Isolation of plant genomic DNA

Modern molecular techniques like PCR and restriction enzyme digestion require isolation of genomic DNA of suitable purity. The DNA extraction process involves separation of DNA from naturally occurring plant cell constituents such as polysaccharides and polyphenolic compounds, followed by deproteinization of the aqueous solution containing the DNA and further precipitation and purification steps. However, a variety of problems are encountered during the isolation and purification of high molecular weight DNA from plant species for example:

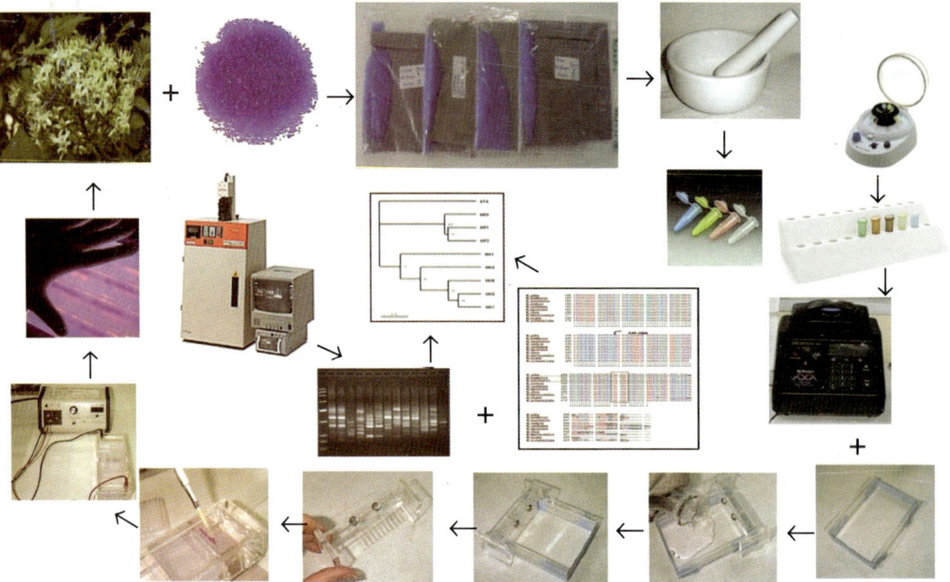

Fig. 7. Different stages of sampling and processing of tissue for molecular systematics

(i) Partial or total DNA degradation due to the presence of endogenous nucleases.

(ii) Co-isolation of highly viscous polysaccharides, which render the handling of samples difficult.

(iii) Co-isolation of polyphenols and other secondary plant metabolites which cause damage to DNA and or inhibit restriction enzymes and *Taq* polymerases.

Considerable variations in biochemical composition of plant species have led to the development of different DNA isolation protocols. One such conventional method used routinely to isolate DNA from a variety of plant species in our laboratory is described below:

3.2.1. Conventional method (CTAB method –Doyle and Doyle, 1990)

Cetyl trimethyl ammonium bromide (CTAB) is a cationic detergent which solubilises membranes and forms a complex with DNA. This general procedure has been used on a wide array of plant genera and tissue types. The protocol is relatively simple, fast, and easily scaled from milligram to grams of tissue and requires no cesium chloride density gradient centrifugation. The protocol is described as follows.

I. Tissue grinding and DNA extraction

- About 1000 mg of the dried sample tissue is macerated to a fine powder by adding liquid nitrogen in pre-cooled mortar and pestle and 100mg PVPP (poly vinyl-polypyrrolidone) added.
- The powdered material thus obtained is transferred to 30 ml centrifuge tubes containing pre- warmed 10 ml of 2% CTAB buffer, and incubated at 65^0 C for 1 hour with intermittent mixing.

  (This step facilitates the breaking or digesting of the cell wall and the membranes in order to release the cellular constituents. The released DNA needs to be protected from endogenous nucleases. EDTA in the extraction buffer chelate magnesium ions which is a necessary co-factor for nuclease).
- After incubation the homogenate is extracted with 10ml (equal amount of the CTAB extraction buffer) chloroform-iso-amyl alcohol (CIA; 24:1 v/v), and mixed up gently 20-30 times and centrifuged at 8000 rpm for 15 min at room temperature.
- The aqueous phase is transferred to fresh tubes very carefully leaving behind the cell debris, and extracted again with CIA solution and centrifuged in the same condition.
- Ice-cold isopropanol (equal volumes of the aqueous phase) and 5M NaCl (10% of the aqueous phase) is added and kept at -20°C overnight. *(This step facilitates the precipitation of the nucleic acid).*

II. DNA precipitation

- Nucleic acid is recovered as a pellet on centrifugation at 12000 rpm for 10 min at 4°C.
- The pellet is washed twice with 70% ethanol in the centrifugation conditions as above. The recovered pellet is air dried and subsequently dissolved in 200µl TE buffer.
- (***Note:*** The nucleic acid obtained from the above procedure is in crude form, and contaminated with RNAse, proteins, phenols, carbohydrates etc. To get rid of these contaminants and to obtain pure and high quality DNA, it is further subjected to the following purification steps).

III. DNA purification

- The nucleic acid obtained is transferred to the fresh 1.5µl tubes and 4µl RNase A (10µg/ml) added and incubated at 37°C with intermittent mixing in a dry bath for 30 min to eliminate the RNA.

- After incubation, the nucleic acid is extracted with CIA 24:1 (equal volumes of nucleic acid), gently mixed and centrifuged at 12000 rpm for 10 min at room temperature and repeated again.

  (This step further eliminates the left over proteins and other contaminants from the crude nucleic acid).
- The final aqueous phase is transferred to a fresh tubes and ice cold isopropanol (equal volumes of the aqueous phase) and 3M sodium acetate (10% of the total volumes of aqueous phase) added and kept at -20^{0}C to precipitate the DNA.
- After 2-3 hours this precipitate is centrifuged to 12000 rpm at 4^{0}C for 10 min to recover the pellet.
- The pellet is washed twice with 70% ethanol; air dried and re-suspended in 500μl 1X TE buffer.

3.2.2. Commercially available kits for DNA isolation

A large number of DNA isolation kits are commercially available in the market, which includes:

(i) Nucleon Phytopure (Amersham Biosciences).

(ii) Fast DNA and Geneclean for Ancient DNA (Qbiogene).

(iii) NucPrep (Applied Biosystems).

(iv) Puregene & Generation DNA Purification Systems (Genera Systems).

(v) AquaPure DNA Tissue Kit (Bio-Rad Laboratories).

(vi) Dynabeads DNA Direct (Dyanal).

(vii) Plant Dnazol reagent & Easy-DNA Kit (Invitrogen Life Technolgies).

(viii) NucleoSpin Plant (Macherey-Nagel).

(ix) peqGOLD DNA pure PT (peqlab).

(x) Wizard Magnetic 96 DNA Plant System (Promega).

(xi) DNeasy Plant kit (Qiagen).

(xii) Extract-N-Amp Plant and Seeds PCR Kits (Sigma-Adrich).

Commercial DNA isolation kits are very convenient, but are usually more expensive, particularly if the costs of the chemicals involved in the isolation of genomic DNA are taken into consideration. These kits are very user friendly and are accompanied by detailed instruction manuals (Weising *et al.* 2005).

3.3. Quantification of Isolated DNA

After the isolation of genomic DNA its quality or purity is tested using spectrophotometric measurements of UV absorption at wavelengths 230, 260 and 280 nm. Measures of DNA purity may be determined by the OD_{260}:OD_{280} ratios. The OD_{260} value provides a measure of concentration (Approx. 1.0 reading at OD_{260} is equivalent to 50µg/ml). A pure DNA has an OD_{260}:OD_{280} ratio of 1.8 ±1.

DNA quality is also checked by agarose gel electrophoresis. The homogenously high molecular DNA preparation appears as a single band (Fig. 8).

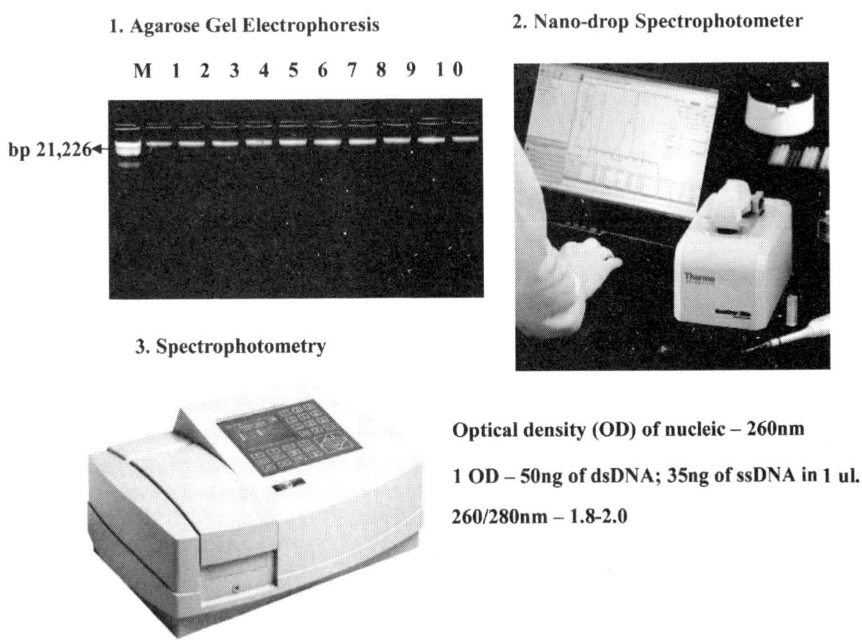

Fig. 8. Determination of the quality and quantity of isolated DNA using agarose gel electrophoresis, Spectrophotometer and nano-drop Spectrophotometer

3.4. Gel electrophoresis

Electrophoresis is a technique commonly used to separate protein, RNA or DNA molecules through agarose or polyacrylamide gels. It is capable of resolving the fragments that cannot be separated adequately by other procedures. Furthermore their location within the gel can be determined directly by staining with low concentrations of fluorescent intercalating dycs such as Ethidium Bromide SYBR Gold and exposing the gel through UV rays.

3.4.1. Agarose gel electrophoresis

Agarose gel electrophoresis is a rapid, simple and standard method used to separate, identify and purify DNA fragments (Sambrook *et al.* 1989). Agarose, though low in resolving power yet has a great range of separation. DNAs from 50bp to several megabases in length can be separated on agarose gels of various concentrations and configurations. Generally 0.8% agarose gels are used for checking the genomic DNA samples, while 1.5% agarose gels are used to separate PCR amplified fragments.

- Agarose gel (0.8 g) powder is boiled in 100 ml 0.5X TBE buffer for 1-2 minutes. Ethidium bromide to a concentration of 0.5µg/ml is added to this agarose solution and allowed to cool.

  Caution: *Ethidium bromide is a carcinogen. Wear gloves while handling this dye. It is also light sensitive and therefore all solutions and gels containing this dye should be kept in dark or covered).*

- Combs are adjusted to about 0.5-1.0mm above the plate with the help of a stand in a gel tray, to form wells for loading the DNA samples.

- Agarose solution when cooled to about a temperature of 40-50^0 C, is poured onto the plate in the gel tray and allowed to polymerize for about 10-15 min *(when ready turns whitish and opaque).*

- Combs are carefully removed once the gel is polymerized. Gel is then placed in the electrophoresis tank containing 0.5X TBE buffer with the wells closest to the cathode (black) end. *(Note: Gel should be completely immersed in the buffer).*

- About 6µl of the genomic DNA is mixed with 6µl sterilized water and 2µl of 10X loading dye and loaded in the wells, against 10µl of the DNA marker Lamda DNA *Hind*III- *Eco*RI double digest as a standard to compare the band intensities.

- Electrophoresis apparatus is then covered with a lid and connected to a power supply. The gel is run for 2-4 hours at a constant voltage (100V). *(DNA molecules along with the loading dye will start moving towards the anode at the start of the current).*

- After the electrophoresis is completed power supply is turned off, gel is removed and visualized and archived using Gel Documentation System).

  *(**Caution:** UV rays are harmful. Do not expose skin and eyes to UV for long durations. Use UV opaque eye glasses and face shield).*

- Presence of highly resolved high molecular weight bands confirms the good quality DNA.

3.5. PCR amplification

Polymerase chain reaction is a versatile technique based on the enzymatic *in vitro* method for the amplification of DNA, discovered by Mullis and Faloona (1987) which allows amplification of large quantities of a targeted region of DNA to generate millions of copies of that region even from a single, initial copy. This technique has become a ubiquitous and powerful tool in diagnostics, forensics and research biology.

PCR amplification is essentially carried out by using a thermostable DNA polymerase and short DNA fragments or "primers" to direct the synthesis of a targeted segment of DNA. The synthesis is repeated numerous times in different cycles, allowing the products of previous synthesis cycles to serve as template for the next cycle, resulting in an exponential amplification of the targeted region of DNA.

Procedure

Master mix is prepared using different PCR components:

- Master mix is aliquot (a quantity of the master mix subtracting the template volume) into each labelled PCR tube. Thereafter template DNA is added into each tube.
- The tubes are vortexed gently and placed in PCR machine.
- Lid is closed slightly tight and PCR programme is run under cycling conditions as per the PCR technique you intend to carry out.
- After the PCR-cycling is completed the amplified DNA fragments are resolved on 1.5% agarose gel.

3.6. Optimization of PCR conditions for multi-locus markers (RAPD, DAMD and ISSR)

Variation in any reaction components as well as any part of the PCR programme can show quite unpredictable effects. This might be because of the various laboratory conditions and different plant taxa in different studies. Therefore, it is important to optimize the PCR reaction components. Optimizations should be carried out progressively for each PCR component in a two-dimensional pattern, *i.e.*, varying one component (low to high) and keeping rest of the parameters constant at a time. Multi-locus techniques such as RAPD, DAMD and ISSR are prone to artefacts and are not always reproducible. Banding patterns may vary in different experiments and among individuals. Such variation sometimes poses a serious problem in genetic diversity analyses. There are several sources of artefacts, and they could be due to primer, template DNA, magnesium and dNTP concentrations, annealing temperature stringency, enzymes and PCR machines. These factors ultimately influence the final amplification profile, and it is therefore, essential to optimize the different parameters.

We optimized the essential reaction conditions for RAPD, DAMD and ISSR in case of pomegranate (Narzary 2010). RAPD reaction which included 25ng template DNA, 0.2µM primer, 100µM each of dATP, dGTP, dCTP and dTTP, 2mM Mg^{2+} and 0.5 units of *Taq* polymerase along with suitable 1x buffer (10mM Tris-Cl, pH 8.3, 50mM KCl, 0.001% w/v gelatine) in 25µl final reaction volumes and amplification in 45 cycles of denaturation at 94°C for 1 min, annealing at 36°C for 1 min and extension at 72°C for 2 min (Williams *et al.* 1990). The different concentrations of template DNA, primer, $MgCl_2$ and dNTPs were tested for RAPD-PCR and optimum results were found at 50ng, 0.4µM, 2.5mM and 200µM of each dNTPs, respectively. Concentrations above or below these values were found either scoring less number of bands or white smearing back grounds in the gel profiles.

In DAMD, the reaction condition of Zhou *et al.* (1997) was optimized for pomegranate DNAs varying the concentrations of PCR components (Narzary 2010). The favourable concentration/amount of the components observed in this optimization was at 60ng template DNA, 0.8µM primer, 2mM $MgCl_2$ and 200µM each of dNTPs in 25µl reaction volume. These reaction conditions were then followed further in the DAMD profile generations with pomegranate accessions.

The optimizations exhibited the most efficient amplification at concentrations of 20ng template DNA, 0.2µM primer, 2mM $MgCl_2$ and 200µM each of dNTPs in case of ISSR-PCR. Further increase in the dNTP concentration inhibits the reaction when $MgCl_2$ was kept constant or vice versa. The other components, such as templates and primer also showed smearing unclear bands beyond these concentrations (Fig. 9).

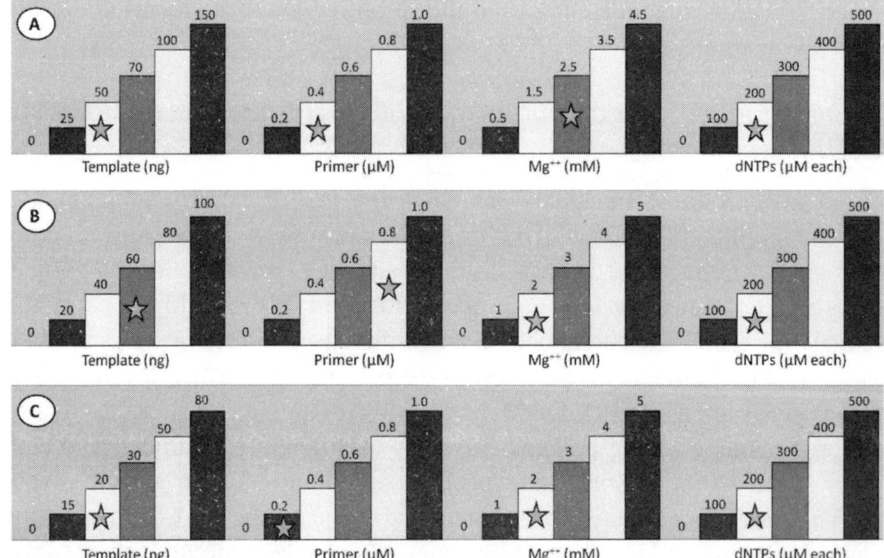

Fig. 9. PCR Optimization for (A) RAPD, (B) DAMD and (C) ISSR. Optimization of PCR reaction conditions was done for pomegranate (Source: Narzary 2010)

The final reaction conditions derived from these optimizations of RAPD, DAMD and ISSR-PCR for pomegranate accessions gave consistent profiles. However, above optimization of PCR conditions for multi-locus markers may vary in different plants, therefore, one has to do the optimization for the plant(s) in question.

4. Data Analysis and Statistical Methods in Plant Molecular Systematics

There are large numbers of computer based softwares available these days for the analyses of molecular systematics data. Some of the softwares and their websites are listed below in the Table 3. Details of the statistical methods have been dealt in chapter 19 of this manual. Programmes mentioned below are the routinely used, and the choice of a particular programme will depend upon the kind of data one would like to analyse.

Table 3. Computer based programme used for the analyses of data in molecular systematics

Programme	Web Site	Operating System	Description
AFLP-SURV	www.ulb.ac.be/sciences/ lagev/aflp-surv.html	Web based	Analyzes dominant data, calculates between populations and between individuals, with specified mating systems
Arlequin v3.5	http://cmpg.unibe.ch/ software/arlequin35/	Windows, MacOS, Linux	Analyzes population genetic data
BioEdit v7.2.3	http://www.mbio.ncsu. edu/bioedit/bioedit.html	Windows	Handles and aligns DNA sequences
ClustalX	http://www.clustal.org/	Windos, MacOS, UNIX	Provides multiple alignment of DNA sequences
FREETREE	www.natur.cuni.cz/~flegr /programs/freetree.html	Windows, MacOS	Analyzes phylogenetic data
GeneScanView	http://www.bmr-genomics. it/seq_i-index.html	Windows	Reads files (ABI and some other brands) for analyzing fragments for AFLP or microsatellite analysis
Genographer	http://hordeum.oscs. montana.edu/genographer/	Windows	Reads files (ABI and some other brands) for analyzing fragments for AFLP or microsatellite analysis
GenAlEx v6.5	http://biology.anu.edu.au/ GenAlEx/Download.html	Windows, MacOS	Analyzes population structure, PCO, Mantel test
MEGA v6	http://www.megasoft ware.net/	Windows, MacOS	Analyzes phylogenetic data
NTSYS v 2.2	http://www.exetersoft ware.com/cat/ntsyspc/ ntsyspc.html	Windows/ Commercial	Performs PCO, Mantel test, also analyzes quantitative data
POPGENE v1.32	http://www.ualberta.ca/~ fyeh/popgene.html	Windows	Analyzes population genetic data
STRUCTURE v2.3.4	http://pritchardlab stanford.edu/structure. html	Windows, UNIX	Analyzes population structure, migration, assignment, hybrid zones

(*Contd.*)

Programme	Web Site	Operating System	Description
TFPGA	http://www.marksgenetic software.net/tfpga.htm	Windows	Analyzes population genetics, dominant /codominant data
PAUP* v4	paup.csit.fsu.edu	Windows, MacOS, UNIX/ Commercial	Analyzes phylogenetic data
PHYLIP v3.695	http://evolution.genetics. washington.edu/phylip.html	Windows, MacOS	Analyzes phylogenetic data
TreeView v1.6.6	www.taxonomy.zoology. gla.ac.uk/rod/treeview. html	Windows, MacOS	Provides visualization of trees, from PAUP, PHYLLIP, ClustalW
FREETREE	www.natur.cuni.cz/~flegr/ programs/freetree.html	Windows, MacOS	Analyzes phylogenetic data

5. Recipes Used in Experimental Protocols

2 X CTAB Buffer (1000 ml)

CTAB powder (20 g) is dissolved in 400 ml double distilled water. 280 ml of NaCl, 100 ml of 1M Tris (pH 8.0) and 40 ml of 0.5 M EDTA (pH 8.0) solutions are added and dissolved thoroughly. Final volume made upto 1000 ml. Sterilized by autoclaving.

1 M Tris-HCl pH 8.0 (1000 ml)

121.1g Tris base is dissolved in 800 ml distilled H_2O, pH is adjusted to 8.0 using concentrated HCl. Final volume made upto 1000 ml with distilled H_2O and sterilized by autoclaving.

0.5M EDTA pH 8.0 (1000 ml)

186.12 g disodium EDTA dehydrate is dissolved in 800 ml ddH_2O. 20 g NaOH pellets added slowly more NaOH pellets are added until pH is adjusted to 8.0. Final volume made up to 1000 ml using ddH_2O and Sterilized by autoclaving.

50 X TE Buffer pH 8.0 (1000 ml)

1M Tris 50 ml (pH 8.0) is added mixed with 10 ml of 0.5M EDTA (pH 8.0). Final volume adjusted to 1000 ml using distilled H_2O, sterilized by autoclaving. 1X concentration is used for dissolving DNA.

3 M Sodium acetate pH 5.2 (1000 ml)

Sodium acetate trihydrate 408.24 g is dissolved in 800 ml distilled H_2O, pH adjusted to 5.2 with glacial acetic acid. Final volume adjusted to 1000 ml and sterilized by autoclaving.

5 X TBE Buffer

54 g Tris base and 27.4 g boric acid are dissolved in 800 ml of distilled water. 20 ml 0.5M EDTA (pH 8.0) added and volume made up to 1000 ml with distilled water and sterilized in autoclave.

(TBE is usually made and stored as a 5× stock solution. p^H adjusted to 8.3. The stock buffer is diluted to 0.5× before using).

5 M NaCl (1000 ml)

Sodium chloride (292.2 g) is dissolved in 800 ml of distilled water thoroughly. Final volume is made to 1000 ml.

CIA 24:1 v/v (1000 ml)

Chloroform (960 ml) is mixed with 40 ml of iso-amyl alcohol and stored below 15^{0}C.

RNase A

RNase 10 mg dissolved in 1 ml distilled H_2O. Mixture is heated to 100^0 C for 10 min to inactivate DNases.

10 X Blue loading dye mix (10 ml)

40 mg of Bromophenol blue and xylene cyanol FF each and 2.5 g of Ficoll are dissolved in 8 ml of distilled water. Final volume is adjusted to 10 ml sterilized by autoclaving. 1X concentration is used when loading DNA samples on the gels.

10 mg/ml Ethidium bromide (25 ml)

Ethidium bromide powder (25 mg) is dissolved in 25 ml water. After completely dissolved, the container is wrapped in aluminium foil or transferred to dark bottle.

References

Adams PJ, Seinfeld JH, Koch DM (1999) Global concentrations of tropospheric sulfate, nitrate, and ammonium aerosol simulated in a general circulation model. J Geophys Res 104:13791-13823

Bakker FT, Olsen JL, Stam WT, van den Hoek C (1992) Nuclear ribosomal DNA internal transcribed spacer regions (ITS1 and ITS2) define discrete biogeographic groups in *Cladophora albida* (Chlorophyta). J Phycol 28:839-845

Baldwin BG (1992) Phylogenetic utility of the internal transcribed spacers of nuclear ribosomal DNA in plants: An example from the Compositae. Mol Phy Evol 1:3-16

Baldwin BG, Sanderson MJ, Porter JM, Wojciechowski MF, Campbell CS, Donoghue MJ (1995) The ITS region of nuclear ribosomal DNA: a valuable source of eviden Ann Missouri Bot Gard 82:247-277

Bebeli PJ, Zhou Z, Somers DJ, Gustafson JP (1997) PCR primed with minisatellite core sequences yields DNA fingerprinting probes in wheat. Theor App Gen 95:276-283

Beebee T and Rowe G (2004) An Introduction to Molecular Ecology, Oxford University Press, Oxford

Blazej RG, Paegel BM, Mathies RA (2003) Polymorphism ratio sequencing: A new approach for single nucleotide polymorphism discovery and genotyping. Gen Res 13:287-293

Botstein D, White RL, Skolnick M, Davis RW (1980) Construction of a genetic linkage map in man using restriction fragment length polymorphism. Am J Hum Gen 32:314-331

Brookes AJ (1999) The essence of SNPs. Gene 234:177-186

Broun P, Tanksley SD (1993) Characterization of tomato clones with sequence similarity to human minisatellites 33.6 and 33.15. Plant Mol Biol 23:231-142

Burow MD and Blake TK (1998) Molecular tools for the study of complex traits. In: Molecular dissection of complex traits. Paterson AH (ed.), CRC Press, Washington, DC, pp. 13-29

Caetano-Anolles G, Bassam BJ, Gresshoff PM (1991) High resolution DNA amplification fingerprinting using very short arbitrary oligonucleotide primers. Bio/Technology 9:553-557

Chase MW, Hills HH (1991) Silica-gel - an ideal material for field preservation of leaf samples for DNA studies. Taxon 40:215-220

Chen LFO, Kuo HY, Chen MH, Lai KN, Chen SCG (1997) Reproducibility of the differential amplification between leaf and root DNAs in soybean revealed by RAPD markers. Theor App Gen 95:1033-1043

Chen X, Sullivan PF (2003) Single nucleotide polymorphism genotyping: biochemistry, protocol, cost and throughput. Pharmacogenomics J 3:77-96

Cho RJ, Mindrinos M, Richards DR, Sapolsky RJ, Anderson M, Drenkard E, Dewdney J, Reuber TL, Stammers M, Federspicl N, Theologis A, Yang WH, Hubbel E, Au M, Chung EY, Lashkari D, Lemieux B, Dean C, Lipshutz RJ, Ausubel EM, Davis RW, Oefner PJ (1999). Genome-wide mapping with biallelic markers in *Arabidopsis thaliana*. Nat Genet 23:203-207

Clifford RJ, Edmonson MN, Nguyen C, Scherpbier T, Hu Y, Buetow KH (2004) Bioinformatics tools for single nucleotide polymorphism discovery and analysis. Ann New York Acad Sci 1020:101-109

Comai L, Young K, Till BJ, Reynolds SH, Greene EA, Codomo CA, Enns LC, Johnson JE, Burtner C, Odden AR, Henikoff S (2004) Efficient discovery of DNA polymorphisms in natural populations by Ecotilling. Plant J 37:778-786

Connell LB (2000) Nuclear ITS region of the alga *Heterosigma akashiwo* (Chromophyta, Raphidophyceae) is identical in isolates from Atlantic and Pacific basins. Marine Biol 136(6):953-960

Cruzan M (1998) Genetic markers in plant evolutionary ecology. Ecology 79:400-412

Deragon JM, Landry BS (1992) RAPD and other PCR-based analysis of plant genomes using DNA extracted from small leaf discs. PCR Methods Appli 1:175-180

Doyle JJ, Dickson EE (1987) Preservation of plant samples for DNA restriction endonuclease analysis. Taxon 36:715-722

Doyle JJ, Doyle JL (1990) Isolation of plant DNA from fresh tissue. Focus 12:13-15

Fakhrai-Rad H, Pourmand N, Ronaghi M (2002) Pyrosequencing (TM): An accurate detection platform for single nucleotide polymorphisms. Hum Mutation 19:479-485

Fang DQ, Roose ML (1997) Identification of closely related citrus cultivars with inter-simple sequence repeat markers. Theor App Gen 95:408-417

Flournoy LE, Adams RP, Pandy RN (1996) Interim and archival preservation of plant specimens in alcohols for DNA studies. Biotechniques 20:657-660

Georges M, Gunawardana A, Threadgill DW, Lathrop M, Olsaker I, Mishra A, Sargeant LL, Schoeberlein A, Steele MR, Terry C, Threadgill DS, Zhao X, Holm T, Fries R, Womack JE (1991) Characterization of a set of variable number of tandem repeat markers conserved in Bovidae. Genomics 11:24-32

Gidoni D, Rom M, Kunik T, Zur M, Izsak E, Izhar S, Firon N (1994) Strawberry-cultivar identification using randomly amplified polymorphic DNA (RAPD) markers. Plant Breed 113:339-342

Godwin ID, Aitken EAB, Smith LW (1997) Application of intersimple sequence repat (ISSR) markers to plant genetics. Electrophoresis 18:1524-1528

Gray M, Charpentier A, Walsh K, Wu P, Bender W (1991) Mapping point mutations in the Drosophila rosy locus using denaturing gradient gel blots. Genetics 127:139-149

Grivet L, Glaszmann JC, Vincentz M, da Silva F, Arruda P (2003) ESTs as a source for sequence polymorphism discovery in sugarcane:example of the Adh genes. Theor App Gen 106:190-197

Gupta M, Chyi YS, Romero-Severson J, Owen JL (1994). Amplification of DNA markers from evolutionary diverse genomes using single primers of simple-sequence repeats. Theor App Gen 89:998-1006.

Hamada H, Kakunaga T (1982) Potential Z-DNA forming sequences are highly dispersed in the human genome. Nature 298:396-398

Hamby RK and Zimmer EA (1992) Ribosomal RNA as a phylogenetic tool in plant systematics. In: Molecular Systematics of Plants. Soltis PS, Soltis DE and Doyle JJ (eds.) Chapman and Hall, New York, pp. 50-91

Heath DD, Iwama GK, Devlin RH (1993) PCR primed with the VNTR core sequences yields species specific patterns and hypervariable probes. Nucl Acids Res 21:5782-5785

Heun M, Helentjaris T (1993). Inheritance of RAPDs in F_1 hybrids of corn. Theor App Gen 85:961-968

Hillis DM, Dixon MJ (1991) Ribosomal DNA: molecular evolution and phylogenetic inference. The Quart Rev Biol 66(4):411-453

Hillis DM, Moritz C, Porter CA, Baker RJ (1991) Evidence for biased gene conversion in concerted evolution of ribosomal DNA. Science 251:308-310

Hilu KW, Borsch T, Muller K, Soltis DE, Soltis PS, Savolainen V, Chase MW, Powell MP, Alice LA, Evans R, Sauquet H, Neinhuis C, Slotta TAB, Rohwer JG, Campbell CS, Chatrou LW (2003) Angiosperm phylogeny based on *mat*K sequence information. Am J Bot 90:1758-1776

Isaac PG (1994) Methods in Molecular Biology: Protocols for Nucleus Acid Analysis by Nonradioactive probes. Humana Press Inc., Totowa, NJ

Jaccoud D, Peng K, Feinstein D, Kilian A (2001) Diversity arrays: a solid state technology for sequence information independent genotyping. Nucl Acids Res 29:25

Jain SK and Rao RR (1977) A Handbook of Field and Herbarium Methods. Today and Tomorrow's Printers and Publications, New Delhi

Jeandroz S, Bastein D, Chandelier A, Du Jardin P, Favre JM (2002) A set of primers for amplification of mitochondrial DNA in *Picea abies* and other conifer species. Mol Ecol Notes 2:389-392

Jeffreys AJ, Neumann R, Wilson V (1990) Repeat unit sequence variation in minisatellites: a novel source of DNA polymorphism for studying variation and mutation by single molecule analysis. Cell 60:473-485

Jeffreys AJ, Wilson V, Thein SL (1985) Hypervariable "minisatellite" regions in human DNA. Nature 314:67-73

Jenczewski E, Properi JM, Ronfort J (1999) Differentiation between natural and cultivated populations of *Medicago sativa* (Legiminosae) from Spain: analysis with random amplified polymorphic DNA RAPD markers and comparison to allozymes. Mol Ecol 8:1317-1330

Jordan B, Charest A, Dowd JF, Blumenstiel JP, Yeh RF, Osman A, Housman DE, Landers JE (2002) Genome complexity reduction for SNP genotyping analysis. Pro Natl Acad Sci USA 99:2942-2947

Julier C, Gouyon DD, Georges M, Guenet JL, Nakamura Y, Avner P, Lathrop GM (1990) Minisatellite linkage maps in the mouse by cross-hybridization with human probes containing tandem repeats. Pro Natl Acad Sci USA 87:4585-4589

Kelly JD, Miklas PN (1998) The role of RAPD markers in breeding for disease resistance in common bean. Mol Breed 4:1-11

Kendall J (1928) Separations by the ionic migration method. Science 67:163-167

Kilian A, Huttner E, Wenzl P, Jaccoud D, Carling J, Caig V, Evers M, Heller-Uszynska K, Cayla C, Patarapuwadol S, Xia L, Yang S, Thomson B (2005) The fast and the cheap: SNP and DArT-based whole genome profiling for crop improvement. In: Proceedings of the international congress "In the wake of the double helix: from the green revolution to the gene revolution", 27 -31 May, 2003, Tuberosa R, Phillips RL, Gale M (eds) Avenue Media, Bologna, Italy, pp 443-461

Konieczny A, Ausubel FM (1993) A procedure for mapping *Arabidopsis* mutations using co-dominant ecotype-specific PCR-based markers. Plant J 4:403-410

Kuhner MK, Beerli P, Yamato J, Felsenstein J (2000) Usefulness of single nucleotide polymorphism data for estimating population parameters. Genetics 156:439-447

Lacou V, Haurogne K, Ellis N, Rameau C (1998). Genetic mapping in pea. 1-RAPD-based genetic linkage map of *Pisum sativum.* Theor Appl Gen 97: 905-915

Laemmli UK (1970) Cleavage of structural proteins during the assembly of the head of bacteriophage T4. Nature 227(5259):680-685

Leberg PL (1992) Effects of population bottlenecks on genetic diversity as measured by allozyme electrophoresis. Evolution 46:477-494

Liston A, Rieseberg LH, Adams RP, Do N, Zhu G (1990) A method for collecting dried plant specimens for DNA and isozyme analyses, and the results of a field test in Xinjiang, China. Ann Missouri Bot Gard 77:859-863

Litt M, Luty JA (1989) A hypervariable microsatellite revealed by *in vitro* amplification of a dinucleotide repeat within the cardiac muscle actin gene. Amr J Hum Genet 44:397-401

Liu Z, Furnier GR (1993) Comparison of allozyme, RFLP and RAPD markers for revealing genetic variation within and between trembling aspen and big tooth aspen. Theor Appl Gen 87:97-105

Lundholm N, Moestrup O, Kotaki Y, Hoef Emden K, Scholin C, Miller P (2006) Inter- and intraspecific variation of the *Pseudo nitzschia delicatissima* complex (Bacillariophyceae) illustrated by r RNA probes, morphological data and phylogenetic analyses. J Phyco 42:464-481

Markert CL, Moller F (1959) Multiple forms of enzymes, tissue, ontogenetic and species specific pattern. Pro Natl Acad Sci USA, 45:753-763

McDermott JM, Brandle U, Dutly F, Haemmerli UA, Keller S, Muller KE, Wolf MS (1994) Genetic variation in powdery mildew of barley: development of RAPD, SCAR and VNTR markers. Phytopathology 84:1316-1321

Meyer W, Mitchell TG, Freedman EZ, Vilgalys R (1993) Hybridization probes for conventional DNA fingerprinting used as single primers in the polymerase chain reaction to distinguished strains of *Cryptococcus neoformans.* J Clinical Microbiol 31:2274-2280

Mullis KB, Faloona FA (1987) Specific synthesis of DNA in vitro via a polymerase-catalyzed chain reaction. Methods Enzymol 155:335-350

Mullis KB, Ferre F and Gibbs RA (1994) The Polymerase Chain Reaction. Birkhauser, Basel, Switzerland

Murphy RW, Sites JWJr, Buth DG and Haufler CH (1996) Proteins: isozyme electrophoresis. In: Molecular Systematics, 2nd Ed. Hillis DM, Moritz C and Mable BK (eds.), Sinauer Associates, Sunderland, MA, pp. 51-120

Murray MG, Pitas J (1996) Plant DNA from alcohol preserved samples. Plant Mol Bio Rep 14:261-265

Myers RM, Maniatis T, Lerman LS (1987) Detection and localization of single base changes by denaturing gradient gel electrophoresis. Meth Enzym155:501-527

Nagaoka T, Ogihara Y (1997) Applicability of inter-simple sequence repeat polymorphisms in wheat for use as DNA markers in comparison to RFLP and RAPD markers. Theor Appl Gen 94:597-602

Nagaraju J, Kathirvel M, Kumar RR, Siddiq EAb, Hasnain SE (2002) Genetic analysis of traditional and evolved Basmati and non-Basmati rice varieties by using fluorescence-based ISSR-PCR and SSR markers. Pro Natl Acad Sci USA 99:5836-5841

Nakamura Y, Carison M, Krapcho K, Kanamori M, White R (1988) New approach for isolation of VNTR markers. Am J Hum Gen 43(6):854-859

Narzary D (2010) The genus *Punica* L. -Assessment of diversity and systematics using PCR based methods (PhD Thesis)

Nicholson G, Smith AV, Jonsson F, Gustaffson O, Stefansson K, Donnelly P (2002) Assessing population differentiation and isolation from single nucleotide polymorphism data. J Royal Stat Soci, Series B, Statistical Methodology 64:695-715

Nickrent DL (1994) From field to film: Rapid sequencing methods for field collected plant species. BioTechniques 16:470-475

Nickrent DL, Soltis DE (1995) A comparison of angiosperm phylogenies from nuclear 18s rDNA and *rbc*L sequences. Ann Miss Bot Gard 82:208-234

Osman A, Jordan B, Lessard PA, Muhammad N, Haron MR, Riffin NM, Sinskey AJ, Rha C, Housman DE (2003) Genetic diversity of *Eurycoma longifolia* inferred from single nucleotide polymorphisms. Pl Physiol 131:1294-1301

Paran I, Michelmore RW (1993) Development of reliable PCR-based markers linked to downy mildew resistance genes in lettuce. Theor Appl Genet 85:985-993

Pyle MM, Adams RP (1989) *In situ* preservation of DNA in plant specimens. Taxon 38:576-581

Rafalski A (2002) Applications of single nucleotide polymorphisms in crop genetics. Curr Opin Plant Biol 5:94-100

Rajapakse S, Belthoff LE, He G, Estager AE, Scorza R, Verde I, Ballard RE, Baird WV, Callahan A, Monet R, Abbott AG (1995) Genetic linkage mapping in peach using morphological, RFLP and RAPD markers. Theor Appl Gen 90:503-510

Rogstad SH (1992) Saturated NaCl-CTAB solution as a means of field preservation of leaves for DNA analyses. Taxon 41:701-708

Rossetto M, McLauchlan A, Harriss FCL, Henry RJ, Baverstock PR, Lee LS, Maguire TL, Edwards KJ (1999) Abundance and polymorphism of microsatellite markers in the tea tree (*Melaleuca alternifolia*, Myrtaceae). Theor Appl Gen 98:1091-1098

Saiki RK, Gelfand DH, Stoffel S, Scharf SJ, Higuchi R, Horn GT, Mullis KB, Erlich HA (1988) Primer-directed enzymatic amplification of DNA with a thermostable DNA polymerase. Science 239:487-491

Sambrook J, Fritch EF and ManiatisT (1989) Molecular Cloning : A Laboratory Manual. Cold Spring Harbor, NY

Sanderson MJ, Doyle JJ (1992) Reconstruction of organismal and gene phylogenies from data on multigene families: Concerted evolution, homoplasy, and confidence. Syst Biol 41:4-17

Savolainen V, Chase MW, Hoot SB, Morton CM, Soltis DE, Bayer C, Fay MF, De Bruijn AY, Sullivan S and Qiu YL (2000). Phylogenetics of flowering plants based on combined analysis of plastid *atp*B and *rbc*L gene sequences. Syst Biol 49:306-362

Skroch P, Nienhuis J (1995) Impact of scoring error and reproducibility of RAPD data on RAPD-based estimates of genetic distance. Theor Appl Gen 91:1086-1091

Soller M, Beckmann JS (1983). Genetic polymorphism in varietal identification and genetic improvement. Theor Appl Gen 67:25-33

Soltis DE, Morgan DR, Grable A, Soltis PAS, Kuzoff R (1993) Molecular systematics of Saxifragaceae *sensu stricto*. Am J Bot 80:1056-1081

Soltis DE, Soltis PS (1997) Phylogenetic relationships in Saxifragaceae *sensu lato:* a comparison of topologies based on 18S rDNA and rbcL sequences. Am J Bot 84:504-522

Soltis DE, Soltis PS Cleggy MT, Durbin M (1990) *rbc*L sequence divergence and phylogenetic-relationships in Saxifragaceae *sensu-lato*. Pro Natl Acad Sci USA 87:4640-4644

Somers DJ, Kirkpatrick R, Moniwa M, Walsh A (2003) Mining single nucleotide polymorphism from hexaploid wheat ESTs. Genome 49:431-437

Somers DJ, Zhou Z, Bebeli PJ, Gustafson JP (1996) Repetitive, genome-specific probes in wheat (*Triticum aestivum* L.) amplified with minisatellite core sequences. Theor Appl Gen 93:982-989

Staub J, Bacher J, Poetter K (1996) Sources of potential errorsin the application of random amplified polymorphic DNAs in cucumber. Hort Sci 31:262-266

Sytsma KJ, Givnish TJ, Smith JF and Hahn WJ (1993) Obtaining and storing land plant samples for macromolecular comparisons. In: Molecular Evolution: Producing the Biochemical Data. Zimmer EA, White TJ, Cann RL and Wilson AC [eds.], Academic Press, pp23-37

Tai TH, Tanksley SD (1990) A Rapid and Inexpensive Method for Isolation of Total DNA from Dehydrated Plant Tissue. Plant Mol Biol Report 8:297-303

Tanksley SD (1983) Molecular markers in plant breeding. Pl Mol Biol Rep 1:3-8

Tautz D, Trick M and Dover GA (1986) Cryptic simplicity in DNA is major source of genetic variation. Nature 322:652-656

Tsumura Y, Suyama Y, Yoshimura K, Shirato N, Mukai Y (1997). Sequence-Tagged-Sites (STSs) of cDNA clones in *Cryptomeria japonica* and their evaluation as molecular markers in conifers. Theor Appl Gen 94:764-772

Vos P, Hogers R, Bleeker M, Reijans M, Van de Lee T, Hornes M, Frijters A, Pot J, Peleman J, Kuiper M, Zabeau M (1995) AFLP: a new technique for DNA fingerprinting. Nucleic Acids Res 23:4407-4414

Wang DG, Fan J-B, Siao C-J, Berno A, Young P, Sapolsky R, Ghandaur G, Perkins N, Winchester E, Spencer J, Kruglyak L, Stein L, Hsie L, Topaloglou T,Hubbell E, Robinson E, Mittmann M, Morris MS, Shen N, Kilburn D, Rioux J, Nusbaum C, Rozen S, Hudson TJ, Lipshutz R. Chee M, Lander ES (1998) Large scale identification, mapping and genotyping of single nucleotide polymorphisms in the human genome. Science 280:1077-1082

Wang L, Liu S, Niu T, Xu X (2005) SNPHunter: A bioinformatic software for single nucleotide polymorphism data acquisition and management. BMC Bioinformatics 6:60 (doi:10.1186/1471-2105-6-60)

Weising K, Nybom H, Wolff K and Meyer W (1995) DNA fingerprinting in plants and fungi. Boca Raton: CRC Press, London

Weising K, Nybom H, Wolff K, Kahl G (2005) DNA Fingerprinting in Plants: Principles, Methods, and Applications, 2nd edn. CRC Press, Taylor & Francis Group, Boca Raton, Florida

Welsh J, McClelland M (1990) Fingerprinting genomes using PCR with arbitrary primers. Nucleic Acids Res 18:7213-7218

Wendel JF, Schnabel A, Seelanan T (1995) Bidirectional interlocus concerted evolution following allopolyploid speciation in cotton (*Gossypium*). Proc Natl Acad Sci USA 92:280-284

Wenzl P, Carling J, Kurna D, Jaccound D, Huttner E, Kleinhofs A, Kilian A (2004) Diversity Arrays Technology (Dart) for whole genome profiling of barley. Proc Natl Acad Sci USA 101(26): 9915-9920

White TJ, Bruns T, Lee S and Taylor J (1990) Amplification and direct sequencing of fungal ribosomal RNA genes for phylogenetics. In: PCR Protocols: A Guide to Methods and Application, Innis MA, Gelfand DH, Sninsky JJ and White TJ (eds.), San Diego, California: Academic Press, pp. 315-322

Wiesner I, Wiesnerova D (2003) Effect of resolving medium and staining procedure on inter-simple-sequence-repeat (ISSR) patterns in cultivated flax germplasm. Genet Res Crop Evol 50:849-853

Williams JGK, Kubelik AR, Livak KJ, Rafalski JA, Tingey SV (1990) DNA polymorphisms amplified by arbitrary primers are useful as genetic markers. Nucleic Acids Res 18:6231-6235

Williams NMV, Pande N, Nair S, Mohan M, Bennet J (1991) Restriction fragment length polymorphism analysis of polymerase chain reaction products amplified from mapped loci of rice (*Oryza sativa* L.) genomic DNA. Theor Appl Gen 82:489-498

Winberg BC, Zhou Z, Dallas JF, McIntyre CL, Gustafson JP (1993) Characterization of minisatellite sequences from *Oryza sativa*. Genome 36:978-983

Wolfe AD and Liston A (1998) Contributions of PCR-based methods to plant systematics and evolutionary biology. In: Molecular Systematics of Plants II. DNA Sequencing. Soltis DE, Soltis PS and Doyle JJ, (eds.), Kluwer, Dordrecht, pp. 43-86

Wolfe AD Xiang Q-Y, Kephart SR (1998a) Assessing hybridization in natural populations of *Penstemon* (Scrophulariaceae) using hypervariable intersimple sequence repeat (ISSR) bands. Mol Ecol 7:1107-1125

Wolfe AD, Xiang Q-Y, Kephart SR (1998b) Diploid hybrid speciation in *Penstemon* (Scophulariaceae). Proc Natl Acad Sci USA 95:5112-5115

Wong Z, Wilson V, Jeffreys AJ, Thein SL (1986) Cloning a selected fragment from a human DNA "fingerprint": isolation of an extremely polymorphic minisatellite. Nucleic Acids Res 14:4605-4616

Zhou Z, Bebeli PJ, Somers DJ, Gustafson JP (1997) Direct amplification of minisatellite-region DNA with VNTR core sequences in the genus *Oryza.* Theor Appl Gen 95:942-949

Zhu J, Gale MD, Quarrie S, Jackson MT, Bryan GJ (1998) AFLP markers for the study of rice biodiversity. Theor Appl Gen 96:602-611

Zhu YL, Song QJ, Hyten DL, Van Tassell CP, Matukumalli LK, Grimm DR, Hyatt SM, Fickus EW, Young ND, Cregan PB (2003) Single Nucleotide Polymorphisms (SNPs) in soybean. Genetics 163:1123-113

Zietkiewicz E, Rafalski A, Labuda D (1994) Genome fingerprinting by simple sequence repeat (SSR)-anchored polymerase chain reaction amplification. Genomics 20:176-183

Chapter – 19

Statistical Treatment of Experimental Data for Biosystematics

S.A. Ranade

1. Introduction

Biological systematics (biosystematics) deals with the grouping of biological organisms in discrete taxonomic units based on their character states. These character states are determined through morphological, anatomical, ecological, biochemical and molecular biological analyses of specific parameters. If these character states are converted into numerical parameters, we can apply several statistical and numerical methods to achieve appropriate biosystematic inferences. In fact this particular application of statistical and numerical methods has given rise to what we may describe as Numerical Taxonomy, a concept first developed by Robert R. Sokal and Peter H. A. Sneath nearly a half century ago. In recent years the application of numerical methods to taxonomy has received a great deal of attention with the development of the computer and computer technology to the extent that even classical taxonomists are now resorting to the use of such numerical taxonomy methods with the help of computers. In this article the various methods commonly used, their application and specific limitations or otherwise are discussed in the context of their usefulness for classification and taxonomic disposition of the plants.

1.1. What is biosystematics?

Biological systematics or Biosystematics is the study of the diversification of living forms, both past and present, and their inter-relationships over a time scale. The inter-relationships are usually visualized as evolutionary trees that are generally described as cladograms or phylogenetic trees or phylogenies. "Systematic biology" and "taxonomy" are often confused and used interchangeably. However, these are defined separately and related to one another (Michener *et al.* 1970) by the following criteria: Systematic biology or systematics is a broader discipline that encompasses (a) taxonomy (scientific names for organisms, their descriptions, collection and preservation, and their classification, identification keys and their distributions; (b) investigation of their evolutionary histories, and (c) adaptation to their environmental niches, both at present as well as in the past. In recent years, systematics has gained a new level of importance with particular emphasis on theoretical considerations, simulations and predictions of ecological as well as evolutionary models and paradigms that explain present day diversity of form and function in the living world. These disciplines are involved with extinct and extant organisms and their relationships through time, and can be synonymous with phylogenetics, that deals with the inferred hierarchy of organisms. Systematics is thus fundamental to biology because it is the foundation for all studies of organisms, by showing how any organism relates to other living things (ancestor-descendant relationships). Similarly, Systematics is also of major importance in understanding conservation issues because it attempts to explain the Earth's biodiversity and this knowledge could be used to assist in allocating available resources to preserve and protect endangered species, by evaluating or assessing the genetic diversity among various taxa of plants or animals and deciding how much of that to preserve.

Biosystematics is divided primarily into phylogenetics and taxonomy where the former deals with an evolutionary account of the relationships while the latter deals with their nomenclature and classification relative to each other. It is greatly influenced by speciation - the origin of new species from previously existing ones and can occur via either anagenesis where one species changes into another over time or cladogenesis where one species splits to make two different species. Evolutionary theory states that groups of similar organisms are descended from a common ancestor. Therefore, phylogenetic systematics is a method of taxonomic classification based on their evolutionary history, and is a method developed by Hennig, a German entomologist (Hennig 1950, 1966).

1.2. How the biosystematic data are generated in experimental systems?

Systematics as has already been stated is a broad discipline. Consequently the systematic data are generated by a wide variety of experimental and empirical approaches. No matter what methods are used for the data generation, these must ultimately yield data that can be subjected to statistical and logical interpretations as well as tests of significance. A schematic description of the strategies available for the generation of systematic data through experimental approaches is given in Fig. 1.

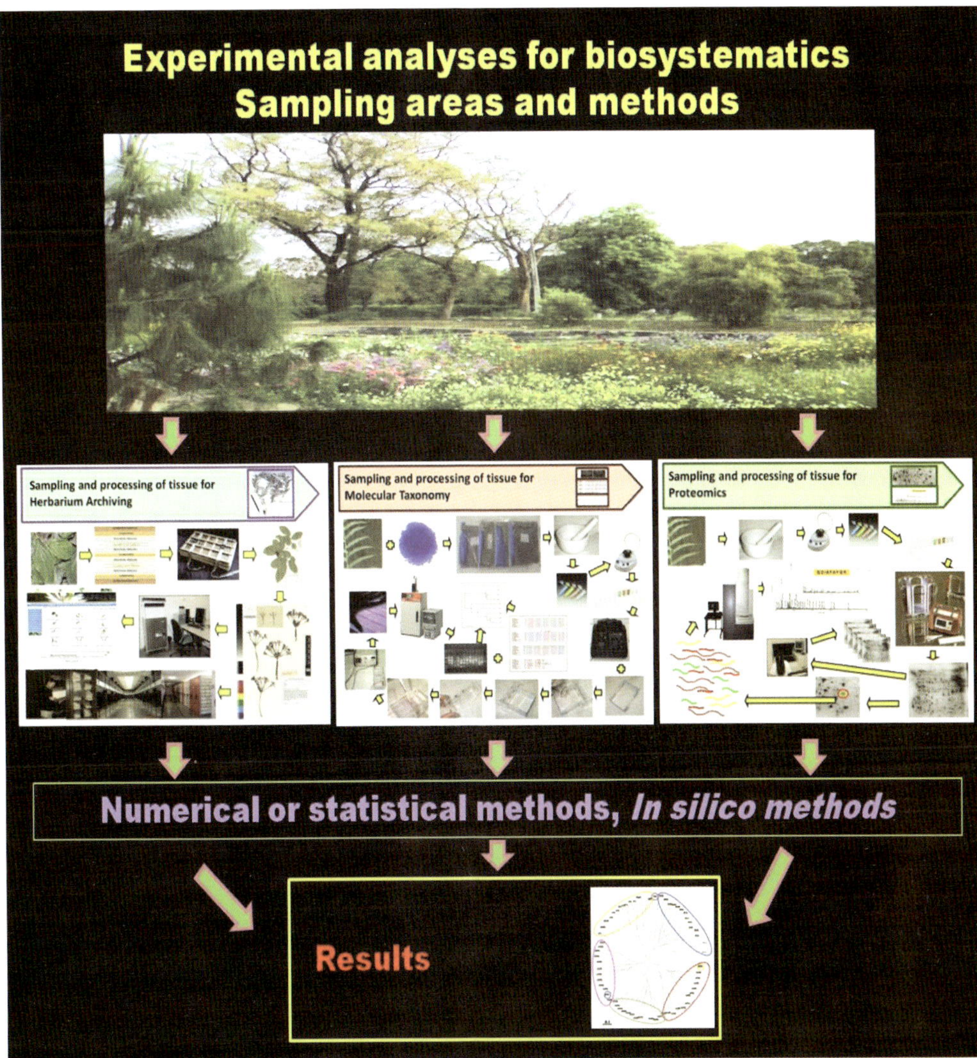

Fig 1. Starting from any habitat where a large number of taxa can be sampled, some of the experimental strategies that are available for the collected samples are depicted in this graphic

The experimental data for phylogenetics and systematic studies in the present times is almost invariably generated through the use of the Polymerase Chain Reaction (PCR) methods. The major rationale for the increasing use of these methods include the need for only low amounts of starting DNA, possibility of isolating DNA from dried and herbarium preserved materials, the suitability of even partially degraded DNA for PCR analyses and overall reliability with which these data are generated. The data from these methods are then subjected to various statistical treatments to test for the robustness of the methods, to test for their statistical significance and to generate the best phylogenetic inter-relationship of the taxa under study. These statistical treatments are described in some details in the following sections of this chapter. Almost invariably, the best method of describing the interrelationships of the taxa to each other is through a diagrammatic representation of what are described as "trees". Various features of the trees and their use in describing phylogenies are detailed below.

2. Phylogenetic Trees

A tree describing phylogenetic relationships between taxa has the following components that are invariably used to describe the tree (Fig. 2).

Operational Taxonomic Unit (OTU) - taxonomic level of sampling selected by the user to be used in a study, such as individuals, populations, species, genera etc.

Node - a branch point in a tree and is also a presumed ancestral OTU

Branch - defines the relationship between the OTU in terms of descent and ancestry

Topology - the branching pattern

Branch length - represents the number of changes that have occurred in the branch

Root - the common ancestor of all OTUs

Clade - a group of two or more OTUs that includes both their common ancestor and all their descendents

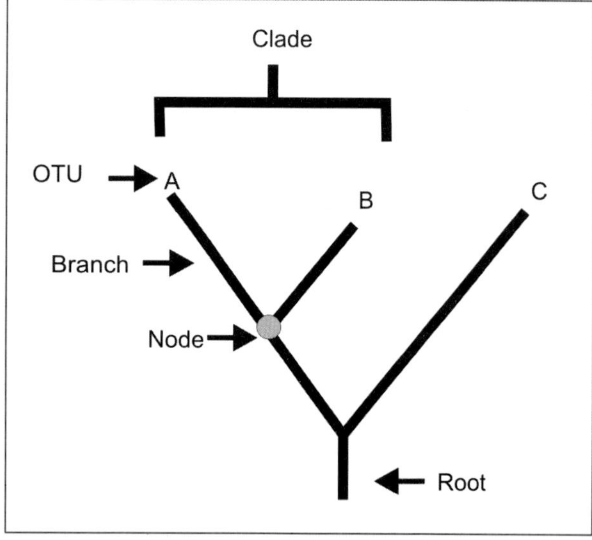

Fig. 2. A phylogenetic tree is described by its components

A phylogenetic tree is characterized by its topology (form) and its length (sum of its branch lengths) such that each node of a tree is an estimation of the ancestor of the elements included in this node. Such a tree can be rooted or unrooted. The numbers of rooted or un-rooted trees for a given number of taxa is calculated from the following relation:

$$N_R = (2n-3)!/2^{n-2}(n-2)!$$

$$N_U = (2n-5)!/2^{n-3}(n-3)!$$

Only one of all possible trees generated from the above relationships (Table 1) can represent the true tree describing the phylogenetic relationships among the OTUs.

Table 1. Possible evolutionary trees, both rooted as well as unrooted, in relation to the numbers of taxa. Even for as few taxa as say 10 taxa, the numbers of possible trees are in millions. This is the reason why most of the phylogenetic tree building is computer resource intensive

Taxa (n)	Rooted (2n-3)!/(2n-2(n-2)!)	Unrooted(2n-5)!/(2n-3(n-3)!
2	1	1
3	3	1
4	15	3
5	105	15
6	954	105
7	10,395	954
8	135,135	10,395
9	2,027,025	135,135
10	34,459,425	2,027,025

While the different trees may topologically appear similar, the nature of the algorithm or data test that underpins the trees ultimately decides the nature of the trees. A Cladogram is a phylogenetic tree that depicts patterns of shared characteristics among taxa grouped as a clade consisting of an ancestral species and its descendants. A clade is monophyletic when it consists of the ancestor species and all its descendants; it is paraphyletic when it consists of an ancestral species and some of the descendants while it is polyphyletic when grouping consists of various species that don't show a common ancestor (Fig.3).

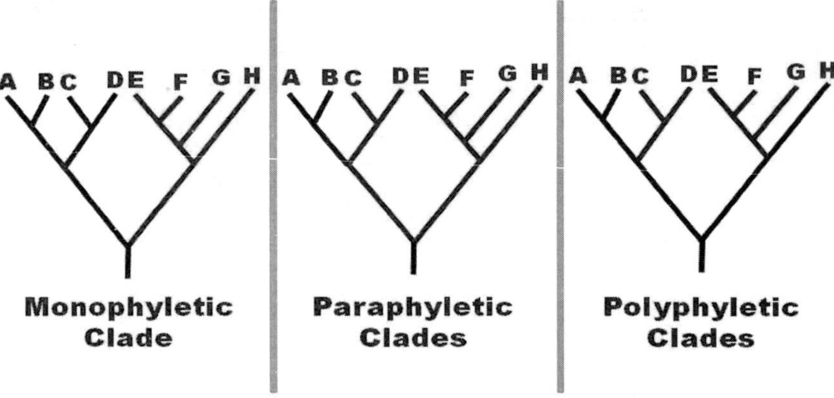

Fig. 3. Diagrammatic representation of the mono-, para- and polyphyletic clades in a phylogenetic tree. The phylogenies are considered to be complete when all taxa can be included in monophyletic clades

3. What are the Various Statistical Treatments Available?

Phylogenetic reconstruction and taxonomy includes two different ways and approaches for reconstructing a phylogeny. In one approach, called as the phenetic approach, a tree is constructed by considering the phenotypic similarities of the species without trying to understand the evolutionary pathways of the species. Such trees do not always reflect evolutionary relationships and are thus called phenograms. In the second approach, called as the cladistic approach, a tree is reconstructed by considering the various possible pathways of evolution and choosing from amongst these the best possible tree. Trees reconstructed via these methods are called cladograms. In phenetic systematic, the objective is to determine the relationships of organisms through a measure of similarity, considering both plesiomorphies (ancestral traits) and apomorphies (derived traits) to be equally informative. For character data where ancestral forms are known and to construct a taxonomic classification the cladistic approach is almost certainly superior. Cladistics, considers plesiomorphies to be uninformative for a phylogeny of various organisms through time. The cladistic methods are also often difficult to implement with assumptions that are not always satisfied by the molecular data. The phenetic approaches are generally faster algorithms and often have better statistical properties for molecular data. Hence, there appears to be a place for both types of methods in the analysis of molecular sequence data. Though the present day systematists rely more on and generally make extensive use of molecular biology and computer programs to study organisms. The use of nucleotide and amino acid sequence for phylogenetic analysis is a common trend these days particularly because of the increasing volume of such data being available in a large number of taxa. For DNA sequences, the four bases of the sequences are the main consideration for the analysis and their frequencies as well as specific positions relative to each other in different taxa generate the phylogenetically useful information. At least five different models have been proposed that give different weightages to frequency as well as change in bases at any position. Based on uniform criteria for sequence comparisons it is essential that any single model for the substitution is used throughout the analysis. Each model can generate a different tree and so the user must select that which best reflects the biological and evolutionary relationships among the taxa analyzed. The various models used are depicted in Fig. 4 and clearly show the essential parameters of the sequence substitutions that have the greatest impacts on the trees to be generated.

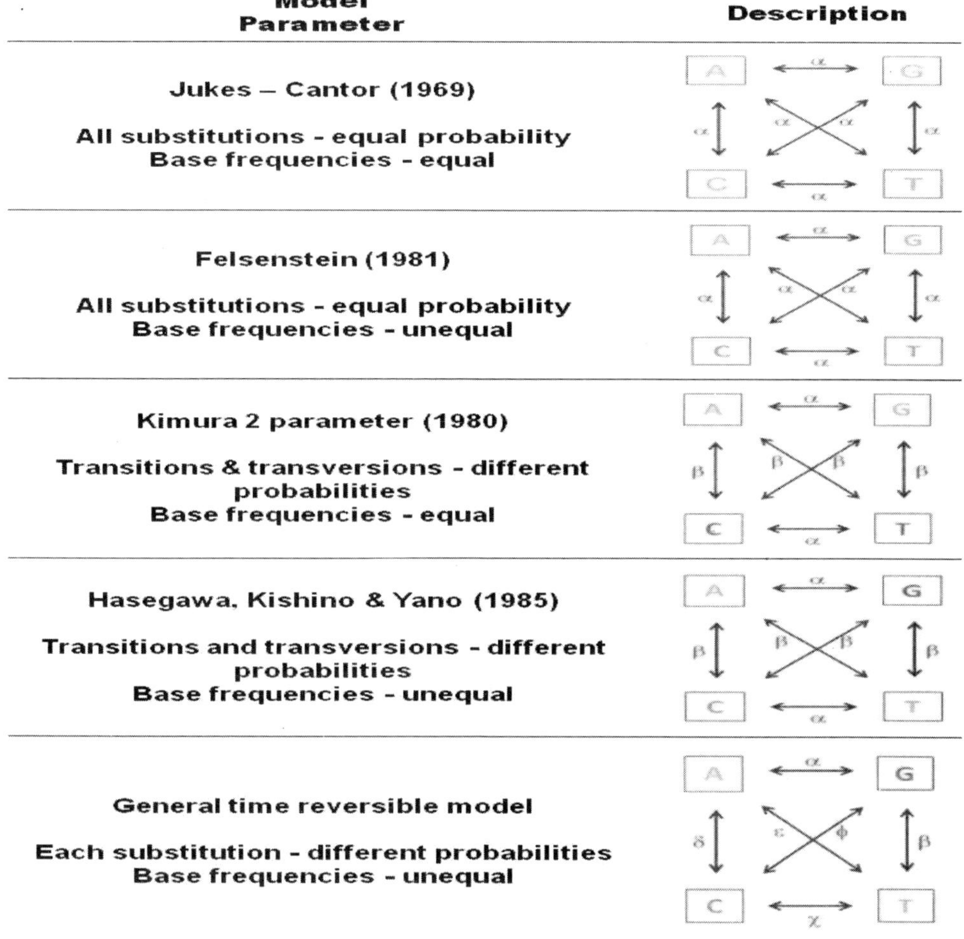

Fig. 4. The various models for consideration of nucleotide substitution to derive phylogenetic information are shown in the figure. The weightage for each type of change as well as the base frequencies is indicated. Experimental data are assesses as per any given model to determine the phylogeny by any selected method

4. Experimental Approaches to Generate Phylogenetically Useful Data

Several experimental approaches are available for generating data useful for phylogenetics and biosystematics. In effect, these approaches determine the laboratory component of the field studies in biosystematics. Starting with habitats where several taxa are present and are sampled from, a series of essential steps are required to be followed in biosystematics as described in Fig. 1. A generalized schema for these steps is described in the sequence starting from "sampling", through "DNA Isolation", "PCR amplification", "Gel Electrophoresis", "Scoring of bands in profiles", "Sequencing of amplified DNA" and ending with "Profile / Sequence Data Analysis". Once the samples are brought to the laboratory, biosystematic studies are enabled through the analysis of nucleic acids (genomics, transcriptomics), proteins (Proteomics) and even metabolites (metabolomics) under the present day "OMICS" scenario. The nucleic acid methods can be based on anonymous DNA profiling methods such as RAPD, ISSR PCR or these can be gene-specific when known genes and their sequences are analyzed. At the level of estimating phylogenetics from nucleotide and amino sequences, the important rationale is that these sequences reflect an evolutionary history relative to each other. Thus, the similarities in sequences can be used to help determine classification and evolutionary relationships. The more similar the DNA or protein sequence is of two species, the more recently they shared a common ancestor, and the more closely they are related in evolutionary terms. Conversely, the more two species have diverged from each other, the less similar their DNA or protein sequences will be. A time scale used to assess the extent of change can now give rise to Molecular Clocks that monitor "neutral" mutations to estimate the length of time two species have been evolving independently. The sequence based phylogeny analysis and its reconstruction is invariably based on one of four methods that are used for the generation of phylogenetic trees (Table 2).

Table 2. Methods for phylogenetic reconstructions are based on algorithms that best fit the experimental data to biological and evolutionary inter-relationships among the taxa

Method	Rationale
Parsimony	Find a phylogenetic tree that explains the data, with as few evolutionary changes as possible
Distance	Find a tree such that branch lengths of paths between sequences (species) fit a matrix of pairwise distances between sequences
Maximum likelihood	Find a tree that maximizes the probability of the genetic data that give the tree
Bayesian	Bayesian posterior probabilities estimate the maximum likelihood topology as well as nodal confidence in the trees

4.1. Maximum parsimony

This method is based on determining the minimum number of changes (substitutions) required to transform a sequence to its nearest neighbor. The basic idea of parsimony analysis was first given by Fitch (1971). Parsimony is a character-based tree estimation method which uses a set of characters to infer one or more trees for the taxa being studied. The various trees are evaluated according to a defined optimality criterion and the tree with the most favorable score (least number of steps) is taken as the best estimate of the phylogenetic relationships of the included taxa. A large number of possible phylogenetic trees (see Table 1) for a set of taxa are evaluated for the best fit to optimality criterion such that the distribution of characters being analyzed directly follow the branching pattern of evolution. Two organisms possess a shared character when they are more closely related to each other than to another organism that lacks this character. The parsimony algorithm searches for the minimum number of genetic events (nucleotide substitutions or amino-acid variation in each position of the sequences) to infer the most parsimonious (the one which needs the fewest changes) tree from a set of sequences. This, results in several trees constructed at random based on the nucleotide changes in iterative calculations until all possible topologies have been examined and the one requiring the smallest number of steps is identified. This final tree is presented as the most likely inferred tree based on the most parsimonious iterations. Inherent in this method is the assumption, not necessarily correct in all cases, that evolution follows the shortest possible route and that the correct phylogenetic tree is therefore the one that requires the minimum number of nucleotide changes to produce the observed differences between the sequences. The method occasionally leads to a difficult situation within practical computational limits, several "equally most parsimonious trees" are generated and it becomes difficult to select a given tree as the "most parsimonious" tree. Necessarily, the procedure involves long computation times to construct a tree.

4.2. Distance methods

Neighbor Joining (NJ) and Unweighted Pair Group Method with Arithmetic mean (UPGMA) are two of the commonly used distance methods for sequence alignments and discrete variable (band) data. Usually used for trees based on DNA or protein sequence data, the distance algorithms require knowledge of the distance between each pair of taxa (e.g., species or sequences) to form the tree. Both NJ and UPGMA are fast methods. Of the two, UPGMA is more affected by variations in rates of evolutionary changes and is therefore, not a reliable method for true phylogenies. Neighbor joining is a clustering method for the creation of phenetic trees (phenograms), developed by Saitou and Nei (1987). UPGMA is a simple agglomcrative or hierarchical clustering method that uses pairwise similarities for the set of taxa being analyzed. The method was first developed by Sokal and Michener (1958). The UPGMA algorithm constructs a tree that reflects the pairwise similarity matrix (or a dissimilarity matrix). At each step, the nearest two taxa are combined into a higher-level cluster until all the taxa are included in the tree.

A measure of how close the tree is to D is given by the least square criterion:

$$\Sigma_{i,j} (D_{i,j} - d_{i,j})^2 / D^2_{ij}$$

where

$D_{i,j}$ = the distance between *i* and *j* sequences;

$d_{i,j}$ = sum of branches on the tree path from i to j

The phylogeny makes an estimation of the distance for each pair as the sum of branch lengths in the path from one sequence to another through the tree. A matrix of distances (or similarities) using a suitable algorithm is thus prepared. The matrix is then used generation of un-rooted trees for the data that can be further subjected to bootstrap analyses.

4.3. Maximum likelihood method

Maximum likelihood is also a character-based analysis like the Parsimony analysis. However, in maximum likelihood, a statistical method tests a specific model of character evolution (Felsenstein 1996). The probability of the data for different taxa given a tree for these taxa can be calculated on the assumption of specific model for incremental changes in the characters in the taxa. Hence, the tree that has the highest probability of producing the observed data is the most likely tree. In this method, the bases (or amino acids) of all sequences at each site are considered separately (as independent), and the log-likelihood of having these bases are computed for a given topology using a probability model. This log-likelihood is added for all sites, and the sum of the log-likelihood is maximized to estimate the branch length of the tree. The procedure is repeated for all possible topologies, and the topology that shows the highest likelihood is chosen as the final tree. This is the best justified method from a theoretical viewpoint and simulations have shown that this method is consistent and superior to other methods in most cases. However, it is a very calculation and computationally intensive method. It is almost always impossible to evaluate all possible trees because of the large numbers. Hence, invariably, a partial exploration of the set of trees is made and the maximum likelihood estimates the branch lengths of the final tree.

4.4. Bayesian inference of phylogenies

Bayesian phylogenetics uses the likelihood function and Bayes' theorem, which relates the posterior probability of a tree to the likelihood of data, and the prior probability of the tree and model of evolution (Yang and Rannala 1997). Here, given *a priori* probabilities of all parameters (topology, branch lengths, relative rates of substitutions), we have to calculate the posterior probability of all possible topologies and the tree for the given sequence alignment.

$\Pr(\tau/\times)\alpha ff_{v'\theta})\ \Pr(\times/\tau,\nu,\theta).\ \Pr_{prior}(\nu,\theta)d\nu d\theta$

where

τ: is the topology of the tree (a priori uniform)

X: are aligned sequences

v: are all branch lengths

è : is parameter substitution model (eg, transition - transversion ratio)

Pr (×/τ,ν,θ):is the likelihood + tree parameters and

$\Pr_{prior}(\nu,\theta)\ d\,\nu\ d\,\theta$: are *a priori* probabilities of parameter values

5. Numerical Taxonomy

Numerical taxonomy is a classification system in biological systematic which deals with the grouping by numerical methods of taxonomic units based on their character states. It aims to create a taxonomy using numeric algorithms like cluster analysis rather than using subjective evaluation of their properties. The concept was first developed by Sokal and Sneath (1963) and later elaborated by the same authors. They divided the field into phenetics in which classifications are formed based on the patterns of overall similarities and cladistics in which classifications based on the branching patterns of the estimated evolutionary history of the taxa, though in recent years many authors treat numerical taxonomy and phenetics as being synonymous. Although intended as an objective classification method, in practice the choice and implicit weighing of characteristics is, of course, influenced by available data and research interests of the investigator. What was made objective was the introduction of explicit steps to be used to create phenograms and cladograms using numerical methods rather than subjective synthesis of data.

6. When Phylogenetics Does Not Work!

The methods suffer difficulties related to the phylogenetic tree construction when the global sequence alignment does not result in a tree that accounts for all observed variations and affinities due to substitution variations between genes in the test organisms or due to incomplete evolutionary histories. Similarly, the facts that the genes do not evolve at the same rate or in the same way in all organisms and only a limited number of genes are shared among all species also limit the successes in phylogenetic tree construction or result in trees that do not account for all parameters of the evolution of the taxa.

References

Felsenstein J (1981) Evolutionary trees from DNA sequences: a maximum likelihood approach. J Mol Evol 17:368–76

Felsenstein J (1996) Inferring phylogenies from protein sequences by parsimony, distance, and likelihood methods. Meth Enzy 266:419-427

Fitch WM. (1971) Toward defining the course of evolution: minimum change for a specified tree topology. Sys Zool 20:406-416

Hasegawa M, Kishino H, Yano T (1985) Dating of the human-ape splitting by a molecular clock of mitochondrial DNA. J Mol Evol 22:160–74

Hennig W (1950) Grundzu¨ge einer Theorie der phylogenetischen Systematik. Deutscher Verlag, Berlin, Germany

Hennig W (1966) Phylogenetic systematics (tr. D. Dwight Davis and Rainer Zangerl), Urbana, IL: Univ. of Illinois Press (reprinted 1979 and 1999), ISBN 0-252-06814-9

Jukes TH and Cantor CR (1969) Evolution of protein molecules. In Munro, H.N. Mammalian protein metabolism. New York: Academic Press. pp. 21–123

Kimura M (1980) A simple method for estimating evolutionary rates of base substitutions through comparative studies of nucleotide sequences. J Mol Evol 16:111–20

Michener CD, Corliss JO, Cowan RS, Raven PH, Sabrosky CW, Squires DS and Wharton GW (1970) Systematics in Support of Biological Research. Division of Biology and Agriculture, National Research Council. Washington, D.C. 25 pp

Saitou N, Nei M (1987) The neighbor-joining method: a new method for reconstructing phylogenetic trees. Mol Biol Evol 4: 406-425

Sokal RR and Michener C (1958) A statistical method for evaluating systematic relationships. Univ Kans Sci Bull 38:1409–1438

Sokal RR and Sneath PHA (1963) Principles of Numerical Taxonomy, San Francisco: W.H. Freeman, 359 pp

Yang Z, Rannala B (1997) Bayesian phylogenetic inference using DNA sequences: a Markov Chain Monte Carlo Method. Mol Biol Evol 14:717–24

Chapter – 20

Ecological Sampling Methods for Biodiversity Assessment

Soumit K. Behera

1. Introduction

If we want to know what kind of plants and animals are in a particular habitat, and how many they are of each species, it is obviously (usually) impossible to go and count each and every one present. It would be like trying to count different sizes and colors of grains of sand on the beach. This problem is usually solved by taking a number of samples from around the habitat, making the necessary assumption that these samples are representative of the habitat in general. In order to be reasonably sure that the results from the samples do represent the habitat as closely as possible, careful planning beforehand is essential.

The usual sampling unit is a quadrat. Quadrats normally consist of a square frame, the most frequently used size being 20×20 m^2 for trees, 5×5 m^2 for shrubs, 1×1 m^2 for herbs. The purpose of using a quadrat is to enable comparable samples to be obtained from areas of consistent size and shape. Rectangular quadrats and even circular quadrats have been used in some surveys. It does not really matter what shape of quadrat is used, provided it is a standard sampling unit and its shape and measurements are stated in any write-up. It may however be better to stick to the traditional square frame unless there are very good reasons, because this yields data that are more readily comparable to other published research. (For instance, you cannot compare data obtained using a circular quadrat, with data obtained using a square quadrat. The difference in shape of the sampling units will introduce variations in the results obtained.)

Choice of quadrat size depends to a large extent on the type of survey being conducted. For instance, it would be difficult to gain any meaningful results when using a 0.5 m^2

quadrat in a study of a forest canopy. The pattern of distribution of species should also be considered when deciding on quadrat size, because different results will be obtained using different quadrat sizes, depending on whether individuals are regularly distributed, randomly distributed, or clustered together in patches. It may often be difficult to decide if there is a pattern of distribution.

Small quadrats are much quicker to survey, but are likely to yield somewhat less reliable data than large ones. However, larger quadrats require more time and effort to examine properly. A balance is therefore necessary between what is ideal and what is practical. As a general guideline, 0.5 - 1.0m quadrats would be suggested for short grassland or dwarf heath, taller grasslands and shrubby habitats might require 5m quadrats, while quadrats of 20X 20 m^2 or larger, would be needed for forest habitats.

2. Methods

2.1. Quadrat sampling

A group of organisms of the same species found within the same environment is called a population. Populations of different species sharing the same area and interacting with each other make up a community. One way of studying the interactions of organisms in a community is by taking an inventory of all the species in the area and comparing the sizes of their populations.

In this investigation, you will observe the abiotic factors of a study site and classify the species of plants in that site. You will then measure randomly chosen quadrats within the site and count the individuals of each population located within each quadrat. With this data, you will estimate the sizes of plant populations within the community.

Before scientists can design an experiment, they must first make observations on which to base their hypotheses. Scientists have many different methods of collecting this data. The task of taking an inventory of the different kinds of plants and their population sizes in an environmental site can be very difficult. Since it would be impractical, if not impossible, to count each individual plant in a large area, ecologists randomly choose small portions of the whole area and classify and count the plant in each small portion. They can then estimate the size of each population in the larger community. This process is called the quadrat method.

The goal of the quadrat method is to estimate the population density of each plant species in a given community. Population density is the number of individuals of each species per unit area. Small square areas, called quadrats, are randomly selected to avoid choosing unrepresentative samples. Once the population densities for all quadrats are determined, the population size within the larger area can be estimated.

There are three main ways of taking samples

- Random Sampling
- Systematic Sampling (includes line transect and belt transect methods)
- Stratified Sampling

2.1.1. *Random sampling*

Random sampling is usually carried out when the area under study is very large, or there is limited time available. When using random sampling techniques, large numbers of samples/records are taken from different positions within the habitat. A quadrat frame is most often used for this type of sampling. The frame is placed on the ground (or on whatever is being investigated) and the plants inside it counted, measured, or collected, depending on what the survey is for. This is done many times at different points within the habitat to give a large number of different samples.

In the simplest form of random sampling, the quadrat is thrown to fall at 'random' within the forest/vegetation site. However, this is usually unsatisfactory because a personal element enters into the throwing and it is rarely completely random. True randomness is an important element in ecology, because statistics are widely used to process the results of sampling. Many of the common statistical techniques used are only valid on data that is truly randomly collected. It must also be noted that this technique would only be possible if quadrats of small size were being used. It would be impossible to throw anything larger than a $1m^2$ quadrat and even this might pose problems.

A better method of random sampling is to map the area and then to lay a numbered grid over the map. A (computer generated) random number table is then used to select which squares to sample in. For example, if we have mapped our habitat, and have then laid a numbered grid over it as shown (Fig. 1), we could then choose which squares we should sample in by using the random number.

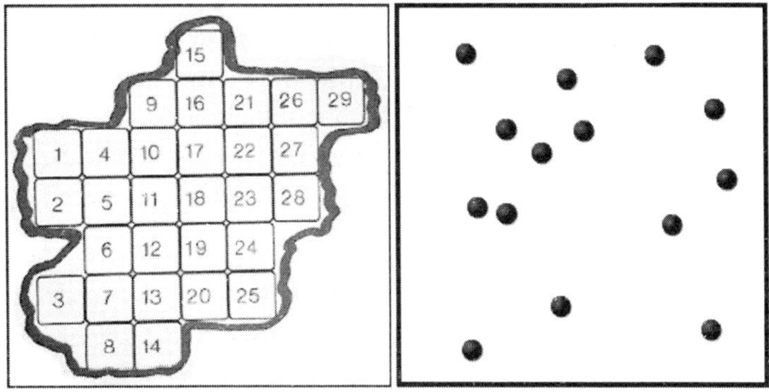

Fig. 1. A numbered grid map of an area to be sampled

2.1.2. Systematic Sampling

Systematic sampling is done when samples are taken at fixed intervals (Fig. 2), usually along a line. This normally involves doing transects, where a sampling line is set up across areas where there are clear environmental gradients. For example you might use a transect to show the changes of plant species as you moved from grassland into forest, or to investigate the effect on species composition of a pollutant radiating out from a particular source. Systematic sampling involves following two types:

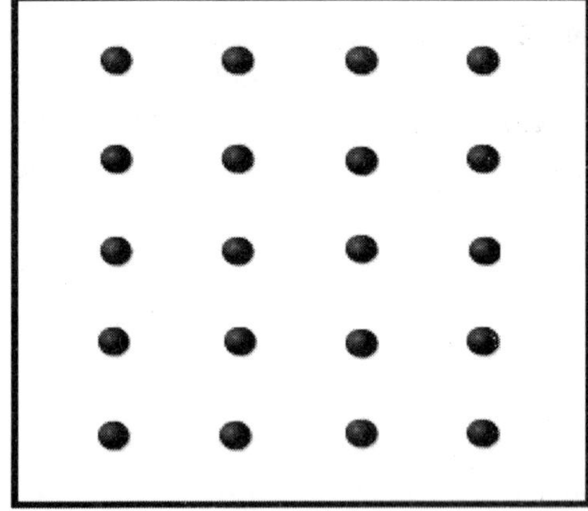

Fig. 2. A layout showing systematic sampling of an area at regular intervals

1. Line Transect Method

A transect line can be made using a nylon rope marked and numbered at 0.5m, or 1m intervals, all the way along its length (Fig. 3). This is laid across the area you wish to study. The position of the transect line is very important and it depends on the direction of the environmental gradient you wish to study. It should be thought about carefully before it is placed. You may otherwise end up without clear results because the line has been wrongly placed. For example, if the source of the pollutant was wrongly identified in the example given above, it is likely that the transect line would be laid in the wrong area and the results would be very confusing.

A line transect is carried out by unrolling the transect line along the gradient identified. The plant species touching the line may be recorded along the whole length of the line (continuous sampling). Alternatively, the presence, or absence of species at each marked point is recorded (systematic sampling).

Fig. 3. Figure showing line transects sampling of an area

2. Belt Transect Method

This is similar to the line transect method but gives information on abundance as well as presence, or absence of species. It may be considered as a widening of the line transect to form a continuous belt, or series of quadrats.

In this method, the transect line is laid out across the area to be surveyed and a quadrat is placed on the first marked point on the line (Fig. 4). The plants inside the quadrat are then identified and their abundance estimated. It is usual to estimate the percentage cover of plant species. Cover is the area of the quadrat occupied by the above-ground parts of a plant species when viewed from above. The canopies of the plants inside the quadrat will often overlap each other, so the total percentage cover of plants in a single quadrat will frequently add up to more than 100%.

Quadrats are sampled all the way down the transect line, at each marked point on the line, or at some other predetermined interval (or even randomly) if time is short. It is important that the same person should do the estimations of cover in each quadrat, because the estimation is likely to vary from person to person. If different people estimate percentage cover in different quadrats, then an element of personal variation is introduced which will lead to less accurate results. The height of plants in the quadrat can be recorded and the biomass of plants can also be measured by harvesting all the plants inside the quadrat and then weighing either fresh, or dry weight in the laboratory. This is obviously a very destructive method of sampling which could not be used too often in the same place. Sampling should always be as least destructive as possible and you should try not to trample an area too much when carrying out your survey.

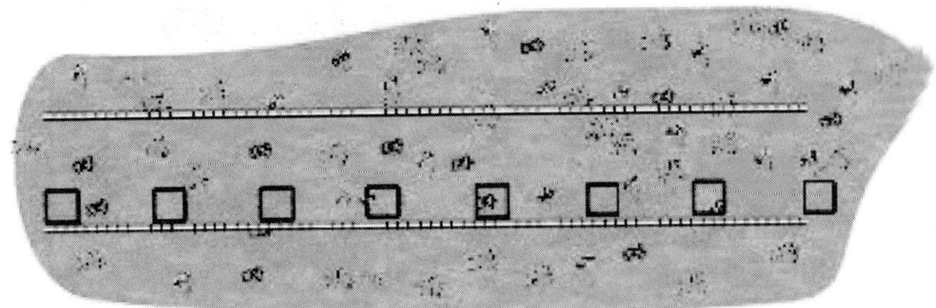

Fig. 4. Figure showing belt transect sampling of an area

2.1.3. Stratified sampling

Stratified sampling is used where there are small areas within a larger habitat which are clearly different (Fig. 5). For example, scrub patches within a forest area, or areas of bracken in a grassland. Sampling would be carried out either randomly, or systematically within each separate 'stratum' identified. This recognizes major differences within communities before sampling begins.

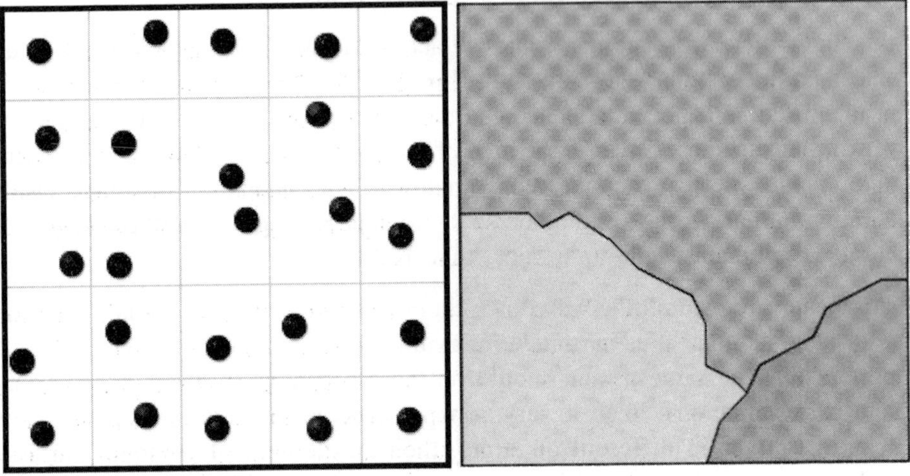

Fig. 5. Figure showing stratified sampling of an area having three different type of vegetation cover or land use shown in three colors

2.2. Sampling unit

By definition, a quadrat is a square, rectangle, circle or other shape area used as a sample unit. The term quadrat and plot are used interchangeably in much of the natural resource literature. In sampling using quadrats, small, manageable areas of known dimensions are designated as the sample unit. A number of quadrats will be selected to provide the data needed to estimate the population parameters of interest.

Usually, everything of interest within the quadrat is counted, collected, weighted, and the responses recorded represent the total for the whole quadrat. Thus, for example, when quadrats are used to sample vegetation, the response might be the total number (count data) of stems of small shrubs in the quadrat, or it might be the total biomass (continuous data) of all shrubs in the quadrat. The response may also relate to the fraction of the total quadrat area covered by the species of interest (coverage), or the fraction of total biomass represented by the species of interest (percentage data).

Unlike sampling using natural individuals, with quadrat sampling one must choose the dimensions and shape of the quadrat prior to implementation. The choice of dimensions and shape will affect the precision and accuracy of the parameter estimates subsequently computed.

When?

Quadrats are used to sample the following populations.

- Vegetation and trees
- Species migration
- Species invasion
- Species associations
- Soil bio-physical characteristics.

Where?

Quadrats are the sample unit of choice where:

- The individual organisms of interest are too numerous to consider separately as sample units.
- It would be impossible to create a sampling frame of individuals prior to unit selection.
- The parameter of interest relates to area coverage or density.

How?

A number of questions must be answered before quadrats can be used in a sample study.

- How large of a quadrat will be used?
- What shape quadrat will be used?

- How will quadrats be oriented?
- Where will quadrats be placed in the study area?
- How many quadrats will be used?
- What will be measured/observed in each quadrat?

Once these questions have been answered, the process of sampling becomes:

- Establish quadrats
- Record observations/measurements in each quadrat
- Summarize the data by quadrat estimate population parameters of interest compute a confidence interval for the estimates

Quadrat size

While the size of the quadrat in vegetation studies will depend on the local conditions of the study site and population being studied. For example:

- Herbaceous vegetation closely spaced - one meter plots (1m x 1m)
- Bushes, shrubs, and saplings up to 4m tall - five to ten meter plots (i.e. 5m x 5m / 10m x 10m)
- Mature forest trees over 4m - twenty to fifty meter plots (i.e. 20m x 20m or 50m x 50m)
- Climbers and lianas - one to five meter plots (i.e. 1m x 1m / 5m x 5m)
- All plots may have to be adjusted to larger sizes when sampling larger plants

Quadrat shape problems

There are two problems relating to quadrat shape

(i) Edge effect - the more edge to the quadrat the more opportunity to miscount individuals

(ii) Habitat homogeneity tends to lead to use of long thin rectangular quadrats

Sample size determination

There is no set percentage that is accurate for every population. What matters is the actual number or size of the sample, not the percentage of the population. Consider a coin toss: the first few times you flip the coin, the average result may be skewed wildly in one direction (say you got 5 tails in a row), but the more times you flip the coin, the more likely that the average result will be an even split between heads and tails.

So, if you survey 20% of a group of 300 participants in a programme, to produce a sample of 60 people, you would under represent the population, as there is a fairly large chance in a small group that the respondents you choose will vary from the whole population. On the other hand, 20% of 30,000 county residents (a sample of 6,000) would be a wastefully large sample, and not significantly more accurate than a sample of 400.

There are 5 steps in deciding a sample size.

(i) Determine goals

(ii) Determine desired precision of results

(iii) Determine confidence level

(iv) Estimate the degree of variability

(v) Estimate the response Rate

Quadrat sampling errors

Quadrat sampling is based on measurement of replicated sample units referred to as quadrats or plots (sometimes transects). This approach allows estimation of absolute density (number of individuals per unit area within the study site). Our challenge is to identify sampling strategies that will provide satisfactory precision with minimum sample effort. Some of the factors that affect precision are:

(i) **Measurement error**. In the real world, it is important to count organisms carefully and lay out plots accurately for good estimates of density. This is not a concern here however, because the computer will be laying out the plots and counting the plants.

(ii) **Total area sampled.** In general, the more area sampled, the more precise the estimates will be, but at the expense of additional sampling effort.

(iii) **Dispersion of the population**. Whether the population tends to be aggregated, evenly spaced, or randomly dispersed can affect precision. Note that the dispersion pattern of the same population may be different at different spatial scales (e.g., 1 x 1 m plots vs 100 x 100 m plots).

(iv) **Size and shape of quadrats**. The size and shape of the plots can affect sampling precision. Often, the optimal plot size and shape will depend on the dispersion pattern of the population.

Sampling with quadrats (plots of a standard size) can be used for most plant communities. A quadrat delimits an area in which vegetation cover can be estimated, plants counted, or species listed. Quadrats can be established randomly, regularly, or subjectively witin a study site. Since plants often grow in clumps, long, narrow plots often include more species than square or round plots of equal area; especially if the long axis is established parallel to environmental gradients. However, accuracy may decline as the plot lengthens because, as the perimeter increases, the surveyor must make more subjective decisions about the placement of plants inside or outside the plot. Round quadrats can be most accurate because they have the smallest perimeter for a given area. Round quadrats are also simple to define in the field, requiring only a center stake and a tape measure (Cox 1990).

The appropriate size for a quadrat depends on the items to be measured. If cover is the only factor being measured, size is relatively unimportant. If plant numbers per unit area are to be measured, then quadrat size is critical. A plot size should be large enough to include significant numbers of individuals, but small enough so that plants can be separated, counted and measured without duplication or omission of individuals (Cox 1990; Barbour *et al.* 1987). Large quadrats with many plants may require two or more people to obtain an accurate census, while one person may be sufficient for smaller plots or those with sparse vegetation.

An accurate estimate of the necessary number of quadrats can be determined by plotting data for a given feature (i.e. percent cover) versus number of quadrats. The appropriate quadrat number will correspond to the point at which the curve plateaus (Barbour *et al.* 1987). Some field researchers sample until the standard error of the quadrat is within a previously decided, acceptable boundary.

3. Plant Community Analysis

Basal cover, density, and frequency are importants aspects structural parameters of the plant community, which can be measured by quadrat sampling.

3.1. Basal cover/basal area

Basal area refers to the ground actually covered by the stems. It is the most important character that determines dominance and nature of community (Misra 1968). The cross section area of the stem or stems of a plant or of all plants in a stand is generally expressed as square units per unit area. Tree basal area is used to determine percent stocking. For shrubs and herbs it is used to determine phytomass. Each herb and shrub species at the ground surface were cut by a sharp scissors and the diameters of the cut ends were measured. Tree basal area was calculated from diameter at breast height (DBH) data.

Dominance (Basal area) = Πr^2 (r = diameter /2)

$$\text{Relative dominance (\%)} = \frac{\text{Total basal area of the species}}{\text{Total basal area of all species}} \times 100$$

3.2. Density

Density is determined by the number of plants rooted within each quadrat. Relative density is the density of one species as a percent of total plant density. Area per plant, or mean area, is plot area per density.

$$\text{Density} = \frac{\text{Total number of individual}}{\text{Total number of quadrats studied}} \text{ x } 100$$

$$\text{Relative density (\%)} = \frac{\text{Number of individuals of the species}}{\text{Number of individuals of all the species}} \text{ x } 100$$

3.3. Frequency

Frequency is the percentage of total quadrats containing at least one rooted individual of a given plant species. Relative frequency of one species defines as a percentage of total plant frequency. Frequency is affected by quadrat size and less meaningful than other measurements.

$$\text{Frequency \%} = \frac{\text{Number of sampling unit in which a species occurred}}{\text{Total number of units studied}} \text{ x } 100$$

$$\text{Relative frequency (\%)} = \frac{\text{Number of occurrences of the species}}{\text{Number of occurrence of all species}} \text{ x } 100$$

3.4. Importance value Index

Plant species vary in their responses to environmental factors. A given species will have a unique set of tolerances to environmental variables, such as light, temperature, moisture, and nutrients. At the community level, these differences in tolerances will cause various species to have competitive advantages, depending on the nature of those environmental factors. The importance value, or the importance percentage, gives an overall estimate of the influence of importance of a plant species in the community. In order to express the dominance and ecological influence of any species, with a single value, the concept of Importance value index has been developed. The index utilized three characters, viz., relative frequency, relative density and relative dominance.

IVI= Relative Frequency + Relative density + Relative dominance (Misra 1968)

3.5. Measuring biodiversity

The simplest measure of biodiversity is the number of species – called species richness.

Usually only count resident species and not accidental or temporary immigrants

Another concept of species diversity is heterogeneity:

	Community 1	Community 2
Species A	99	50
Species B	1	50

Heterogeneity is higher in a community where there are more species and when the species are more equally abundant.

3.6. Species richness

- Ecologists recognise that the diversity of an ecological system has 2 facets:
  - species number (=richness)
  - evenness of distribution

Both these systems have 30 individuals and 3 species.
Which is more diverse?

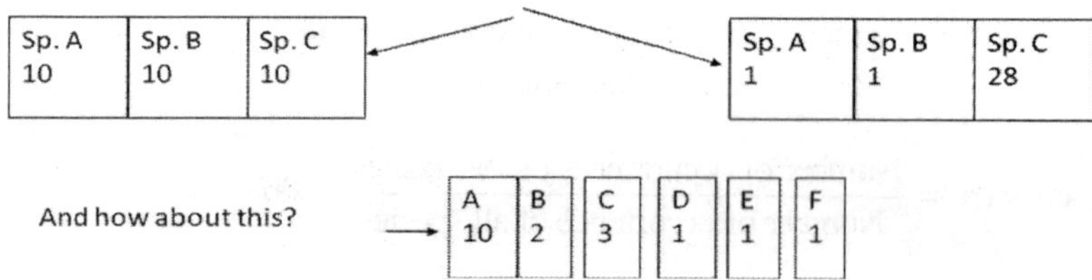

This is what people often (not always) mean when they talk of biodiversity. The value of species richness is, in principle, easy. List all the species in the habitat and count them.

But this principle has flaws:

1. This index is usually criticised as being too sensitive to the occurrence of rare species.
2. Are you sure of your list? The more samples you take the more species you find this links to rank abundance curves. And that's assuming your taxonomic basis is solid.

3.7. Dispersion patterns

Dispersal is the movement of individuals away from their area of origin or from centers of high population density, resulting in a spacing pattern within specified boundaries. Dispersion usually takes place at the time of reproduction. Populations within a species are translocated through many methods, including dispersal by people, wind, water, and animals. Dispersal range refers to the distance a species can move from an existing population or the parent organism. An ecosystem depends critically on the ability of individuals and populations to disperse from one habitat patch to another. Therefore, biological dispersal is critical to the stability of ecosystems.

4. Types of Distribution

4.1. Clumped distribution

Clumped distribution is the most common type of dispersion found in nature (Fig. 6). In clumped distribution, the distance between neighboring individuals is minimized. This type of distribution is found in environments that are characterized by patchy resources. Clumped distribution is the most common type of dispersion found in nature because plants need certain resources to survive. When these resources become rare during certain parts of the year, plants tend to "clump" together around these crucial resources.

Fig. 6. Clumped distribution of an individual species

Individuals might be clustered together in an area due to social factors such as selfish herds and family groups. Organisms that usually serve as prey form clumped distributions in areas where they can hide and detect predators easily. Clumped distribution can be beneficial to the individuals in that group.

4.2. Uniform distribution

Uniform distributions are found in populations in which the distance between neighboring individuals is maximized (Fig. 7). The need to maximize the space between individuals generally arises from competition for a resource such as moisture or nutrients, or as a result of direct social interactions between individuals within the population, such as territoriality.

Fig. 7. Uniform distribution of an individual species

4.3. Random distribution

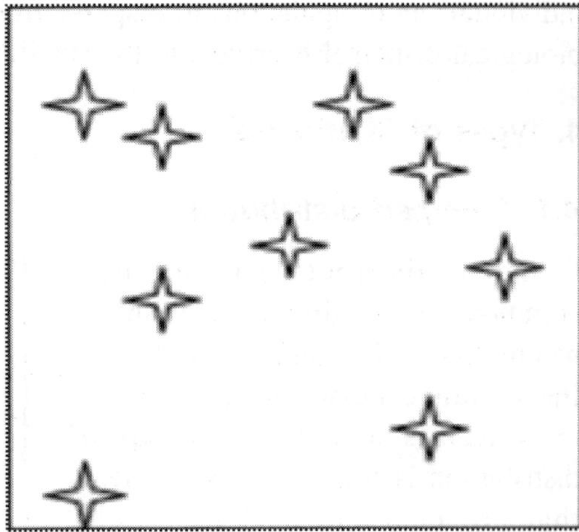

Fig. 8. Random distribution of an individual species

Random distribution (Fig. 8), also known as unpredictable spacing is the least common form of distribution in nature and occurs when the members of a given species are found in homogeneous environments in which the position of each individual is independent of the other individuals: they neither attract nor repel one another. Random distribution is rare in nature as biotic factors, such as the interactions with neighboring individuals, and abiotic factors, such as climate or soil conditions, generally cause organisms to be either clustered or spread apart. Random distribution usually occurs in habitats where environmental conditions and resources are consistent. This pattern of dispersion is characterized by the lack of any strong social interactions between species.

Tropical fig trees exhibit random distribution as well because of wind pollination. Although random is thought to be unpredictable, it is the only dispersion that has a mathematical equation to represent it. This is due to the individualistic characteristics of random dispersion based on the idea that every species has equal opportunity and access to resources.

5. Diversity Indices

5.1. Simpson index

Simpson's index considered a dominance index because it weights towards the abundance of the most common species. For example, the probability of two trees, picked at random from a tropical rainforest being of the same species would be relatively low, whereas in the boreal forest would be relatively high.

Simpson's dominance index (C) was calculated following Simpson (1949):

$C = \Sigma (ni/N)^2$

Where C = Dominance index.

ni = Density of each species.

N = Total density for all species.

5.2. Shannon Weiner Index

The widely used Shannon-Wiener's Diversity Index (H') is a mimic of the so called information theory formula that combines the variety and evenness components as one overall index of diversity. This index is one of the best for making comparisons where one is not interested in separating out diversity component; it is also the best because of reasonable independent sample size.

Shannon-Wiener's diversity index (H') is calculated based on the formula (Magurran, 1988)

$$H' = - \sum pi \ln pi$$

$$= - \sum (ni/N) \ln (ni/N)$$

Where ni = density of each species.

N = Total density of all species.

Pi = Importance probability for each species = ni/N.

5.3. Similarity Index

Similarity index between two forest stands is calculated using the community coefficient (Whittaker, 1975). Community coefficient or index of similarity (IS) is equal to the number of common species as expressed in percent of total number of species in both forest stands. Thus index of similarity (IS) is calculated as follows:

$$IS = \frac{c}{a+b+c} \times 100$$

Where, c = number of species common to both the forest stands

a = number of species unique to forest stand 1

b = number of species unique to forest stand 2

5.4. Species-Area Curve

Species-area curve (Fig. 9) is a relationship between the area of a habitat, or of part of a habitat, and the number of species found within that area. Larger areas tend to contain larger numbers of species, and empirically, the relative numbers seem to follow systematic mathematical relationships. The species-area relationship is usually constructed for a single type of organism, such as all vascular plants or all species of a specific trophic level within a particular site (Colinvaux 1973).

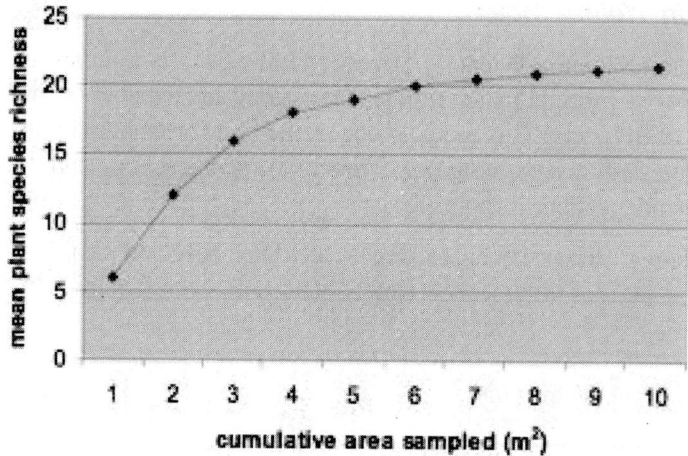

Fig. 9. Species-area curve of a sampled area

References

Barbour MG, Burk JH and Pitts WD (1987) Terrestrial Plant Ecology. Chapter 9: Method of sampling the plant community. Menlo Park, CA: Benjamin/Cummings Publishing Co

Colinvaux, Paul A (1973). Introduction to Ecology. Wiley. ISBN 0-471-16498-4

Cox G (1990) Laboratory manual. of general ecology (6th Edition). Dubuque, Iowa

Magurran AE (1988) Ecological Diversity and its Measurement. Princeton University Press, New Jersey, pp. 145-146

Misra R (1968) Ecology work book. Oxford and IBH Publishing Company, New Delhi, pp. 224

Simpson EM (1949) Measurement of diversity. Nature 163:688

Whittaker, RH (1975) Communities and Ecosystems (2nd Edition). MacMillan Publishing Co, New York

Chapter – 21

Remote Sensing and Geographic Information System (RS - GIS): Important Tools for Assessment of Plant Diversity

Manoj Semwal

1. Introduction

Plant diversity is vital for the survival and well-being of humanity. A number of domesticated plant species are critical to food security. In addition to the cultivated species, many wild plants also play an important role in meeting local needs for food, fuel, medicine and construction materials; crop wild relatives are also of special interest for crop breeding programmes. There are currently hundreds of underutilized plant species and varieties displaying traits of interest to meet present and future needs, while the value of many other plant species is yet to be discovered (Stohlgren *et al.* 1997). Researchers have confirmed the importance of the contributions and interdisciplinary understanding in relation to taxonomy, physiology, reproductive biology, conservation biology, hydrology, soils, as well as socioeconomic and climate change for holistic understanding of biodiversity patterns (Dale 1999; Myers *et al.* 2000). It is thus necessary that the professionals studying plant diversity require a synoptic view on the regulators of biodiversity and the tools through which the processes could be understood.

Spatial analysis can contribute significantly for improved understanding and monitoring of plant diversity. Results obtained from spatial analysis make it possible to formulate and implement targeted conservation strategies, which can be more effective (Escudero *et al.* 2003; Dale 1999). Outputs from spatial studies can provide critical information like evaluation of the current status of plant species and prioritize arcas for conservation on the plant diversity, in particular geographic region. Spatial information, combined with available characterization and/or evaluation data, is very useful for effective genebank management (e.g. definition of core collections, identification of collection gaps, etc.). This type of analysis is conducted using Geographic Information System (GIS) tools, which allows one to carry out complex analyses combining different (spatial) data sources and generate maps, to facilitate the development and implementation of conservation policies (Goodchild 1992; Crosetto *et al.* 2000; McCarthy 2001).

2. Components for Assessment of Plant Diversity

Statements like '75 % of diversity have been lost' or 'this site is rich in diversity' are frequently made in articles and reports focusing on conservation. However, to use such in the policy and decision-making process, they must be credible and based on well-established methodologies.

The first challenge while conducting any type of diversity analysis is to determine the appropriate level at which to work. Plant diversity is studied at community level, species level and genetic level (Soberón and Peterson 2004). At the community level, ecosystems are the units of study. At species level, the observed unit of diversity is the species, which may be present or absent in a certain location. In terms of genetic diversity, the observed unit of diversity may either be a phenotypic trait (the product of gene/genes expression) or a DNA base pair composition (analyzing neutral markers or known functional DNA, based on sequences or molecular weights) (Hijmans *et al.* 2001a). The plant diversity studies assess the status of species or genetic diversity at a specific point in time.

The holistic understanding of the complex mechanisms controlling the plant diversity, as well as their spatial and temporal dynamics, requires synergetic adoption of measurement approaches, sampling designs and technologies (Anderson *et al.* 2003). The data requirements for plant diversity assessment are both of spatial and non-spatial nature and also of various time scales (Joshi *et al.* 2004) (Fig. 1). Thus, the use of integrative tools (Remote Sensing, GIS and Ground Data) are important complimentary system assisting the ground-based studies. Geographic Information System(GIS) and remote sensing can be integrated to provide regional-scale data products across time for use in strategic and management level policymaking (Kozak *et al.* 2008).

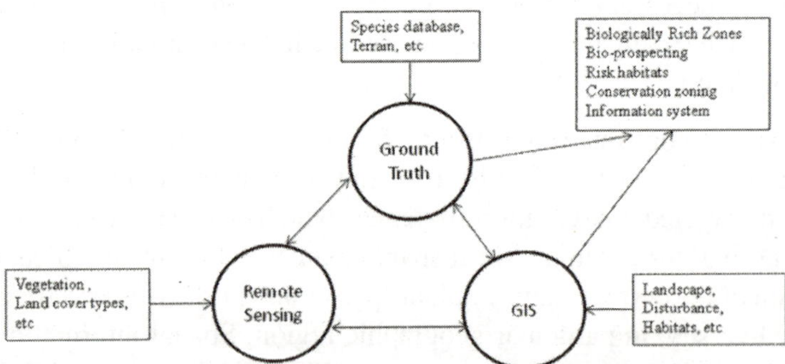

Fig. 1. Components for assessment of plant diversity

2.1. Remote sensing

On global to local scales, the only feasible way to monitor the Earth's surface to prioritize and assess the success of conservation efforts is through remote sensing (Turner *et al.* 2003) (Fig. 2A, B). Currently a suite of remote sensing satellites, having various resolutions, are available to generate spatial information on vegetation and land-cover from global to local level (Nagendra 2001).

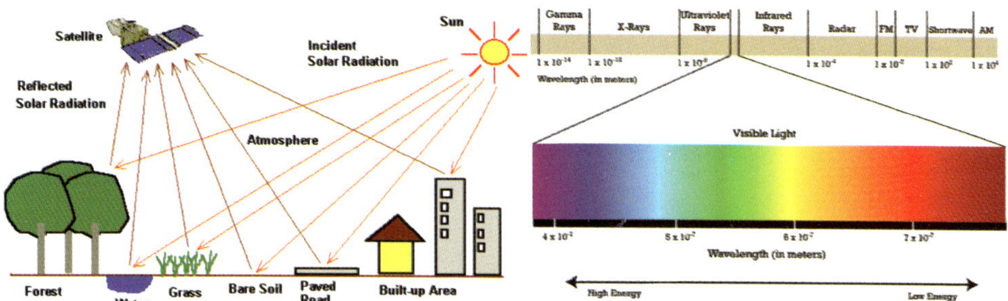

Fig. 2. (A) Remote Sensing Principle Source: www.crisp.nus.edu.sg, (B). Electromagnetic Spectrum *Source*: www.pion.cz

The remote-sensing-based information on vegetation and land cover provides a potential spatial framework and works as one of the vital input layers for the following (Turner *et al.* 2003):

i. Vegetation, land cover losses and conversion.
ii. Stratification base for optimal ground sampling and assessment of diversity.
iii. Fragmentation and neighborhood analysis.
iv. Delineation of broader vegetation types and analysis of species assemblages along with ancillary data.
v. Identification of gregarious and ecological important species.
vi. Inputs for species habitat models.
vii. Spatial delineation of biologically rich zones.
viii. Developing conservation strategies.

2.2. Ground truth data

Ground-based measurements of various parameters are a vital input for effective plant diversity assessment. In this regard, Global Positioning Systems (GPS) provide powerful tools for acquiring accurate positiond information. Precise point-location data on microclimate, topography, and soil in association with species distribution helps in understanding the interrelationships and controls of biotic and abiotic factors on species distribution pattern. GPS-aided survey in terms of presence/absence data is useful in various studies to understand the habitat distribution patterns (Peterson 2003). The use of differential GPS surveys will help in estimating the accurate change in area and the kind of species alternation with pinpoint coordinates, demarcation of permanent plots, and patches for temporal monitoring.

2.3. Geographic Information Science

'GIS' can be defined as Geographic Information Science. GIS provides the means to visualize and analyze masses of information about the state of the world's plant life and allows us to reveal new relationships, patterns and trends in a rapidly changing environment (Hijmans *et al.* 2001b) (Fig. 3A). By mapping vegetation, we can also analyze the information we collect alongside other environmental data to find specific relationships among the plant diversity and surrounding landscape (Fig. 3B). GIS tools can also be used for species conservation assessment, management and planning, monitoring plant diversity on the ground temporally and forecasting climate change (Laffan *et al.* 2010).

The translation of the ground-measured data into a spatial domain, or linking with any other spatial or non-spatial parameters to analyze the relationships and understand the trends, precise locations, and areas under consideration, is done using GIS software. The recent development of GIS technologies are ideally suited to conservation effort because they empower ecologists to expeditiously acquire, store, analyze, and display spatial data on organisms and their environment.

The integrated analysis of spatial data generated both from remote sensing and other sources, becomes critical in order to generate information on several aspects of plant diversity. Landscape processes, habitat evaluation, characterization of disturbance regimes, analysis of forest structure and function in spatial domain, are the areas where GIS works as a powerful tool to perform integrated analysis. Geostatistical techniques are the modern tools to perform interpolate sample data to create a map of plant/tree species distribution, estimate the possible occurrence, disturbance and effective prediction by spatial information on local variability (Guarino *et al.* 2002).

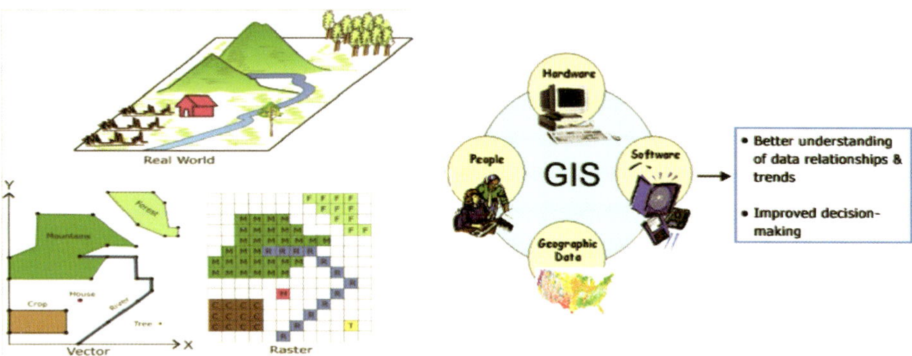

Fig. 3. (A) GIS Principle (Source:www.rst2.edu), (B) Representing the real world using GIS (*Source*: http://www.gis.com/whatisgis)

3. Methods / Techniques

Scientists and professionals who work with biodiversity data can use spatial analysis using the GIS applications to bring out new insights in the biodiversity data. In this manual, examples have been provided on the inter-specific and intra-specific diversity analyses using different types of data like species presence, morphological characterization data (phenotypic data) and molecular marker data (data of DNA base pair compositions or molecular weights) based on free and publically available software: DIVA-GIS, a GIS programme specifically designed to undertake spatial diversity analysis; and Maxent, a species distribution modeling programme (Hijmans *et al.* 2001a).

The basis for spatial plant diversity analysis is observation data. Observation data are snapshots of species, trait or allele presence in time and space. To analyze observation data, users can use their own data, publically available data provided through international platforms such as the GBIF (http://www.gbif.org).

The analyses outlined in this manual are based on two components of observation data such as passport data and associated data. Basic passport data consists of an identification code (ID), a taxonomic identification of an observed individual plant (or group of individuals) and the location of the collection or observation site. Presence points may be associated with additional data describing the collection or observation sites (e.g. land cover, soil type, etc.), the date of collection, source of coordinates, name(s) of collector(s) and the institution housing/conserving the specimen. Information about the collection date provides a time dimension to the analyses and can be used to analyze trends in species distribution (e.g. to monitor possible genetic erosion).

After following the instructions and completing the analyses outlined in this manual, it is anticipated that the reader will have the capacity to carry out basic spatial diversity analyses to address common questions in conservation biology and plant genetic resources research.

4. Applications

4.1. Spatial analysis of species richness based on point data

Many diversity analyses focus on diversity at the species level. Richness is the most straightforward method to evaluate (*alpha*) diversity. This section outlines how to undertake this type of analysis.

The analysis below uses data from a diversity study of the *Vasconcellea* genus (Scheldeman *et al*. 2007).

Before starting the analysis, it is important to remember that a sufficient number of observations within a raster cell are necessary in order to undertake reliable diversity analyses. The credibility and accuracy of the final result will depend on the quality of the sampling strategy, although the choice of the cell size used for the analysis (ideally, the cell size should be defined while formulating the sampling strategy) will also influence the quality of the final result. If an analysis is run with cells that are too small, the resulting raster will generate a high resolution map with results of limited value, as each cell will most likely contain too few presence points (often only one) to detect a spatial pattern of species diversity. On the contrary, if raster cells are too large, they will have a sufficient number of observations, but the map will be of poor resolution, complicating its interpretation and use.

The following analysis will use the number of species as the measured unit of diversity. Here, we would use DIVA-GIS to conduct spatial analysis of the species richness of *Vasconcellea* genus.

Steps

i. Start by visualizing two layers: *Vasconcellea data* and *countries data*. Then, select the layer for the *Vasconcellea* species. Go to *Analysis* → *Point to Grid* → *Richness*.

ii. Next, define the properties of the raster that will be used for the analysis. The dimensions of the study area as well as the resolution (cell size) must be determined. In the *Point to Grid* window, go to the *Define Grid* option and select *Create a New Grid* (Fig. 4).

iii. Click on the *Options* window to define the raster properties: origin and extent of the study and the resolution (cell size).

iv. The study area can be defined using one of the following options:

   a. In the *Options* window, manually enter the values for the X-axis and Y-axis (you may wish to select the default options); or

   b. Draw the extension on the map using the *Draw Rectangle* tool.

v. The values used for this analysis are: Min X: -100; Max X: -45 and Min Y: -34; Max Y: 21 (Fig. 4).

vi. The cell size used in this analysis will be one (1) degree (default value under *Cell Size*).

vii. Click *OK* to accept the values.

viii. Return to the *Point to Grid* window to select the type of analysis to be carried out. Here, we will undertake a species richness analysis.

  a. First, select the Output Variable: *Number of different classes (Richness).*

  b. Click on the *Parameters* tab to select the units.

ix. In the *Parameters* tab, under the *Field* option, indicate the parameter you wish to analyze. For this analysis, the parameter will be *Species*, in order to analyze the species richness (Fig. 4). You are given the option to exclude specific species from the analysis.

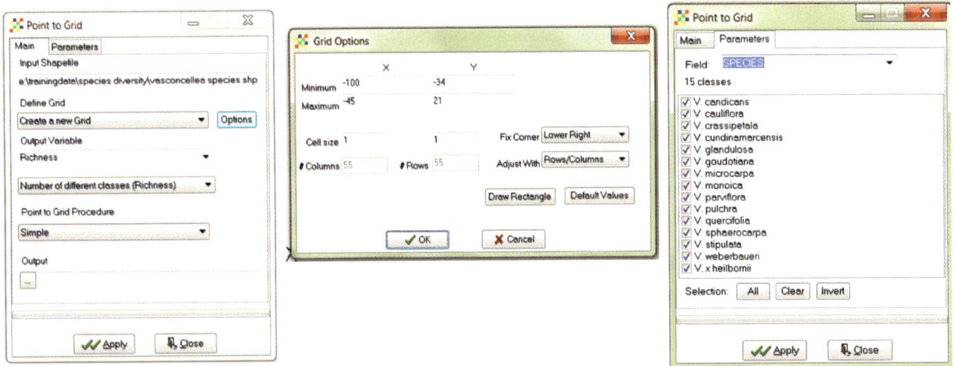

Fig. 4. Species richness steps in DIVA – GIS

Fig. 5. (A) Output file path in DIVA – GIS, (B) Species richness map generated

x. Click on the *Main* tab and select the button to the left of the *Output* box (Fig. 5A). Enter the name of the file to be saved and its file path. Finally, click on *Apply*. The resulting raster will show the number of species observed in each cell (Fig. 5B).

4.2. Spatial analysis for Intra-specific diversity based on phenotypic data

Genetic diversity studies, including the analysis of spatial patterns in genetic diversity, are frequently based on molecular marker data. However, phenotypic data and, particularly morphological data, can be another indirect source of genetic diversity information (Heywood 1991). Phenotypic data from a single individual varies as a function either of the genotype (G), the environment (E) or a combination of both (the GxE effect). When using data based on *in situ* characterization, conducted at environmentally heterogeneous locations, it is recommended to focus on the traits that are not influenced by environment.

When working on other traits, in order to minimize variation determined by environmental influences, one option is to conduct the characterization outside the original collection site and under controlled, uniform environmental conditions (*ex situ* characterization), be it in the same geographic location (e.g. experimental fields) or in a controlled environment (e.g. greenhouse) (Hijmans *et al.* 2001a).

The following analysis outlines how to carry out a spatial analysis based on phenotypic data resulting from *ex situ* (in the same experimental field) morphological characterization, combined with passport data (information about the site where material was originally collected). In this section, we would use DIVA-GIS to undertake an intra-specific diversity analysis based on morphological data.

Steps

i. Add the layer with the characterization data (Cassava: *Manihot ex situ.shp*).

Explore the *ex situ* characterization values for the exercise. Select the cassava characterization layer and go to *Analysis* → *Point to Grid* → *Statistics* and different type of statistics can be generated for the phenotype characterization data (Fig. 6).

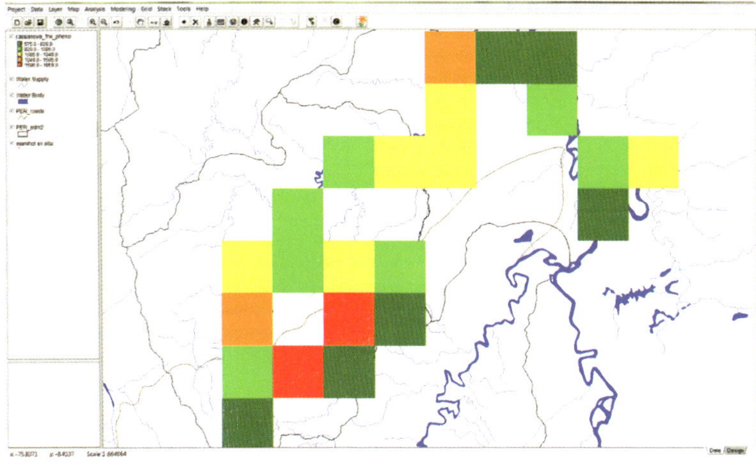

Fig. 6. Intra-specific diversity map based on phenotypic data generated

4.3. Spatial analysis of Intra-specific diversity based on molecular marker data

Phenotypic/morphological data can be used to conduct intra-specific diversity analyses; however, the influence of the environment where the characterization was conducted will always affect the results (Dale 1999). Allelic composition, or gene sequences of plant individuals, are not influenced by such environmental factors (a change in environmental conditions, e.g. wet vs. dry year, does not alter DNA base pair composition but will alter phenotypic appearance, e.g. leaf size or growth). Therefore, molecular markers are the measurements of choice to carry out an analysis of intra-specific diversity (Hijmans *et al.* 2001a).

Although molecular markers are, for the most part, not directly related to functional genes, it can be anticipated that a high diversity based on molecular markers also indicates a high abundance of useful genes. The following section outlines a basic spatial analysis of molecular marker data, whereby microsatellites (SSRs), a widely used co-dominant marker, are utilized. It should be noted, though, that any type of molecular marker data, e.g. AFLPs, can be used to conduct a diversity analysis, provided a unique identity can be given to each variation in the DNA composition (Turner *et al.* 2003).

Principles of spatial diversity analysis using molecular marker data are very similar to those of an analysis at the species level. SRR analysis looks at differences in length of microsatellite regions (usually not associated to functional genes, i.e. neutral). These differences in length, further referred to as different alleles, are the observed units of diversity in this analysis.

To carry out a spatial analysis, the molecular marker data must be formatted in such a way that each allele includes georeferenced information. In the following analysis, each allele is formatted according to the following: microsatellite code + weight of base pairs (e.g. SSR1-293).

Uneven distribution of observations can have a significant impact on the richness analysis. The rarefaction method compares the richness among cells that have a dissimilar number of observations or samples. It recalculates the richness measured in the different cells as if a standard number of observations were made in each cell. Only cells with an equal or higher number of observations than the standardized number are included in the analysis; cells with fewer observations are excluded. The choice of the standardized number of observation is a trade-off between the number of cells included in the analysis and the maximum richness to be calculated.

Steps

i. Select the layer with the molecular marker data and go to *Analysis/Point to Grid/ Richness*. Under *Output Variable,* select *Rarefaction* (Fig. 7).

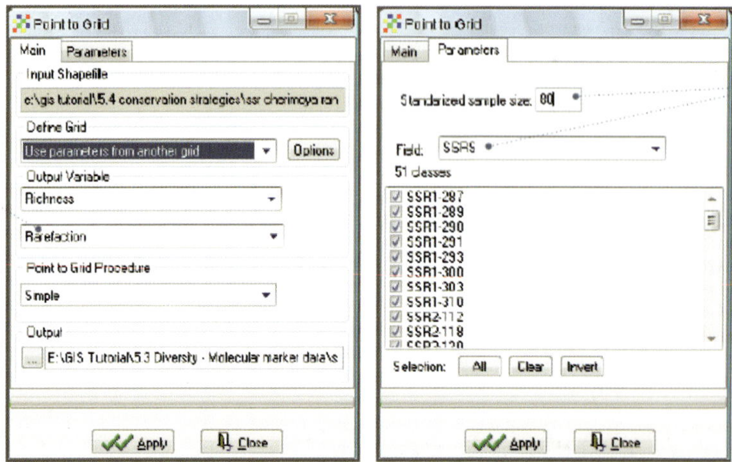

Fig. 7. Steps for spatial analysis for intra-specific diversity on molecular marker data

Under the *Parameters* tab, mark the SSRs using a *Standardized Sample Size* of 80.

As can be observed, the difference in diversity between northern Peru and Bolivia (where much sampling was carried out) becomes even more evident in this analysis (Fig. 8).

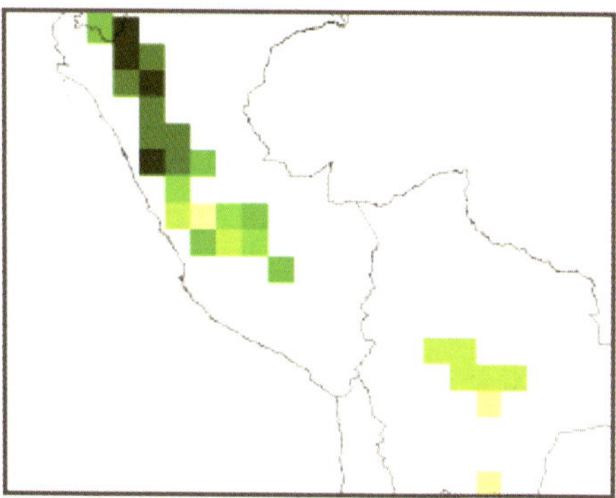

Fig. 8. Intra-specific diversity map based on molecular marker data generated

5. Prediction of Species Niche due to Climate Change Using Maxent

With the recent and rapid spread of ecological niche modeling (ENM) and Geographic Information Sciences (GIS), the need for a detailed dataset of environmental characterization has increased substantially (Phillips *et al.* 2006, Anderson *et al.* 2003). The WorldClim dataset (Hijmans *et al.* 2005, available at http://www.worldclim.org) provides high resolution (i.e. nearly 1 km) climatic surfaces derived from historical records obtained from a number of weather stations across the globe. WorldClim provides high resolution monthly maximum (tmax), minimum (tmin), and mean temperatures (tmean), and monthly precipitation (prec); and a set of 19 bioclimatic variables can be derived (Busby 1991).

Species distribution models (or SDM's) are used to explore how the occurrence of a species is related to the environment, and how a species might respond to changes in its environment. This can help find new locations where a rare species might be found, or understand the potential threats to a species due to urban encroachment, climate change, or other causes. The maximum entropy algorithm for species distributions modeling is one of the most accurate and globally used ecological niche models (Phillips *et al.* 2006; Phillips and Dudík 2008).

Many modelers currently use the set of bioclimatic variables available at the WorldClim website when modeling a certain species geographic distribution using Maxent. This is a relatively easy task when the user works with current conditions (interpolations historical observed data, representative of 1950-2000 climates) since the bioclimatic variables needed for the analysis can be directly downloaded from the WorldClim website. However, often when working with future conditions (i.e. climate change), these 19 bioclimatic layers must be derived from the three basic climatic variables (i.e. tmin, tmax, prec).

1. Download the tmin, tmax and prec variables from the climate model (e.g. CCCMA), emission scenario (e.g. A2A), year (e.g. 2020) and spatial resolution (e.g. 30 arc-seconds) of interest (http://www.ccafs-climate.org/data/) (Fig. 9 A).
2. Open DIVA-GIS and Import each of the .ASC file into a .GRD format (Fig. 9 B)
    a. Data → import to Grid file → multiple files
    b. On "type", select "Arc ASCII" Click on "add file", browse the path where you saved your ASCII files and then add all of them
    c. In "output folder" you may browse for a folder in which you want to save your new .GRD files. Otherwise, select "same as input" and the program will save the files inside the folder from which you loaded the input ASCII files.

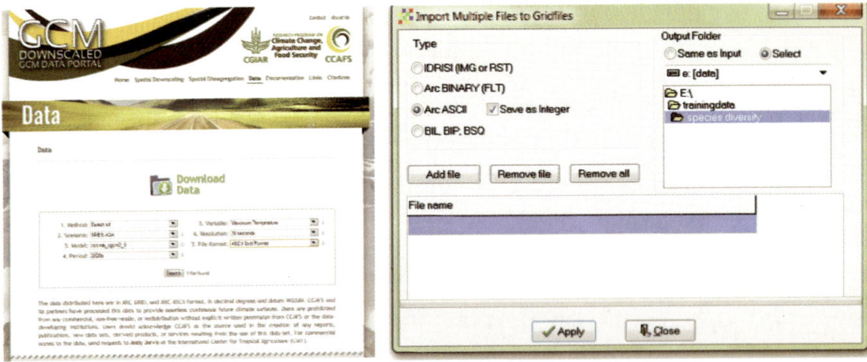

Fig. 9. (A) Downloading the climatic data for future prediction *Source*: http://www.ccafs-climate.org/, (B) Importing the climate data to DIVA – GIS for analysis

3. This process will produce a .CLM file to be used as input for DIVA-GIS in order to generate the bioclimatic variables.
    a. Data → Climate → Make CLM Files
    b. Browse to the folder where all the .GRD files are located.
    c. In "File prefixes" add a "_" after "tmin" and "tmax", and change the word "rain" by "prec_".
    d. Leave the "suffix" and other boxes blank.
    e. Click on "Files".
    f. Click "OK".
4. Load the CLM File in DIVA-GIS.
    a. Tools → Options
    b. Click on "Folder" and browse the folder containing the CLM files.

c. Click on "Apply" and verify that all the boxes are now filled with some names, metadata, or names of the files.

d. Click on "OK" Close the window.

5. Load the CLM File in DIVA-GIS

   a. Tools → Options.

   b. Click on "Folder" and browse the folder containing the CLM files.

   c. Click on "Apply" and verify that all the boxes are now filled with some names, metadata, or names of the files.

   d. Click on "OK" Close the window.

6. Create the bioclimatic variables

   a. Select "Bioclim" from the "Output" several choices.

   b. Check the "all" variables box, Check the "Add to map" box to visualize each of the 19 derived bioclimatic variables.

   c. Click on "File", browse for an output folder, and put any name you may want.

   d. Click on "Apply.

7. Export the 19 variables (BIO1 to BIO19) to an ASCII format (ASC).

   a. Data → Export Gridfile → Multiple files.

   b. Select "ESRI ASCII" on "File type".

   c. Click on "add file" and browse for your bioclimatic files in. GRD format.

   d. Choose an output folder if desired. Otherwise, select "Same as input".

8. Load the ASCII files (.ASC) of interest into the Maxent interface (Fig. 10 A) to produce the predicted climate change outcomes (Fig. 10 B).

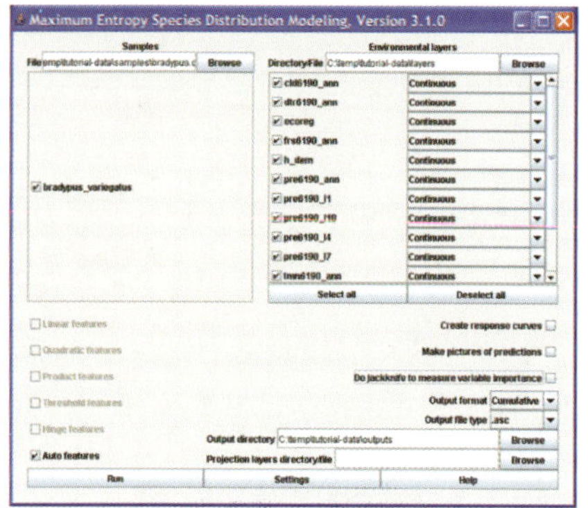

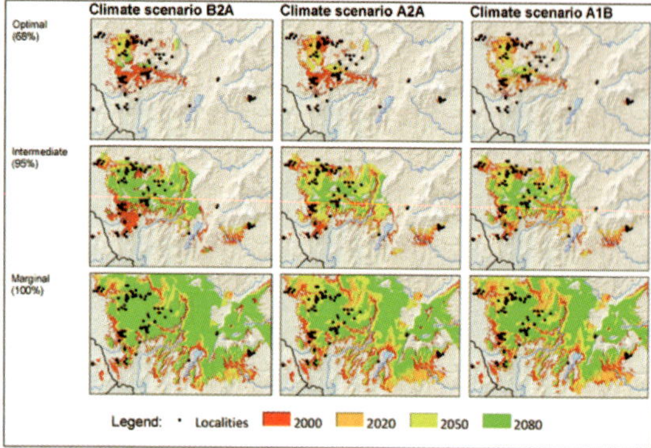

Fig. 10. (A) Maxent software interface for running the prediction model, (B) Predicted climate change outcomes using MaxEnt Modeling for indigenous Arabica for the year intervals 2020, 2050 and 2080. *Source*: http://www.plosone.org/article/ info%3Adoi%2F10.1371%2Fjournal.pone.0047981

References

Anderson RP, Lew D, Peterson AT (2003) Evaluating predictive models of species' distributions: criteria for selecting optimal models. Ecol Model 162:211–232

Busby J (1991) BIOCLIM-a bioclimate analysis and prediction system. Plant Prot. Q. 6

Crosetto M, Tarantola S, Saltelli A (2000) Sensitivity and uncertainty analysis in spatial modelling based on GIS. Agric Ecosyst Environ 81:71–79

Dale MR (1999) Spatial pattern analysis in plant ecology. Cambridge University Press, United Kingdom.

Escudero A, Iriondo JM, Torres ME (2003) Spatial analysis of genetic diversity as a tool for plant conservation. Biol Conserv 113:351–365

Goodchild MF (1992) Geographical information science. Int J Geogr Inf Syst 6:31–45

Guarino L, Jarvis A, Hijmans RJ, Maxted N (2002) 36 Geographic Information Systems (GIS) and the Conservation and Use of Plant Genetic Resources. Manag Plant Genet Divers Electron Resour 387

Heywood JS (1991) Spatial analysis of genetic variation in plant populations. Annu Rev Ecol Syst 22:335–355

Hijmans R, Cruz M, Rojas E and Guarino L (2001a) DIVA-GIS version 1.4: A geographic information system for the analysis of biodiversity data

Hijmans RJ, Guarino L, Cruz M, Rojas E (2001b) Computer tools for spatial analysis of plant genetic resources data: 1. DIVA-GIS. Plant Genet Resour Newsl 15–19

Joshi C, de Leeuw J and van Duren IC (2004) Remote sensing and GIS applications for mapping and spatial modelling of invasive species. Proc. ISPRS. p B7

Kozak KH, Graham CH, Wiens JJ (2008) Integrating GIS-based environmental data into evolutionary biology. Trends Ecol Evol 23:141–148

Laffan SW, Lubarsky E, Rosauer DF (2010) Biodiverse, a tool for the spatial analysis of biological and related diversity. Ecography 33:643–647

McCarthy JJ (2001) Climate change: Impacts, adaptation, and vulnerability: contribution of Working Group II to the third assessment report of the Intergovernmental Panel on Climate Change. Cambridge University Press, Cambridge

Myers N, Mittermeier RA, Mittermeier CG *et al.* (2000) Biodiversity hotspots for conservation priorities. Nature 403:853–858

Nagendra H (2001) Using remote sensing to assess biodiversity. Int J Remote Sens 22:2377–2400

Peterson AT (2003) Predicting the geography of species' invasions via ecological niche modeling. Q Rev Biol 78:419-433

Phillips SJ, Anderson RP, Schapire RE (2006) Maximum entropy modeling of species geographic distributions. Ecol Model 190:231–259

Phillips SJ, Dudík M (2008) Modeling of species distributions with Maxent: new extensions and a comprehensive evaluation. Ecography 31:161–175

Soberón J, Peterson T (2004) Biodiversity informatics: managing and applying primary biodiversity data. Philos Trans R Soc Lond B Biol Sci 359:689–698

Stohlgren TJ, Coughenour MB, Chong GW, *et al.* (1997) Landscape analysis of plant diversity. Landsc Ecol 12:155–170

Turner W, Spector S, Gardiner N, *et al.* (2003) Remote sensing for biodiversity science and conservation. Trends Ecol Evol 18:306–314

Chapter – 22

Identification of Herbal Drugs Using Pharmacognostical Tools

A.K.S. Rawat and M.M. Pandey

1. Introduction

In the last few decades there has been worldwide revival on the use of herbal drugs/ phytochemical for diverse purposes including medicinal, nutritional and as cosmetic. The revival of interest in natural drugs and the herbal products was due to the widespread belief that 'green' medicine is healthier than synthetic products. This has led to the rapid spurt of demand for health products like herbal tea, ginseng and such products of traditional medicine during the 1980s. The health promotion and disease prevention strategy in treatment is widely prevalent in oriental systems, especially the Indian ('Ayurveda', 'Siddha', 'Unani and 'Amchi') and the Chinese systems of medicine, which are finding increasing popularity and acceptance in the world over. Because of this sweeping 'greenwave' a large number of herbal drugs and the plant derived herbal products are sold in the health food shops all over the developed countries. Export–Import Bank reports revealed that the global trade of plant-derived and plant originated products is around US $60 billion (with growth of 7% per annum) where India holds stake of US $1 billion which is expected to reach 3 trillion US$ by the end of 2015 (Pandey *et al.* 2008). These targets can be achieved by providing scientifically validated, safe and standardized herbal products in domestic and international markets.

Standardization is an important step for the establishment of a consistent biological activity, a consistent chemical profile, or simply a quality assurance programme for production and manufacturing of herbal drugs. It is now well known that the pharmacological activity of a medicinal plants is due to the presence of certain bioactive phytochemicals, which are either primary or secondary metabolites. Amount of these compounds are controlled and conditioned by a variety of factors such as its genetic predisposition, habitat of the plant agro climatic conditions, season and also the stage of growth and development of the plant etc. Therefore, it is desirable to study the effect of these on secondary metabolites to get the right/ optimum amount of secondary metabolites in the plant part/ raw drug. The formation of herbal drug based on the traditional knowledge using modern tools is the requirement of the present day. WHO's specific guidelines for the assessment of the safety, efficacy and quality of herbal medicines as a prerequisite for global harmonization are of utmost importance. Considering the importance of the herbal drugs/ products, quality assurance of raw drug/ formulation and developing botanical and chemical parameters for assuring their quality, have become major riders on the pharma or drug research and industry.

2. Need of Standardization in Herbal Drug/Products

The need of the hour is to evolve a systematic approach and to develop well-designed methodologies for the standardization of herbal raw materials and herbal formulations. Numerous technologies of phytochemical standardization, such as preliminary phytochemical screening, fingerprint profiling, and quantification of marker compound(s) with reference to herbal raw materials and polyherbal formulations have been developed well. But these technologies should consider and possibly use the fact that the biological activity of plant extracts often results for additive or synergistic effects of its components. The herbal market is facing difficulties for keeping batch to batch consistency in quality products. Most of these drugs do not have well defined and characterized composition. The three pillars of ideal herbal drug and their rational use are quality, safety and efficacy (Fig. 1).

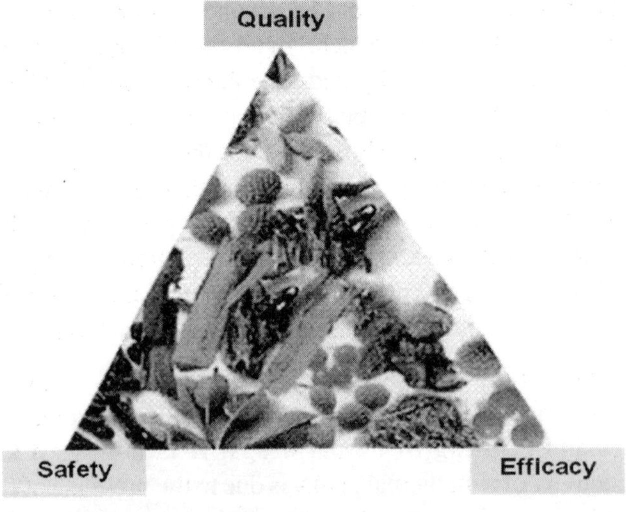

Fig. 1. Three pillars of herbal drug development

3. Pharmacognostical Study of Raw Herbal Drugs

The quality evaluation as per WHO guidelines has great significance in herbal drug sector. Development of parameters for quality control of raw herbal drugs and standardization of polyherbal formulations is utmost important. In the Indian Systems of Medicine (comprising of Ayurveda, Unani and Siddha), drugs of plant, animal and mineral origin, are used in their natural or so called 'Raw' forms singly or in their mixture or in combination, to make a compound preparation of formulation. Nearly 90 per cent of the Raw Drugs are obtained from the plant sources while about 10 per cent of the drugs are derived from animal and mineral sources. The drugs of plant origin especially of herbaceous nature frequently used as whole plant; otherwise their parts such as Root, Stem, Leaf, Flower, Seed, Fruit and modifications of Stem and Root, Bark of a Stem or Root, Wood, and their Exudates or Gums etc. constitute single drugs in the Indian Systems of Medicine. These vegetable drugs are either used in dried forms or some times as whole fresh or their juice. The study of these crude drugs made with a view to recognize them is called Pharmacognosy (Pharmakon = Drug; Gignosco = to acquire knowledge of), meaning the knowledge or science of Drugs coined by Seydler in 1815 (Trease and Evans 2009).

The parameters used for standardization of plant based drugs are classical pharmacognostical, physico-chemical, phytochemical and molecular approaches. In Pharmacognosy, a complete and systematic study of a drug is done, which comprises of (i) origin, common names, scientific nomenclature and family, (ii) geographical source (and history), (iii) cultivation, collection, preservation and storage, (iv) Macroscopical, Microscopical and sensory (organoleptic) characters, (v) Chemical compostition wherever possible, (vi) Identity, Purity, Strength and Assay, (vii) substitute and adulterants etc. such systematic study of a drug as complete as possible, is claimed to be the scientific or pharmacognostical evaluation (Fig. 2). The detail study includes correct taxonomic identification of plants and its parts, organoleptic characterization, macro-microscopical details and histochemical analysis of the plant parts used as medicine; physico-chemical standards *viz.* foreign matter, ash and extractive values, total volatile matter, estimation of phyto-constituents e.g., alkaloids, phenolics, tannins, sugars, starch, protein etc.; microbial load; heavy metal estimation; chemical markers for the identification, quality assurance and batch to batch consistency with the help of HPTLC, HPLC and GC, identification of biomarkers and their estimation etc.

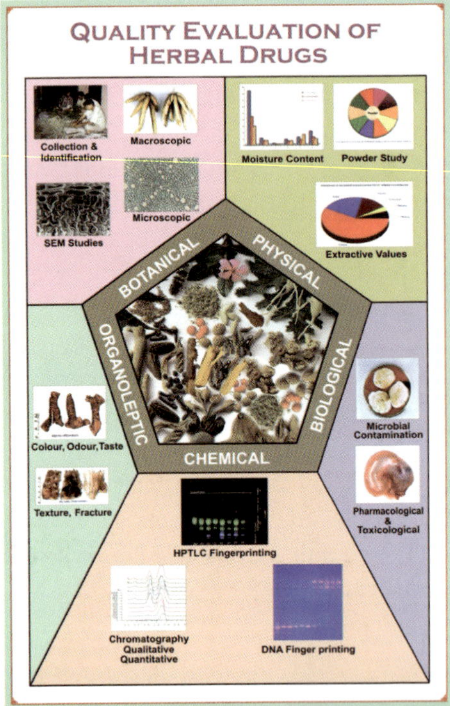

Fig. 2. Quality parameters of herbal drug standardization

3.1. Morphological studies or organoleptic characters

The organoleptic characters like, colour, ornamentation, texture, fracture and odour of the raw drug sample is helpful to identify the drug and its morphology. Many drugs also have some specific aroma/fragnance, i.e. *Hemidesmus indicus, Withania somnifera, Nardostachys jatamansi* etc. These are the characteristic features used in the morphological study (Fig. 3).

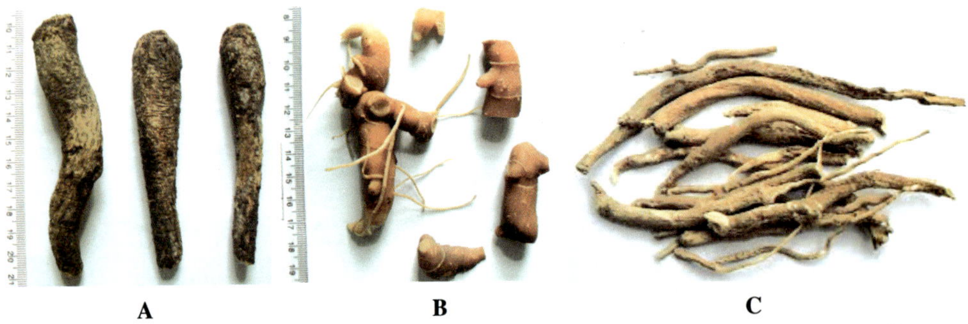

Fig. 3. Morphological structure of the roots of; (a) *Saussurea costus,* (b) *Polygonatum verticillatum,* (c) *Withania somnifera*

3.2. Microscopic identification of raw herbal drugs

Methods of preparing specimens of raw materials of herbal drugs for microscopical studies vary, depending on the morphological groups of drugs to be examined and also on the nature of the material i.e., entire, cut or powdered (Fig. 4).

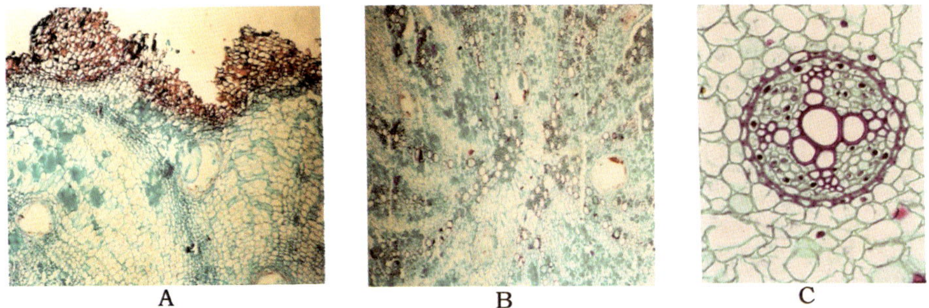

Fig. 4. Anatomical structure (TS) of the root; (A,B) *Saussurea costus,* (C) *Polygonatum verticillatum* (Pandey et al. 2007).

3.2.1. Section and slide preparation

The fixed samples should be properly washed in running water before sectioning. The sectioning can be done in transverse and longitudinal planes at a thickness of 10-15 µm. The best sections are picked out for mounting after the staining and dehydration is completed, the best results can be obtained using the method described in Plant microtomy by Johanson, (1940).

3.2.2. Maceration technique

To study the microscopic details of the individual elements of the drug pieces are macerated using suitable reagent. These slices are boiled with 40% Nitric acid solution for 3-5 min or the treatment continued till the cells of the slices become sufficiently loose to allow the separation of the individual elements to soften the tissues. Then wash with distilled water, stain with safranin and transfer on a slide after a slight pressure over the cover slip after mounting in glycerine (Fig. 5).

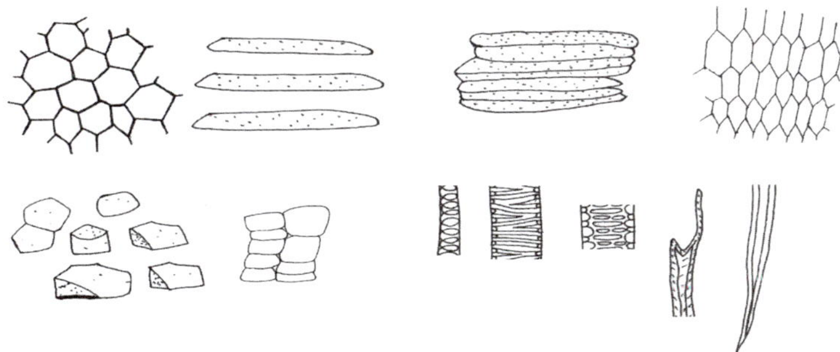

Fig 5. Powdered microscopic characters of the root of *Saussurea costus*

3.2.3. Powder studies

The powder study help us to examine the of different characters of powdered drugs like organoleptic characters viz., colour, tastes, odour, fineness and degree of uniformity of the particles, sensation of smoothness. Fluorescence tests of powders (under UV and visible light) is useful to identify the elements (Chase and Pratt 1949; Kokoski *et al.* 1958).

3.2.4. Study of leafy drug/ stomatal study

There are several types of stomata, distinguished by the form and arrangement of the surrounding cells. The type of stomata present in the leafy drug is one of the important parameter to study the drug and its identification. The following types of stomata are present in the plants (Kokate 2004) (Fig. 6).

Fig 6. Some example of stomata types in the leaf surface

3.2.4.1. Anomocytic (irregular-celled)

Previously known as ranunculaceous. The stoma is surrounded by a varying number of cells in no way differing from those of the epidermis generally.

3.2.4.2. Anisocytic (unequal-celled)

Previously known as cruciferous or solanaceous. The stoma is usually surrounded by three subsidiary cells, of which one is markedly smaller than the others.

3.2.4.3. Diacytic (cross-celled)

Previously known as caryophyllaceous. The stoma is accompanied by two subsidiary cells whose common wall is at right angles to the guard cells.

3.2.4.4. Paracytic (parallel-celled)

Previously known as rubiaceous. The stoma has each side one or more subsidiary cells parallel to the long axis of the pore and guard cells.

3.2.4.5. Actinocytic

The stomata surrounded by epidermal cells uniformly arranged along the radii of a circle with the stomata at the sample.

Fig. 6. Some example of stomata types in the leaf surface

3.2.5. Study of herbs and flowers

For examining leaves, herbs and flowers (entire or cut) under microscope, following methods are employed for clarification.

3.2.5.1. Entire material

When examining entire leaves, herbs and flowers, take pieces of leaf (margin and vein of leaves only), herbs (only leaf) and flowers (only calyx and corolla) in test tube. Add a solution of caustic alkali or nitric acid to the test tube and boil for 1-2 minutes, pour the contents into a porcelain dish, drain off the liquid, wash the material with water and leave for sometimes. Remove the pieces of the material from the water with a spatula and put on the slide, add a few drops of the solution of glycerol or chloral hydrate. Crush the material with scalpel and cover with slip before examining.

3.2.5.2. Cut material

For examining cut leaves, herbs and flowers, take several pieces in a test tube and employ the same methods as described for entire material.

3.2.6. Study of fruit and seeds

Entire material

For microscopical examination of fruit and seed take the specimens or outer coat of seed or fruit and examine as described below:

Outer coat- For examining the outer coat boil 3 or 4 seeds or fruits in caustic alkali solution in a test tube for 1-2 minutes (outer coat specimens with intensive pigmentation are boiled for longer period). After boiling, place the pieces on slide, remove the layers of the coat and examine them after mounting in glycerol solution.

Section- If fruits or seeds are too hard to cut then boil them for 15-30 minutes or more depending on their hardness or keep them in moistening chamber or absorb in water and chloroform solution or soften them with stem and then cut the specimen for examining purpose. For cutting small, flat seeds (which are difficult to hold) place them in a pith or potato slit for section cutting. For this, a block of paraffin (0.6 x 0.5 x 1.5 cms. in size) is made and the seed in embedded in the block by making a cavity or a pit in the block with a hot teasting needle. Cut the section with a sharp razor (through the object) together with the paraffin, place them on to the slide, remove paraffin with a needle or wash it with xylene and examine the section in chloral-hydrate solution.

3.2.7. Study of bark drugs

Lignified elements – For testing lignin add several drops of phloroglucinol and a drop of concentrated hydrochloric acid to the section on a slide then draw off the liquid, and cover with a cover slip (the specimen should not be heated); the lignified elements are pink. The excessive stain can be washed out with acidified alcohol.

Starch- Starch is detected by treating with iodine solution.

Tannin- Tannin is detected by treating with ferric ammonium sulphate solution (blue-black or green black colour shows the presence of Tannin) or with potassium-di-chromate solution (brown colour indicates the presence of Tannin).

Anthraquinone derivatives – Anthraquinone derivatives are detected by treating with alkali solution (blood-red-colour shows the presence of anthraquinone derivatives).

Cut material of bark drug

Prepare small pieces or scraping of bark and boil them for 3-5 minutes in a solution of caustic alkali or potassium hydroxide or in nitric acid solution and then mount in glycerin for examination on a slide covered with a cover slip.

3.2.8. Study of Roots and rhizomes

3.2.8.1. Entire material

For anatomical examination of entire roots and rhizomes cut transverse and longitudinal sections. For this soften small pieces of roots without heating in glycerol solution for 1-3 days, depending on their hardness. The softened roots are straightened with the help of a scalpel in the right direction and then cut a section with the razor. First, cut thicker entire slices and then make thin, smaller sections. Stain the entire slices with phloroglucinol and concentrated hydrochloric acid or with safranin examine the specimen under a dissecting microscope. For micro-chemical test the small and thin sections are examined under microscope, as follows:

Starch- Starch is detected with iodine solution. For this, prepare specimen with water to measure the granule of starch with an ocular micrometer.

Inulin- Inulin is detected with Molish's reagent. For this place a little powder on a slide and apply 1-2 drops of naphthol and drop of concentrated sulphuric acid, if inulin is present, the powder will appear reddish-violet coloured. Starch also gives this test, so the test for inulin can be done in the absence of starch.

Lignified elements- Lignified elements (fibrovascular bundles, mechanical tissue etc.) are detected with phloroglucinol and concentrated hydrochloric acid or safranine solution as mentioned above for barks.

Fixed oil- For fixed oil detection use Sudan IV, as mentioned above for fruits and seeds.

If required for tannin, anthraquinone derivatives, test as mentioned above.

3.2.8.2. Cut material of roots and rhizomes

Make small pieces or scrapping of roots or rhizomes and boil them for 3-5 minutes in caustic alkali, or in nitric acid and then make pressed specimen and immerse them in glycerol. Microchemical tests can be performed with scrapings for various chemicals as mentioned above.

3.3. Quantitative microscopy

3.3.1. Determination of stomatal index

The stomatal index is the percentage ratio of the number of stomata formed by the total number of epidermal cells, including the stomata, each stoma being counted as one cell.

3.3.2. Determination of palisade ratio

Palisade ratio is the average number of palisade cells beneath one epidermal cell.

3.3.3. Determination of vein-islet number

The mesophyll of a leaf is divided into small portions of photosynthetic tissue by anastomosis of the veins and veinlets; such small portions or areas are termed "Vein-Islets". The number of vein-islets per square millimeter is termed the "Vein-Islet number". This value has been shown to be constant for any given species and, for full-grown leaves, to be unaffected by the age of the plant or the size of the leaves. The vein-islet number has proved useful for the critical distinction of certain nearly related species. The determination is carried out as follows:

3.3.4. Determination of stomatal number

Place leaf fragments of about 5x5 mm in size in a test tube containing about 5 ml of chloral hydrate solution and heat in a boiling water-bath for about 15 minutes or until the fragments become transparent. Transfer a fragment to a microscopic slide and prepare the mount the lower epidermis uppermost, in chloral hydrate solution and put a small drop of glycerol-ethanol solution on one side of the cover glass to prevent the preparation from drying. Examine with 40X objective and a 6X eye piece, to which a microscopical drawing apparatus is attached. Mark on the drawing paper a cross (x) for each stomata and calculate the average number of stomata per square millimeter for each surface of the leaf.

4. Physico-chemical Parameters for the Standardization of Crude Drugs/ Products

4.1. Foreign matter

Medicinal plant materials should be entirely free from visible sign of contamination by moulds or insects, and other animal contamination, including animal excreta. No abnormal odour, discoloration, slime or signs of deterioration should be detected. It is seldom possible to obtain marketed plant materials that are entirely free from harmful foreign matter or residue. Macroscopic examination can conveniently be employed for determining the presence of foreign matter in whole or cut plant materials. However, microscopy is indispensable.

Foreign material is material consisting of any or all of the following:

- Parts of the medicinal plant material or materials other than those named with the limits specified for the plant material concerned.
- Any organism, part or product of an organism, other than that named in the specification and description of the plant material concerned.
- Mineral admixtures not adhering to the medicinal plant materials, such as soil, stones, sand, and dust.
- Drugs should be free from moulds, insects and animal fecal matter.

4.2. Determination of moisture content (Loss on drying)

Procedure set forth here determines the amount of volatile matter (i.e., water drying off from the drug). For substances appearing to contain water as the only volatile constituent (Anonymous 1998).

4.3. Determination of total ash and acid insoluble Ash

Ash values are determined to estimate the total amount of the inorganic salts present in the drug. This includes Total Ash, and Acid Insoluble Ash. The ash remaining following ignition of medicinal plant materials is determined by two different methods, which measure total ash, and acid insoluble ash.

4.3.1. The total ash

Method is designed to measure the total amount of material remaining after ignition. This includes both 'physiological ash', which is derived from the plant tissue itself, and 'non-physiological' ash, which is the residue of the extraneous matter (e.g., sand and soil) adhering to the plant surface.

4.3.2. Acid insoluble ash

The residue obtained after boiling the total ash with dilute hydrochloric acid, and igniting the remaining insoluble matter. This measures the amount of silica present, especially as sand and siliceous earth. That may be added on at the time of collection.

4.4. Extractive values

These are used to determine the amount of the matter, which is soluble in the solvents used, which include alcohol soluble extractives and water-soluble extractives. Percentage of alcohol and water-soluble extractives are calculated and used as standards.

4.5. Determination of foaming index

Many medicinal plant materials contain saponins that can cause a persistent foam when an aqueous decoction is shaken. The foaming ability of an aqueous decoction of plant materials and their extract is measured in terms of a foaming index.

5. Qualitative Analysis

5.1. Preliminary phytochemical screening

The preliminary phytochemical studies is performed for testing the different chemical groups present in plant extracts. 10% (w/v) solution of extract is taken unless otherwise mentioned in the respective individual test. The chemical group tests is performed.

General screening of the alcoholic and aqueous extracts of the plant material is carried out for qualitative determination of the groups of organic compounds present in them.

5.1.1. Alkaloids

5.1.1.1. Dragendorff's test

Dissolve a few mg of alcoholic or aqueous extract of the drug in 5 ml of distilled water, add 2 M hydrochloric acid until an acidic reaction occurs, then add 1 ml of Dragendorff's reagent, an orange or orange-red precipitate produced immediately indicate the presence of alkaloid.

5.1.1.2. Hager's test

Take 1 ml of alcoholic extract of the drug take in a test tube, add a few drops of Hager's reagent. Formation of yellow precipitate confirms the presence of alkaloids.

5.1.1.3. Wagner's test

Acidify 1 ml of alcoholic extracts of the drug with 1.5% v/v of hydrochloric acid and add a few drops of Wagner's reagent. A yellow or brown precipitate formed confirms presence of alkaloids.

5.1.1.4. Mayer's test

Add a few drops of Mayer's reagent to 1 ml of acidic aq extract of the drug. White or pale yellow precipitate indicates the presence of alkaloids.

5.1.2. Carbohydrates

5.1.2.1. Anthrone test

Take 2 ml of anthrone solution, add 0.5 ml of aqueous extract of the drug. A green or blue colour indicates the presence of carbohydrates.

5.1.2.2. Molisch's test

In a test tube containing 2 ml of aqueous extract of the drug add 2 drops of a freshly prepared 20% alcoholic solution of á-naphthol and mix, pour 2 ml conc sulphuric acid through the side of the test tube so as to from a layer below the mixture. Carbohydrates, if present, produce a red-violet ring, which disappears on the addition of an excess of alkali solution

5.1.2.3. Benedict's test

Take 0.5 ml of aqueous extract of the drug add 5 ml of Benedict's solution and boil for 5 minutes. Formation of a coloured precipitate is due to the presence of carbohydrates.

5.1.2.4. Fehling's test

In 2 ml of aqueous extract of the drug add 1 ml of a mixture of equal parts of Fehling's solution 'A' and Fehling's solution 'B' and boil the contents of the test tube for few minutes in a water bath. A red or brick red precipitate is formed.

5.1.3. Flavonoids

Schinoda test. In a test tube containing 0.5 ml of alcoholic extract of the drug, add 5-10 drops of diluteed hydrochloric acid followed by a small piece of magnesium. In the presence of flavonoids a pink, reddish pink or brown colour is produced.

5.2. Triterpenoids

Lebermann-Burchard's test. Add 2 ml of acetic anhydride solution to 1 ml of petroleum ether extract of the drug in chloroform followed by 1 ml of conc sulphuric acid through the sides. A violet coloured ring is formed which indicates the presence of triterpenoids.

5.3. Proteins

5.3.1. Biuret's test

Take 1 ml of hot aqueous extract of the drug add 5-8 drops of 10% w/v sodium hydroxide solution followed by 1 or 2 drops of 3% w/v copper sulphate solution. A red or violet colour is obtained.

5.3.2. Millon's test

Dissolve a small quantity of aqueous extract of the drug in 1 ml of distilled water and add 5-6 drops of Millon's reagent. A white precipitate is formed which turns red on heating, indicates protein.

5.4. Resins

Dissolve the extract in acetone and pour the solution into distilled water. Turbidity indicates the presence of resins.

5.5. Saponins

In a test tube containing about 5 ml of an aqueous extract of the drug add a drop of sodium bicarbonate solution, shake the mixture vigorously and leave for 3 mnts. Honeycomb like froth formed indicates saponins.

5.6. Steroids

5.6.1. Liebermann-Burchard's test

Add 2 ml of acetic anhydride solution to 1 ml of petroleum ether extract of the drug in chloroform followed by 1 ml of conc sulphuric acid. A greenish colour is developed which turns to blue.

5.6.2. Salkowski reaction

Add 1 ml of conc sulphuric acid to 2 ml of chloroform extract of the drug carefully, from the side of the test tube. A red colour is produced in the chloroform layer.

5.7. Tannins

Take 1 – 2 ml of extract of the drug add a few drops of 5% FeCl3 solution. A green colour indicates the presence of gallotannins while brown colour tannins.

5.8. Starch

Dissolve 0.015g of Iodine and 0.075g of Potassium iodide in 5 ml of distilled water and add 2 – 3 ml of an aq extract of drug. A blue colour is produced.

6. Chromatographic Analysis

Chromatography is a physical method of separation in which the components to be separated are distributed between the two phases; one of these is a stationary phase bed and the other is a mobile phase which percolates through this bed. Separations occur as a result of repeated adsorption/desorption during the movement of the sample components along the stationary bed, and are due to differences in distribution constants of an individual sample component. The stationary phase includes solid and liquid coated on a solid support. The mobile phase includes liquid and gas.

Based on the stationary and mobile phase, the chromatographic technique can be of the following types.

6.1. Thin-Layer Chromatography (TLC) and High Performance Thin Layer Chromatography (HPTLC)

Thin-layer chromatography is a technique in which a solute undergoes distribution between two phases, stationary phase acting through adsorption and a mobile phase in the form of a liquid. The adsorbent is a relatively thin, uniform layer of dry finely powdered material applied to a glass, plastic or metal sheet or plate. Glass plates are most commonly used. Separation may also be achieved on the basis of partition and adsorption, depending on the particular type of support, its preparation and its use with different solvent.

Identification can be effected by observation of spots of identical Rf value and about equal magnitude obtained, respectively, with an unknown and a reference sample chromatographed on the same plate co-TLC. A visual comparison of the size and intensity of the spots usually serves for semi-quantitative estimation.

TLC is used for the separation of simple mixtures where speed, low cost, simplicity is required. HPTLC is an advanced versatile chromatographic technique for quantitative analyses with high sample throughput and is complementary to HPLC. It provides a chromatographic drug fingerprint. It is therefore, suitable for monitoring the identity and purity of drugs. In HPTLC the various steps involved are as follow.

- Application of sample
- Chromatographic development
- Detection of spots
- Quantitation
- Documentation

a. **Application of sample.** An automatic applicator (Linomat IV) is used for sample application. A known quantity of sample is dissolved in a known volume of solvent and the sample applied on precoated TLC plate either in the form of a spot or a band.

b. **Chromatographic development (separation).** Development of the chromatogram is affected after the solvent of the applied sample is completely evaporated. Rectangular glass chambers or twin trough chambers are commonly used for TLC development.

c. **Detection of spots.** For densitometric scanning, detection under UV light is generally preferred. But post chromatographic derivatisation reactions are essentially required for detection when individual compounds does not respond to UV light or do not have intense fluorescence.

d. **Quantitation and documentation.** Densitometry is *in situ* instrumental measurement of visible, UV absorbance, fluorescence quenching directly. The scanner converts the spot/band on the layer into a chromatogram consisting of peaks similar in appearance to HPLC (Fig. 7).

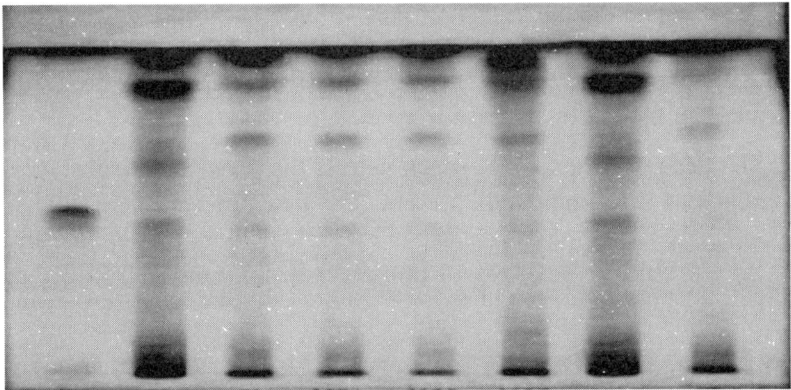

Fig. 7. HPTLC profile of the drug samples (under UV 254)

The portion of the scanned peaks on the recorder chart is related to Rf values of the spots on the layer and the peak height or area is related to the concentration of the substance on the spot.

6.2. High Performance Liquid Chromatography (HPLC)

HPLC is one of the latest analytical techniques, which is very essential for both quantification and standardization of the herbal materials. The technique is based on the same modes of separation as of column chromatography, i.e., adsorption, partition, ion exchange and gel permeation, but it differs from column chromatography in that the mobile phase in HPLC is pumped through the packed column under high pressure. The principle advantages of HPLC when compared to classical column chromatography are improved resolution of the separated substances, faster separation times and the increased accuracy, precision and sensitivity with which the separated substances may be quantified.

The HPLC technique involves the following:

Mobile phase. The HPTLC technique is generally an isocratic elution technique, in which the same solvent system, which may be single or a combination of solvents are used throughout the separation procedure. But the highest advantage of HPLC over HPTLC is that gradient elution can be used in HPLC, wherein, the composition of the mobile phase can be changed during the course of the separation. The selection of the mobile phase is one of the crucial steps in HPLC.

Pump. There may be one or more pumps, which are essentially used in pumping the mobile phase through the packed column under pressure.

Injection. The sample is injected (in ml) into the injection port in the load position and then after injection the lever is pulled back to inject position.

Column. The column used can be of two types, normal phase and reversed phase. Generally reverse phase column is used. The column is chosen based on the sample to be analysed in composition of the mobile phase. If polar mobile phase is used then reverse phase column is used of which C18 or RP18 is generally used. If the mobile phase is nonpolar then normal phase is used.

Detector. The detector is a device, which detects the presence of substance and gives the absorbance of the compounds present. The detector may be a dual wavelength detector, which measures the absorbance at 2 wavelengths, or Photodiode array detector for scanning the entire wavelength, or fluorescence detector or an RI detector of non-UV sensitive compounds.

Recorder. Recorder is a computer, which records the absorbance of the substances and gives out the results in the form of peaks.

Interpretation. The peaks are then integrated (mainly valley to valley) and the report contains the retention time, height of the peak and the percentage area under the peak. With the Retention Time (RT), we can identify the substances, as the RT, is characteristic of the substance under the specific conditions. With the area and the height, the amount of the substances can be calculated (Fig. 8).

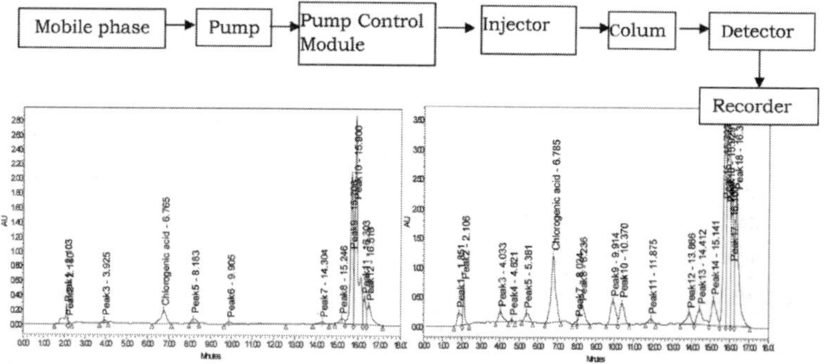

Fig. 8. HPLC profile of the drug samples

6.3. Gas Chromatography

Gas Chromatography has an entirely different field of application to that of HPTLC and HPLC. In general, gas chromatography is used for the separation of volatile materials. GC can separate volatile and semivolatile compounds with great resolution, but it cannot identify them. Without the use of gas chromatography the analysis of essential oils would be extremely difficult.

General uses

- Identification and quantitation of volatile and semivolatile organic compounds in complex mixtures.
- Determination of molecular weights and (sometimes) elemental compositions of unknown organic compounds in complex mixtures.
- Structural determination of unknown organic compounds in complex mixtures both by matching their spectra with reference spectra and by a priori spectral interpretation.

Mobile phase

In the GC system the first important component is the carrier gas or mobile phase. For a basic GC system, extremely pure nitrogen or helium is usually used, and hydrogen is less frequently encountered due to its explosive nature. Helium is used due to its inertness, non-reactive nature, and the shape of its van Deemter curve that allows for a relatively wide range of optimum mobile phase linear velocities.

Columns

There are two general types of column, packed and capillary (also known as open tubular). Packed columns contain a finely divided, inert, solid support material (commonly based on diatomaceous earth) coated with liquid stationary phase. Most packed columns are 1.5 - 10m in length and have an internal diameter of 2 - 4mm. Capillary columns have an internal diameter of a few tenths of a millimeter. They can be one of two types; wall-coated open tubular (WCOT) or support-coated open tubular (SCOT).

Detectors

There are many detectors which can be used in gas chromatography. Different detectors will give different types of selectivity. Detectors can also be grouped into concentration dependant detectors and mass flow dependant detectors. Detectors include Flame ionization (FID), Thermal conductivity (TCD), Electron capture (ECD), Nitrogen-phosphorus, Flame photometric (FPD) and Photo-ionization (PID). The signal from a concentration dependant detector is related to the concentration of solute in the detector, and does not usually destroy the sample dilution of with make-up gas which will lower the detectors response (Fig. 9).

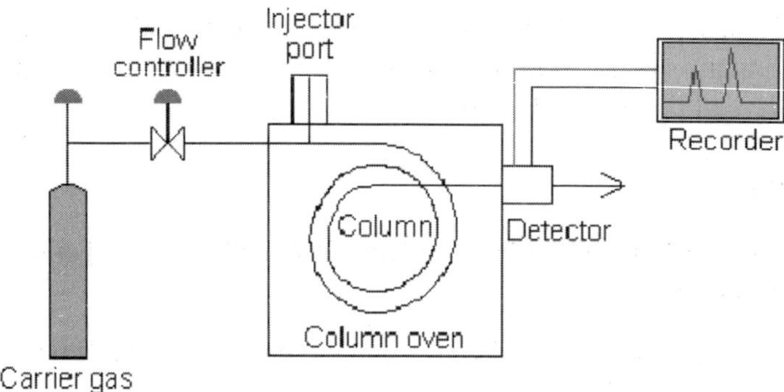

Fig. 9. Flow diagram of Gas Chromatography

References

Anonymous (1998) Quality control methods for medicinal plant materials World Health Organisation Geneva

Chase CR, Pratt RJ (1949) Fluorescence of powered vegetable drugs with particular reference to development system of identification. J Am Associ (Sci.edn.) 38: 324-331

Johanson DA (1940) Plant Microtechnique (5th Edition). McGraw-Hill Book Company, INC. New York pp: 523

Kokate CK (2004) Practical Pharmacognosy, Nirali Prakashan New Delhi

Kokoski CJ, Kokoski RJ, Slama FJ (1958) Fluorescence of powdered vegetable drugs under ultraviolet radiation. J Am Pharm Assoc 47:715-717

Pandey MM, Govindarajan R, Khatoon S, Rawat AKS, Mehrotra S (2007) Comparative pharmacognostical evaluation of Polygonatum species viz. *Polygonatum cirrhifolium* and *Polygonatum verticilatum.* J Herbs Spices Med Plants 12(1/2):37-48

Pandey MM, Rastogi S, Rawat AKS (2007) Evaluation of pharmacognostic characters and comparative morphological study of *Saussurea costus* (Falc.) Lipchitz and *Articum lappa* L. roots. Nat Prod Sci 13(4):304-310

Pandey MM, Rastogi S, Rawat AKS (2008) Indian Herbal Drug for General Healthcare: An Overview. Internet J Alternative Med 6(1):3

Trease and Evans (2009) Pharmacognosy (15th edition). Saunders: Elsevier (A Division of Reed Elsevier India Pvt. Limited), 2002

Chapter – 23

Characterization of New Cultivars of Ornamentals

R.K. Roy, A.K. Goel, Shilpi Singh and Rameshwar Prasad

1. Introduction

Ornamentals include plants which have either attractive foliage or flowers. They are used for various purposes like bedding plants in the garden or as cut-flowers and have huge commercial importance in terms of nursery business. New cultivars of ornamentals, which have been developed by various plant improvement techniques (hybridization, mutation breeding and colchiploidy) are always an added attraction to the plant lovers and a highly prized commodity for commercialization.

New cultivars of ornamentals usually inherit characters of both the parents singly or in combination and produce an entirely new trait. The progenies, thus, express a range of phenotypic characters. One has to screen these plants and identify elite plant(s) having desired phenotypic characters.

Second part of this process is characterization of the selected plants. Various characters of these plants are studied as per laid down procedures, which include vegetative, floral, seed and rhizome characters. The recorded observations are then compared with the parental characters to establish their distinctiveness. Subsequently, these plants are subjected to field trials for studying their uniformity and stability of the desired phenotypic characters.

New cultivars of ornamentals have enough potential for commercialization and good source of earning on the part of breeders who may be farmers, nurseryman, and researchers. However, for claiming right over the new cultivars developed, it is now mandatory to get those new varieties registered with appropriate agencies. The registration authorities ask for every morphological details as well as parental history. Therefore, characterization is a must for the newly developed cultivars / varieties.

2. Various Methods for Development of New Varieties

There are several methods used by the plant breeders for development of new varieties.

2.1. Bud Sport / Spontaneous mutation

It is most easy and popular method to sport these natural mutants out of the whole germplasm collections. There are a lot of changes in the foliage and flowering also. Bud sports are isolated, multiplied vegetatively and performance is evaluated in the subsequent generations with regard to the stability of changed morphological characters and its distinctiveness in comparison to the parental characters. *Bougainvillea* varieties developed by this method are – 'Abhimanyu', 'Archana', 'Bhabha', 'Jawahar Lal Nehru', 'Manohar Chandra Variegata', 'Parthasarthy', 'Scarlet Queen Variegata', 'Surekha' and 'Thimma'.

2.2. Hybridization

This involves crossing between selected inter-specific parents having desirable characters as flower size, shape and colour, length of inflorescence, flowering season, vase life, fruit size, shape and colour, height of plant, etc. The inter-specific hybrids exhibit morphological intermediary for various vegetative characteristics but produce flowers with distinct colours. Backcrossing of the F1 hybrids result in a number of flower colours and shapes enhancing the ornamental value of the plant. This method is very applicable for the ornamental crops like: *Bougainvillea, Gladiolus*, Rose and *Canna*. Examples of new hybrid varieties developed in *Bougainvillea* are – 'Begum Sikander', 'Chitra', 'Mary Palmer Special' and 'Wajid Ali Shah' and in *Gladiolus* new varieties developed are – 'S. Percy Lancaster', 'Neelima', 'Suverna', 'Urvashi', 'Roshni' etc.

2.3. Mutagenesis

Induced mutagenesis by application of Gamma Ray (Cobalt 60) for genetic manipulation leading to new cultivars. Induced mutagenesis is an established method for plant improvement, and the technique has been successfully exploited for development of new or novel ornamental varieties in general. Stem cuttings (soft/semi hardwood/ hardwood), seeds, bulbs, rhizomes, bud-wood are irradiated in various doses and planted for root regeneration and growth. Varieties developed by gamma radiation are – 'Arjuna, Los Banos Variegata', Los Banos Variegata 'Silver Margin', Mahara Variegata 'Abnormal Leaves', and 'P.V. Sane' (Datta and Banerji 1990; Datta 2009). A large number of new flower colour/shape mutant varieties have been developed in chrysanthemum through sports. The main bottleneck of induced mutagenesis with vegetatively propagated plants is that the mutation appears as a chimera after treatment with physical and/or chemical mutagens. Sports are also chimeric in nature. Mutated cells are present along with normal cells in chimeric tissues. The size of the mutant sector varies from a narrow streak on a petal to the entire flower head and from a portion of a branch to the entire branch. The mutated tissue can be isolated in pure form when a portion of a branch or an entire branch is mutated. However, a small sector of a mutated branch or flower cannot be isolated using the available conventional propagation techniques. Therefore, many mutagen-induced new flower colour/shape mutants or spontaneously developed mutants are lost due to lack of microtechnique for management of such chimeric tissues either *in vivo* or *in vitro* (Datta and Chakrabarty 2009).

2.4. Chemical mutagenesis

Chemical mutagenesis is the application of chemical mutagens for development of new cultivars. The most commonly used mutagens are Ethyl Methane Sulphonate (EMS) (0.01-0.03%) and Methyl Methane Sulphonate (MMS) for induction of somatic mutations (foliage/flower). Aqueous solution of various concentration of these chemical are used on vegitative part for creating genetically variability in the selected plant materials. Examples of varieties developed through chemical mutagenesis are: *Bougainvillea* 'Los Banos Variegata' 'Jayanti' and Bougainvillea 'Pixie Variegata' (Jayanti and Datta 2006).

3. Characterization of New Cultivars

Characterization is a study of morphological characters, both vegetative and floral characters of ornamentals and recording of observations carefully. It is a type of horto-taxonomical study for documenting vegetative and floral characters for the selection of elite plant (s), correct identification and establishing authenticity of the new cultivars

There are standard methods and approaches for characterization of new cultivars. Depending upon the type of plant *viz.,* trees, shrubs, herbs, bulbous and rhizomatous, the characters to be studied are decided. Morphological characters are mainly considered for characterization is as follows (Sharma and Goel 2006).

3.1. Vegetative characters

Plant height; stem (diameter and colour); leaf (length, width, shape and colour); spines (length, width, shape and colour); rhizomes (weight, internode length, colour and roots); corms (number of corms, diameter, weight, number of cormels, cormel diameter) and bulbs (weight, diameter and colour).

3.2. Floral characters

Type of inflorescence (length of inflorescence, no. of flowers per inflorescence); flower (sessile or pedicelled, bracteate or ebracteate, length, width, shape and fragrance); calyx/ sepals (no.of sepals, aestivation, shape, size and colour); corolla/ petals (no, of petals, aestivation, shape, size, colour as per international colour chart); androecium/ stamens (no. of stamens, free or united, nature of insertion in the flower, size, exserted or inserted and colour); gynoecium/carpel (shape, size and insertion of ovary, placentation, nature of style and stigma and colour); and vase-life, etc.

3.3. Fruits and seeds

Fruits (type of fruits, weight, shape, size and colour); seeds (number of seeds per fruit, shape, size and colour).

4. Naming New Cultivars

The plants which were identified having desired phenotypic characters and validated with respect to vegetative and floral characters and found clearly distinct, uniform and stable are recognized as new cultivars. Documented characters justify and establish their individuality as new cultivar as a part of recognition process.

Naming of the new cultivars is generally done on various ways *viz.* on the basis of any particular diagnostic characters, may be flower colour, shape, etc. or after any personality for commemoration purpose. The following examples illustrate the important morphological traits to be taken into account for assessing commercial value of these ornamental crops.

Bougainvillea

Bougainvillea, an ornamental climbing shrub, is a native of Brazil and highly popular all over the world for their colourful bracts. Excellent R&D work has been done at CSIR-National Botanical Research Institute, Lucknow. Several novel varieties have been evolved in the institute, which are highly popular and in high demand in the floricultural trade. Some of the *Bougainvillea* cultivars developed in the recent past are: 'Dr. P.V. Sane' (2011), 'Abhimanyu' (2010), 'Pixie Variegata' (2009), 'Aruna' (2008), 'Los Banos Variegata 'Jayanthi' (2006), etc. (Banerji 2012; Banerji and Datta 1986; Banerji *et al.* 1983; Jayanthi *et al.* 1999). The important morphological traits of these cultivars are: Plant height, plant habit, thorns (size, shape); leaves (colour of mature leaf and young leaf, texture, size, shape); flowering (profuse, sparse, medium, blooming season); and inflorescence/ bracts (colour at flowering, range of colour change from young to old bracts, size, margin, base shape, persistent or non-persistent after flowering), floral tube and star (Fig. 1A, B, C).

Fig. 1. Bougainvillea: (A) Different types of spines, (B) Leaves, and (C) Flower with bracts

Canna

Canna is a perennial rhizomatous plant grown for their colourful flowers and foliage. They are extensively used for bedding purpose in all types of garden (Roy *et al.* 2008). The important morphological characteristics of Canna cultivars are: Plant height (pre flowering); stem characteristics, (shape, colour, diameter); leaf characteristics (size, colour, variegated/ non variegated, shape, leaf apex, total number of leaves/plant at flowering); rhizome characteristics (fresh weight/10 cm, internodes/10 cm length rhizome, diameter of rhizome, no. of eyes/10 cm length, no. of roots/10 cm length, color) and flower characteristics (type of inflorescence, length of inflorescence, no. of flowers/inflorescence, size of flower, staminoide-shape, size, color) (Fig. 2 A,B,C, D).

Fig. 2. Canna: (A) Rhizomes of different cultivars (B) Leaves, (C) Flower and (D) Seeds

Gladiolus

Gladiolus is a popular bulbous plant grown for their magnificent flowers in various colour. Cuts spikes of *Gladiolus* are commercially grown as an open field crop all over India. Owing to its potentiality as a commercial floriculture crop, R&D work was initiated for development of new varieties (Roy 2003). The important characters taken into consideration while assessing *Gladiolus* cultivars are: Plant habit; corms and cormels (no. of corms produced/plant, corm diameter, corm weight, no. of cormels produced/plant, cormel diameter and corms weight) and floral characters such as blooming period, spike length, flower colour, flower size, number of florets per spike (Fig. 3A, B).

Fig. 3. Gladiolus: (A) Floral characteristics (B) L.S. of flowers

The Protection of Plant Varieties and Farmer's Rights Act, 2001 enacted by Government of India have made it compulsory for all (individuals, nurserymen, farmers and institutes) that any new varieties of agricultural crops and ornamentals must be registered with the authority for getting recognition, rights and benefits. Therefore, characterization of new cultivars of ornamentals is an essential tool for establishing distinctiveness, uniformity and stability of the newly developed genetic material.

References

Banerji BK (2012) Improvement of Bougainvilleas: Approaches and Methods. Indian Bougainvillea Ann 24:17-22

Banerji BK, Datta SK (1986) Improvement of ornamental plants by induced mutation: Bougainvillea. Floriculture Today 3:4-9

Banerji BK, Nath P and Gupta MN (1983) Gamma ray induced somatic mutation in Bougainvillea cv. 'Roseville's Delight' I: induction of chlorophyll mutation. National Symposium on Recent Development in Nuclear and Allied Techniques and their application in Agriculture, Biology & Animal Science' May 5-7, G.B. Pant Univ Agric Tech, Pantnagar pp 41

Data SK, Banerji BK (1990) 'Los Banos Variegata'-A new double bracted chlorophyll variegated bougainvillea induced by gamma rays. J Nucl Agric Biol 19:134-36

Datta SK (2009) A Report on 36 Years of Practical Work on Crop Improvement Through Induced Mutagenesis: Induced Plant Mutations in the Genomics Era. Food and Agriculture Organization of the United Nations, Rome pp 253-256

Datta, S K and Chakrabarty D (2009) Management of Chimera and In Vitro Mutagenesis for development of new flower colour/shape and chlorophyll variegated mutants in *Chrysanthemum* In: Q.Y. Shu (ed.), Induced Plant Mutations in the Genomics Era. Food and Agriculture Organization of the United Nations, Rome, pp 303-305

Jayanthi R, Datta SK, Verma JP (1999) Effect of gamma rays on single bracted Bougainvillea. J Nucl Agric Biol 28:228-33

Jayanti R and Datta SK (2006) Improvement of Bougainvillea varieties using Chemical Mutagens. In: National Conference on *Bougainvillea* April 12-13, NBRI Lucknow pp 17

Roy RK (2003) Studies on the genetic variability of androecium and gynoecium amongst the cultivars of Gladiolus. Ecoprint. Intl J Ecol, Nepal 10(1):31-34

Roy RK, Banerji BK and Verma TS (2008) Effect of gamma irradiation on *Canna generalis* 'Confetti' with particular reference to induction of somatic mutation in foliage/flower colour. In: Proceedings of the DAE-BRNS National Symposium on Landscaping for Sustainable Environment (Nov.20-21), at BARC, Mumbai, pp181-184

Sharma SC, Goel AK (2006) Horto-taxonomy of ornamentals: Case study of *Bougainvillea*. J Econ Tax Bot 30(1):71-78

Chapter – 24

Cultivation and Evalution of Medicinal Plants

S.K. Tewari, R.C. Nainwal, Shweta Singh and S.K. Sharma

1. Introduction

Medicinal plants play an important role in human life to combat various diseases. The rural folks and tribals in India even now depend largely on the surrounding plants/ forests for their day-to-day needs. Majority of the medicinal and aromatic plants are still collected from wild for preparation of herbal drugs and perfumes/cosmetics. Rapid population growth and rising popularity of herbal drugs and natural essential oils have brought into focus the acute scarcity in availability of some of the plants due to indiscriminate and unregulated collection, habitat destruction through expanding agricultural lands, deforestation and urbanization. Fall in the supply of good quality, genuine raw material has resulted in price rise and deterioration in the quality of formulations. With growing demand and use of medicinal and aromatic plants, the important species have to be introduced into commercial agriculture.

The Indian annual turnover of herbal material has crossed Rs. 4000 crores mark, of which Rs. 300 to 400 crores worth is being the export market. The share of Indian trade in medicinal and aromatic plants is only 1.5 percent of the world market. India is one of the major exporters of crude drugs mainly to the developed countries like the USA, the UK, Germany, Japan, France, Switzerland, etc. The main crude drugs having good export opportunities are Aconite, Aloe, Dioscorea, Ephedra, Digitalis, Bach, Belladona, Cincona, Ergot, Isabgol, Opium, Vinca, Senna, Sarpagandha, etc. Of these senna leaves, isabgol husk/seed, and *Cassia tora* seeds are in maximum demand. China is the major producer of herbal plant material in the world. China earns US$ 5 billion per year from herbal trade besides meeting its domestic requirements. Though India has more potential of production of number of medicinal and aromatic plants than China, the contribution of India in the world market is very low because of the production of poor quality produce (Chadha and Gupta 2005).

2. Current Scenario of Availability of Raw Material of Herbal Drugs

About 90% of medicinal plants used by the industries are collected from the wild. While over 800 species are used in production by industry, less than 20 species of plants are under commercial cultivation. Over 70% of the plant collections involve destructive harvesting because of the use of parts like roots, bark, wood, stem and the whole plant in case of herbs. This poses a definite threat to the genetic stocks and to the diversity of medicinal plants, if biodiversity is not sustainably used. Crude drugs are usually the dried parts of medicinal plants (roots, stem wood, bark, leaves, flowers seeds, fruits, and whole plants, etc.) that form the essential raw materials for the production of traditional remedies of Ayurveda, Siddha, Unani, Homeopathy, Tibetan and other systems of medicine including the folk, ethno or tribal medicines. The crude drugs are also used to obtain therapeutically active chemical constituents by specialised methods of extraction, isolation, fractionation and purification and are used as phytochemicals for the production of modern allopathic medicines or herbal/ phytomedicines. An approximate estimate of area under cultivation of medicinal plants in India is provided in Table 1.

Table 1. Area under major medicinal plants in India (Kumar 1997)

S. No.	Common name	Botanical name	Producing states	Estimated area (ha)
1.	Psyllium	*Plantago ovata*	Rajasthan and Gujarat	55,000
2.	Opium poppy	*Papaver somniferum*	Madhya Pradesh, Uttar Pradesh and Rajasthan	20,000
3.	Senna	*Cassia senna*	Tamil Nadu, Rajasthan and Uttar Pradesh	20,000
4.	Coleus	*Coleus forskohlii*	Tamil Nadu, Karnataka and Andhra Pradesh	450
5.	Cinchona	*Cinchona* spp.	Darjeeling (West Bengal) and Tamil Nadu	8,000
6.	Ashwagandha	*Withania somnifera*	Madhya Pradesh, Rajasthan and Uttar Pradesh	5,000
7.	Safed musali	*Chlorophytum* sp.	Madhya Pradesh, Gujarat and Uttar Pradesh	5,000
8.	Periwinkle	*Catharanthus roseus*	Andhra Pradesh, Karnataka, Tamil Nadu and Maharashtra	4,000
9.	Khai katari	*Solanum* spp.	Maharashtra	4,000
10.	Sarpagandha	*Rauvolfia serpentina*	Madhya Pradesh	2,500
11.	Ipecac	*Cephaelis ipecacuanha*	Darjeeling (West Bengal)	100

3. Indian Medicinal Plants: Poor Global Competitiveness

Several medicinal plants have been assessed as endangered, vulnerable and threatened due to over harvesting or unskillful harvesting in the wild. Habitat destruction in the form of deforestation is an added danger. The other main source of medicinal plant is from cultivation. Cultivated material is infinitely more appropriate for use in the production of drugs. Indeed, standardisation whether for pure products, extracts or crude drugs are critical and while become increasingly so, as quality requirements continue to become more stringent. Given the higher cost of cultivated material, cultivation is often done under contract. In the majority of cases, companies would cultivate only those plant species which they use in large quantity or in the production of derivatives and isolates, for which standardisation is essential and quality is critical. More recently growers have set up cooperatives or collaborative ventures in an attempt to improve their negotiating power and achieve higher price. According to Farooqi and Sreeramu (2001), some of the constraints associated with the processing of medicinal plants, which may result in reducing their competitiveness in global markets and which have to be remedied are:

- Lack of research on development of high-yielding varieties, domestication etc.
- Poor propagation methods.
- Poor agricultural practices.
- Poor harvesting and post-harvest processing.
- Poor quality control procedures.
- Lack of current good manufacturing practices.
- Lack of R&D on product and process development.
- Difficulties in marketing.
- Lack of local market for primary processed products.
- Lack of trained personnel and equipments.
- Lack of extension service for latest technologies and market information.

The World Health Organization (WHO) has released the guidelines for good agricultural and collection practices for medicinal plants for industry. The guidelines are intended for national governments to ensure production of herbal medicines is of good quality, safe, sustainable and poses no threat to either people or the environment. The WHO guidelines on good agricultural and collection practices (GACP) for medicinal plants are an important initial step to ensure good quality, safe herbal medicines and ecologically sound cultivation practices for future generations. In an easy-to-understand style, they cover the spectrum of cultivation and collection activities, including site selection, climate and soil considerations and identification of seeds and plants. Guidance is also given on the main post-harvest operations and includes legal components such as national and regional laws on quality standards, patent status and benefits sharing (Handa and Kaul 1996).

4. Methods/Techniques

4.1. Prospects of cultivation

Small farmers on marginal lands are generally cultivating medicinal and aromatic plants with low input in resources. Besides this, information on several aspects of their agricultural productivity is also not easily available to the farmers. Though we have been practicing the theory of organic farming for centuries, the modern science based chemicalized agriculture has pushed it in background. The theory and practices of organic cultivation of medicinal and aromatic plants for stable production of good quality raw material without harmful effects on soil health and environment have been described here (Handa and Kaul 1997).

4.2. Selection of site

The selection of site for cultivation of medicinal and aromatic plants mainly depends upon the space available, physical and chemical properties of the soil, medicinal plant species and the micro-climatic conditions (availability of sunlight, shade, etc.). The major area of the site selected should get sufficient sun light, especially in the morning. The shady area may be used for growing shade-loving plants. The area should be safe from grazing by cattle.

4.3. Soil

The majority of medicinal and aromatic plants prefer moderately well drained soil. The soil should preferably be neutral in reaction (pH between 6.5-7.5) and fertile with balanced water retention and drainage capacities. The soil should be rich in organic matter. During recent years, several recommendations have emerged for undertaking cultivation of medicinal and aromatic plants on marginal soils like saline, alkaline soils or wastelands. The cultivation on these soils requires special agronomic management and amelioration before starting the cultivation. The plants have varying degree of tolerance to such soils and the cropping systems shall be adopted accordingly.

4.4. Improved varieties of medicinal plants

Genetic improvement through selection, classical breeding and biotechnological tools is important for production of high quality raw material and herbal drugs. The plant genetic resources for medicinal plants represent the basis for health security through herbs. During past few decades, efforts have been made by various researchers for germplasm collection and characterization, describing the variations in growth and yield, content of the active constituents etc. Based on these efforts, several varieties have been released for some important medicinal plants. Recently, the National Medicinal Plants Board (NMPB) in the department of AYUSH has been supporting the cultivation of medicinal plants through several schemes. Considering the importance of quality of raw material as one of the important factors for quality of the herbal products, the NMPB and National Bureau of plant Genetic Resources, New Delhi (NBPGR) have compiled the details of improved/new varieties of medicinal plants, developed and released by various institutions (Table 2).

The medicinal plants are valued for the content of secondary metabolites, and hence, their content and composition are often more important than the total yield of the economic part(s). In this context, cultivation of seeds/planting material of known genetic origin, with higher productivity and optimum quality in a particular edapho-climatic condition is very important.

5. Propagation and Nursery Management

Medicinal and aromatic plants are propagated by a wide variety of methods. For propagation, the most suited method for the plant should be selected. For multiplication through seed, healthy and vigorous seeds may be sown either in containers or in prepared soil in open ground. Some seeds require special treatment for breaking their dormancy or improving germination. Propagation through cuttings is another popular method of propagation, suitable for woody perennial herbs. Some plants that form clump can be divided into smaller sections and replanted. Some plants can be sown directly in the field, while others are first sown in nursery-beds where seedlings are raised and then transplanted. Proper nursery management for raising seedlings and transplanting them is important. It is advisable to have soil sterilization of the nursery bed before sowing. The soil can be sterilized through the process of solarization. When seeds are sown in extremely cold weather, rainy season or hot weather, a low tunnel of semi-circular shape may be erected over the seed beds to protect the seedlings from adverse weather conditions.

6. Field Preparation

The soil of the field should be prepared thoroughly before transplanting. The root crops and small seeded crops require fine tilth. The field should be leveled properly and beds of convenient size shall be prepared with ease in irrigation and drainage. Deep ploughing in summers and green manuring are beneficial before the final field preparation.

6.1. Direct sowing/transplanting

The seedlings should be transplanted as early as possible after carefully removing from the nursery beds without damaging their roots. For better establishment of the seedlings, it is advisable to transplant them in the evening so that they establish themselves within the night and may recover from the shock of transplanting before the sunrise. Transplanting should immediately be followed by irrigation. Seedlings that fail to establish or not doing well should be removed and replaced by new ones.

Table 2. Crop wise improved varieties and their characters (Anonymous 2009)

S.No.	Name of varieties	Main characters	Breeders seed production sites	Region suitable for cultivation
1.	Aswagandha (*Withania somnifera*)			
	Poshita	Tall green leaves and long roots. Average yield of roots.	CIMAP, Lucknow	Northern Indian Plains
	Jawahar Asgandh –20 (JA-20)	Dry root yield (6-8 q/ ha), Total alkaloid content 1.20-1.28%.	JNKVV, Mandsaur	Northern-Central Plains, Madhya Pradesh and Chhattisgarh
	Jawahar Asgandh-134 (JA-134)	Dry root yield 8-10 q/ha.Total alkaloid yield 1.02-1.09%.	JNKVV, Mandsaur	Central Indian Plains and Madhya Pradesh
	Ashwagandha	Dry root yield 8-10 q/ha Quality parameters: Withaferin A, Withanolide A, Withnine	IIIM, Jammu	Jammu and Kashmir
2.	**Senna (*Cassia angustifolia* syn. *Cassia senna*)**			
	Sona	High leaf yield and high sennoside content	CIMAP, Lucknow	Northern Indian plains, Rajasthan, Gujarat and south India
	Anand Late Selection Maharashtra	Late flowering type, high foliage yield with 75% higher dry matter	AAU, Anand	Gujarat, Rajasthan, and Madhya Pradesh
	AFLT-2	Late flowering, 2.3% sennoside content in leaves, leaf and pod yield 1.2–1.5 t/ha.	AAU, Anand	Northern Indian plains, Rajasthan, Gujarat and south India
	KKM-1	High leaf yielding variety suited for rainfed cultivation with higher dry leaf yield (918 kg/ha), sennoside content is 2.5%.	TNAU, Killikulum unit Tamilnadu	Trinnel valley and Ramnathpuram district, Tamilnadu, Karnataka, Gujarat, Rajasthan and Delhi
3.	**Isabgol (*Plantago ovata*)**			
	Niharika	Long panicle, 120 days maturity period, high seed yield.	CIMAP, Lucknow	Northern Indian plains, Gujarat and Rajasthan
	Jawahar Isabgol-4(JI -4)	High seed yield-1300-1500 kg/ha	JNKVV, Mandsaur	Gujarat, Maharashtra, some parts of Madhya Pradesh and Rajasthan

(*Contd.*)

S.No.	Name of varieties	Main characters	Breeders seed production sites	Region suitable for cultivation
	Gujarat Isabgol-1(GI-1)	Dwarf and erect plant, 1 ton/ ha seed yield and early maturing,	GAU, Gujarat	Gujarat, Maharashtra, parts of Madhya Pradesh and Rajasthan
	Gujarat Isabgol-2(GI-2)	1 ton/ha seed yield, moderately resistant to downy mildew	GAU, Gujarat	Gujarat
	Haryana Isabgol-5(HI-5)	1-1.2 ton/ha seed yield, maturity period 140-145 days	CCS-HAU, Hissar	South western Haryana, some parts of Madhya Pradesh, Maharshtra, Gujarat and Rajasthan
4.	**Opium Poppy *(Papaver somniferum)***			
	Kirtiman(NOP-4)	Latex yield 45-50 kg/ha, Morphine content is 11.95%, moderately resistant to downy mildew.	NDUAT, Faizabad	Eastern Uttar Pradesh
	Jawahar Aphim -539 (MOP-539)	Latex yield 65 kg/ha, Morphine content 14.85%).	JNKVV, Mandsaur	Central Western plains and Madhya Pradesh
	Jawahar Aphim–540 (MOP-540)	75 kg/ha latex yield, Morphine content 13%	JNKVV, Mandsaur	Central Plains and Madhya Pradesh
	BROP-1	Average opium yield 54 kg/ha, seed yield 10-12 q/ha, Morphine content 13% more than local check.	NBRI, Lucknow	Northern Indian Plains
	NBRI-1	Average opium yield 52 kg/ha, seed yield 10 q/ha, Morphine content is 12-13% more than parent plants.	NBRI, Lucknow	Northern Indian Plains
	NBRI-2	Average opium yield 52 kg/ha, seed yield 12 q/ha, Morphine content is up to 15% more than local check.	NBRI, Lucknow	Northern Indian Plains
	NBRI-3	High latex yield (47-58 kg/ ha)	NBRI, Lucknow	Central Eastern Uttar Pradesh
	NBRI-6	Average opium yield 55 kg/ha, seed yield 12 q/ha.	NBRI, Lucknow	Northern Indian Plains
	NBRI-9	Larger capsules, seed yield 14 q/ha, average opium yield 52 kg/ha	NBRI, Lucknow	Northern Indian Plains

(*Contd.*)

S.No.	Name of varieties	Main characters	Breeders seed production sites	Region suitable for cultivation
	Madakini	Multiple disease resistant, opium yield up to 64 kg/ha, high morphine content up to 15%.	NBRI, Lucknow	Northern Indian Plains
	Sweta, Sampada, Shyama, Rakshit	Medium tall, dark green leaves and white flowers. The characters of these four varieties are different and give optimum yield of morphine content.	CIMAP, Lucknow	Northern Indian Plains
	JA-16(Jawahar Aphim)	Early-maturing (150-110 days), Latex yield (60-70 kg/ha); 10-12.5% morphine content.	JNKVV, Mandsaur	Northern Central India, Madhya Pradesh and Chhattisgarh
	Trishna (IC-42)	Medium dwarf, Latex yield-65-70 kg/ha); 14.78% Morphine	NBPGR, New Delhi	Throughout India (Northern Central India)
	Chetak 17 (UO 285)	High latex yield-65-70 kg/ha; High Morphine content (12-13%)	RAU, Udaipur	Central eastern Uttar Pradesh and Rajasthan
5.	**Tulsi (*Ocimum tenueflorum* syn. *O. sanctum*)**			
	CIM-Ayu	Tall with green leaves	CIMAP, Lucknow	Northern Indian Plains, Madhya Pradesh, Rajasthan, Gujarat, Haryana and Punjab
	CIM-Angana	Tall with purple leaves	CIMAP, Lucknow	Northern Indian Plains, Madhya Pradesh, Rajasthan, Gujarat, Haryana and Punjab
	CIM-Kanchan	Tall plants with light green leaves, high herbage yield	CIMAP, Lucknow	Lower Himalayan Region
6.	**Mulhatti *(Glycyrrhiza glabra L.)***			
	Hariyana Mulhati No.1 (HM-1)	Root yield 60-75 q/ha after 2.5-3 years, Glycyrrhizin content 6-7% in rest at 30m age	HAU, Hissar	Northern Central Plains, Haryana and Punjab
7.	**Brahmi *(Bacopa monerii)***			
	CIM-Jagarti	Creeping/floating type plants with light green leaves & purple flowers.	CIMAP, Lucknow	Northern Indian plains, Gujarat, Haryana, Karnataka, Rajasthan, Madhya Pradesh, Punjab and Central-eastern India

(Contd.)

S.No.	Name of varieties	Main characters	Breeders seed production sites	Region suitable for cultivation
8.	**Mandukparni (*Centella asiatica* Syn. *Hydrocotyl asiatica*)**			
	Zandu Brahmi (Sel.)	Creeping herb, preferably cultivated as under crop in orchid, produces 1.0 to 1.2 t/ha., leaf crop in 3 harvests in a year. Tri-terpinoid content (1.0%).	Zandu Foundation for Health Care, Vapi, Gujarat	Gujarat (Proper moist climate in high rainfed tracts)
9.	**Periwinkle (*Catharanthus roseus* syn. *Vinca rosea*)**			
	Prabhat	Dry root yield- 15-18q/ha), Dry leaf yield- 20-25 q/ha after 2-3 cuts.	CCS-HAU, Hissar	Northern Central Plains
	IC- 49581	High biomass yield & high alkaloid content; serpentine (0.41% more).	NBPGR, New Delhi	Central plains
	EC-120835	Total alkaloid content is 1.2% more in leaves.	NBPGR, New Delhi	Central plains
	EC-120837	Alkaloids are 2.49% more in root.	NBPGR, New Delhi	Central plains
10.	**Kalmegh (*Andrographis paniculata*)**			
	Anand Kalmegh-1(AK-1)	High biomass yield- 3.5 t/ha, Andrographolide content-1.20%.	AAU, Anand & ZFHC, Vapi	Middle Gujarat
	CIM-Megha	Medium height, bushy, light green leaves.	CIMAP, Lucknow	All over India
	IC-111286,IC-111287, IC-111289	High biomass yield and 1-2%. Andrographolide content.	NBPGR, New Delhi	Northern plains
	KI-5	30-35q/ha dry herb yield, 2.15% Andrographolide content.	College of Agriculture, Indore	Middle Gujarat, Madhya Pradesh and Chhattisgarh
11.	**Safed Musli (*Chlorophytum borivilianum*)**			
	Anand Safed Musli-1 (ASM-1)	Blunt end and whitish attractive colour, high root yield fresh (3-5 ton / ha), sapogenine content 0.62%.	AAU, Gujarat	Northern Central Plains and middle Gujarat
	Jawahar (JSM-405)	Fresh root yield: 22-24 q/ha, tuber length- 5-7.5 cm, 2-8% Saponin.	JNKVV, Mandsaur	Northern Central Plains, Madhya Pradesh and Chhattisgarh.
12.	**Kewanch (*Mucuna pruriens*)**			
	CIM-AJAR	Fast growing, early maturing, lint less pod	CIMAP, Lucknow	Northern Plains, Rajasthan, Madhya Pradesh, Gujarat and south India

(*Contd.*)

S.No.	Name of varieties	Main characters	Breeders seed production sites	Region suitable for cultivation
	IC-2533	High seed yield, high L-dopa content (3.82%).	NBPGR, New Delhi	Northern Plains, Rajasthan, Madhya Pradesh, Gujarat and south India
	Zandu Kewanch-1	High food yield, pods devoid of stinging hairs, high L –dopa (5-6% more).	ZFHC, Vapi, Gujarat	Northern plains, Rajasthan, Madhya Pradesh, Gujarat and south India
	IIHR MP 10, IIHR MP 11	Short trichomes, high seed yielding lines with high L-Dopa content (4.4-4.6%) with an average duration of 180 days.	IIHR, Banglore	Northern plains, Rajasthan, Madhya Pradesh, Gujarat and south India
13.	**Aloe/Ghritkumari (*Aloe barbadenis*)**			
	IC-111271	High aloin content (>20%).	NBPGR, New Delhi	Throughout India
	IC-111280	High gel yield (2.45 mg/ml).	NBPGR, New Delhi	Throughout India
	CIM-Sheetal	High aloin content (>20%), High gel yield (2.45 mg/ml).	CIMAP, Lucknow	Throughout India
	RLAV-18	Average leaf yield: 50t/ha, Aloin 0.02%.	IIIM, Jammu	Sub-tropical and arid areas
14.	**Sarphgandha (*Rauvolfia serpentina*)**			
	RS-1	Erect, evergreen, roots dull yellow with specific aroma.	NDUAT, Faizabad	Northern Indian plains
	CIM Sheel	Medium height plant, long roots and light green leaves.	CIMAP, Lucknow	Northern Indian plains
	RI-1	75-90 cm height, seeding to flowering 115-145 days, 1.5% total alkaloid content.	JNKVV, Indore	Madhya Pradesh and Chhattisgarh
	Sarpgandha	75 cm height, dry root yield 800-1000 kg/ha.	IIIM, Jammu	Tropical and sub-tropical regions of India
15.	**Shatavar (*Asparagus racemosus*)**			
	Shatavar	Dry tuber yield 10 t/ha in 2 years, four types of saponins.	IIIM, Jammu	Tropical and sub-tropical regions of India

The direct sowing of crops should preferably be done in rows spaced at appropriate distance. The optimum depth of sowing may be decided on the basis of the seed size. For direct sowing, pre-sowing irrigation sowing is provided. When the field is at optimum moisture level, the rows are opened by cultivator or hand-hoe. These can be opened in the evening and left overnight to conserve the dew moisture. The seeds can be sown and covered with the soil in early morning. Widely spaced and large seeds can be dibbled into the soil. After the seed emergence, excess plants can be removed (thinning) to maintain the optimum plant population at the time of first weeding/hoeing after first irrigation.

6.2. Nutrient management

While cultivating medicinal plants, one has to be very careful with application of chemical fertilizers, as this may reduce the quality of the herbs. Organic materials such as vermi-compost, farm yard manure, bio-gas slurry, compost, neem cake and other oilseed cakes, biofertilizers, green manure and cover crops can substitute the inorganic fertilizers. Nutrient management in integrated manner has beneficial effect on soil organic matter and available plant nutrients resulting in sustainable crop production. The application of organic manures including vermicompost and biofertilizers has been found beneficial in several medicinal plants. However, research work on nutrient requirements of medicinal and aromatic plants is quite meagre. Systematic studies need to be initiated to exploit full potential productivity by proper fertilization of crops.

6.3. Irrigation

Watering the plants is very critical as many plants produce medicinally active constituents in dry conditions. The object of irrigation is to keep adequate water available to the crops, or, if water is in short supply, to use it most effectively. The plants should be watered well after planting and later irrigations shall depend upon visual observation of the plant. These are based on early signs of dryness, slight temporary wilts, foliage colour and dryness of the soil. Stagnation of water due to over-irrigation or rains should be avoided and the general principle shall be 'light and frequent irrigation'.

6.4. Weeding

The plants growing out of the place compete with the medicinal and aromatic herbs for space, nutrients and water. The field should be kept free from weeds by hand weeding (removing them by *khurpi*) and hoeing which increases aeration for the plant roots. Use of herbicides should be strictly avoided, using alternative weed management techniques leading to minimum loss in crop production and least disturbance to the ecosystem. Use of mulches smother perennial weeds and prevent the germination of annual weeds. Mulches also conserve moisture and lower the surface temperature.

6.5. Insect- pest and disease

Use of synthetic chemicals for the control of insects and diseases should be avoided due to their great hazard to humans, animals and also to the active principle of the medicinal herb. Non-chemical, biological methods should be used to control insects-pests and diseases. These include conservation and augmentation of natural enemies of insect-pests and adoption of all cultural, physical, mechanical and biological methods. Oils and soaps and botanical pesticides are fast emerging tools for the control of insect pests.

6.6. Harvesting and processing

Harvesting of herbs requires careful planning so as to retain their active ingredients. To prevent the crushing and deterioration of the plant material, wooden tray or an open basket should be used for collecting herbs. The cuts should be made with a sharp knife or scissors to minimize damage. The material should be collected from healthy plants that are free from disease and insect damage. It is important to discard any damaged plants as they can lead to disease or decay in dried plant material. Only single herb should be collected at one time and the harvested plant material should not be mixed to avoid mistakes in identification. The herbs should be harvested in dry weather, preferably on a sunny morning after the dew has evaporated. Herbs can be preserved in a number of ways, the most common and simple being air or oven drying. A warm, dry place is ideal for storage and processing. Plain paper should be used for drying herbs and not the printed newspaper. Dried herbs can be stored for many months in a dark glass jar or a brown paper bag. The underground parts of the plant are usually collected in January when the plants become inactive. After removing the required amount, the remaining underground part should be replanted.

Medicinal plants should be harvested during the optimal season or time period to ensure the production of medicinal plant materials and finished herbal products of the best possible quality. The time of harvest depends on the plant part to be used. Detailed information concerning the appropriate timing of harvest is often available in national pharmacopoeias, published standards, official monographs and major reference books. However, it is well known that the concentration of biologically active constituents varies with the stage of plant growth and development. This also applies to non-targeted toxic or poisonous indigenous plant ingredients. The best time for harvest (quality peak season/time of day) should be determined according to the quality and quantity of biologically active constituents rather than the total vegetative yield of the targeted medicinal plant parts. During harvest, care should be taken to ensure that no foreign matter, weeds or toxic plants are mixed with the harvested medicinal plant materials. Medicinal plants should be harvested under the best possible conditions, avoiding dew, rain or exceptionally high humidity. If harvesting occurs in wet conditions, the harvested material should be transported immediately to an indoor drying facility to expedite drying so as to prevent any possible deleterious effects due to increased moisture levels, which promote microbial fermentation and mould. Cutting devices, harvesters, and other machines should be kept clean and adjusted to reduce damage and contamination from soil and other materials. They should be stored in an uncontaminated, dry place or facility free from insects, rodents, birds and other pests, and inaccessible to livestock and domestic animals.

Effect of cultivation and harvesting practices on yield and quality of some important medicinal plants is illustrated below.

6.6.1. Chamomile

Common name: *German Chamomile, Gul-e-baboonah;* Botanical name: *Matricaria recutita* L. *(Asteraceae)*

It is the most popular source of the herbal product chamomile, although other species are also used as chamomile. The flowers of *Matricaria* contain the blue essential oil from 0.2 to 1.9%, which finds a variety of uses. Chamomile is used mainly as an anti-inflammatory and antiseptic, also antispasmodic and mildly sudorific. It has a wide range of adaptability in sodic soils up to 9.5 pH and 50-60 ESP. The yield of fresh flowers varies from 2 to 4 t/ha in sodic soils corresponding to soil and water management practices and fertilizer applications. The fresh flowers are shade-dried which is reduced to 20-25% of the fresh weight. It yields about 0.5-0.6% blue oil on distillation. M. *recutita* was successfully grown on partially reclaimed sodic soils at Banthara. The flower yield, essential oil yield and its composition were comparable with its production in normal soils. The flowers harvested in different flowering flushes during February-March showed variations in their number, weight, oil content and chemical constituents. The flowers harvested in middle of March recorded higher flower yield as well as the oil content. The chamazulene content was highest in the first flush of flower and showed declining trend in later flushes (Tewari *et al.* 1998).

6.6.2. Kalmegh

Common name: Kalmegh, Kirayat; Botanical name: *Andrographis paniculata* Nees (Acanthaceae)

Kalmegh is an erect annual herb, extremely bitter in taste, and sometimes known as "king of bitters". As an official drug in Indian pharmacopoeia, *A. paniculata* is extensively used in Ayurvedic, Unani, and Siddha medicine as a home remedy for various diseases. The herb is reported to possess astringent, anodyne, tonic, alexipharmic, antibacterial, antifungal, antidiabetic, antioxidant, anticancer, antimalarial, anti-HIV, anti-pyretic, hepatoprotective, and immunological activities plus helping to control dysentery, cholera, influenza, bronchitis, swellings, itches, piles, gonorrhea, colds, and fevers. Tewari *et al.* (2010) studied the edaphoclimatic conditions that affect the yield and qualities of *Andrographis paniculata*. Under seasonal and soil pH variations, andrographolide content ranged between 2.25–3.85% in aerial parts of the plant. Total phenolic content, antioxidant activity and reducing power were 20.3 mg g^{-1} GAE, 66.8% and 2.8 ASE mL^{-1}. FRSA measured by DPPH showed an IC50 of 0.34 mg mL^{-1}, EC50 of 14.7 mg mg^{-1} DPPH, and antiradical power of 6.76. AOA in different assays expressed as IC50 ranged from 0.31 to 1.29 mg mL^{-1}. The extract showed significant protective effect against Fenton's reaction on super-coiled pUC18 DNA.

6.6.3. Mucuna

Common name: Kewanch; Botanical name: *Mucuna pruriens* (L.) DC. (Leguminosae)

Mucuna is a leguminous genus comprising, twining annual plant found growing almost in all regions of India. This crop can be grown on a wide range of soils but it does well on sandy to clay loam type of soils, provided drainage is adequate. The seeds of *Mucuna* contain L-DOPA (L-3,4-Dihydroxyphenylalanine) which is useful for treatment of Parkinson's disease. Three species of *Mucuna* were cultivated on partially reclaimed sodic soils at Banthra Research Station. The seed weight, yield (26.4 q ha^{-1}) and protein content (27.5%) were highest in *M. utilis*, while the L-DOPA content in dry mature seeds ranged between 3.6 to 4.2% and was maximum in *M. nivea*. The maximum number of pods per plant and highest seed yield were obtained with application of 30 kg N and 60 kg P ha^{-1} in *M. nivea* (Prakash *et al.* 2001). In another trial, application of 40 kg N and 80 kg P produced higher seed yield in *M. pruriens*. Application of 20 kg N and 40 kg P ha^{-1} was the optimum dose to get the maximum L-DOPA content (4.31 %). Additional dose of 20 kg Zn ha^{-1} increased the seed yield with lower L-DOPA content. In both the species (*M. nivea* and *M. pruriens*), the highest seed yield was recorded at 90 x 60 cm spacing (18519 plants ha^{-1}). The numbers of pods and seed yield per plant were higher at wider spacings of 120 x 90 and 120 x 120cm. The plants of *M. pruriens* showed more drastic reduction in seed yield at small plant populations of less than 10, 000 plants ha^{-1}, as compared to *M. nivea* (Tewari *et al.* 2001).

6.6.4. Shatawar

Botanical name: *Asparagus racemosus* Willd. (Liliaceae)

Shatavari is a perennial much branched climbing herb found all over India. The fleshy root of shatavari has been used as one of the most powerful nutritive and spermatogenic herbs in Ayurvedic system of medicine. Studies on satawari at Banthra Research Station showed better development of roots with increased proportion of sand Tewari and Misra (1996). The plant, though, commonly grown from its roots, was successfully propagated through seeds. Irrespective of more number of roots and their higher fresh weight per plant under wider spacing, the per hectare yields were highest in the closer spacing of 30x30 cm. Harvesting the crop after two years provided higher root yield than annual harvests in pots as well as in field experiments. Application of 60 kg nitrogen and 30 kg phosphorous significantly increased root number. The second year crop, when planted in the same soil showed reduction in growth and yield. Mishra *et al.* (1996) did comparative study on chemical composition of *Asparagus racemosus* cultivated under sodic and normal soil.

References

Anonymous (2009) Status of High-yielding varieties developed in Medicinal Plants with general guidelines for Seed Production and Certification related to new Seed Act, National Bureau of Plant Genetic Resources, New Delhi & National Medicinal Plants Board, Govt. of India, New Delhi

Chadha KL and Gupta R (2005) Advances in Horticulture Medicinal and Aromatic Plants Volume 11, Malhotra Publication House, New Delhi

Prakash D, Niranjan A, Tewari SK (2001). Some nutritional properties of the seeds of three *Mucuna* species. Int J Food Sci Nutrition 52:79-82

Farooqi AA and Sreeramu BS (2001) Cultivation of Medicinal and Aromatic crops, Universities Press (India) Limited, Hyderabad

Handa SS and Kaul MK (1996) Supplement to Cultivation and utilization of Medicinal Plants, Regional Research Laboratory, Jammu-Tawi

Handa SS and Kaul MK (1997) Supplement to Cultivation and utilization of Aromatic Plants, Regional Research Laboratory, Jammu-Tawi

Kumar S (1997) CIMAP Records, Central Institute of Medicinal Aromatic Plants, Lucknow

Tewari SK, Balak Ram, Naqvi AA, Misra PN (1998) Growth, flowering and oil yield of German chamomile on partially reclaimed sodic lands. J Med Arom Plant Sci 20:386-387

Tewari SK, Prakash D, Gupta, RC, Misra PN (2001) Cultivation trials on kewanch (*Mucuna* species). J Med Arom Pl Sci 23:59-62

Tewari SK, Misra PN (1996) Cultivation trials on Satavari (*Asparagus racemosus*) in alluvial plains. J Med Arom Plant Sci 18(2):270-73

Tewari SK, Niranjan A, Lehri A (2010) Variations in yield, quality and antioxidant potential of Kalmegh (*Andrographis paniculata* Nees) with soil alkalinity and season. J Herbs, Spices Med Pl 16 (1): 41-50

Chapter – 25

Role of Botanic Gardens in Plant Systematic Studies

A.K. Goel and R.K. Roy

1. Introduction

Present day botanic gardens have their origins from medicinal plant gardens established during the middle of the 16th century meant for the cultivation of medicinal plants. Therefore, the roots of botany are traced from the medicine. The modern botanic gardens are considered as an ordered collection of plants, assembled primarily for scientific and educative purposes. They serve as the living repositories and offer excellent opportunities for *ex-situ* conservation, taxonomy, and studies on various other botanical aspects as mentioned below:

- A living repository and unique facility as one stop destination to study a wide range of plant groups for systematic studies and commercial exploitation.
- Live specimens of little known species for providing the taxonomic description and updating the incomplete descriptors.
- Comparative studies of fresh and dried plant specimens (preserved in a herbarium) for better and complete understanding of a plant species.
- Experimental stdies on taxonomic complexes grown in the iso-climatic regions by applying cytogenetic, molecular, chemical and other parameters.
- Revisionary studies of wild relatives of diverse taxa and ornamental crops for better understanding of the affinities, relationships at inter and intra-specific and varietal levels.
- Taxonomic and floristic studies with the help of living fresh material which can reveal several unknown mysteries of the life cycle of a taxon.

- Understanding the phenotypic plasticity of the indigenous, introduced as well as ornamental and economically significant taxa.
- Centres for the systematic studies on ornamental crops having larger collections.
- Besides, the botanic gardens play a very important role in generating public awareness and imparting environmental education for the masses. CSIR-NBRI Botanic Garden offers the scope for the systematic studies by fulfilling above requirements.

2. CSIR-NBRI Botanic Garden

Botanic Garden of CSIR-National Botanical Research Institute, Lucknow is well known all over the world. It is the third largest and one of the oldest Botanic Gardens in India. Spread over in 65 acres of area, it is located in the heart of Lucknow city, the capital of Uttar Pradesh. It has four main functions *viz.*, conservation, education, scientific research and display of plant diversity in plant houses and arboreta. Botanic Garden is reputed for organised display of plant wealth in a well-designed landscape (Goel 2002). Botanical strength of this Botanic Garden can be gauged by the conservation of diverse living collections of various groups of plants comprising 5000 taxa/cvs. distributed under 212 families. Considering the rich plant genetic diversity in the Botanic Garden, the Institute has been designated as a Living National Repository under the Govt. of India notification during 2007 by National Biodiversity Authority, Chennai under the *Biological Diversity Act - 2002*. Besides, it has also been recognized as a Lead Botanic Garden by the Ministry of Environment & Forests, New Delhi for enhancing ex-situ conservation activities. Moreover, it has also been designated as a 'DUS Test Centre' for three ornamental crops viz.: Bougainvillea, Canna and Gladiolus by the Protection of Plant Varieties and Farmer's Rights Authority, Ministry of Agriculture, Government of India, New Delhi during 2012 (Roy 2006a,b).

Botanic Garden covers within its limits, historical 'Sikander Bagh' laid out around 1800 AD as a royal garden by 'Nawabs' of Oudh region. It was named by Wajid Ali Shah, last 'nawab' of Lucknow, after the name of Begum Sikander Mahal, one of his favourite queens. After 1857 war, it was taken over by the state of Oudh as a Horticultural Garden in 1869. Later on after independence, it was taken over by Government of Uttar Pradesh in 1948 as National Botanic Garden and finally by Council of Scientific & Industrial Research (CSIR), New Delhi in 1953 (Goel and Sharga 2001).

2.1. Germplasm collections

Plant wealth in this Botanic Garden has been displayed in Arboretum, Bonsai House, Conservatory, Cacti and Succulents House, Cycad House, Fern House, Moss House, and Palm House, besides the mandated ornamental crops.

2.2. Arboretum

It covers seven hectares of area representing nearly 300 species of trees, shrubs and climbers and located on the left side of main entrance where trees are systematically planted and labelled, providing botanical/common name, family and nativity. Some of the notable indigenous and exotic tree species in the arboretum are: *Adansonia digitata, A. rubrostipa, A. za, Alstonia macrophylla, Aphanamixis polystachya, Bauhinia variegata, Bligha sapida, Boswellia serrata, Butea superba, Castanospermum australe, Chorisia × insigniosa* (NBRI Cultivar), *Cinnamomum camphora, Coccoloba uvifera, Crateva nurvula, Dalbergia lanceolaria, Diospyros kaki, D. malabarica, Ficus benghalensis* 'Krishnae', *Jacaranda mimosaefolia, Khaya senegalensis, Koelreuteria apiculata, Mitragyna parviflora, Oroxylum indicum, Parmenteira cerifera, Protium serratum, Santalum album, Saraca declinata, Strychnos potatorum,Tabebuia palmeri, Tecomella undulata, Tectona grandis, Terminalia muelleri, Triplaris surinamensis, Vitex pinnata* and *Wrightia tinctoria*. Plant resources of gymnospermous taxa like: *Agathis, Biota, Ephedra, Ginkgo, Pinus* and *Podocarpus* are also conserved here (Roy *et al.* 2013). (Fig. 1)

Fig. 1. (A) *Tecomella undullata*, (B) *Triplaris surinamensis* (Male) in Flowering

2.3. Plant houses

2.3.1. Bonsai House

'Bonsai' is a unique technique and an art of growing plants in dwarf form in a shallow container with the help of pruning and training. Bonsai House covers 450 sq m area and is equipped with automated misting systems and shade nets. A collection of 350 Bonsai specimens, trained in different styles, has been exhibited. Plants are well labelled. Few important species worth mentioning are: *Acacia gotezei, Achras sapota, Adansonia digitata, A. za, Bambusa ventricosa, Callistemon lanceolatus, Citrus microcarpa, Cycas revoluta, Drypetes roxburghii, Ficus benjamina* 'Nuda', *F. indica, F. infectoria, Ficus* 'Long Island', *F. tesiela, Portulacaria afra, Psidium guajava* and *Punica granatum,* besides a large number of *Bougainvillea* cultivars in a wide range of flower colours.

2.3.2. Cacti and Succulents House

This house is a pagoda shaped glass house meant for conservation of germplasm collection of cacti and succulents. It is centrally located covering 284 sq m area and shelters about 300 species/varieties. Some of the interesting species displayed are: *Adenium obesum, Agave parviflora, A. stricta, Cereus grandiflorus, C. peruvianus, Cotyledon orbiculata, Dudleya virens, Dyckia remotifolia, Echinocactus grusonii, Euphorbia tirucalii, Gasteria maculata, Gymnocalycium mihanovichii, Haworthia fasciata, Hesperaloe parviflora, Mammillaria echinata, Notonia grandiflora, Opuntia argentina, O. microdasys* 'Albida', *O. vulgaris* 'Variegata', *Pereskia aculeata, P. bleo, Senecio pendula, Stapelia gigantean, Yucca aloifolia* and *Y. filamentosa.*

Welwitschia mirabilis, a unique gymnosperm, known as "Tree Tumbo" has been introduced from NBG, Kirstenbosch, Republic of South Africa. This species has only two leaves throughout its life-span (over 500 years) which elongate continuously in opposite direction. It is very important from educational and evolutionary point of view and considered a bizarre in plant kingdom. It grows wild in Namib Desert of South Africa along the coastal line. Among the SAARC nations, only CSIR-NBRI Botanic Garden possesses this extremely rare plant (Sharma and Goel 2003). (Fig.2)

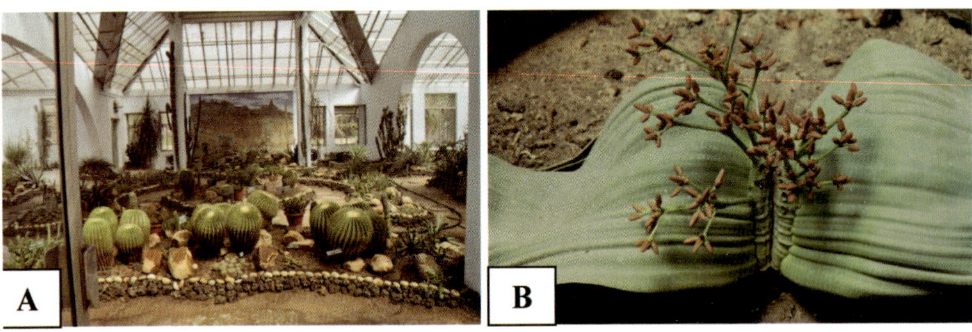

Fig. 2. (A) Inside view of Cacti & Succulents House, (B) *Welwitschia mirabilis*–Native to South Africa

2.3.3. Conservatory

An arch-shaped plant house, spread in 1370 sq m area, is meant for the conservation of house plants from tropical and subtropical climates. There are nearly 350 species/ cultivars of house plants in the conservatory. Some of the novel and interesting plants are: *Alocasia* × *amazonica, Bambusa ventricosa, Dracaena marginata* 'Tricolor',

Encephalartos transvenosus, E. villosus, Fatsia papyrifera, Ficus 'Long Island', *Ginkgo biloba, Heliconia rostrata, Hoya carnosa* and *H. wightii.* Besides, a large collection of *Aglaonema, Alocasia, Anthurium, Asparagus, Calathea, Chlorophytum, Codiaeum, Dieffenbachia, Dracaena, Maranta, Peperomia, Philodendron, Pandanus* and *Syngonium* is being conserved. Besides, *Microcycas calocoma* is an extremely rare cycad, introduced from Florida. It is designated as the National Plant of Cuba- Besides, some economic plant species viz., *Cinnamomum camphora, Coffea arabica* and *Piper betel* have also been showcased here. (Fig. 3)

Fig. 3. (**A**) View of Conservatory, (B) *Microcycas calocoma*-Rare Cuban Cycad

2.3.4. Cycad House

Cycads have 45 species and eight genera (*Cycas, Dioon, Encephalartos, Lepidozamia, Macrozamia, Microcycas, Stangeria* and *Zamia*), from various phyto-geographical regions of Australia, India, South Africa and South America are being conserved in Cycad House covering 300 sq m area for educative and aesthetic display. This is a maiden effort and first of its kind in India, that *ex-situ* conservation and creating awareness about cycads have been taken-up (Kumar *et al.* 2006; Roy and Goel 2006) (Fig. 4A).

Fig. 4. (A) Views of Cycad House, (B) Fern House

2.3.5. Fern House

Ferns and fern allies are widely used as ornamentals for interior plantscaping. This fernery is pyramidal in shape and occupies 400 sq m area. It is covered with shade net to cut-off excessive sun light and heat and is kept humid by sprinklers, mist irrigation and water-filled channels all around. A germplasm collection of 60 species of ferns and fern allies is maintained in fern house. Few notable ferns are: *Adiantum capillus-veneris, A. hispidulum, Blechnum occidentale, Bolbitis heteroclita, Diplazium esculentum, Drynaria quercifolia, Equisetum debile, Lygodium flexuosum, Microsorium alternifolium, Nephrolepis cordifolia, N. cordifolia* 'Duffii', *N. tuberosa, Ophioglossum reticulatum, Psilotum nudum, Pteris cretica* 'Albolineata', *P. vittata,* and *Selaginella bryopteris.* (Fig. 4B)

2.3.6. Moss House

A Moss House is established in 270 sq m area, which is covered with netlon and fitted with water sprinklers, misting system and water filled channels around to simulate the microclimatic conditions. It is first of its kind in India which conserves bryophytes under local tropical conditions. A collection of nearly 20 species of liverworts and mosses is being maintained here. Some of the noteworthy species conserved are: *Cyathodium cavernarum*, *Marchantia linearis*, *Marchantia paleacea*, *Plagiochasma appendiculatum*, *Riccia billardieri* and *Targionia hypophylla* and mosses like *Barbula indica*, *Semibarbula ranuii* and *Vesicularia montagnei* are conserved here. It facilitates demonstration of variety of bryophytes for educative purpose and general awareness for the public. (Fig. 5)

Fig. 5. (A) Outer, (B) Inner view of Moss House

2.3.7. Palm House

Palm House with an area of 285 sq m, is meant for conserving and showcasing the palm collections. The germplasm comprises over 60 species displayed in pots and in ground. It includes some fine specimens of: *Areca catechu, Arenga pinnata, Bentinkia nicobarica, Caryota mitis, C. urens, Chamaedorea elegans, C. stolonifera,*

Chrysalidocarpus lutescens, Cocos nucifera, Daemonorops kunstleri, Elaies guineensis, Licuala grandis, L. spinosa, Livistona chinensis, L. cochinchinensis, Phoenix reclinata, P. robelinii, P. rupicola, Ptychosperma macarthurii, Sabal palmetto, Thrinax barbadensis, T. excela, Trachycarpus takil and *Washingtonia filifera*. Several palm species can be grown in pots are useful as house plants for interior decoration. (Fig.6)

Fig. 6. (A) Outer, (B) Inner view of Palm House

2.4. Ornamental crops

2.4.1. Bougainvillea

Bougainvillea (Nyctaginaceae) and has wider adaptability and grown as shrubs, climbers, hedges, topiary, ground covers, standards, hanging baskets and bonsai. A rich collection of 200 cultivars is maintained in Bougainvillea garden and as potted plants. Institute has developed 24 new varieties like: 'Arjuna', 'Aruna', 'Begum Sikander', 'Chitra', 'Hawain Beauty', 'Los Banos Variegata', 'Mahara Variegata', 'Mary Palmer Special', 'Pallavi', 'Shubhra' and 'Wajid Ali Shah', which are highly popular in the horticultural trade. *Bougainvillea* 'Dr. P.V. Sane' was developed in 2011 as, Gamma ray mutant. Two new varieties viz.: 'Los Banos Variegata – Jayanthi and 'Pixie Variegata' have been developed by using the chemical mutagen (Ethyl Methane Sulphonate). Some of these newly developed cultivars, have been registered with International Bougainvillea Registration Authority, IARI, New Delhi. The existing germplasm collection is also enriched by introducing Bougainvillea varieties like: 'Royal Daupline', 'Himani', and 'B.T. Red' from botanic gardens outside the country (Roy *et al*. 2007; Sharma and Goel 2006). (Fig. 7)

Fig. 7. (A) Bougainvillea 'Chitra', (B) Bougainvillea 'Dr. PV Sane'

2.4.2. *Canna*

Canna (Cannaceae) is native to tropical America and is represented by about 12-15 species. Its magnificent flowers in different colour combinations (bicoloured, spotted, blotched, margined) are quite attractive and remain in bloom throughout year in tropical and sub-tropical regions. Germplasm collection of 50 varieties is conserved here, which has been planted in formal beds in accordance with height and flower colour for educative and aesthetic display. Few notable varieties are: 'Assault', 'Black Knight', 'Butter Cup', 'King Alfred', 'King Humbert', 'Lucifer', 'President', 'Striatus', 'New Red', 'Orange King' and 'Trinacria variegata'. Recently three new *Canna* varieties have been developed at CSIR-NBRI viz.: *Canna generalis* 'Kanchan', 'Agnisikha' and 'Raktima' under ongoing improvement programme. (Fig. 8)

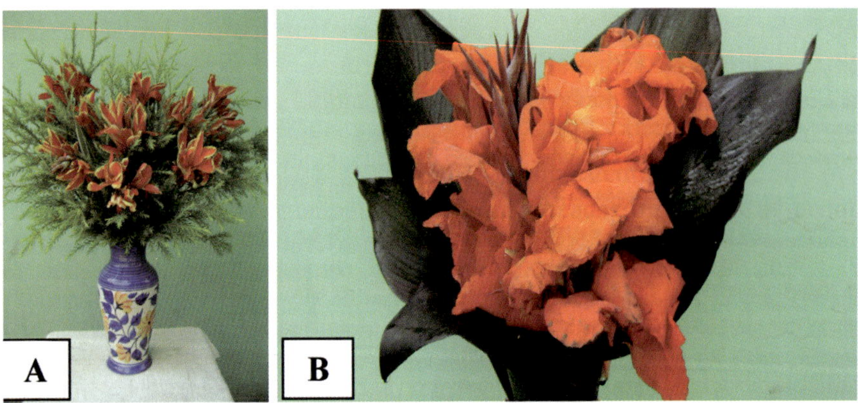

Fig. 8. (A) *Canna generalis* 'Agnisikha', (B) 'Raktima' developed in the Institute

2.4.3. *Chrysanthemum*

A rich collection of Chrysanthemum with 200 varieties is being showcased and conserved. A large number of 'Dwarf No Pinch No Stake' type cultivars have been selected which are suitable for mini-pot culture. Some of the varieties, *viz.*, 'Himanshu', 'Jwala', 'Jyoti' and 'Phuhar' bloom profusely during summer and rainy seasons. A mini chrysanthemum 'Mother Teresa', developed by NBRI, got US Patent (Patent No.PP-13678) in 2003. Recently, Institute developed new cultivars of Chrysanthemum viz., Vijay Kiran (Bud sport of Vijay), NBRI Little Hemant, NBRI Little Pink, NBRI Little Orange, NBRI Little Kusum (Seedling selection), NBRI Himanshu (Seedling selection); NBRI Kaul, NBRI Khoshoo (Hybrid) and NBRI-Pushpangadan. (Fig. 9A, B)

Fig. 9. NBRI cultivars of Chrysanthemum: (A) 'Kaul', (B) 'Pushpangadan', (C) Polyhouse cultivation of *Gerbera*

2.4.4. Gerbera

For the diversification of floricultural crops for commercial purpose, a 'Pilot facility' demonstrating agro-technology for cut-flower production of *Gerbera* (Asteraceae) in a Poly-house covering 560 sq m area has been established. Initially seven Gerbera cultivars *viz.* 'Danaellen', 'Goliath', 'Rosalin', 'Salvadore', 'Silvester', 'Sunway' and 'Zingaro' in various colours are cultivated. Techno-economics is worked out by taking into account the cost of Poly House construction and total income generated during a span of three years. The estimated net profit comes to Rs.7-10 lakhs in 3 years if properly marketed at a sale price of Rs.5-6 per flower. (Fig. 9C)

2.4.5. Gladiolus

Gladiolus (Iridaceae) is native to South Africa and an important floricultural crop grown throughout the world. Institute initiated R & D work for developing new and novel Gladiolus varieties. Botanic Garden has developed agro-technology for commercial cultivation in north Indian plains, besides identification of suitable cultivars *viz.*: 'Aldebaran', 'American Beauty', 'Big Time Supreme', 'Deciso', 'Friendship Pink', 'Oscar', 'Red Beauty', 'Tropic Sea', 'White Prosperity' and 'Yellow Stone'. Germplasm collection of 100 varieties has been conserved and maintained, besides development of new and novel varieties. Notable ones are: 'Rashmi', 'Surekha', 'Classic White', 'Neelima', 'Roshni', 'Jamuni' ('Amethyst'), Suverna, 'Sydney Percy Lancaster' and 'Urvashi'. Recently a new Gladiolus variety 'Acharya Jagadish Chandra Bose' was released during Annual Rose and Gladiolus Show held on 19 January 2013. (Fig. 10)

Fig. 10. Gladiolus, (A) 'Urvashi', (B) 'S. Percy Lancaster', (C) 'Acharya Jagadish Chandra Bose'

2.4.6. *Hippeastrum*

Hippeastrum (Amaryllidaceae), native to tropical and subtropical America, is an important genus among the bulbous ornamentals. A germplasm collection of 10 varieties is conserved. Several new hybrids in attractive colour combinations and forms, adaptable to North Indian plains, have been developed through hybridization, involving Royal Dutch hybrids and local strains. Notable ones are: 'Garima', 'Man-Mayura', 'Kiran Rekha', 'Smriti', 'Jwala', 'Niharika', 'Prakash', 'Dhruva' , 'Agni' and 'NBRI Kiran'.

2.4.7. (Lotus) *Nelumbo nucifera*

Nelumbo (Nelumbonaceae) is represented by two species namely: *N. lutea* 'Yellow Lotus', indigenous to eastern North America and *N. nucifera* 'Kamal', native to tropical and subtropical Asia. *Nelumbo nucifera* Gaertn., is an elegant creation of nature. Besides, it is an important ornamental in floriculture and landscaping. Germplasm collection of Indian and exotic races in different shades of pink and white is being maintained in the aquatic bodies in Botanic Garden (Goel *et al.* 2003; Sharma and Goel 2001). Besides, the germplasm collection of *Nymphaea* cultivars and *Euryale ferox* (Makhana) is also conserved in aquatic bodies. (Fig.11)

Fig. 11. *Nelumbo nucifera* in (A) Pink, (B) White colours

2.4.8. Rose

Rose is an important floricultural crop all over the world and ranks first in top ten cut-flowers in the international floricultural trade. In India, roses are used in various forms *viz.,* cut and loose flowers and in the extraction of essential oils. Besides, there is a good demand of roses in horticultural trade. Considering its commercial importance, germplasm collection of 225 rose varieties (both Indian and exotic) is being maintained, comprising all groups *viz.,* hybrid tea, polyantha, floribunda, miniature and climbling roses. Some important varieties are: 'City of Lucknow' (a thornless variety), 'Viridiflora' (clusters of green flowers), 'Coctail', 'Shyamlata' (climbing type), and *Rosa clinophylla.*

2.5. Other important R & D activities

2.5.1. Plant Introduction

It is the most important activity in the Botanic Garden for enrichment of germplasm collection and creating wider genetic base for further use by researchers and plant lovers (Roy 2009; Roy *et al.* 2013). Seeds and plant materials are procured from over 50 Botanic Gardens from within and outside the country on exchange. It also provides authentic plant material to other sister organizations in India and abroad, specifically for R & D work under MTA. Some of the important species, recently introduced are: *Adansonia za, A. rubrostipa, Adenanthera microsperma, Billbergia alfonsi-joannis, Caesalpinia cacalaco, C. mexicana, Clerodendrum speciosissimum, Crescentia mirabilis, Cycas media, Dasilirion glucophyllum, Dracaena draco, Hesperaloe parviflora, Jacaranda cuspidifolia, Khaya senegalensis, Livistona carnavron, Pavetta revoluta, Phoenix canarensis,* and *Triplaris surinamensis* (Roy 2011; Roy and Datta 2005). (Figs. 12)

Fig. 12. New Introductions: (A) *Clerodendrum speciosissimum,* (B) *Jacaranda cuspidifolia*

2.5.2. Ex-situ *Conservation*

Botanic Garden plays an important role for conservation of plant diversity and acts as a centre of excellence for the conservation of wild and introduced plant species. The germplasm under conservation includes *Adhatoda beddomei, Alectra chitrakutensis, Allium hookerii, Anogeissus sericea* var. *nummularia, Commiphora wightii, Curcuma pseudomonatana, Cycas beddomei, Dischidia benghalensis, Ephedra foliata* var. *ciliata, Erythrina resupinata, Frerea indica, Hoya lanceolata, H. wightii, Hyphaene dichotoma, Isonandra villosa, Indopiptadenia oudhensis, Phoenix rupicola, Sophora mollis, Tecomella undulata, Trachycarpus takil* and *Vanilla walkeri* (Goel and Mamgain 2001; Goel 2002, 2003; Goel and Kulshreshtra 2005; Goel and Sharma 2005; Goel *et al.* 2007; Roy 2000, 2003). (Fig. 13)

Fig. 13. RET species: (A) *Alectra chitrakutensis*, (B) *Isonandra villosa*

2.5.3. Touch-'n'-Smell Garden

Visually impaired people are totally deprived of enjoyment associated with natural resources. Institute has created a unique garden meant for visually impaired and disabled persons enabling them to learn about the diversity plants by touch and smell. The garden is first of its kind in Indian subcontinent and seventh in the world (Pushpangadan *et al.* 2002). It is spread in 1000 sq m area. Plant labels and legends are written in Braille on the aluminium sheets mounted on stands which provide botanical and common name, family, origin and other relevant information in English and Hindi. Prerecorded information through audio system has been provided. The plant species either having fragrant flowers, aromatic / coarse leaves have been planted. The notable ones are: *Buddelia madagascariensis, Cestrum diurnum, C. nocturnum, Cinnamomum camphora, Crinum* spp., *Cymbopogon martinii, Ficus krishnae, Ixora parviflora, Jasminum* spp., *Lantana camara* 'Flava', *Nyctanthes arbortristis, Ocimum* spp., *Polianthes tuberosa* and *Rosa* cv. 'City of Lucknow' (Roy 1999, 2005). (Fig. 14A)

2.5.4. *Vertical Gardening*

Botanic Garden has taken-up a unique initiative to display different styles of Vertical Gardening. Eight different styles of Vertical Gardens are displayed at the entrance gate. This provides an excellent opportunity for the garden lovers to enjoy the visual beauty of the plants displayed in different styles on a wall and also to study the technical aspects and further application at suitable places. Some suitable plant species for this garden style are: *Adiantum capillus-veneris, Aptenia cordifolia, Phalaris minor, Pilea cadierei, Pteris vittata, Callisia repens, Sedum morganianum, Chlorophytum comosum* 'Variegatum', *Cryptanthus bivittatus* 'Minor', *Setcreasea purpurea* 'Purple Heart', *Stenotaphrum secundatum* 'Variegatum', *Epipremnum aureum* 'Marble Queen', *Epipremnum aureum* 'Wilcoxii', *Tradescantia albiflora* 'Variegata', *Tradescantia navicularis, Tradescantia sillamontana, Pellionia daveauana, Peperomia obtusifolia, Zebrina pendula, Zebrina pendula* 'Discolor', etc. (Figs. 14B,C)

Fig. 14. (A) Touch-'n'-Smell Garden, (B,C) Different views of Vertical Garden

2.5.5. *Annual Flower Shows*

Institute organizes every year two Annual Flower Shows *viz.* Rose and Gladiolus Show in January and Chrysanthemum and Coleus Show in December for the last 45 years. The purpose of organizing these flower shows is to create aesthetic sense and to develop awareness among the masses for keeping their surroundings and environment clean/ healthy and green besides generating self-employment. They also provide a common platform and an ideal opportunity to the scientists, researchers, horticulturists, gardeners and general public, for interaction on problems of mutual interest, well identified ornamental varieties and quality planting material. (Fig. 15A)

Fig. 15. (A) Prize winning entries at Annual Chrysanthemum and Coleus Show and, (B) view of articles prepared by using floral dehydration techniques

2.5.6. Dehydration of Flowers and Floral Craft

Dehydrated flowers and floral craft is one of the important components of value added products. Dehydrated flowers and floral craft is used for making various useful products like the greeting cards, landscapes, photo frames, table mats, coasters, wall hangings and potpourri items. Institute has a pioneering role for development of dry flower technology. Methodologies for the dehydration of different ornamental plants/flowers have been standardized which have added a new dimension. Many training and awareness programmes and exhibitions on dehydration of flowers and floral craft have been organized for the benefit of cross section of society. (Fig. 15B)

2.6. Botanic Gardens as excellent facility for taxonomic studies

With such diversified germplasm resources in this Botanic Garden, it offers a great scope of conducting the taxonomic studies on wild as well as the cultivated plants. The available fresh materials can be used for dissection purpose and making the detailed taxonomic descriptions. Many new plant species are described every year from various Botanic Gardens world over thus providing a unique opportunity and scope to discover and describe them. Even the complex taxa can be grown under *ex-situ* and their correct identities may be established at the later stages. Identification of introduced species has been found to be quite interesting and cumbersome task and requires well trained taxonomic eyes as well as expertise.

The students at junior and senior levels can be trained in plant identification techniques properly and more precisely with the help of fresh plant material. The phenotypic plasticity can be well explained to the students and the researches at one place. Studies with the live plant material make plant taxonomy more lively and interesting. Even the floristic studies of a district may not reveal such a wide phenotypic as well as the genotypic diversity as available in the Botanic Gardens.

References

Goel AK, Mamgain SK (2001) *Ex-situ* Conservation of some endemic and threatened plant species of India, Indian For 127(5):552-562

Goel AK and Sharga AN (2001) NBRI Botanic Garden - A National Facility: Centre for Education and Conservation. In: The Power for Change: Botanic Gardens as Centres of Excellence in Education for Sustainability, Sutherland LA, Abraham T.K., Thomas J (eds.), – Proceedings of the 4th International Congress on Education in Botanic Gardens, 8-12 November 1999. TBGRI, Kerala, India, pp 65-73

Goel AK (2002) Botanic Gardens - Potential source for conservation of plant diversity, J Econ Tax Bot 26(1):67-74

Goel AK (2002) *Ex-situ* Conservation studies on some rare, endangered and endemic plant species at NBRI Botanic Garden, Indian J For 25(1):67-78

Goel AK (2003) *Ex-situ* Conservation of plant genetic resources in NBRI Botanic Garden, In: Natural Resource Management & Conservation, Singh J, Pandey Girish (eds.), Kalyani Publishers, Ludhiana, pp 131-142

Goel AK, Swarup Vishnu and Roy RK (2003) Lotus (11), In: Commercial Flowers, Bose TK, Yadav LP, Pal P, Parthasarthy VA, Das P (eds.),. 2nd Edition, Noya Udyog, Kolkata, pp 419-428

Goel AK, Kulshreshtra K (2005) *Ex-situ* conservation of *Isonandra villosa* (Sapotaceae) at the National Botanical Research Institute, Lucknow, India, BGjournal (UK) 2(2):28-29

Goel AK and Sharma SC (2005) CITES and conservation of plant diversity in India. In: Plant Response to Environmental Stress. Proceedings of ICPAP-II, Tripathi RD *et al.* (eds.), International Book Distributing Co., Lucknow, pp 85-94

Goel AK, Roy RK, Sharma SC (2007) NBRI Botanic Garden – A Centre of Excellence for conservation, Education and Bio-aesthetics, Environews 13(3):11-14

Kumar S, Chauhan P, Goel AK (2006) Double male cone formation- A rare and interesting phenomenon in *Cycas circinalis* L. (Cycadaceae), Vatika 3: 3-5

Pushpangadan P, Sharga AN, Roy RK, Kulshreshtra K (2002) Touch-N- Smell Garden: Opening a new garden for the visually impaired and disabled in Lucknow, India. BGCI News (United Kingdom) 3(8):51-52

Roy RK (1999) Access to the People. 'Roots' BGCI Education Newsletter, (United Kingdom) 18:27-29

Roy RK (2000) NBRI: A national institute for conservation and exploitation of plants in India. Botanic Garden Conservation News (United Kingdom) 3(4):37-39

Roy RK (2003) Conservation of Plant genetic diversity in India: The Role of Botanic Gardens in new millennium. J Non-Timber For Prod 1/2):61-64

Roy RK (2005) Environmental education in Botanic Gardens. In: Challenges and opportunities in the new millennium, Tripathi RD *et al.,* (eds.), Proceedings of ICPAP-II, International Book Distributing Co., Lucknow, pp 391-396

Roy RK, Datta SK (2005) Botanic Gardens: A gateway of plant introductions. Ind J Plant Genet Resour 18 (1):153

Roy RK (2006a) The role of botanic gardens in conservation of plants and environment. In: Proceedings of the 'Recent Advances in Plant Sciences – Environment and Plants: Glimpses of research in South Asia, Ecological Society, Jha P K *et al.,* (eds.), Nepal

Roy RK (2006b) Role of Botanic Gardens in education and public awareness: NBRI Botanic Garden – A case study. Proceedings BGCI's 6th Education Congress, Oxford University Botanic Garden, United Kingdom, www. bgci.org.edncongress

Roy RK (2009) Save these rare ornamental trees. Curr Sci 97(11):1536-38

Roy RK, Goel AK (2006) Cycads: An excellent group of plants for indoor and outdoor decoration. Vatika 4:11-16

Roy RK, Banerjee BK, Goel AK (2007) *Bougainvillea* – Germplasm collection at NBRI, Indian Bougainvillea Ann 20:10-13

Roy RK (2011) Introduction and Migration of Ornamental Trees to India, Chronica Horticulturae, Int'l Soc. for Hort Sci Belgium 51(1): 29-33

Roy RK, Kumar S, Goel AK (2013) History, Myth and Conservation Threat of the African Baobab Tree in India, Chronica Horticulturae, Int'l Soc Hort Sci Belgium 53(1): 26-30

Sharma SC, Goel AK (2001) Conservation of Lotus (*Nelumbo nucifera*), Indian Horticulture 46(1):18, 22

Sharma SC, Goel AK (2003) *Welwitschia mirabilis*- a rare xerophytic South African gymnosperm in National Botanical Research Institute, Lucknow, Curr Sci 84(1):11

Sharma SC, Goel AK (2006) Horto-taxonomy of ornamentals: Case study of *Bougainvillea,* J Econ Tax Bot 30(1):71-78

About the Editors

Dr. T. S. Rana is presently working as a Senior Principal Scientist at CSIR-National Botanical Research Institute, Lucknow. He has over 23 years of research experience in Angiosperm Taxonomy and Biodiversity assessment at the species, ecosystem and genetic levels. Dr. Rana has published one book and about sixty research papers in peer reviewed National and International journals, and have supervised/guided four PhD and ten MSc (Biotech.) students. He is a life member and fellow of various renowned scientific bodies. Dr. Rana was bestowed with the prestigious BOYSCAST Fellowship (2001-2002) by the Department of Science and Technology, Ministry of Science and Technology, Government of India, New Delhi, for his outstanding contributions in the field of plant sciences.

Dr. K. N. Nair presently works as a Senior Principal Scientist in the Plant Diversity, Systematics & Herbarium Division at the CSIR- National Botanical Research Institute, Lucknow. He has about 27 years of experience in the field of plant taxonomy, nomenclature and plant conservation. His special areas of research include systematic studies and genetic diversity assessment of Indian *Citrus* and its wild relatives using morpho-taxonomic and molecular taxonomic approaches. Dr. Nair has to his credits 24 peer reviewed research papers published in reputed National and International journals, two patents, 3 books (co-authored), and 10 contributed chapters to different books.

Dr. Dalip K. Upreti is presently working as a Chief Scientist and Area Coordinator in Plant Diversity, Systematics and Herbarium Division at CSIR-National Botanical Research Institute, Lucknow. Dr. Upreti has discovered more than 100 new species, and about 200 new records of Lichens in India. Dr. Upreti has carried extensive researches on various aspects like taxonomy, pollution monitoring, medicinal activities, *in vitro* culture, antibiotic and bio-prospection studies of Lichens. He is the first Indian Lichenologist, who participated in 11th Indian Antarctic Expedition to carry out research studies on Antarctic Lichens. Dr. Upreti has five books and more than 300 research papers in peer reviewed National and International journals to his credits. Dr. Upreti has given a new dimension to the Lichen studies in India, and recently, he has been nominated as a *Fellow of National Academy of Sciences* (FNASc), Allahabad for his outstanding contributions in the Lichen biology.